D1395531

PLANT COLD HARDINESS
and FREEZING STRESS

MECHANISMS *and* CROP IMPLICATIONS

Volume 2

Academic Press Rapid Manuscript Reproduction

*Proceedings of an International Seminar on Plant Cold Hardiness
Held at the Sapporo Educational and Cultural Hall,
Sapporo, Japan, August 11–14, 1981*

*Sponsored by
The United States National Science Foundation
The Japan Society for the Promotion of Science*

PLANT COLD HARDINESS
and FREEZING STRESS

MECHANISMS
and CROP IMPLICATIONS
Volume 2

Edited by

P. H. LI

Laboratory of Plant Hardiness
Department of Horticultural Science and Landscape Architecture
University of Minnesota
St. Paul, Minnesota

A. SAKAI

Laboratory of Frost Injury in Plants
The Institute of Low Temperature Science
Hokkaido University
Sapporo, Japan

1982

ACADEMIC PRESS
A Subsidiary of Harcourt Brace Jovanovich, Publishers

New York   London
Paris   San Diego   San Francisco   São Paulo   Sydney   Tokyo   Toronto

ACADEMIC PRESS, INC.
111 Fifth Avenue, New York, New York 10003

United Kingdom Edition published by
ACADEMIC PRESS, INC. (LONDON) LTD.
24/28 Oval Road, London NW1   7DX

Library of Congress Cataloging in Publication Data
Main entry under title:

Plant cold hardiness and freezing stress.

   Proceedings of an international plant cold hardiness
seminar held in St. Paul, Minn., Nov. 2-4, 1977 and
sponsored by United States National Science Foundation, Japan Society for the Promotion of Science,
and College of Agriculture, University of Minnesota.
   Vol. 2 based on the proceedings of the 2nd
international seminar on plant cold hardiness, held
at the Sapporo Educational and Cultural Hall,
Sapporo, Japan, Aug. 11-14, 1981.
   Includes bibliographies and indexes.
   1. Plants--Frost resistance--Congresses.
2. Plants, Effect of cold on--Congresses. 3. Crops
and climate--Congresses. I. Li, P. H. (Paul H.),
1933-   .   II. Sakai, A. (Akira), 1920-   .
III. National Science Foundation (U.S.) IV. Nippon
Gakujutsu Shikōkai, V. University of Minnesota,
College of Agriculture.
QK756.P53   582' .019165   78-7038

ISBN 0-12-447602-3 (v. 2)

Contents

PART III   FREEZING STRESS

PART IV   STRATEGIES FOR IMPROVING FREEZING SURVIVAL

PART V   OTHER TEMPERATURE-RELATED STRESSES

Contributors

Numbers in parentheses indicate the pages on which the authors' contributions begin.

P. K. ANDREWS (529), IAREC, Washington State University, Prosser, Washington

T. ASAHI (671), Laboratory of Biochemistry, Faculty of Agriculture, Nagoya University, Nagoya, Japan

M. R. BECWAR (307), National Seed Storage Laboratory, Colorado State University, Fort Collins, Colorado

M. J. BURKE (211, 307), Department of Fruit Crops, University of Florida, Gainesville, Florida

J. V. CARTER (169, 379), Laboratory of Plant Hardiness, Department of Horticultural Science and Landscape Architecture, University of Minnesota, St. Paul, Minnesota

H. H. CHEN (5), Crop Development Center, University of Saskatchewan, Saskatoon, Saskatchewan, Canada

L. CHRISTERSSON (605), Department of Ecology and Environmental Research, The Swedish University of Agricultural Science, Uppsala, Sweden

H. A. DONG (243), Institute of Botany, Academia Sinica, Peking, China

M. F. DOWGERT (459), Department of Agronomy, Cornell University, Ithaca, New York

B. ELFMAN (129), Department of Plant Sciences, University of Western Ontario, London, Canada

N. ESTRADA R. (615), Tuberous Crops Program, Colombian Institute of Agriculture, Bogota, Colombia

R. Y. EVANS (459), Department of Agronomy, Cornell University, Ithaca, New York

B. J. FINKLE (643), USDA-SEA, Western Regional Research Center, 800 Buchanan Street, Berkeley, California

D. B. FOWLER (23), Crop Development Center, University of Saskatchewan, Saskatoon, Saskatchewan, Canada

L. H. FUCHIGAMI (93), Department of Horticulture, University of Hawaii, Honolulu, Hawaii

H. FUCHINOUE (499), Saitama Prefectoral Experiment Station of Tea, Iruma Saitama, Japan

M. F. GEORGE (367), School of Forestry, University of Missouri, Columbia, Missouri

W. J. GORDON-KAMM (459), Department of Agronomy, Cornell University, Ithaca, New York

M. GRIFFITH (129), Department of Botany, University of British Columbia, Vancouver, Canada

L. V. GUSTA (23, 93, 297), Crop Development Center, University of Saskatchewan, Saskatoon, Saskatchewan, Canada

C. L. GUY (169, 561), Laboratory of Plant Hardiness, Department of Horticultural Science and Landscape Architecture, University of Minnesota, St. Paul Minnesota

S. HATANO (145, 157), Department of Food and Technology, Faculty of Agriculture, Kyushu University, Fukuoka, Japan

D. B. HAYDEN (129), Department of Plant Sciences, University of Western Ontario, London, Canada

T. HOLUBOWICZ (541), Department of Pomology, Academy of Agricultural Science, Poznan, Poland

S. G. HONG (341), Department of Forestry, Kon-Kuk University, Seoul, South Korea

W. G. HOPKINS (129), Department of Plant Sciences, University of Western Ontario, London, Canada

N. P. A. HUNER (129), Department of Plant Sciences, University of Western Ontario, London, Canada

S. ICHIKI (181), Aomori Apple Experiment Station, Kuroishi, Aomori, Japan

I. IKEDA (575), Akitsu Branch, Fruit Tree Research Station, Ministry of Agriculture, Forestry and Fisheries, Akitsu, Hiroshima, Japan

M. ISHIKAWA (325), The Institute of Low Temperature Science, Hokkaido University, Sapporo, Japan

M. IWAYA (357), College of General Education, Kyushu University, Fukuoka, Japan

K. G. JENSEN (221), Department of Botany, University of Iowa, Iowa City, Iowa

K. B. JEON (357), Department of Biology, Jeonbug National University, Jeonju, Korea

L. C. JIAN (243), Institute of Botany, Academia Sinica, Peking, China

K. KABATA (145), Department of Food and Technology, Faculty of Agriculture, Kyushu University, Fukuoka, Japan

A. KACPERSKA (261), Institute of Botany, University of Warsaw, Warsaw, Poland

S. KAKU (357), College of General Education, Kyushu University, Fukuoka, Japan

M. A. KHAMIS (541), College of Moistour of Agricultural University of Zagazig, Egypt

R. J. KLOSSON (55), Institute of Botany, University of Düsseldorf, Düsseldorf, West Germany

K. KOBAYASHI (93), Department of Horticulture, Oregon State University, Corvallis, Oregon

G. H. KRAUSE (55), Institute of Botany, University of Düsseldorf, Düsseldorf, West Germany

W. LARCHER (417), Institute of Botany, University of Innsbruck, Innsbruck, Austria

J. S. LI (285), Shanghai Institute of Plant Physiology, Academia Sinica, Shanghai, China

P. H. LI (5, 221, 379), Laboratory of Plant Hardiness, Department of Horticultural Science and Landscape Architecture, University of Minnesota, St. Paul, Minnesota

S. E. LINDOW (395), Department of Plant Pathology, University of California, Berkeley, California

M. MAESHIMA (671), Laboratory of Biochemistry, Faculty of Agriculture, Nagoya University, Nagoya, Japan

M. MATSUOKA (671), Laboratory of Biochemistry, Faculty of Agriculture, Nagoya University, Nagoya, Japan

R. Q. MI (285), Shanghai Institute of Plant Physiology, Academia Sinica, Shanghai, China

T. NIKI (189), The Institute of Low Temperature Science, Hokkaido University, Sapporo, Japan

S. OOHATA (437), Kyoto University Forest in Hokkaido, Department of Agriculture, Kyoto University, Kyoto, Japan

T. OSHIMA (661), Mitsubishi-Kasei Institute of Life Sciences, Tokyo, Japan

J. P. PALTA (221), Department of Botany, University of Iowa, Iowa City, Iowa

D. M. PATON (77), Department of Botany, The Australian National University, Canberra, Australia

J. PIENIAZEK (541), Department of Pomology, Academy of Agricultural Science, Poznan, Poland

E. L. PROEBSTING (529), IAREC, Washington State University, Prosser, Washington

C. RAJASHEKAR (211, 379), Laboratory of Plant Hardiness, Department of Horticultural Science and Landscape Architecture, University of Minnesota, St. Paul, Minnesota

H. SADAKANE (157), Department of Food and Technology, Faculty of Agriculture, Kyushu University, Fukuoka, Japan

A. SAKAI (199, 325, 427, 437, 487, 635), The Institute of Low Temperature Science, Hokkaido University, Sapporo, Japan

K. A. SANTARIUS (475), Institute of Botany, University of Düsseldorf, Düsseldorf, West Germany

T. SATO (447), The Institute of Low Temperature Science, Hokkaido University, Sapporo, Japan

E. SIKORSKA (261), Institute of Botany, University of Warsaw, Warsaw, Poland

D. SIMINOVITCH (117), Chemistry and Biology Research Institute, Agriculture Canada, Ottawa, Ontario, Canada

P. L. STEPONKUS (459), Department of Agronomy, Cornell University, Ithaca, New York

E. SUCOFF (341), Department of Forest Resources, University of Minnesota, St. Paul, Minnesota

D. L. SUN (243), Institute of Botany, Academia Sinica, Peking, China

L. H. SUN (243), Institute of Botany, Academia Sinica, Peking, China

R. TIMMIS (93), Weyerhaueser Company, Tacoma, Washington

B. TISSERAT (643), Fruit and Vegetable Chemistry Laboratory, USDA-ARS, Pasadena, California

T. TOYAO (591), Laboratory of Genetics and Physiology, National Research Institute of Tea, Kanaya, Shizuoka, Japan

U. TRÖSTER (55), Institute of Botany, University of Düsseldorf, Düsselforf, West Germany

T. I. TRUNOVA (41), Institute of Plant Physiology, The USSR Academy of Science, Moscow, USSR

N. J. TYLER (23), Crop Development Centre, University of Saskatchewan, Saskatoon, Saskatchewan, Canada

M. UEMURA (487, 635), The Institute of Low Temperature Science, Hokkaido University, Sapporo, Japan

J. M. ULRICH (643), USDA-SEA, Western Regional Research Center, 800 Buchanan Street, Berkeley, California

I. URITANI (671), Laboratory of Biochemistry, Faculty of Agriculture, Nagoya University, Nagoya, Japan

H. C. WANG (285), Shanghai Institute of Plant Physiology, Academia Sinica, Shanghai, China

Y. Q. WANG (285), Shanghai Institute of Plant Physiology, Academia Sinica, Shanghai, China

C. J. WEISER (1, 93), Department of Horticulture, Oregon State University, Corvallis, Oregon

S. C. WIEST (511), Department of Horticulture, Kansas State University, Manhattan, Kansas

H. YAMAYA (181), Aomori Apple Experiment Station, Kuroishi, Aomori, Japan

G. YELENOSKY (561), Horticultural Research Laboratory, USDA-SEA-AR, 2120 Camden Road, Orlando, Florida

S. YOSHIDA (273, 297, 487), The Institute of Low Temperature Science, Hokkaido University, Sapporo, Japan

F. YOSHIE (427), The Institute of Low Temperature Science, Hokkaido University, Sapporo, Japan

Symposium participants and contributors (left to right). First row: L. C. Jian, H. Fuchinoue, S. Yoshida, S. Hatano, J. P. Palta, B. J. Finkle, A. Kacperska, D. Siminovitch, A. Sakai, P. H. Li, T. I. Trunova, T. Holubowicz, S. C. Wiest, P. L. Steponkus, W. Larcher, E. Ashina. Second row: T. Niki, N. Estrada, M. Uemura, S. Oohata, I. Ikeda, T. Asahi, M. Iwaya, E. L. Proebsting, M. N. Westwood, S. E. Lindow, H. Sadakane, G. Yelenosky, N. Huner, S. Kaku, E. A. Ashby. Third row: T. Sato, F. Yoshie, T. Hashimoto, T. Toyao, M. J Burke, J. V. Carter, L. Christersson, D. M. Paton, W. R. Breidenbach, G. H. Krause, J. Y. Guo, K. A. Santarius. Fourth row: Y. Oshima, M. F. George, L. V. Gusta, L. H. Fuchigami, O. Shibata, S. Ichiki, M. Ishikawa, H. C. Wang, Y. C. Zhu, S. G. Hong, T. Oshima, I. Nishiyama, N. Murata.

Preface

This volume is based on the proceedings of the second international seminar on plant cold hardiness, which was held at the Sapporo Educational and Cultural Hall, Sapporo, Japan, August 11-14, 1981. It contains a series of valuable articles on the studies of plant cold hardiness since the first seminar was held in 1977.*

The seminar was again jointly sponsored by the United States National Science Foundation and the Japan Society for the Promotion of Science under the auspices of the United States Japan Cooperative Science Program. The primary focus was to update the knowledge of the fundamental phenomena of plant cold acclimation and freezing behavior, to examine the hypotheses and new ideas that were thought to be important in plant cold hardiness research, and to review the application of research findings for improving quality of life. It was also our intention to utilize this forum to introduce younger investigators and newcomers to the scientific community interested in plant cold hardiness.

It is hoped that the information reported in this volume will make an additional, significant contribution to researchers involved in understanding and planning research strategies for plant cold hardiness and for attenuating crop losses by frosts and severe winters.

We would like to express our appreciation to Dr. Mikio Arie, the President of Hokkaido University; Dr. Seiiti Kinosita, the Director of the Institute of Low Temperature Science, Hokkaido University; and Dr. Malcolm Purvis, the Director of International Agriculture Programs, College of Agriculture, University of Minnesota, for their interest and support.

*P.H. Li and A. Sakai, editors (1977). "Plant Cold Hardiness and Freezing Stress. Mechanisms and Crop Implications." Academic Press, New York.

We would also like to express our thanks to the Japan Society of the Promotion of Science; the United States National Science Foundation (Tokyo Regional Office), especially to Dr. Ebert A. Ashby, the Head of Japan and Southeast Asia Programs; and the staff and the graduate students of the Section of Frost Injury in Plants, the Institute of Low Temperature Science, for their help in organizing the seminar. A special word of thanks goes to Mrs. E. Ueno at the Institute of Low Temperature Science, for her impeccable secretarial work for more than two years.

Finally, we would like to acknowledge Ms. Virginia M. Dahm, Department of Horticultural Science, for her preparation of the camera-ready manuscript, and the contributions of the staff of Academic Press in bringing the volume to rapid publication.

Contents
of Previous Volume

Part I
Introduction

COLD HARDINESS AND STRESS RESEARCH: AN EVOLVING AGRICULTURAL SCIENCE

C. J. Weiser

Department of Horticulture
Oregon State University
Corvallis, Oregon

It is constructive to project future research developments and potentials (1). In plant cold hardiness and freezing stress research one wonders whether breeding crops for freeze and chill resistance will become as effective and commonplace as selecting for disease resistance? Will primary gene pools of crops be systematically assessed to identify and utilize genetic traits for stabilizing productivity and reducing losses in stressful environments? Will valid selection methods be perfected for screening large seedling populations for cold resistance? Will the focus on breeding crop cultivars for broad environmental adaptability shift towards development of fine-tuned cultivars for specific environments and locales? Will centralized national and international hardiness testing centers be established to assist breeders evaluate genotypes for stress resistance as well as for disease resistance, insect resistance, or processing quality? Will germplasm repositories routinely store sexual and vegetative genetic resources for long periods at ultra-low temperatures?

In crop physiology and production research will we soon elucidate the basic mechanisms of injury and resistance at cellular and organismal levels? Will the non-lethal effects of low temperature on plant growth and productivity be more fully investigated? Will identification of vulnerable stages of development and stresses that limit survival and productivity of crops in major production areas become an organized and significant area of study? Will crop models, based on

Technical Paper No. 6132, Oregon Agricultural Experiment Station.

physiological responses, be developed to select sites and crops, predict losses, and make sound management decisions? Will perennials be grown as annuals or biennials in stressful environments? Will physiologists, breeders, and genetic engineers collaborate closely rather than simply exchanging interesting plant materials and methodologies? Will chemical cryoprotectants be found that work in the field as well as in the laboratory? Will it be possible to regulate the timing of crop development to attenuate or avoid frost damage at particularly sensitive stages? Will supercooling of plants in nature be extended to avoid freezing in sensitive crops, or will nucleation be induced to reduce nonequilibrium freezing stress and damage in tolerant crops?

The answer to these questions will hopefully be yes and soon as man seeks ways to double world food productivity in the next few decades. It is timely and essential for environmental stress biology to evolve and emerge as a well-defined branch of biological and agricultural science because hardiness and environmental stress research is destined to play a pivotal role in mankind's success or failure in balancing crop productivity with resources and population growth.

Consider that average yields of most crops fall 3 to 7-fold short of potential yields (2); that for all major crops there is a 10-fold range among countries in their average yields (3); that an estimated 90% of the shortfall between potential and average yield is due to environmental factors and weed competition, 6% to diseases, and 4% to insects (2); that a relatively small drop in world mean temperature 0.5° to 1.0°C) would have a devastating impact on crop productivity--i.e. a 45% decrease in world rice production (4); and that temperature and precipitation are the least predictable of the major climatic variables (5).

The most common cause of crop failure in new environments is lack of synchrony of a crop life-cycle with the sequence of seasonal conditions, and the greatest major gains in productivity have been achieved by selecting crops for environmental adaptation, rather than yield potential (3). These gains have been highly significant, but they have also been unnecessarily slow because the essential collaboration between stress physiologists and plant breeders has lagged (1). There is excellent potential for rapid and sustained progress. A top priority should be selection of crops for adaptation to localized environments (3).

Breeding for climatic adaptation and stress resistance has the potential of achieving recurring genetic gains in productivity without commensurate recurring costs of protection and production. Crop physiology and production practices also play a vital role in increasing productivity and reducing losses-- particularly in long life-cycle perennial crops where genetic strategies face severe time constraints. Even in annual crops the major gains in productivity often result from improved crop production and protection. In Mexico, for example, 4.3% of the average annual increase in wheat yields since 1950 is attributed to improved cultural practices, and 0.9% to genetic improvement in yield potential (3).

Future increases in crop productivity will, of necessity, become increasingly science-based as we seek ways to conserve and optimize the utilization of finite land, water, and petroleum resources. Knowledge is our most valuable resource, and the one catalyst that can balance the seemingly unrealistic food-resource-population equation. The generation and prompt application of knowledge about plant responses to environmental stresses has unique application and urgency in science, and for society as a whole, because environmental stresses are the primary limiting factor to crop productivity, climatic change may impose new constraints, and because there is great potential for genetically and physiologically increasing productivity and reducing crop losses due to environmental stress.

REFERENCES

1. Howell, G.S., Bartholic, J., Breidenback, W., Dilley, D., Evert, D., Frey, K., George, M., Hanson, A., Kanemasu, E., Knezek, B., Olien, R., Parsons, L., and Weiser, C., Alleviation of Environmental Stress on Renewable Resource Productivity (G.S. Howell, ed.). Report to DOE-AAAS Climate Project. 39 pp. (1980).
2. Boyer, J.S. Environmental Stress: Improving Crop Production in Unfavorable Environments. USDA, SEA Research Proposal. 23 pp. (1978).
3. Evans, L.T., Amer. Scientist 68:388-397 (1980).
4. Stansel, J.W. Climatic Impact Assessment Program--Rice. Texas A and M University. (1973).
5. Christianson, M.N., in "Stress Physiology in Crop Plants" (H. Mussel and R.C. Staples, eds.). John Wiley & Sons, New York, 1979.

Part II
Cold Acclimation

POTATO COLD ACCLIMATION

Hwei-Hwang Chen[1] and P. H. Li

Laboratory of Plant Hardiness
Department of Horticultural Science and Landscape Architecture
University of Minnesota
St. Paul, Minnesota

I. INTRODUCTION

Many plant species increase in frost hardiness when exposed to environmental conditions such as shortening daylengths and/or lowering temperatures. The characteristics of cold acclimation have been studied extensively in plant species (22), including tuber-bearing Solanum species (3-6,37), but the physiological mechanisms remain a mystery. Shortening daylengths and decreasing environmental temperatures are physical changes and the resultant frost hardiness induced in plants involves biological response. We do not know how plants sense and translate environmental signals into biological responses that result in cold acclimation? An understanding of this process is needed to interpret and manipulate plant cold adaptation.

Levitt proposed (22) that metabolites accumulate during cold acclimation as photosynthetic light reactions and carbon assimilation proceeds efficiently as environmental temperatures drop. He suggests that the resulting metabolic changes lead to plant adaptation to freezing stress. How such changes may specifically lead to the cold adaptation is still not yet answered. Kacperska-Palacz (18), on the other hand, proposed that environments which induce acclimation do so by triggering changes in hormonal balance which lead to growth cessation and result in metabolic shifts. It is still not clear how the environmental factors induce changes in the hormonal balance, and which hormones are likely to be involved in cold acclimation.

---

[1] Present address: Crop Science Department, University of Saskatchewan, Saskatoon, Sask., Canada, S7N 0W0.

PLANT COLD HARDINESS AND FREEZING STRESS

5

Observations of seasonal changes in metabolites such as carbohydrates, proteins, nucleic acids, lipids, etc. in woody plants (21,24,27,32,34), and herbaceous plants (3,9,16) relative to the changes in frost hardiness have led to the conclusion that the drastic alterations in metabolites during the fall enable plants to survive winter cold.

Generally, the development of frost hardiness is always accompanied with the changes in carbohydrate content and composition (25,32,34), osmotic concentration (23,25,32,34), hormonal balance (15,16,18,30,39,40), nucleic acids (3,26,33,34) and the protein quality (2,20,31) and quantity (3,11,19,24,27,34,38). The possible significance of specific biochemical changes in relation to cold acclimation has been discussed extensively (22), but causal relationship between metabolic changes and cold acclimation are difficult to establish.

II. REVIEW OF POTATO COLD ACCLIMATION

Characteristics of frost hardiness and cold acclimation have been investigated among a number of tuber-bearing Solanum species (4,5). Species differ in frost hardiness (4,5,6) and cold acclimating capabilities when subjected to low temperatures (4,5). Species such as S. acaule and S. commersonii acclimate in response to low temperatures, whereas, others like S. tuberosum fail to cold acclimate (4,5). The level of frost hardiness achieved, for instance, in S. acaule and S. commersonii, are dependent on acclimating temperatures (4,5). A day/night temperature of 2°C, regardless of photoperiod, was found to be the optimum acclimating temperature and 15 days of cold exposure was needed to achieve full frost hardiness potential (4).

Under such optimum acclimating temperature leaves of all tested potato species showed an increase in total sugar and starch content in spite of their frost hardiness levels and/or acclimation capabilities (3). An increase in leaf soluble protein content, however, was observed only in those species which were able to cold acclimate, i.e. S. acaule and S. commersonii (3). The net increase of soluble proteins was significantly and positively correlated with the increases of frost hardiness in those species.

Plasma membrane alterations were observed in callus tissue of S. acaule during cold acclimation (36). Membrane protein particles were aggregated in this species to a peak at the 5th day after the initiation of 2°C acclimation treatment, and subsequently redistributed almost to the same level as observed in control callus after 15 days. No such changes were observed in S. tuberosum callus under similar experimental conditions. The change in protein particle aggregation pattern in S. acaule is interpreted as indicating the presence of an adaptive fluidity control mechanism in that species via protein metabolism.

Potato stem cuttings can be easily rooted and propagated in a nutrient agar medium (5). Such stem cultured plants and undifferentiated callus tissue can both be acclimated to similar frost hardiness levels as normal tuber propagated plants grown in pots (5). In stem-cultured plants, we observed that abscisic acid (ABA) could substitute for low temperature in inducing frost hardiness in S. commersonii (5). Surprisingly, ABA also induced frost hardiness in S. tuberosum (5). Gibberellic acid (GA$_3$) did not affect the cold acclimation.

III. HYPOTHESIS

Increases of sugars have been shown to accompany increases in frost hardiness in many woody and herbaceous plants (22), but relationships are not directly parallel; and plants which do not cold acclimate also increase in sugar content (3), when exposed to environmental conditions that elicit acclimation in responsive species. This leads one to ask whether sugar content has a role in cold acclimation besides being a source of energy.

The parallel relationship between the carbohydrate accumulation and the increases in osmotic concentration (25,32) observed during cold acclimation implies that carbohydrates may be a major contribution to the observed changes in osmotic concentration. It has been shown that an increase in osmotic concentration to a critical level can increase the endogenous ABA concentration (29,30) in plants. Exposure of plants to cool temperatures has also been shown to raise the endogenous level of ABA (8,40), to induce the protein synthesis (2,20,29), and to increase the frost hardiness. Specifically, it was found that ABA could preferentially induce the synthesis of particular proteins (10,17,36) and at the same time increase the resistance to low temperature stress (1,5,28,30). It is not known as yet whether ABA is directly involved in the cold acclimation via the induction of synthesizing particular proteins in the potatoes, but the causal relationship as hypothesized in Fig. 1 may exist. Experiments to test this hypothesis were conducted and are described.

IV. MATERIALS AND METHODS

Two tuber-bearing Solanum species which differ in levels of frost hardiness and in cold acclimation capabilities served as a model system for examining changes in sugar content, osmotic concentration, endogenous levels of free ABA and protein

LOW TEMPERATURE

ACCUMULATION OF SUGARS

ELEVATION OF OSMOTIC CONCENTRATION

INCREASE OF FREE ABA CONTENT

INDUCING PROTEIN SYNTHESIS

INCREASING FROST HARDINESS

FIG. 1. A hypothesis of the sequence of events as S. commersonii plants convert the signal from inductive low temperature into biological adjustments which result in cold adaptation.

synthesis in leaves in relation to the induction of frost hardiness during cold acclimation. One is S. commersonii which is capable of acclimating to 12°C at low temperature (2°C) (4). It survives at -5°C when it grows in a normal environment. The other is S. tuberosum which is incapable of acclimating (4) and is killed at -2.5°C. Potted plants of both species, grown from tubers, were grown in a uniform environment chamber for two months in a regime of 14 hr. photoperiod and 20/15° day/night temperature. Plants of each species were then divided into two groups. One group (controls) was maintained in the same chamber. The second group (treated) was exposed to low temperatures in a chamber with 14 hr. photoperiod and 2/2°C day/night temperature. Frost hardiness (4) of fully expanded leaves was determined after 0, 1, 2, 3, 4, 5, 10 and 15 days of treatment.

At the same time, fully mature leaves were collected for analysis of sugars (3), proteins (3), ABA (35). A basis of fresh weight of leaf tissues was used to express chemical changes during cold acclimation since we observed that the leaf moisture changed insignificantly and was indifferent between control and treated plants during the treatment. The determination of osmotic concentration was carried out as follows:

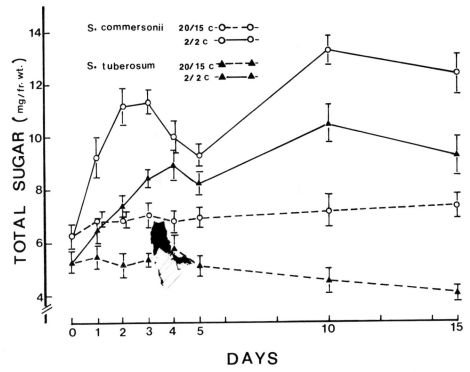

FIG. 2. Changes in total sugar content in potato leaves during cold acclimation.

Fresh leaves were infiltrated in spring water for about 5 min and sliced to 65-75 μm thick. The slices were then transferred to different concentrations of manitol solution ranging from 0.2 to 1.0 M with 0.05 M interval for 2 hr. The concentration of manitol solution at which 50% cells showed incipient plasmolysis was regarded as being isotonic to the osmotic concentration of the cells.

The effects of exogenously applied ABA and cycloheximide on the frost hardiness of stem-cultured S. commersonii were evaluated by adding these chemicals to the medium. The ABA effects on frost hardiness were studied by transferring plants from old medium to fresh medium containing 0.3% agar and various concentration of ABA (mg/liter). Plants transferred to fresh medium without ABA served as control. Plants were acclimated at 2°C (14 hr daylength), and the frost hardiness was evaluated after 15 days of treatment (4). To determine the effect of cycloheximide on the induction of frost hardiness, a $10^{-5}M$ concentration of the chemical was added to the medium just prior to transferring plants onto it, and subsequently, plants were treated as usual (4), and the frost hardiness was measured after 15 days of treatment.

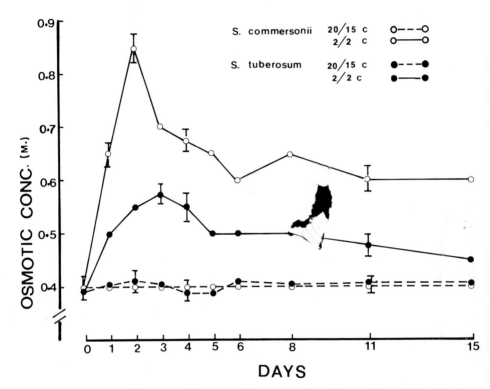

FIG. 3.   Changes in osmotic concentration of potato leaf cells during cold acclimation.

V.   RESULTS

A.   Sugar Content

Figure 2 illustrates the changes in sugar content in leaves of pot-grown S. commersonii and S. tuberosum plants during cold acclimation.   In both species, plants grown in the control showed little change in sugar content.   Under the 2/2°C acclimating regime, the sugar content in S. commersonii increased dramatically and reached a peak on the second day of treatment.   In S. tuberosum, sugar content also increased but at a slower rate, and peaked on the 4th day at a lower concentration.   On the 5th day, sugar contents in both species declined slightly and increased again to a considerably higher level which was approximately maintained during the course of the cold treatment.

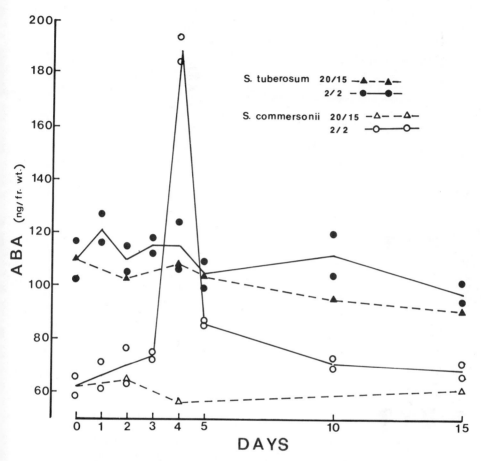

FIG. 4.   Changes in endogenous ABA concentration in potato leaves during cold acclimation.

B.   Osmotic Concentration

Changes in osmotic concentrations in leaf cells of S. commersonii and S. tuberosum are shown in Fig. 3. For the controls of both species, the osmotic values were about 0.4 M and remained at this level during the 15 day sampling period. Treated S. commersonii plants under acclimating condition showed a dramatic increase in osmotic concentration that peaked at 0.85 M on the second day, and declined thereafter. The values, however, never fell below the highest osmotic concentration (0.55 M) observed for S. tuberosum during treatment period. The highest value for S. tuberosum was observed on the third day of the treatment.

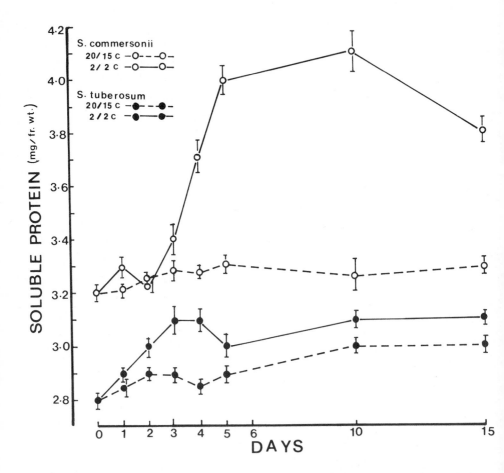

FIG. 5. Changes in soluble protein contents in potato leaves during cold acclimation.

C. Free ABA Content

Changes in the free ABA content of leaf tissues of S. commersonii and S. tuberosum are shown in Fig. 4. Leaves of control and treated plants of S. tuberosum showed a more or less constant and similar level of free ABA during 15 sampling days. Treated (2°C) S. commersonii plants, however, had free ABA contents which increased to a very high peak on the 4th day and then declined to its initial level thereafter. The increase was about 3-fold, and was not observed in leaves of S. commersonii control plants.

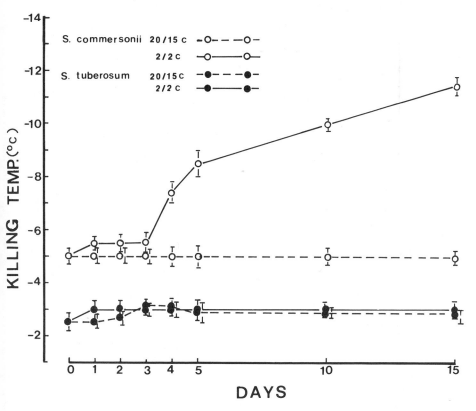

FIG. 6. Changes in frost hardiness in potato leaves during cold acclimation.

D. Soluble Protein Content

Figure 5 illustrates changes observed in soluble protein content of leaves of S. commersonii and S. tuberosum during cold acclimation. Treated S. tuberosum plants, grown at 2°C, showed slightly higher leaf soluble protein content than control plants. In S. commersonii, leaves of control plants maintained a relatively constant level of soluble proteins, but leaves of treated plants showed a dramatic increase. This significant increase was initiated on the 3rd or 4th day, peaked on the 5th day, and remained at this high level during the 15 days of sampling period.

E. Frost Hardiness

Figure 6 shows daily levels of frost hardiness in S. commersonii and S. tuberosum plants. S. tuberosum leaves typically failed to acclimate and were always killed at -3°C.

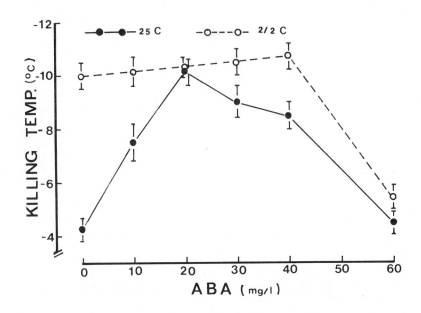

FIG. 7. The effects of various ABA concentrations on the frost hardiness of S. commersonii stem-cultured plants. Abscisic acid (mg/l) was added to the medium just prior to transferring plants on to the medium for 2°C or 25°C temperature conditioning. Frost hardiness (killing temp) was evaluated after 15 days of conditioning.

Table I. The effects of low temperature, ABA (20 mg/l) and cycloheximide (10⁻⁵M) on the acclimation of S. commersonii stem-cultured plants. Frost hardiness was evaluated after 15 days of treatment.

Treatment	Killing temperature (°C)
Control	−4.0 ± 0.3
LTa	−10.0 ± 0.5
LT + cycloheximide (added at 0 day)	−4.0 ± 0.3
LT + cycloheximide (added at 5th day)	−8.0 ± 0.5
ABA	−10.0 ± 0.5
ABA + LT	−10.0 ± 0.4
ABA + cycloheximide (added at 0 day)	−4.0 ± 0.3

a LT = low temperature (2°C)

Control plant leaves of S. commersonii were frost hardy to -5°C, and leaves of treated plants began acclimating on the 4th day, and ultimately became hardy to -12°C (killing temperature) by the 15th day.

F.   Abscisic Acid and Cycloheximide Treatments

As shown in Fig. 7, ABA was capable of inducing frost hardiness in leaves of S. commersonii whether plants were grown under a warm (25°C) or cold (2°C) temperature regime. The hardiness of leaves from stem-cultured plants grown at 25°C after 15 days of growth increased from a killing temperature of -4°C to a maximum of -10°C when treated with 20 mg per liter concentration of ABA. Leaves of plants grown at 2°C after 15 days growth increased in hardiness to -10°/-11°C when treated with a concentration up to 40 mg/l of ABA. Plants in media treated with high concentrations of ABA did not acclimate as well, and showed senscense, growth cessation, and in some case, tuberization. Plants grown in the low temperature regime tolerated higher concentrations of added ABA without showing a decrease in frost hardiness.

When cycloheximide, a cytoplasmic protein synthesis inhibitor, was added to the medium of stem-cultured plants at the beginning of low temperature (2°C) acclimation, or at the beginning of ABA treatment in the warm temperature (25°C) regime, it completely inhibited the development of frost hardiness. However, when cycloheximide was added on the 5th day to plants in the low temperature (2°C) regime, the development of frost hardiness was not inhibited. Table 1 summarizes the results of this study which suggests that protein synthesis is essential for cold acclimation, that ABA induces protein synthesis and subsequent frost hardiness, and that these ABA effects occur within 5 days of low temperature treatments, i.e. note that endogeneous ABA levels peaked within 4 days (Fig. 4), and that addition of cycloheximide to the medium after 5 days of low temperature treatment did not inhibit cold acclimation (Table 1).

VI. DISCUSSION

Sugar accumulation during cold acclimation have been observed in numerous plant species (22). Table 2 cites references of studies which have indicated that osmotic concentration, ABA status and protein synthesis also increase in plants during low temperature conditioning.

Table II. List of references concerning the studies of osmotic
          concentration, and the status of ABA and proteins
          during low temperature conditioning.

Parameters	Observations	Plants	Ref.
Osmotic concentration	Increase	Cabbage & Clover	23
		Fatsia japonica	25
		Mulberry	32
		Black locust	34
	Causing ABA accumulated	Cucumber (seedlings)	29
		Xanthium	41
Abscisic acid	Low temp causing ABA accumulated	Tomato	8
		Kharkov wheat	40
	Increasing cold tolerance	Tobacco (callus)	1
		Potato	5
		Cotton	28
		Cucumber	29
		Alfalfa	30
	Inducing new protein synthesis	Sugar cane (internodal tissue)	11
		Barley (aleurone layer)	17
		Bean (embryo) and	
		Rape (embryo)	36
Proteins	Increase	Potato (leaf)	3
		Alfalfa (root)	11
		Winter rape (leaf)	19
		Spruce (needle)	21
		Dogwood (bark)	24
		Red pine (needle)	27
		Black locust (bark)	34
		Wheat (leaf)	38,40
	New protein synthesized	Black locust	2
		Dogwood	7
		Winter rape	20
		Wheat	31
	Preventing hardiness induction by cycloheximide	Chlorella	12
		Winter rape	19
		Wheat	38

A.   The Role of Sugar

According to Levitt (22) the accumulation of sugars during cold acclimation is due to: 1) the excess of photosynthesis over respiration and growth, and 2) the starch-sugar conversion.   In leaf tissue of tuber-bearing Solanum species, starch and sugars increased simultaneously during acclimation (3). The observed sugar accumulation (Fig. 2) appears to be more likely due to the excess of photosynthesis over respiration and growth rather than a starch-sugar conversion.   In addition the rapid increase in sugars in 3 days in S. commersonii at 2°C makes a starch-sugar conversion unlikely.   Huner and MacDowell (13) found that cold adapted rye had an increased RuBPCase stability under freezing stress.   This could enable plants to retain photosynthetic capability at low temperatures.   It was also observed that the configuration of RuBPCase from S. commersonii has fewer exposed SH groups which make it structurally more stable under stress than the enzyme from S. tuberosum (14).   At about 0°C, RuBPCase from S. commersonii showed a higher activity of CO_2 fixation than the same enzyme from S. tuberosum (14).   These observations help to explain why sugar accumulated more quickly and abundantly in leaves of the former than of the latter.

During cold acclimation, sugar accumulation can account for the major proportion of the increase in osmotic concentration observed (27).   In potato leaves, at the early stage of acclimation (up to 3rd day), the sugar content (Fig. 2) and osmotic concentration (Fig. 3) showed a parallel relationship.   However, as acclimation continued, the osmotic concentration declined while total sugar content remained at a high level. Qualitative analysis of sugars in plant species has shown that sucrose and glucose are the dominant sugar constituents in leaves prior to low temperature treatment, but that high molecular weight sugars such as raffinose and starchyose tended to accumulate with prolonged exposure to low temperatures.   It may be possible that a transition of smaller molecular weight sugars (glucose, sucrose, etc.) to larger sugar molecular occurs during a prolonged cold temperature treatment.   Therefore, although high total sugar levels were observed at the later stage of acclimation, the osmotic concentration might actually decrease due to the lower osmotic activity of bigger molecules.

It is generally assumed that an increase in cellular sugar content lowers the freezing point of cell sap (22), thus lowering the killing temperature of the cells.   However, we observed no correlation between the level of frost hardiness and cell sap concentration in the potato (6).   Furthermore, changes in osmotic concentration (Fig. 3), did not parallel the changes in frost hardiness (Fig. 6) during acclimation.   For example, at the 3rd and 4th days of low temperature treatment, although a significantly high osmotic concentration was

observed in S. tuberosum, it did not exhibit any increase in
frost hardiness. In S. commersonii, the highest osmotic concentration occurred before any increase in frost hardiness. We
conclude that the increase of sugar content makes no significant or direct contribution to the increase of frost hardiness
in the potato by lowering the freezing point of cell sap or by
protecting cells from freezing injury.

Sugar accumulation with resultant high osmotic concentrations during cold acclimation usually occurred prior to any
measurable increase of frost hardiness (21,25,27,32,34). The
phenomenon of sugar increase in plants during cold acclimation
is an universal event. It seems illogical that this ubiquitous
increases in sugar content during cold acclimation is causally
unrelated to the development of frost hardiness in plants;
except as a source of energy.

It has been demonstrated that an increase in osmotic concentration due to water stress can trigger an elevation in
endogenous ABA levels (29,41). It has also been shown that ABA
can   induce   frost   hardiness   in   many   plant   species
(1,5,28,29,30). Therefore, one might hypothesize the sugar
accumulation during cold acclimation might play a key, but
indirect, role in the development of frost hardiness by
triggering the endogenous increase in ABA levels due to the
resulting elevation of osmotic concentration, i.e. a process by
which plants may convert the signals provided by a physical
change in environment into biological adjustments.

The osmotic concentration required to trigger ABA build-up
is species-specific (29,41). Perhaps, S. tuberosum failed to
develop frost hardiness in response to low temperature treatment because its sugar increase was not sufficient to raise the
osmotic concentration to the critical threshold level needed to
trigger the endogenous ABA build-up. This line of speculation
is supported by an observation (5) that exogenously supplied
ABA did indeed increase the frost hardiness of stem-cultured S.
tuberosum plants after 5 days of treatment.

B.   The Role of Abscisic Acid

Changes in hormonal balance have been shown to be associated with cold acclimation (15,18,30,39,40). The so called
"translocatable hardiness promotor" is thought to be a hormone
(9,15,16). In winter rape leaves, a decrease in a GA-like substance and an increase in an inhibitor (possibly ABA) were
observed during cold acclimation (19). In alfalfa seedlings,
Waldman et al. (39) concluded that ABA/GA balance regulates the
cold acclimation capacity. Exogenously applied ABA, or other
treatments which induce an increase in endogenous ABA levels,
have been observed to increase the tolerance of plants to low
temperature (1,5,28,29,30), and an increase in endogenous ABA
level is observed when plants are exposed to low temperature
(8,18,40).

Because of the similar involvement of ABA in many environmental stresses, ABA has been proposed as a common mediator for plant stress responses (8), although its role is not fully understood. The involvement of ABA metabolism in potato cold acclimation seems very likely, and the elevation of ABA level observed in potato during acclimation (Fig. 4) is likely induced by increased sugar content and the resulting elevation of osmotic concentration.

The requirement of protein synthesis for cold acclimation has been demonstrated in many plant species (12,19,38). Specific proteins are preferentially synthesized during cold acclimation (2,7,20,31). Interestingly, it has been demonstrated that ABA is able to induce synthesis of certain protein species (10,17,36). During seed maturation, for example, the accumulation of major storage proteins is controlled by the ABA metabolism (36). Evidence from this laboratory also indicates that an elevation of ABA during cold acclimation may induce protein synthesis (new?) responsible for the increase of frost hardiness. Specifically it has been observed that: 1) Increase in frost hardiness was initiated after endogenous ABA levels peaked (Fig. 4), 2) The increase in soluble protein content paralleled and followed the ABA peak (Fig. 5), 3) ABA can substitute for exposure of plants to low temperature in inducing frost hardiness (Fig. 7), 4) ABA and low temperature did not induce an increase of frost hardiness when plants were exposed to a protein synthesis inhibitor at beginning of acclimation (Table 1), 5) The protein synthesis inhibitor did not inhibit cold acclimation (Table 1) if it was supplied after (i.e. at the 5th day) the endogenous ABA in plants had peaked (Fig. 4) and the protein synthesis has been initiated (Fig. 5).

Although the ABA content in stem-cultured plants of S. commersonii was not assayed, it seems to be a reasonable assumption that the change in ABA content in stem-cultured plants during cold acclimation is similar to that in pot-grown plants (Fig. 4) because of their similar hardening responses under identical environmental conditions. The significant correlation between the net increase of soluble protein and frost hardiness in acclimatable potato during cold acclimation has been documented (3). At the 5th day during a 2°C acclimation, a peak in plasma membrane protein particle aggregation was observed in S. acaule. Solanum acaule is capable of acclimation (4), and the synchrony of timing of protein synthesis, membrane protein particle aggregation, and cold acclimation supports the view that the occurrence of protein synthesis at the early stage of low temperature treatment and induction of frost hardiness are causally related.

In leaves of plants grown at warm temperatures, an elevated ABA level was observed within 5-6 hr after an increase in leaf water potential (41). At low temperature treatment conditions it is not so clear what the time lag between the increase in osmotic concentration and the increase in endogenous ABA.

There was a time-lag of about 48 hr between the peak of osmotic concentration (2nd day at 2°C) and the peak of ABA (4th day at 2°C) in S. commersonii plants, but the infrequency of sampling could give rise to an error. Assuming an average of Q_{10} of about 2 for most physiological reactions, and a temperature difference of 23°C between most physiological studies (25°C) and our low temperature 2°C acclimating treatment, theoretically, there should be about a 45 hr time lag between the peak of osmotic concentration and the peak of ABA concentration in the potato. This is quite close to the observed time of 48 hr (Fig. 3 and 4).

VII.   SUMMARY

When potato plants are exposed to low temperatures, sugars increase (Fig. 2) resulting in rapid increase in cellular osmotic concentration. When the increase of osmotic concentration (Fig. 3) reaches a critical level ABA is released from a bound form (?) or synthesized (?) (Fig. 4). This may be a crucial step in a plant's translation of an environmental signal into biological adjustments during cold acclimation. The elevation of endogenous ABA appears to induce the synthesis of specific proteins which are responsible for the increase of frost hardiness. In non-photosynthetic organs, such as roots (11) and bark (24,32,34), the starch-sugar conversion may contribute to the accumulation of sugars which results in a high osmotic concentration (32,34). In species which increase in sugar content at low temperatures but fail to cold harden such as S. tuberosum species specific (29,41) increase in sugars and osmotic concentrations may not be high enough to trigger the increase in ABA content needed to induce frost hardiness via protein synthesis.

ACKNOWLEDGMENT

The authors wish to thank Dr. C. J. Weiser, Professor of Horticulture at the Oregon State University for critical reading of the manuscript.

REFERENCES

1. Bornman, C.H., and Janson, E., Physiol. Pl. 48:491-493 (1980).
2. Brown, G.N., and Bixby, J.A., Physiol. Pl. 34:187-191 (1975).

3. Chen, H.H., and Li, P.H., Pl. Physiol. 66:414-421 (1980).
4. Chen, H.H., and Li, P.H., Pl. Physiol. 65:1146-1148 (1980).
5. Chen, H.H., Gavinlertvatana, P., and Li, P.H., Bot. Gaz. 140:142-147 (1979).
6. Chen, P.M., Burke, M.J., and Li, P.H., Bot. Gaz. 137:313-317 (1976).
7. Craker, L.E., Gusta, L.V., and Weiser, C.J., Can. J. Plant Sci. 49:279-286 (1969).
8. Dale, J., and Campbell, W.F., Pl. Physiol. 67:26-29 (1981).
9. Fuchigami, L.H., Evert, D.R., and Weiser, C.J., Plant Physiol. 47:164-167 (1971).
10. Gayler, K.R., and Glasziou, K.T., Planta 84:185-194 (1969).
11. Gerloff, E.D., Stahmann, M.A., and Smith, D., Pl. Physiol. 42:895-899 (1967).
12. Hatano, S., Sadakane, H., Tutumi, M., and Watanabe, T., Pl. & Cell Physiol. 17:643-651.
13. Huner, N.P.A., and MacDowall, F.D.H., Biochem. Biophys. Res. Commun. 73:411-420 (1976).
14. Huner, N.P.A., Palta, J.P., Li, P.H., and Carter, J.V., Can. J. Biochem. 59:280-289 (1981).
15. Irving, R.M., Pl. Physiol. 44:801-805 (1969).
16. Irving, R.M., and Lanphear, F.O., Pl. Physiol. 43:9-13 (1968).
17. Jacobsen, J.V., Higgens, T.J.V., and Zwar, J.A., in "The Plant Seed: Development, Preservation and Germination" (I. Rubenstein, R.L. Phillips, C. Green and G.G. Gengenbach, eds.), pp. 241-262. Academic Press, New York, 1979.
18. Kacperska-Palacz, A., in "Plant Cold Hardiness and Freezing Stress" (P.H. Li and A. Sakai, eds.), pp. 139-152. Academic Press, New York, 1978.
19. Kacperska-Palacz, A., Dlugokecka, E., Brietenwald, J., and Weislinska, B., Biologia Planta 19:10-17 (1977).
20. Kacperska-Palacz, A., Jasinska, M., Sobczyk, E.A., and Weislinska, B., Biologia Planta 19:19, 18-26 (1977).
21. Kandler, O., Dover, C., and Ziegler, P., Ber. Deutch. Bot. Ges. Bd. 92:225-241 (1979).
22. Levitt, J., Responses of Plants to Environmental Stresses, Vol. 1. Academic Press, New York, 1980.
23. Levitt, J., and G.W. Scarth. Can. J. Res. C14:267-284 (1936).
24. Li, P.H., and Weiser, C.J., Plant & Cell Physiol. 7:475-484 (1966).
25. Nishiyama, I., J. Facul. of Sci., Univ. of Tokyo, Sec. III, Vol. IX, Part 6-9, pp. 219-239 (1966).
26. Paldi, E., and Devay, M., Phytochem. 16:177-179 (1977).
27. Pomeroy, M.K., Siminovitch, D., and Wightman, F., Can. J. Bot. 48:953-967 (1970).
28. Rikin, A., Atsmon, D., and Gitler, C., Pl. & Cell Physiol. 20:1537-1546 (1979).
29. Rikin, A., Blumenfeld, A., and Richmond, A.E., Bot. Gaz. 137:307-312 (1976).

30. Rikin, A., Waldman, M., Richmond, A.E. and Dovrate, A., J. Exp. Bot. 26:175-183 (1975).
31. Rochat, E., and Therrien, H.P., Can. J. Bot. 53:2411-2416 (1975).
32. Sakai, A., Inst. Low Temp. Sci., Ser. B 11:1-40 (1962).
33. Sarhan, F., and Daoust, H.J., Physiol. Pl. 35:62-65 (1975).
34. Sminovitch, D., Can. J. Bot. 41:1301-1308 (1963).
35. Setter, T.L., Brun, W.A., and Brenner, M.L., Plant Physiol. 67:774-779 (1981).
36. Sussex, I.M., Dal, R.M.K., and Creuch, M.L., in "Genome Organization and Expression in Plants" (C.J. Leaver, ed.), pp. 283-289. Plenum Press, New York, 1980.
37. Toivio-Kinnucan, M.A., Chen, H.H., Li, P.H. and Stushnoff, C., Pl. Physiol. 67:478-483 (1981).
38. Trunova, T.I., and Zvereva, G.N., Soviet Pl. Physiol. 24:311-316 (1977).
39. Waldman, M., Rikin, A., Dovrate, A., and Richmond, A.E., J. Exp. Bot. 26:853-859 (1975).
40. Wightman, F., in "Plant Regulation and World Agriculture" (T.K. Scott, ed.), pp. 327-377. Plenum Press, New York, 1979.
41. Zeevaart, J.A.D., in "Plant Growth Substances" (N.B. Mandava, ed.), pp. 99-114. Amer. Chem. Soc., Washington, D.C., 1979.

FACTORS INFLUENCING HARDENING AND SURVIVAL IN WINTER WHEAT

L. V. Gusta, D. B. Fowler and N. J. Tyler

Crop Development Centre
University of Saskatchewan
Saskatoon, Saskatchewan Canada

I.   INTRODUCTION

A species' ability to tolerate low temperatures is one of the primary factors determining its area of adaptation and distribution. Survival is accomplished primarily through the process of cold acclimation or hardening (24). In winter wheat the process of cold acclimation is under the control of a genetic system induced by low temperature (12,29,35). The energy required for the process is supplied either from seed reserves or from photosynthesis. Any form of stress, e.g. salinity, nutrient deficiencies, drought, etc. may interfere with the hardening process, thereby resulting in the plant not being able to reach its full cold hardiness potential.

Winter wheat plants have a distinct killing temperature. Upon exposure to a critical temperature, one of the first signs of damage is the loss of membrane semipermeability (23). Recent results have established that injury occurs during freezing upon passing through the critical temperature and suggest that the primary site of injury is the plasma membrane (32).

This report is primarily concerned with the hardening process, how it is affected by various stresses and the role of membranes in relation to freezing and injury in winter cereals.

II.   SINGLE PLANT SELECTION TECHNIQUES

Field survival is considered to be the ultimate test of a cultivar's winter-hardiness, but is often inconclusive due either to complete winterkill or a lack of it (13). Variation in stress levels within field trials mask small but important

TABLE I. Simple correlation coefficients among several variables measured on cold acclimated plants from 36 winter wheats (Study I)[a] and 14 winter wheats (Study II).

	Study I			Study II	
	FSI	LT_{50}		FSI	LT_{50}
LT_{50}	-95		LT_{50}	-98	
Crown H_2O	-88	89	Uronic acid	-71	75
Crown F.W.	72	68	Proline	21	-18
Crown D.W.	-42	37	Osmotic potential	50	-52
Shoot D.W.	-61	64	Cell buffering	71	-71
Leaf H_2O	-89	89	Crown H_2O N_2	-76	87
Leaf F.W.	-71	71	Crown N_2	-76	83
Leaf D.W.	-58	58	Viscosity	84	82
Leaf No.	-22	16	ATP	2	4
No. tillers	-41	32			
No. coleoptiles	-19	15			
No. crown tillers	-47	41			
Height	-68	66			
Mesocotyl length	-23	21			
Crown depth	38	-29			
Erectness	-85	81			
Heading date	30	-26			
Crown P	-81	80			
N	-70	76			
K	-66	63			
S	-47	59			
Mn	-64	75			
Mg	-63	66			
Fe	-69	65			
Total sugars	80	-81			
Glucose	67	-64			
Fructose	72	-69			

FSI - Field Survival Index
Values greater than 33 and 43 (Study I) and 51 and 64 (Study II), sign ignored, are significant at the 0.05 and 0.01 level respectively; decimals omitted.

[a]   Fowler, D.B., Gusta, L.V., and Tyler, N.J., Crop Sci. 21:896-901 (1981).

differences among cultivars even when differential winterkill does occur (11).

Due to the limitations inherent in field trials, rapid and efficient methods for predicting cold hardiness have been sought. Controlled freeze tests overcome many of the difficulties associated with field tests, but are of little value in selecting for superior genotypes in a segregating population. Large populations are required and the method may be destructive.

Many biochemical, morphological, and physiological changes are known to occur during cold acclimation (24,42), however it is difficult to determine which are causal. Thirty-four biochemical, physiological and morphological characters of thirty-six winter wheat cultivars were evaluated to determine their usefulness in winter survival prediction tests (Table 1) (14). Close linear associations were found among many of the characters indicating that there are a number of possible screens to supplement field survival trials. Crown LT_{50} gave the highest correlation with field survival. Since the LT_{50} method is destructive and requires a large number of plants, it is limited to non-segregating populations. Tissue water content combined with a measure of plant erectness provided nearly as much information on field survival as did LT_{50}. Both leaf water content and plant erectness are nondestructive to the crown and therefore should provide a method to the breeder in selecting for hardy genotypes.

A close correlation (0.9) between the LT_{50} of leaf segments and field survival was found among thirty winter wheat cultivars (unpublished results). This also provides the breeder with a simple, fast, nondestructive method of comparing the cold hardiness of segregating populations.

III.   INDUCTION OF COLD HARDINESS

The process of cold acclimation in winter wheat is under the control of a genetic system induced by low temperature (12,29,35). Unlike in certain deciduous trees (42) photoperiod is not involved other than to supply photosynthates. It is generally considered that temperatures below 10°C are effective in inducing the system. Under controlled conditions six weeks of cold exposure (temperatures above 0°C) are required to develop the cultivars' full hardiness potential. However a continuous frost of -3°C will shorten the hardening period considerably (20). Under controlled environment conditions winter wheat crowns of cold hardy cultivars can readily acclimate up to 10°C in 7 days. Norstar seedlings hardened in petri dishes at 2°C/0°C 12h photoperiod acclimated from -3°C to -7°C in 24 hours and from -3°C to -11°C within 4 days.

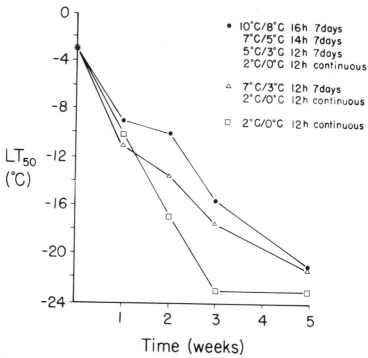

FIG. 1.   Cold hardening of Norstar crowns under different hardening regimes.

Experiments were conducted to determine how rapidly crowns of Norstar winter wheat would acclimate under different temperature regimes.   Plants transferred from a growing regime of 15°C, 16h photoperiod directly to 2°C/0°C 12h photoperiod cold hardened faster and attained a higher level of hardiness as compared to similarly grown plants transferred either to 10°C/8°C 16h photoperiod or 7°C/3°C 12h photoperiod (Fig. 1).

Winter wheat crowns acclimated for two weeks will attain a higher level of hardiness if subjected to a continuous frost of −3°C (20).   Further studies have shown that plants acclimated for as little as 2 days at 7°C/3°C and then transferred to a constant −3°C acclimate and maintain a similar level of hardiness as plants held at 7°C/3°C.   To determine the minimum period of hardening prior to storage at −3°C, winter wheat plants growing at 15°C, 16h photoperiod were hardened at 7°C/3°C for 0, 3, 6, 12 and 24h.   Hardiness of the crowns was determined at weekly intervals over a five week period (Fig. 2).   Plants transferred after 0, 3, and 6h of hardening could tolerate approximately −9°C to −12°C of frost after 7 days at −3°C.   However after 14 to 21 days at −3°C these plants lost whatever hardiness they attained and eventually succumbed to

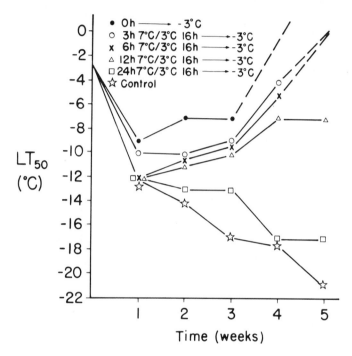

FIG. 2. Cold hardening of Norstar crowns at -3°C after pretreatment at 7°C/5°C, 16 h photoperiod for either 1, 2 or 3 days.

the -3°C treatment. Plants hardened for 12 hours did not lose their hardiness as rapidly as the plants exposed to a shorter hardening period, and were still alive after 5 weeks at -3°C. It appears that plants hardened for 24h at 7°C/3°C and transferred to -3°C are able to harden comparable to plants maintained at 7°C/3°C, 14h photoperiod continuously. These results suggest that cold hardening is cumulative, requiring a certain level of hardiness factors to maintain a high degree of hardiness at subzero temperatures. The results also suggest that the process can occur very effectively at temperatures close to zero and to a limited degree at temperatures just below zero. Preliminary tests suggest that a longer hardening period is required for plants stored at -5°C.

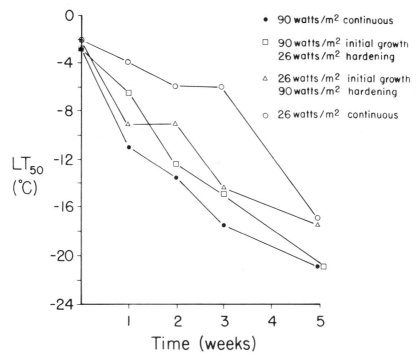

FIG. 3. The effect of light on the hardening of Norstar crowns.

IV.   THE EFFECT OF CERTAIN STRESS FACTORS .ON HARDENING

A.   Energy Source

The level of cold hardiness of winter wheat plants has been shown to be directly related to light intensity and photoperiod (4,29,36). Hardiness may also be induced by feeding plants sucrose at low temperatures (37). Germinating seeds may also attain a high degree of hardiness in the dark, by using the endosperm as an energy source (2,3).

The hardening patterns of plants grown and hardened under high light were compared for plants: a) grown under high light and hardened under low light; b) grown under low light and hardened under high light; and c) grown and hardened under low light (Fig. 3). As expected, plants grown and hardened under low light did not harden as fast or reach the same level as plants grown and hardened under high light. Plants grown under high light and hardened under low light were comparable to plants grown and hardened under high light.

Winter wheat plants maintained at -3°C in the dark longer than 90 days tended to be a few degrees less hardy than plants

TABLE II. The effect of cooling rate on the frost tolerance of winter wheat crowns.

Freezing rate	LT_{50} (°C)*
-3°C 12 hr, 2°C/hr	-21
-3°C 12 hr, 5°C/hr	-18
-3°C 12 hr, 10°C/hr	-17
-3°C 12 hr, 15°C/hr	-13
-3°C 2 hr, 2°C/hr to -6°C, 10°C/hr	-21
-3°C 2 hr, 2°C/hr to -6°C, 15°C/hr	-18

* mean of seven cultivars. Kharkov, Ulianovka, Alabaskaja, Winalta, Norstar, Sundance and Yogo.

maintained under low light conditions at -3°C (unpublished results).

B. Freezing Rate

Ice in hardy plants under slow freezing conditions is confined to the extracellular spaces. The plasma membrane, which is probably the main barrier to water efflux, partitions ice away from the protoplasm so that the protoplast remains unfrozen. The water in plants freezes as is expected for a dilute salt solution. As long as the plasma membrane remains intact the extracellular ice is essentially solute-free. Over 60% of the total freezable water in plant cells is frozen by -5°C (19). Under natural freezing conditions it is highly unlikely that water efflux would limit the freezing process. However, for wet soil conditions, a sudden drop in temperature may produce some unusually high freezing rates. Also this aspect of freezing is important in controlled environment freeze tests.

Since such a high percent of the total water crystallizes in the first few degrees of a frost the initial freezing rate is critical. If freezing is initiated at -3°C in a sample which is held for several hours at this temperature to allow ice growth throughout the tissue and then cooled at 13°C/h, the killing temperature is much warmer as compared to plants cooled at 2°C/h. However, a freezing rate of 13°C/h has little effect if the plants after being frozen initially at -3°C are cooled at 2°C/h to -6°C and held at this temperature till freezing reaches equilibrium (Table 2).

TABLE III. Effect of varying crown water content on LT_{50} (°C).

Cultivar	A		B		C	
	LT_{50}	H_2O(%)	LT_{50}	H_2O(%)	LT_{50}	H_2O(%)
Sundance	-24	73	-22	-	-21	78
Winalta	-23	72	-24	-	-15	77
Norstar	-24	73	-24	-	-17	77
Alabaskaja	-23	71	-24	-	-17	76

A - Frozen in moist sand.
B - Held in excessively wet sand for 15 min prior to freezing.
C - Held for 24 hours in excessively wet sand prior to freezing.

C. Tissue Water Content

With cold hardening the crown water content drops several percent and is highly correlated with survival. However, a drop in tissue water content does not mean the tissue has a water deficit. Tyler et al. (38) could not detect any difference in crown water potential between tender and hardened winter wheat. In all probability the decrease in total water content is due to an increase in dry matter or loss of water in the extracellular spaces.

Although there is a gradual decrease in percent crown water during cold acclimation of winter cereals there is not a reduction in the weight of water per crown. During acclimation dry matter accumulates at a faster rate then water with the result that percent water decreases.

Gullord et al. (18) found that by increasing the water content of cereals by placing the crowns in wet sponges, the killing temperature occurred at a warmer temperature. Studies in our laboratory confirmed these findings. If water content of wheat crowns was increased by less then 5%, the LT_{50} was reduced approximately 6°C (Table 3). It is assumed that the additional water is confined to the extracellular spaces. Therefore, as suggested by Olien (28), extracellular ice may play a role in frost injury.

D. Temperature

It has been shown that frost injury increases with the length of exposure to cold as the killing temperature is approached (1,17,20). Fully hardened winter wheat crowns can tolerate -23°C for 12h, but can only tolerate -18°C for 24h and -15 for 5 to 6 days (Table 4). The effect of exposure time

TABLE IV. Duration of freeze effect on survival temperatures for winter wheat crowns.

Exposure time (days)	Survival temperature (°C)
0.04	-23
0.5	-23
1	-18
6	-15
15 plus	-12

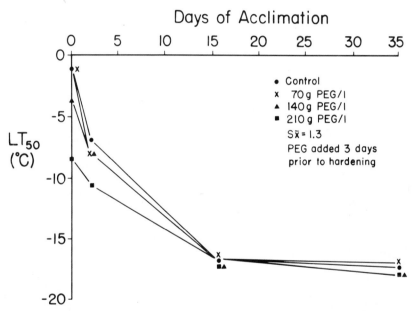

FIG. 4. The effect of a water stress on the cold hardening of Kharkov winter wheat crowns grown in hydroponics.

appears to be more associated with herbaceous material than with hardy woody material. Eight different species of hardy woody perennials were kept at -75°C for up to two weeks without any measureable effect on hardiness (unpublished results).

E. Water Deficit

Cold acclimation in cereal plants is paralleled by a decrease in tissue water to dry weight ratio (7,19,26,27).

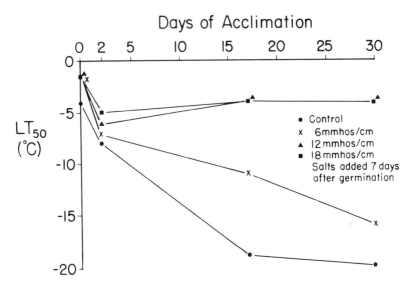

FIG. 5.   The effect of salinity on the cold hardening of Kharkov winter wheat crowns grown in hydroponics.

Artificial desiccation has been shown to increase cold hardiness in red osier dogwood (5,6,25), cabbage (8) and in winter cereals (9,33).

When winter wheat plants were water stressed with PEG-20,000 MW prior to hardening, both tender crowns and crowns hardened for 2 days were hardier than control plants (Fig. 4, 41). After 16 days of hardening there were no differences in hardiness. When the water stress was extended into the acclimation period to provide different periods of exposure, the stressed plants were hardier only in the tender state. In both studies, the crown moisture content of the stressed plants was lower than that of the controls before acclimation. In the long-term study, the moisture content of the stressed plants remained lower than the controls throughout the acclimation period. As hardiness increased, both the crown moisture content and osmotic potential decreased. Most of the variability in LT_{50} could be explained by changes in leaf osmotic potential and crown moisture content.

F. Salinity

Field observations have indicated that there is a reduction in cold hardiness of winter wheat and rye under conditions of salt stress (15). Reductions in cold hardiness of lilac and ash trees exposed to highway de-icing salts have also been reported (34). The presence of salts ($MgSO_4$ and Na_2SO_4) in the nutrient solution of winter wheat reduced hardiness in

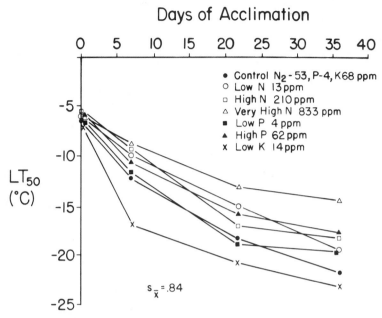

FIG. 6. The effect of N, P, and K on the cold hardening of Kharkov winter wheat crowns grown in hydroponics.

tender plants and decreased the rate of hardening in acclimating plants (Fig. 5, 39). The presence of salts also reduced crown moisture content prior to cold acclimation.

G.   Nutrition

The addition of nitrogen fertilizer to plants prior to acclimation has been found to decrease cold hardiness (10,16,31), although there have been reports where the addition of nitrogen had no effect on cold hardiness (31). Applications of phosphorus and potassium are generally associated with increased cold hardiness (22,31,43).

The rate of cold acclimation and the maximum level of hardiness of winter wheat grown in modified Hoagland's solutions was influenced by the mineral composition of the solution (Fig. 6, 40). Plants grown in the control solution or in those low in K or P acclimated the fastest and were the most hardy. Plants grown in high and very high levels of N and high P were the least hardy. Over the acclimation period, crown dry weight increased, crown moisture content and leaf osmotic potential decreased. Leaf water potential did not change over the acclimation period. Crown moisture content was most closely related to LT_{50}.

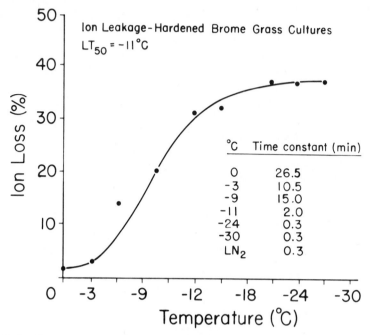

FIG. 7. Percent ion leakage and rate of ion loss from cold hardened brome grass cell aggregates.

V.   MEMBRANES AND FROST DAMAGE

Upon exposure of a tissue to a critical temperature, one of the first measurable events of injury is the loss of membrane semipermeability (23).   Ions start to leak from the tissue immediately upon thawing with no measurable lag period. Depending upon how homogenous the tissue is, there is little or no ion leakage prior to exposure to the critical temperature (Fig. 7).   It appears from time constant data for ion leakage that the cells are severely lacerated upon passing through the critical temperature.   For example the time constant for ion leakage for control brome grass cell cultures was 26.5 min. The time constant for cells frozen to -9°C was 15 min as compared to 0.3 minutes for cells exposed to a critical temperature of -24°C.   The increased ion leakage at -9°C as compared to the control may be due to a few cells being killed at -9°C or the effect of a frost on membrane permeability.

Under conditions of slow cooling and freezing rates in cold hardened winter wheats, ice first forms in the extracellular spaces and is partitioned away from the cell interior by the plasma membrane.   The rate of freezing in tissue is dependent upon many things including the following: the degree of

TABLE V. First order time constants for freezing from -3°C to -5°C and thawing from -5°C to -3°C of fully hardened winter wheats and woody perennials.

Species	Cultivar	LT$_{50}$ (°C)	Time constant (sec)					
			Freezing			Thawing		
			Alive		Dead	Alive		Dead
Triticum aestivum L.	Cappelle	-12	68(39*)	243(61)	64(100)	66(18)	210(82)	100(100)
Triticum aestivum L.	Dawsons Golden Chaff	-15	70(14)	130(86)	100(100)	69(24)	194(75)	108(100)
Secale cereale L.	Puma	-28	60(25)	176(75)	57(100)	95(100)		79(100)
Abies lasiocarpa Nutt.	Alpine Fir	-75	23(11)	177(89)	115(100)	33(9)	135(91)	
Thuja occidentalis L.	Cedar	-75	20(11)	175(89)	112(100)	42(11)	135(89)	117(100)
Cornus stolonifera Michx.	Red-osier Dogwood	-196	42(33)	222(67)	62(100)	38(25)	208(75)	91(100)

* Fraction of water (%) associated with rate constant.

TABLE VI. Effect of temperature and dehydration on ion loss from winter wheat leaves.

Cultivar	LT_{50} (°C)	Water g H_2O/g D.Wt.	Temp. (°C)	Time (min)	Ion loss (%)
Cappelle-Desprey	-3	1.5	0		26
			-12	10	48
			-12	25	66
Norstar	-3	2.8	0		16
			-10	10	25
			-10	15	25
			-10	25	29
			-10	150	51
Norstar	-7	2.0	0		15
			-13	25	37
			-14.2	15	39

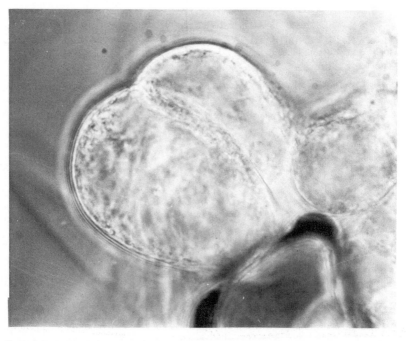

Fig.8a

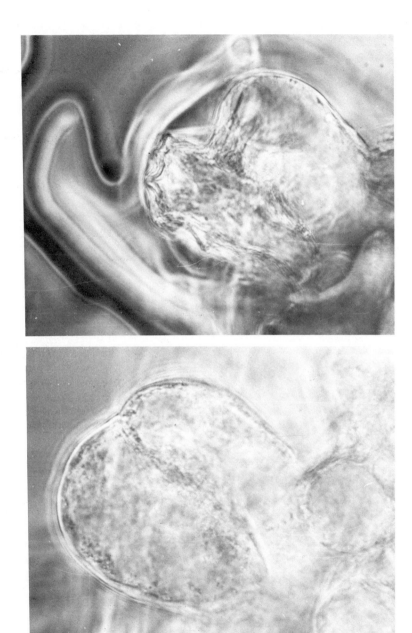

FIG. 8. Freezing and thawing of brome grass cell suspension culture cells a) cells in the non frozen state, b) cells
frozen to -7°C, c) thawed cells frozen to their critical temperature (-20°C).

supercooling, the rate of water efflux (membrane permeability), the rate of heat removal and the rate of ice front growth. To determine the role of membranes in freezing, samples were pre-frozen to -3°C to avoid some of the above factors. The time constant for freezing and thawing was determined from -3°C to -5°C to -3°C on a range of tissues differing greatly in hardiness (Table 5). Also the time constants for live and freeze killed tissues were compared. In these experiments the temperature was changed quickly (approximately 20°C/min) to ensure that the primary factor limiting freezing would be the contribution of the membrane. In living tissue, two fractions of water, each with a different time constant, were observed during freezing and thawing. The time constants for the major fractions of water of frost killed crowns were markedly smaller than those for living crowns. Also only one fraction was observed in frost killed tissue. There was not a good correlation between LT_{50} and the time constants for freezing and thawing.

There is convincing evidence to indicate that in winter wheat injury occurs during freezing immediately upon passing through a critical temperature and that the primary site of injury is the plasma membrane. Although the cells are severely dehydrated by freezing it appears that dehydration alone is not the sole cause of injury. Nearly all of the plants' freezable water is frozen by -10°C (19). Thus there is little additional dehydration of the cells below this temperature, yet many of our plants are killed at much lower temperatures, e.g. Puma rye -30°C, bentgrasses -50°C.

It has been suggested that one form of frost injury in winter wheat is due to a membrane structural transition (32). This evidence is based on the findings that upon passing through the critical temperature there is a change in the environment of unfrozen cellular water; Mn^{2+} ion becomes permeable to the plasma membrane and there is a marked reduction in the rate constant for water movement. These determinations were all done on tissue in the frozen state.

In a recent study of phospholipids extracted from cereal crowns, the membrane lipid phase transition did not correspond to the LT_{50} of the tissue (21). However, there on the average a 20°C difference in the phase separation temperature of hardened and tender tissue. Further work is required to establish if the plasma membrane has the same phase separation temperature as the total lipids.

If tender or hardy leaf segments are supercooled below their respective critical temperatures and then shaken in distilled water, there is no difference in ion loss as compared to the control. However, if these segments are dehydrated to 50% of their original water content at 0°C and then supercooled below their critical temperature, enhanced ion leakage is observed (Table 6, unpublished results). The results suggest that freezing injury to the membranes are manifested in the following manner. First, with dehydration of the cell due to water efflux to extracellular ice, the cells shift, twist and collapse resulting in a strain on the membrane (Fig. 8). As the critical temperature is approached the plasma membrane starts to undergo a phase transition. The combination of dehydration and change in membrane structure results in complete loss of the semi-permeability properties. As shown in Fig. 8 brome grass cells from suspension cultures prior to freezing have a smooth cell wall (Fig. 8a), however upon freezing the cell wall collapses (Fig. 8b) resulting in a strain on the cell membrane-cell wall complex. If the cell is frozen to its critical temperature and then thawed the cell wall does not return to its original shape suggesting loss of turgor (Fig. 8a). In these hardy grass species injury appears to be due to a combination of both dehydration and temperature. This process is irreversible in contrast to chilling cell injury where there is no ice present to cause dehydration. These findings are consistent with the fact that large cells and cells with a high water content are not as hardy as small or undifferentiated cells and cells with a low water content.

REFERENCES

1. Ahring, R.M., and Irving, R.M., Crop Sci. 9:615-618 (1969).
2. Andrews, J.E., Can. J. Plant Sci. 28:1-7 (1958).
3. Andrews, C.J., Pomeroy, M.K., and de la Roche, I.A., Can. J. Bot. 52:2539-2546 (1974).
4. Barta, A.L., and Hodges, H.F., Crop Sci. 19:535-538 (1970).
5. Chen, P., Li, P.H., and Weiser, C.J., HortSci. 10:372 (1975).
6. Chen, P., and Li, P.H., Plant Physiol. 59:240 (1977).
7. Chen, P., and Gusta, L.V., in "Plant Cold Hardiness and Freezing Stress: Mechanisms and Crop Implications" (P.H. Li and A. Sakai, eds.). Academic Press, New York, 1978.
8. Cox, W., and Levitt, J., Plant Physiol. 57:553 (1976).
9. de la Roche, A.I., Plant Physiol. 59:S-36 (1977).
10. Dexter, S.T., Adv. Agron. 8:203 (1956).
11. Fowler, D.B., Crop Sci. 19:773-775 (1979).
12. Fowler, D.B., and Gusta, L.V., Can. J. Plant Sci. 57:751-755 (1977).
13. Fowler, D.B., and Gusta, L.V., Crop Sci. 19:769-772 (1979).
14. Fowler, D.B., Gusta, L.V., and Tyler, N.J., Crop Sci., 21:896-901 (1981).
15. Fowler, D.B., and Hamm, J.W., Can. J. Soil Sci. 60:439 (1980).
16. Freyman, S., and Kaldy, M.S., Can. J. Plant Sci. 59:853 (1979).

17. Greenham, C.G., and Daday, H., Aust. J. Agric. Res. 11:1-15 (1960).
18. Gullord, M., Olien, C.R., and Everson, E.H., Crop Sci. 15:153-157 (1975).
19. Gusta, L.V., Burke, M.J., and Kapoor, A.C., Plant Physiol. 56:707-709 (1975).
20. Gusta, LV., and Fowler, D.B., in "Stress Physiology in Crop Plants" (N. Mussel and R.C., Staples, eds.). Wiley-Interscience, New York, 1979.
21. Harvey, G.W., Gusta, L.V., Fork, D.C., and Bery, J.A., Plant, Cell and Environment 5: (1982).
22. Jung, G.A., and Smith, D., Agron. J. 51:585 (1959).
23. Levitt, J., "The Hardiness of Plants", 278 pages. Academic Press, New York (1956).
24. Levitt, J., "Responses of Plants to Environmental Stresses", 697 pages. Academic Press, New York (1972).
25. Li, P.H., and Weiser, C.J., Cryobiol. 8:108 (1971).
26. Metcalf, E.L., Cress, C.E., Olein, C.R., and Everson, E.H., Crop Sci. 10:362-365 (1970).
27. Newton, R., J. Agric. Sci. XII:1-17 (1922).
28. Olien, C.R., J. Theoret. Biol. 39:201-210 (1973).
29. Paulsen, G.M., Crop Sci. 8:427-430 (1968).
30. Pellet, N.E., J. Am. Soc. Hort. Sci. 98:82 (1973).
31. Ragan, P., and Nylund, R.E., HortSci. 12:320 (1977).
32. Rajashekar, C., Gusta, L.V., and Burke, M.J., in "Low Temperature Stress in Crop Plants" (J.M. Lyons, D. Graham, J.K. Raison, eds.). Academic Press, New York, 1979.
33. Siminovitch, D., and Cloutier, Y., Plant Physiol. 567:Report 345.
34. Sucoff, E., Hong, S.G., and Wood, A., Can. J. Bot. 54:2268-2274 (1976).
35. Svec, L.V., and Hodges, H.F., Can. J. Plant Sci. 52:165-175 (1972).
36. Trunova, T.I., Soviet Plant Physiol. 12:70-77 (1965).
37. Tumanov, I.I., and Trunova, T.I., Soviet Plant Physiol. 4:379-388 (1957).
38. Tyler, N.J., Masters Thesis, University of Saskatchewan, Saskatoon, Sask. (1979).
39. Tyler, N.J., Fowler, D.B., and Gusta, L.V., Can. J. Plant Sci. 61:543-548 (1981).
40. Tyler, N.J., Gusta, L.V., and Fowler, D.B., Can. J. Plant Sci. 61:879-885 (1981).
41. Tyler, N.J., Gusta, L.V., and Fowler, D.B., Can. J. Bot. 59:1717-1721 (1981).
42. Weiser, C.J., Science 169:1269-1278 (1970).
43. Zurawicz, E., and Stushnoff, C., J. Am. Soc. Hort. Sci. 102:342-346 (1977).

MECHANISM OF WINTER WHEAT HARDENING AT LOW TEMPERATURE

T. I. Trunova

K. A. Timiryazev Institute of Plant Physiology
USSR Academy of Sciences
Moscow, USSR

The problem of plant frost resistance is one of the theoretical problems of plant physiology which is intimately allied to the practical needs of agriculture, to the practical and theoretical objectives of selection and genetics. Frost resistance of plants can be governed if we know the physiological mechanism of hardening, its biological and molecular bases. The plants wintering in the conditions of very low temperatures, for example, winter cereals, have developed, in the process of evolution, the most perfect protective mechanisms against the action of ice.

Frost hardiness is an inherent quality, characteristic of all winter plants, which is realized in the process of ontogenesis during hardening under specific conditions and which is expressed, ultimately, in the rearrangement of metabolism and protoplast structure interferring with extracellular ice formation and increasing intracellular ice resistance.

The characteristic of frost hardiness is determined by two major factors: 1) genetic feature of the organism to develop a certain degree of resistance to negative temperatures; and 2) external conditions in which the indicated feature is made manifest.

It is known that frost hardiness is not a permanent characteristic of plants, but it is expressed under specific environmental conditions. The realization of genetically determined property of frost hardiness under specific environmental conditions is called the process of hardening. Hardening of winter cereals proceeds in two stages: first, in the exposure to low positive temperatures, and subsequently, in the exposure to negative temperatures in order of -4 and -5°C (34). In our studies we concentrated attention on the processes taking place in the conditions of low positive temperatures. In this period increased frost resistance is associated with the functional and structural rearrangement of the cell.

TABLE I. Frost hardiness of the winter wheat Ulyanovka, its sugar content after hardening at 2°C, in the light and in the dark in 12% saccharose solution.

Variant	Frost hardiness, % of plants that survived					Sugars in the tillering nodes, % per dry weight
	-13°	-16°	-20°	-23°	-26°	
1. Before laboratory hardening	100	100	62	20	0	24.7
2. After hardening; in the electric light	100	100	100	100	100	57.2
3. In the dark, on 12% saccharose solution	100	100	100	100	100	55.1
4. In the dark, on water	100	100	40	8	0	23.4

Low temperature is known to inhibit growth and main physiological processes; the inhibition, in its turn, results in sugar accumulation (18,35). However, such changes are observed both in frost resistant, and cold resistant plants and even in warm-requiring plants. These processes are of tremendous importance, but in the final analysis, frost hardiness is determined by the mechanisms characteristic of frost resistant forms alone. Consequently, it is necessary to single out those processes (of all variety of processes occurring at low temperature) which would be the decisive ones for frost hardiness.

We have developed a laboratory method of hardening for winter cereals by the intake of exogenous sugars, from the solution (36). This method is convenient for modeling the conditions of hardening and exposure to different substances, either stimulating or inhibiting the development of frost hardiness. Winter wheat plants were shown to increase frost resistance after being in the dark for 7 days at a temperature of 0 to +2°C, with the roots and tillering nodes plunged in 12% saccharose solution followed by successive freezing (Table I). It was shown experimentally that sugars were absolutely necessary for increasing frost resistance of plants.

What is the role of sugars in frost hardiness? Plant hardening by keeping them in solutions of different sugars at +2°C showed exogenous sugars to contribute to frost hardiness if they possess the following properties: 1) to permeate into the cells; 2) to participate actively in the metabolism in the conditions of low temperature; 3) to transform into another

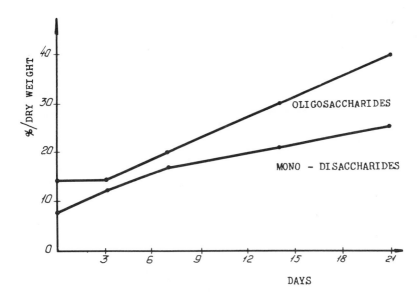

FIG. 1. Dynamics of various forms of sugar in the tillering nodes of Ulyanovka winter wheat during hardening at 2°C in 12% saccharose solution.

form of sugar (Fig. 1); and 4) to accumulate in the form of different sugars in great amounts (22,23,27,37).

These data verify the fact that effect of sugars on frost hardiness is not restricted to an osmotic factor (18). At low positive temperatures sugars are necessary as a substrate which can easily be used for energy and synthesis processes.

This idea is also confirmed by the results of studies on the dynamics of different sugars content in the period of hardening. The experiments showed an elevated content of glucose, fructose, saccharose and other polymeric oligosaccharides in the process of the plants hardening at 2°C. Elevated content of the total sugars over the period of the first six days can be attributed mainly to the increase in mono- and disaccharides (Fig. 1). Within the following days of hardening sugars were conserved in the form of oligosaccharides. As for wheat cereals, whose starch level is not high, their oligosaccharides serve as a reserve form of carbohydrates which can hydrolize at low negative temperatures (-3, -6, -9°C) and supply the plants with mono- and disaccharides in winter periods (27,28).

Study of sugar localization in the cells of frost resistant plants is of great importance for elucidating their role at low positive temperatures and for understanding the mechanism of their protective effect at negative temperatures. Sugars were shown to accumulate in chloroplasts at low positive temperature

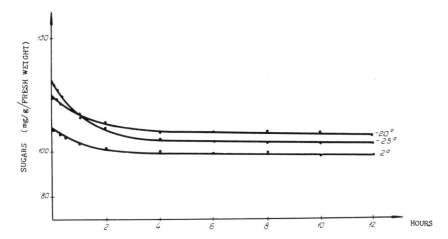

FIG. 2. Kinetics of washing off sugars from the tillering
nodes of Ulyanovka winter wheat after its hardening at 2°C and
freezing at -20 and -25°C.

(9,10,29). Maximum sugar content amounted to 15%, on conversion to absolutely dry chloroplast mass. As was shown in vitro
studies, this concentration is quite sufficient for preserving
photophosphorylation activity of organellas following their
freezing (30).

Intracellular sugar localization of hardy plants was
studied by compartmentation analysis and graphic calculation
technique (14). This technique permits dividing the main sugar
localization zones and finding out how securely sugars are
retained in the cells of hardy plants.

The obtained evidence has shown that about 90% of sugars
accumulated in the process of hardening is localized in the
compartments beyond the membranes which are difficult for them
to permeate (probably, in vacuoles, and partially, in cyto-
plasma); approximately 10% of sugars are released from the tissues relatively easily. Easily washed-off sugars are likely to
be localized in free tissue space. These sugars play a significant part in protecting plasmalemma from the action of negative temperatures (31). It should be noted that sugars are not
released from the cells even after freezing, unless the cells
were affected by frost (Fig. 2).

These experiments have shown that protoplasma and frost
sensitive organellas are protected by sugars. However, much
sugar is also accumulated in the vacuole which is of great
importance for enhancing water-retaining capacity of the cells

TABLE II. Amount of ice in the killed Ulyanovka winter wheat plants with different sugar content (% of total water).

°C	Plants with sugar content of 24.3%	50.6%	Difference
-10	80.2	73.7	6.5
-15	83.8	76.7	7.1
-20	86.6	79.0	7.6
-25	88.3	81.4	6.9
-30	89.2	83.4	5.8

and decreasing the degree of their dehydration by extracellular ice (24).

Ice quantity was determined in the plants of winter wheat with different sugar content by microcalorimetric method (15). The plants with a high sugar level pre-killed by steam were shown to retain osmotically, in the course of freezing, a considerable amount of water in the non-frozen state. As can be seen from Table II, two-fold sugar content increase resulted in 6.5% decrease of ice amount at -10°C in the pre-killed plants. Approximately the same values are characteristic of lower temperatures.

The data permit deduction that sugars as an osmotic factor play a greater role in the development of winter wheat rather than woody plants, for frost hardiness of the former is based, to a high degree, on the water-retaining capacity of the cells, while frost hardiness of the latter is based on the timely water yield from the cell. For example, in one of the experiments, the winter wheat Ulyanovka plants had 57.2% of non-frozen water at -10°C, while the apple tree Antonovka had -59.6%. These facts can probably account for a higher sugar content in the hardy plants of winter cereals compared to the woody plants.

The data obtained verify the stabilizing effect of sugar on the cellular membrane structures. Freezing was shown to induce acid phosphatase yield in the external medium, for the increases of acid phosphotase activity (2,3). Following the irreversible cell damage and the cell destruction, the yield of acid phosphatase reaches its maximum (Fig. 3). Consequently, acid phosphatase yield can be an indicator of the degree of the membrane damage, particularly, plasmalemma along whose side it is localized.

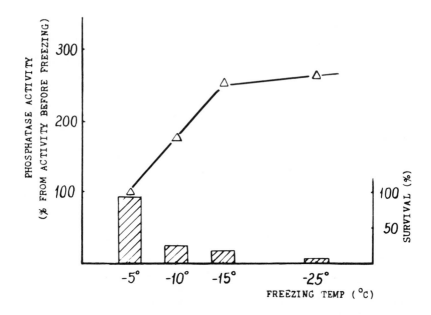

FIG. 3. Acid phosphatase activity after freezing the tillering nodes of nonhardy Mironovskaya 808 and Mironovskaya 808 winter wheat.

Hardening of plants stabilizes the membrane structure. Acid phosphatase yield from the hardy tissues starts at lower negative temperatures than from the non-hardy tissues. Fig. 4 shows the data on phosphatase yield from the cells of nonhardy and hardy plants (in saccharose solution) after freezing at -5 and -15°C for 6 hours at each temperature. The experimental results have shown that nonhardy plants were considerably damaged even at -5°C; phosphatase yield at this temperature was higher than at 2°C. Following the frost of -15°C which is lethal for nonhardy plants, phosphatase yield reached its maximum. At the same time, the frost of -5°C did not cause any damage to hardy plants, with phosphatase yield being approximately the same as before freezing. After the frost of -15°C, phosphatase yield increased, but to a much lesser degree than in nonhardy plants. The data show that hardening produces a great effect on the membrane stability, and that sugars play a significant part in this process.

Both literature data and the evidence that we have obtained experimentally indicate that the processes proceeding in the course of hardening reduce a harmful effect of frost by enhancing water-retaining capacity and decreasing cell dehydration. The role of sugars on these processes is quite significant. At the same time, there is an increase of cell

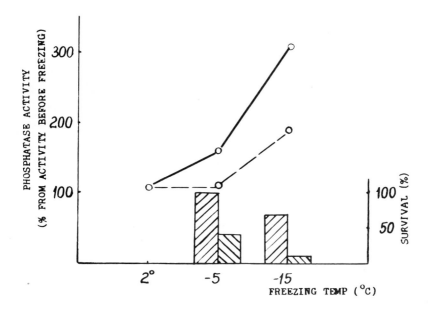

FIG. 4. Yield of acid phosphates in the external solution from the tillering nodes after freezing nonhardy and hardy (at 2°C in saccharose solution) winter wheat: 1 - hardy; 2 - nonhardy.

resistance to frost. This is achieved by the functional and structural rearrangement of the cell.

To assess the significance of sugar accumulation and other processes proceeding at low temperature we conducted a study in which these two factors were differentiated in time (Table III).

The experiments showed that 42% sugar accumulation alone at the expense of intake from the solution at 15°C increased frost hardiness of winter wheat by 3°C (variant 2). A long-term exposure to low temperature alone in the conditions excluding sugar accumulation did not result in increased frost hardiness (variant 5). Increase in frost hardiness by 10° and more was observed either in the simultaneous exposure to sugar and low temperature factors (variant 4), or in a successive procedure: the plants were first enriched with sugars at 15°C, then exposed to low temperature (variant 3). It should be noted that in the latter case frost hardiness increase was 3°C less. This is explained by the unfavorable effect of etiolation the plants are subjected to when they are kept in the dark at 15°C (39).

It was shown that the effect of low temperature should be both long-term (7-10 days) and continuous. Temperature increase from 2 to 15°C for only 2 hours a day was quite sufficient to interfer with raising frost hardiness (40). The

TABLE III. The effects of low temperature and sugar accumulation on frost hardiness of winter wheat (Ulyanovka).

Variant	Frost hardiness, % of plants that survived					Percent of sugar per absolutely dry weight
	-10°	-13°	-16°	-20°	-23°	
1. Before hardening After the exposure in the dark:	10	0	0	–	–	13.9
2. 2 days at 15°C in 20% saccharose solution	100	100	14	0	0	42
3. 2 days at 15°C in 20% saccharose then 6 days at 2°C in water	100	100	100	100	85	37.3
4. 8 days at 2°C in 20% saccharose	100	100	100	100	100	47.2
5. 8 days at 2°C in water	0	0	0	0	11.2	

TABLE IV. The effect of temperature on the content of soluble protein, sugars and frost hardiness of the winter wheat Mironovskaya 808 plants.

Variant	Percent of sugar per dry weight	Soluble protein, mg/g dry weight	Frost hardiness, % of plants that survived			
			-10°	-13°	-16°	-18°
1. Before hardening	18.9	11.2	90	0	0	0
2. After 2 days at 18°C in 10% saccharose	50.1	12.5	100	0	0	0
3. The same as 2 + 7 days at 2°C in the dark in water	47.2	38.9	100	100	100	85

limiting factor in raising frost hardiness was heat interruption; sugar accumulation was not inhibited.

Consequently, the experiments showed that low temperature at hardening was necessary not only for sugar accumulation, but it induced a number of other processes connected with increased frost hardiness, for example, with protein synthesis (32).

In one of the experiments the plants were kept in 10% saccharose solution at 18°C for two days (Table IV). Over this period of time about 50.1% of sugar accumulated in the tillering nodes. There was almost no change in frost hardiness; soluble protein increase was only from 11.2 to 12.5 mg/g per dry weight. After that the plants were placed in a dark chamber, +2°C, for 7 days. The exposure to cold resulted in frost hardiness increasing almost by 8°. Soluble protein content amounted to 38.9 mg/g per dry weight, that is, increased almost by 3.5 times. As is known, period of hardening at low positive temperatures is associated with the production of gelated structures (41), increased volume of protoplasts (25), rearrangement of organella structure (20), endoplasmatic reticulum expansion (16). The obtained data make it possible to suggest that low temperature induces soluble protein synthesis.

In literature there are data available concerning the increased water soluble protein fraction after hardening, for example Black locust (25), box-wood (7), alfalfa, cereals (1,6). Soluble proteins are characterized by thermostability and a high capacity for gel production. Protoplast gelification is known to considerably enhance their water-retaining capacity. Certain proteins act as protective substances, much more effective than sugars (11). There are also data available on the changes in the number of sulfhydril protein groups during hardening (17). Changes in the chloroplast protein structures and mitochondria have been observed (13) in the course of frost hardiness. Summing up the available literature data, we can suggest that protein thermostability is achieved as a result of conformation changes and synthesis of proteins more resistant to frost.

The analysis of soluble protein fraction in the tillering nodes of hardy and nonhardy winter wheat plants showed great differences in their protein content (32). During hardening in the light soluble protein content increased by 1.5-2 times (Fig. 5). The analysis by electrophoretic protein partition showed the soluble protein fraction released from the tillering nodes to be heterogenous, with heterogeneity increasing after hardening. In hardy plants the proteins with Rf 0.125 and 0.5 disappeared, at the same time new proteins with Rf 0.35 and 0.55 appeared. A new component with Rf 0.82 was observed in the zone of highly mobile proteins. Though indirectly, these data can be used to support a hypothesis that protein synthesis 'de novo' takes place in the period of hardening. Such an assumption is quite feasible, as hardening is a fairly long process, lasting days and even weeks. This idea is also

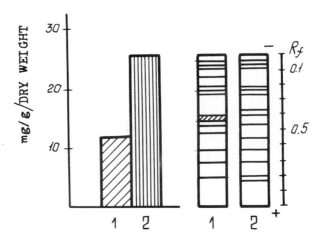

FIG. 5. Content and electrophoretic spectrum of soluble proteins, extracted from the tillering nodes of hardy and non-hardy Ulyanovka winter wheat: 1 - hardy; 2 - nonhardy.

confirmed by the observations over frost resistance of separate parts of the plant. Young leaves appearing in the period of hardening are more stable than the old ones, with their growth terminated (4,33).

Quite likely, the process of retarded growth is associated with the production of qualitatively new structures better adapted to dehydration and mechanical deformations brought about by freezing. The hypothesis of protein synthesis 'de novo' can also be confirmed by the research studies on the multiple molecular forms of enzymes-mechanisms of subtle regulation of metabolic processes in the conditions of temperature changes. There is evidence on considerable alterations in the isoenzyme composition of plants during hardening (19,26), as well as the properties of enzymes synthesized at low temperature (21). An important feature of these enzymes is the low temperature coefficient of their activity.

Reduction of temperature minimums and a long-term lag-period of enzyme activity (when there is a shift from low to high temperatures) can also testify to temperature enzyme adaptation of plants during hardening. If follows that numerous indirect data point to a significant role of protein synthesis in the hardening of plants for frost resistance.

To study the relationship between the development of frost resistance property and protein-synthesizing system we have used the method of inhibition analysis (32). Chloramphenicol (100-200 mg/l) and cycloheximid (2.5-5 mg/l) were used to inhibit protein synthesis during frost hardiness. The roots, tillering nodes and part of the leaves were immersed in the

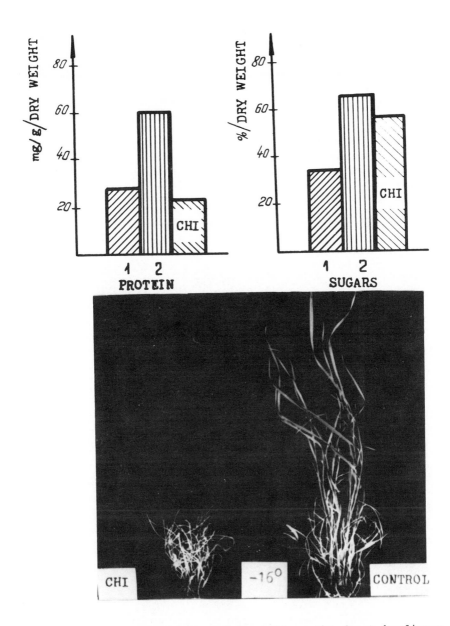

FIG. 6.   The effect of cycloheximid on the frost hardiness of Ulyanovka winter wheat and on the content of soluble protein and sugars: 1 - hardy; 2 - nonhardy.

solutions of antibiotics and treated with them during the whole period of hardening (8 days), in the light at 2°C; every 2 days there was a new antibiotic solution.

It was shown experimentally that the development of frost hardiness was inhibited in the treated plants, with cyclo-heximid having a greater effect than chloramphenicol. The frost hardiness of the plants treated with cycloheximid was equivalent to that of nonhardy plants (Fig. 6). Similar results were obtained by other researchers (8,12).

Soluble protein synthesis was inhibited by antibiotics during hardening, as could be expected. The analysis showed that cycloheximid-induced soluble protein level was the same as in nonhardy plants or even lower. Chloamphenicol inhibited protein accumulation only partially.

Since frost hardiness in plants is largely determined by the number of protective substances, it was quite natural to consider a question of the effect of protein synthesis inhibitors on sugar content. The analysis showed that their content was somewhat reduced in the plants treated with antibiotics; however, it was sufficiently high to ensure a high degree of front resistance (more than 50% per dry mass).

To prove that hardening of plants is associated with low temperature induced protein synthesis 'de novo', we have carried out additional experiments: cycloheximid was added to the solution on the seventh day of hardening, two days before freezing. As is known, the major processes related to hardening are normally terminated by that time. Despite the fact that two days prior to freezing 80S ribosomes were blocked by cycloheximid the frost hardiness of plants did not change. It was equivalent to that of hardy plants which were not treated with the antibiotic.

The obtained results can be interpreted in the following way. At low temperature of hardening (about 0°), induced protein synthesis is carried out in addition to sugar accumulation; proteins participate in the processes specifying a hardy state of the cells. In case protein synthesis is blocked by antibiotic exposure, no hardening takes place, and the plants get frozen at the same temperature as nonhardy plants, despite a high sugar content. Antibiotic exposure at the end of hardening period does not affect frost hardiness; this fact signifies protein synthesis completion. Probably, the major protein synthesis associated with hardening proceeds in 80S ribosomes, as indicating that the cycloheximid reduces frost hardiness to a much greater degree than chloramphenicol.

Based on our data, we can conclude that development of frost hardiness at low positive temperature is determined by both sugar and protein synthesis. The former induces increasing water-retaining capacity of the cell, establishing cell structure. The latter seems to take part in function and structure rearrangement so that the plants not only function at low temperature but increase frost resistance.

RE FERENCES

1. Babenko, V.I., Biryukov, S.V., and Inkina, A.G., Fisiol. Rast. 16:921 (1969).
2. Bolduc, R., Zvereva, G.N., and Trunova, T.D., Fisiol. Rast. 28:601 (1981).
3. Bolduc, R., Rancourt, L., Dolbec, P., and Chouinard-Lavoie, L., Canad. J. Teant. Sci. 58:1007 (1978).
4. Cox, W., and Levitt, J., Plant Physiol. 44:923 (1969).
5. Garber, M., and Steponkus, P., Cryobiology 9:313 (1972).
6. Gerloff, E.D., Stahman, M.A., and Smith, D., Plant Physiol. 42:895 (1967).
7. Gusta, L.V., and Weiser, C.I., Plant Physiol. 49:91 (1972).
8. Hatano, S., Sadakane, H., Tutumi, M., and Watanabe, T., Plant Cell Physiol. 17:643 (1976).
9. Heber, U., Protoplasma 51:284 (1959).
10. Heber, U., in "The Cell and Environmental Temperature". "Nauka", M.-L., 1964.
11. Heber, Y., and Volfer, H., Biochim. et Biophys. Acta 412:335 (1975).
12. Kasperska-Palacz, A., Dlugolercka, E., Breitenwold, J., and Weislinska, B., Biol. Plantarum 19:10 (1977).
13. Khokhlova, L.P., Eliseeva, N.S., Stupishina, E.A., Bondar, I.G., and Suleimanov, I.G., Fisiol. Rast. 22:831 (1975).
14. Kholodova, V.P., Fisiol. Rast. 14:444 (1967).
15. Krasavtsev, O.A., "Calorimetry of Plants at Temperatures Below Zero". "Nauka", M., 1972.
16. Krasavtsev, O.A., and Tutkevich, G.I., Fisiol. Rast. 18:607 (1971).
17. Levitt, I., and Dear, J., in "The Frozen Cell". p. 149 Aciba Found. Sympos., London, 1970.
18. Levitt, J, "The Hardiness of Plants". pp. 278. Academic Press, New York, 1956.
19. McCown, B.H., McLeester, R.C., Bech, G.H., and Hall, T.C., Cryobiology 5:410 (1969).
20. Parber, J., and Philpott, D., Protoplasma 53:575 (1961).
21. Roberts, D.W.A., Canad. J. Bot. 45:1347 (1967).
22. Sakai, A., Inst. Low Temp. Sci. Ser. B 11:1 (1962).
23. Sakai, A., and Joshida, S. Cryobiology 5:160 (1968).
24. Samigin, G.A., "Causes for Freezing of Plants". "Nauka", 1974.
25. Siminovitch, D., Rheaume, B., Pomeroy, K., and Lepage, M., Cryobiology 5:3 (1968).
26. Suleimanov, I.G., Ramasanova, L.H., and Alexeeva, V.Y., Report of USSR Academy of Sciences 202:718 (1972).
27. Trunova, T.I., Fisiol. Rast. 10:588 (1963).
28. Trunova, T.I., Fisiol. Rast. 12:85 (1965).
29. Trunova, T.I., Fisiol. Rast. 17:902 (1970).
30. Trunova, T.I., and Zvereva, G.N., Fisiol. Rast. 21:1000 (1974).

31. Trunova, T.I., Fisiol. Rast. 16:658 (1969).
32. Trunova, T.I., and Zvereva, G.N., Fisiol. Rast. 24:395 (1977).
33. Trunova, T.I., and Smirnova, N.A., Selchoz Biol. 12:534 (1977).
34. Tumanov, I.I., in "Nauka", Moscow, 1979.
35. Tumanov, I.I., in "Selkhozgiz", 1940.
36. Tumanov, I.I., and Trunova, T.I., Fisiol. Rast. 10:176 (1963).
37. Tumanov, I.I., and Trunova, T.I., Fisiol. Rast. 4:408 (1957).
38. Tumanov, I.I., Krasavtsev, O.A., and Trunova, T.I., Fisiol. Rast. 16:907 (1969).
39. Tumanov, I.I., Trunova, T.I., Smirnova, N.A., and Zvereva, G.N., Fisiol. Rast. 22:1231 (1975).
40. Tumanov, I.I., and Trunova, T.I., Reports of USSR Academy of Sciences 175:1168 (1967).
41. Tumanov, I.I., Fisiol. Rast. 14:520 (1967).

ON THE MECHANISM OF FREEZING INJURY
AND COLD ACCLIMATION OF SPINACH LEAVES

G. H. Krause, R. J. Klosson and U. Tröster

Botanisches Institut der Universität Düsseldorf
Düsseldorf, Germany

I.   INTRODUCTION

The leaves of many plants are killed by freezing and thawing when they are in an unacclimated state, but they can withstand freezing after acclimation to low temperatures. However, even after cold acclimation, the temperature range in which plants survive freezing is limited. It seems well-founded that freezing injury primarily consists of membrane damage (see ref. 17). Membrane systems isolated from leaves such as thylakoids (6) and mitochondria (25) can be readily inactivated by freezing and thawing. Freezing injury of thylakoid membranes has been extensively studied in vitro. Obviously, there are different types of "cryotoxic" substances, i.e., compounds that damage the membranes when their concentration is raised in a membrane suspension due to ice formation. The adversory action of inorganic salts during freezing is distinctly different from that of amphiphilic organic compounds (5). The presence of inorganic salts during freezing and thawing primarily leads to increased permeability of the membranes to protons and thereby to uncoupling of photophosphorylation from electron transport (see ref. 6). Presumably, this is due to an interaction of the electrolytes with the charged surfaces of the membranes. Inactivation of electron transport is observed only under extreme conditions. In contrast, amphiphilic molecules, for instance phenylpyruvate, cause less uncoupling but strong inhibition of photosynthetic electron transport, possibly by interacting with the lipid bilayer and intrinsic membrane proteins.

In the present report, the question as to the primary mechanism of freezing inactivation of photosynthesis in vivo is raised. We have tried to relate the pattern of damage observed in cold acclimated and nonhardened spinach leaves after defined

frost treatment to the known freezing inactivation of thylakoids in vitro in the presence of different cryotoxic substances. Freezing injury occurring in vivo was examined by
testing the activities of various reactions in whole leaves as
well as in thylakoid membranes isolated from the frost-treated
leaf blades (see also refs. 9 and 10). As will be shown, a
very complex picture of freezing injury emerges, which does not
completely correspond to either damage observed in vitro.

Freezing injury of isolated membranes is diminished or
fully averted by cryoprotective substances (6). Sugars are
particularly well known for their cryoprotective action (see
K.A. Santarius, this volume). Increase in sugar contents of
plant tissue in relation to frost hardening has been reported
for numerous instances (15), but little is known about the distribution of the sugars within the cold acclimated cells
(3,8,21). We have therefore analyzed the accumulation and compartmentation of the most abundant sugars between chloroplasts
and extraplastidic space during the process of cold acclimation. The results confirm a cryoprotective role of sugars in
spinach leaves.

II.   COLD ACCLIMATION OF SPINACH LEAVES

There are several advantages to use spinach (Spinacia
oleracea L.) for model studies on freezing injury and frost
tolerance of herbaceous plants. Spinach plants are easy to
culture. When grown in a greenhouse at 20°C (day)/15°C (night)
and a photoperiod of 9 h, spinach leaves virtually do not
exhibit tolerance to extracellular ice formation in the tissue
(9). Freezing starts at about -4 to -5°C. Strong damage to
the leaves as seen by lowered rates of photosynthesis and
respiration (Fig. 1a) and by water infiltration of the tissue
was usually observed after exposure to about -5 to -7°C.
Within 8 days, spinach leaves can be cold acclimated in a
reproducible manner by a stepwise decrease of the day and night
temperature to 1°C at a photoperiod of about 8 h (9). Notably,
the temperature at which ice formation begins is not lowered by
cold acclimation. Fig. 1b shows, however, that in acclimated
leaves, damage is seen only after exposure to considerably
lower temperatures (-11 to -13°C). Evidently, these leaves
possess an increased but limited tolerance to freezing. When
the growth temperature is raised again to 20/15°C, the leaves
lose frost tolerance almost completely within 6-8 days. It is
not necessary to prolong the photoperiod for de-hardening.

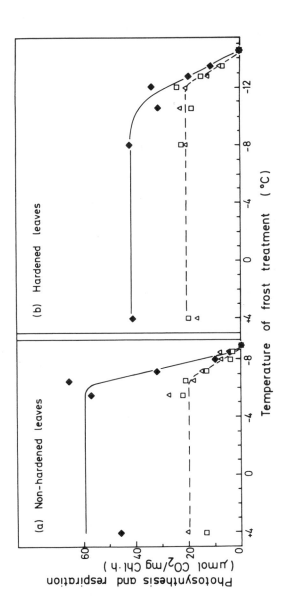

FIGURE 1. Rates of photosynthesis and respiration of nonhardened (a) and hardened (b) spinach leaves as a function of cold treatment. Detached leaves were cooled from +4°C at a rate of 6°C/h to a minimum temperature as given on the abscissa, kept at this temperature for 2 h and warmed up to +4°C at the same rate. Controls were kept at +4°C. For measurements, leaf samples were brought to room temperature. After preillumination for 10 min in the presence of 300 µl·l⁻¹ CO_2 in air, respiratory CO_2 evolution (□) and apparent photosynthetic CO_2 uptake (◆) were determined with an infrared analyzer during a 3 min dark and subsequent 2 min light period. Actinic light (half band width 630-680 nm, 45 W·m⁻²) was about half saturating for photosynthesis. Respiration was also measured in CO_2-free air (△).

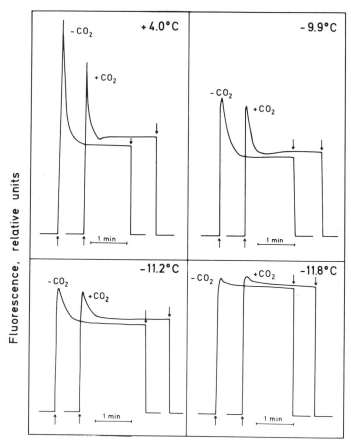

FIGURE 2.   Effects of frost treatent on chlorophyll a
fluorescence signals of hardened spinach leaves. Experimental
conditions were as for Fig. 1. Chlorophyll fluorescence in the
presence   and   absence   of   300 $\mu l \cdot l^{-1}$ CO_2 in   air   was
recorded at room temperature after 10 min preillumination and
3 min darkness as described previously (11,12). Upward arrows
denote actinic light on; downward arrows, actinic light off.
Minimum temperatures of frost treatment are given in the
graph: +4°C denotes the unfrozen control.

III.   NATURE OF FREEZING DAMAGE

A.   Effects of Freezing and Thawing on Whole Leaves

Inhibition of photosynthetic CO_2 assimilation, measured
at room temperature after frost treatment of the leaves, as
depicted in Fig. 1, proved to be a reliable measure of freezing

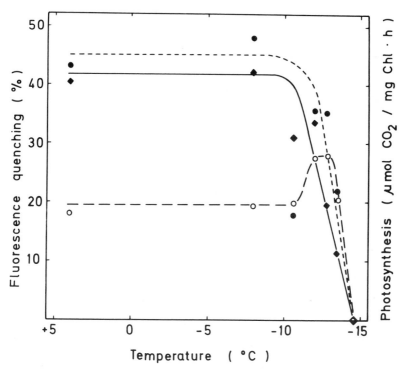

FIGURE 3. Effects of freezing and thawing on chlorophyll a fluorescence quenching in hardened spinach leaves. Fluorescence quenching is defined as the difference between fluorescence emission in the peak and in the steady-state (in % of maximal emission) reached after 2 min of illumination (see Fig. 2). Quenching in the presence (O) and absence (●) of CO_2 in air is plotted as a function of minimum temperature of frost treatment. Rates of apparent photosynthesis (◆) are given for reference.

injury. It should be noted that this inactivation was irreversible; if partial damage occurred, this was aggravated by storing the leaves for prolonged time at room temperature. Within limits of error, respiration was inactivated in parallel with photosynthesis, indicating damage of the mitochondria which has been studied in more detail by Thebud and Santarius (26).

Damage of the photosynthetic apparatus in the leaves can be detected by physical parameters such as chlorophyll a fluorescence emission and light scattering reactions at 535 nm. Fig. 2 depicts long-term fluorescence signals of cold acclimated spinach leaves recorded in the presence and absence of CO_2 after frost treatment. It has been described before (11)

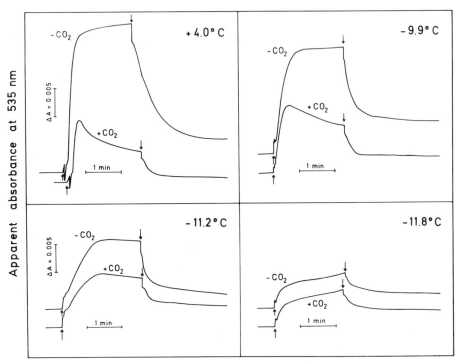

FIGURE 4.   Light-induced apparent absorbance changes of
hardened spinach leaves; effects of frost treatment.   Slow
changes in apparent absorbance at 535 nm are caused by altered
light-scattering (4,11), whereas fast phases represent membrane
potential-dependent absorbance changes.   Minimum temperatures
of treatment are given in the figure.   For experimental details
see legends to Figs. 1 and 2.   Upward arrows, actinic light on;
downward arrows, light off.

that both the peak and the steady-state level of fluorescence
emission are affected by CO_2-fixing activities of the leaves,
as   shown   in   the   control   samples   (+4°C)   of   Fig. 2.   With
progressing frost damage, these differences between the signals
in the presence and absence of CO_2 become smaller and finally
disappear.   In addition, the fluorescence peak is drastically
lowered.   As   shown   before   (10),   this   is   due   to   a   strong
decrease in the variable part of fluorescence emission, indi-
cating damage in the region of Photosystem II.   In Fig. 3, the
course of these fluorescence changes during freeze-inactivation
of leaves is expressed in terms of fluorescence quenching,
i.e., the extent of fluorescence decline from the peak to the
stationary level.   Two effects of frost treatment can clearly
be   distinguished.   In   the   controls,   quenching   is   strongly

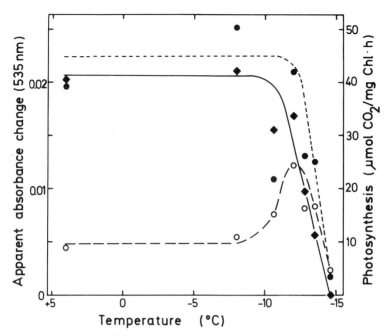

FIGURE 5.   Photoinduced light scattering changes of
hardened spinach leaves as a function of the minimum temperature of frost treatment. The increase in light scattering
(slow phase of apparent absorbance increase) upon illumination
of the leaves in the presence (O) and absence (●) of
300 $\mu l \cdot l^{-1}$ CO_2 in air was recorded after thawing at
room temperature (see Fig. 4). Rates of photosynthesis (◆)
are also given.

diminished by CO_2 fixation (lower curve, presence of CO_2);
quenching increases, when CO_2 fixation is partially inhibited
by frost treatment.   In the absence of CO_2, fluorescence
quenching is strong in the undamaged samples and gradually
decreases with progressing damage.   In the last phase of
inactivation, both curves decline in parallel.   Quenching of
chlorophyll fluorescence has recently been shown (13,28) to
consist predominantly of two components, determined by photosynthetic electron transport and by the photoinduced proton
gradient across the thylakoid membranes, respectively.   The
general decline of quenching in relation to freezing injury
thus denotes an inactivation of the photosynthetic apparatus
located in the thylakoids.   However, the intermediary increase
of quenching in the presence of CO_2 suggests that inactivation of CO_2 fixation starts at a point when the energy-
conserving apparatus of photosynthesis is still largely

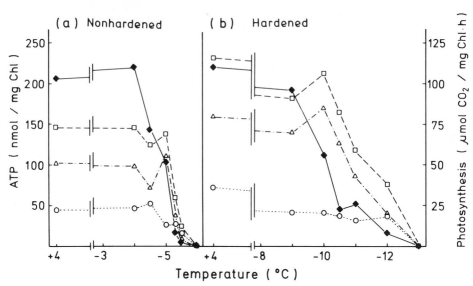

FIGURE 6. Steady-state ATP levels in frost-treated non-hardened (a) and cold acclimated (b) spinach leaves. Experimental conditions were as for Fig. 1. In the steady-state (light or dark), reactions were stopped with liquid N_2. Samples were extracted with $HClO_4$, and ATP was determined by the luciferin/luciferase test with an LKB Wallac luminometer 1250 and purified enzyme and reagents supplied by LKB. Control values (+4°C) are means of two measurements. Symbols: □ , ATP levels in the light, Δ, ATP levels in the dark; O, light/dark differences in ATP levels; ◆ , rates of photosynthesis.

intact. This is supported by the photoinduced light scattering changes at 535 nm (Figs. 4 and 5). In the signals depicted in Fig. 4, slow changes of apparent absorbance seen upon illumination and darkening denote light scattering changes that are caused by the build-up of the ΔpH at the thylakoid membranes and its decay in the dark (4,11,12). In the presence of CO_2 (see control samples kept at +4°C), the light scattering signal is small, because the proton gradient is kept at a low level due to fast consumption of photosynthetic ATP by carbon metabolism (14). Like the chlorophyll fluorescence signals, the changes in light scattering observed in the presence and absence of CO_2 are affected differently by frost treatment. Partial damage of the leaves is characterized by increased signals in the presence and decreased signals in the absence of CO_2. This is clearly seen in Fig. 5, depicting the light scattering changes observed upon illumination as a function of minimum temperature of frost treatment. Notably, in the presence of CO_2, the increase of the light scattering change coincides with beginning inactivation of CO_2 fixation. It appears that at an intermediary stage of damage, photosynthetic energy cannot be fully utilized by carbon metabolism.

To prove this hypothesis, ATP levels in the leaf tissue were determined. In the experiment of Fig. 6, nonhardened and cold acclimated leaves were subjected to frost treatment with different minimum temperatures. After thawing, the leaves were kept in the presence of CO_2 at room temperature either in the light or in darkness. When steady-state was reached, the leaf samples were fast frozen with liquid nitrogen, and subsequently their ATP content was measured. The results clearly show a delay in the decrease of ATP levels, as compared to the inhibition of CO_2 fixation by frost treatment. When CO_2 assimilation was significantly diminished, ATP levels still remained high. But when the minimum temperature was decreased further, ATP levels both in the light and in the dark broke down and dark-light differences disappeared, indicating severe damage to the energy-conserving systems in chloroplasts and mitochondria.

From these data and the changes in physical parameters dipicted in Figs. 2-5 we conclude that CO_2 fixation by the Calvin cycle is slightly more sensitive to freezing stress than the energy-conserving apparatus of photosynthesis. Inhibition of CO_2 fixation is, in fact, the earliest detectable sign of freezing injury so far. The mechanism of this inactivation is presently unknown. Frost treatment does not cause stomatal closure (9). However, the intercellular spaces of injured tissue are partly filled with water that has not been reabsorbed during thawing. This may slow down CO_2 diffusion to the chloroplasts and thereby CO_2 fixation. On the other hand, it is possible that the complex light activation of the Calvin cycle is disturbed by failure of compartmentation in the cell. It was indeed shown (9) that the chloroplast envelope is affected by freezing. Finally, it cannot be excluded that there is a more direct effect of freezing on the activity of ribulose-1,5-bisphosphate carboxylase, which is regulated by a complicated mechanism (16). A number of other soluble enzymes localized in the chloroplast stroma, however, have been shown to be rather insensitive to freezing (20,27).

B. Effects of Freezing and Thawing on Thylakoid Membranes in vivo and in vitro

Functionally intact chloroplasts and thylakoid membranes can routinely be isolated from spinach leaves. In thylakoids prepared from frost-treated leaf tissue, the most conspicuous sign of injury is inhibition of photosynthetic electron transport. This is shown in Fig. 7 for nonhardened and cold acclimated leaves. Rates of coupled electron transport under phosphorylating conditions, as well as of uncoupled electron transport (denoting the maximum rate), decline with decreasing temperature of frost treatment. The differences in the rates

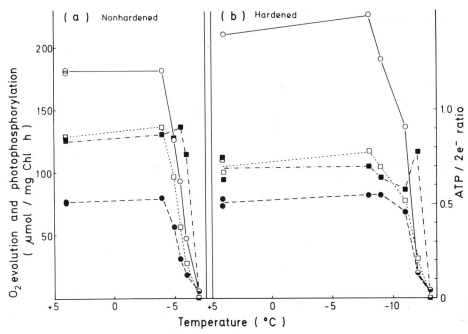

FIGURE 7.   Inactivation of noncyclic electron transport and photophosphorylation by freezing and thawing of nonhardened (a) and hardened (b) spinach leaves.   After frost treatment (minimum temperatures given on the abscissae), thylakoid membranes were isolated and photosynthetic reactions under light saturation measured at 20°C and pH 7.6 as described before (9). Rates of O_2 evolution and ATP formation, respectively, are given in μmol/mg chlorophyll·h.
Symbols: O, electron transport, uncoupled with 5 mM NH_4Cl; 1.2 mM ferricyanide present as electron acceptor; ●, electron transport in the presence of 2 mM ADP, 10 mM phosphate and 1.2 mM ferricyanide; □, photophosphorylation during noncyclic electron transport; ATP determined enzymatically; ■, ATP/2e⁻ ratio.

of the two reactions disappear.   This decline occurs simultaneously with the changes in physical signals of Figs. 3 and 5 and therefore cannot be viewed as a mere artifact of thylakoid isolation.   Notably, photophosphorylation declines in parallel with electron transport.   There is no indication of uncoupling of photophosphorylation from electron transport, i.e., in thylakoids from partially damaged leaves, ATP/2e⁻ ratios remain constant.   Only in thylakoids isolated from fully damaged leaves, where always a low residual electron transport rate was observed, photophosphorylation was absent.   Table I demonstrates that the O_2 evolution then was the most strongly

TABLE I. Inhibition of partial reactions of photosynthetic electron transport by freezing and thawing of non-hardened spinach leaves (minimum temperature of frost treatment, -8°C; see Fig. 1a). For experimental details see refs. 5 and 9).

Electron transport reaction tested	Region of electron transport system	Inhibition (% of control)
1) H_2O to DCIP	Photosystem II plus water splitting system	75
2) 1,5-diphenylcarbazide to DCIP	Photsystem II	24
3) Ascorbate/DCIP to O_2 (via methylviologen)	Photosystem I	31

Abbreviation: DCIP, 2,6-dichlorophenol indophenol.
Control rates (in μequiv./mg chlorophyll·h): 1) 146; 2) 84: 3) 198.

inhibited partial reaction of electron transport, but the regions of Photosystem II and I were also considerably affected. However, as chlorophyll fluorescence spectra at 77°K indicate (9), the photosynthetic pigment systems did not seem to be substantially damaged. The pattern of damage was identical in unhardened and hardened leaves.

Fig. 8 shows that in nonhardened and in acclimated leaves the light-induced proton gradient across the thylakoid membranes is not more sensitive to freezing stress than other photosynthetic reactions (see also ref. 9). The proton gradient was estimated from the quenching of 9-amino-acridine fluorescence (24) in the isolated thylakoids and was found to be only slightly affected in samples showing partially inactivated electron transport. Moreover, the half time of dark decay of the 9-amino-acridine signal does not decrease in the course of freezing injury (9); in other words, the permeability of the membranes to protons does not increase.

These results are clearly in disagreement with effects of freezing on thylakoids in vitro in the presence of high concentration of inorganic salts (see refs. 5 and 6). In the experiment of Fig. 9, various photosynthetic parameters were studied after freezing of thylakoid membranes in vitro. Samples were frozen to -10°C in the presence of 50 mM NaCl and varying concentration of sucrose. At low levels of the cryoprotectant, uncoupling is evident from lowered rates of photophosphorylation in conjuction with increased rates of electron transport, decreased ΔpH values and decreased half times of the decay of

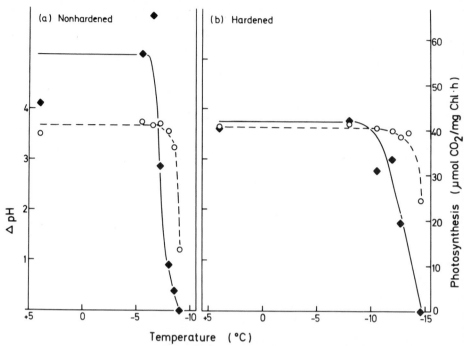

FIGURE 8. Effects of frost treatment of nonhardened and hardened spinach leaves on the photoinduced proton gradient across the thylakoid membranes. Thylakoids were isolated after thawing as for Fig. 7. ΔpH values (O), which should be viewed as relative, not absolute measures of the proton gradient, were calculated from quenching of 9-amino-acridine fluorescence (9,24) at 20°C and pH 7.6; the thylakoid volume was assumed to be 10 µl per mg chlorophyll. In the reaction medium, 25 µM methylviologen and 7.5 µM 9-amino-acridine were present. For reference, rates of photosynthesis of the frost-treated leaves (◆) are given.

the 9-amino-acridine signal. On the other hand, even in the absence of sucrose, when photophosphorylation is fully abolished, there is no inhibition of electron transport. This suggests that thylakoid inactivation by freezing in vivo is primarily not caused by the damaging effects of increased electrolyte concentrations. The question arises, whether freezing injury to chloroplast membranes in vivo may then arise from the action of amphiphilic substances. A potent compound of this type is phenylpyruvate (5). In the experiment of Fig. 10, isolated thylakoids were frozen to -20°C in the presence of increasing concentration of sodium phenylpyruvate. The pattern of damage emerging after thawing does indeed show similarities to freezing injury in vivo: Coupled and uncoupled electron

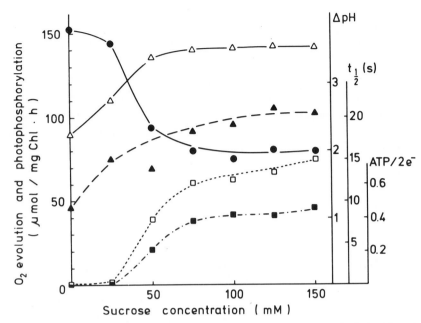

FIGURE 9.   Pattern of freezing injury to thylakoids in
vitro in the presence of NaCl, and protection of the membranes
by sucrose.   Thylakoids were isolated from nonhardened spinach
leaves in NaCl media as described by Santarius and Ernst (23)
but with 40 mM HEPES buffer, adjusted to pH 8.0.   Freezing
occurred in media of 50 mm NaCl, 40 mM HEPES (pH 8.0) and
sucrose as given on the abscissa; samples were cooled from 0°C
with 6°C/h to -10°C, kept there for 1 h and warmed up to 0°C at
the same rate.   Photosynthetic activities were measured after
thawing at 20°C.   Symbols denote: ●, electron transport in
the presence of ADP (see Fig. 7); □, noncyclic photophosphorylation   (see Fig. 7); ■, ATP/2e$^-$ ratio; Δ, ΔpH values
(see Fig. 8); ▲, half time (t$_{1/2}$) of the decay of the
9-amino-acridine fluorescence quenching observed upon darkening.

transport are effectively inhibited.   However, in contrast to
freezing in vivo, distinct uncoupling effects are also visible
(cf. ref. 5).   Non-cyclic photophosphorylation declines more
strongly than electron transport, as expressed by decreased
ATP/2e$^-$ ratios;   the   lowered   half   times   of   the   decay   of
9-amino-acridine   fluorescence   quenching   indicate   enhanced   proton leakage of the membranes.
     Thus, present evidence confirms the earlier observation by
Heber et al. (5), that freezing injury of the thylakoid membranes in vivo is similar, although not identical with injury
occurring in vitro in the presence of amphiphilic substances.
Presently, it is puzzling, what compounds in the cells might be

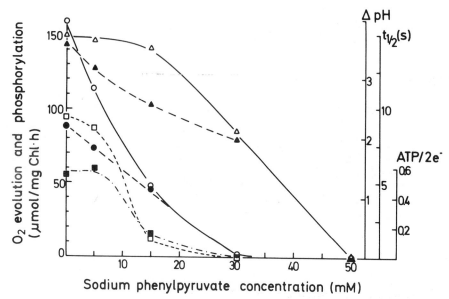

FIGURE 10.   Pattern of freezing inactivation of thylakoids in the presence of phenylpyruvate.   Thylakoids were isolated as for Fig. 9 and suspended in media containing 0.1 M sucrose, 40 mM HEPES (pH 8.0), 5 mM $MgCl_2$ and sodium phenypyruvate as denoted on the abscissa.   These samples were cooled from 0 °C to -20 °C and kept at this temperature for 1 h.   Rates of cooling and warming were 12°C/h.   After thawing, activities were measured at 20 °C; O, uncoupled electron transport (see Fig. 7); other symbols as for Fig. 9.

responsible for the damage observed in vivo.   It seems questionable whether high levels of cryotoxic amphiphilic compounds reside in the chloroplasts.   For instance, phenylalanine has been detected there in small amounts (1).   However, unphysiologically high concentrations of this substance are needed in freezing experiments to achieve substantial inhibition of electron transport (data not shown).   We have speculated (9) that cryotoxic compounds residing in other compartments of the cell, particularly in the vacuole, may flood the chloroplasts when cell compartmentation breaks down in the course of freezing and thawing.   Therefore, experiments were designed to demonstrate the effects of crude cell sap on the activities of thylakoid membranes in vitro (R.J. Klosson, unpublished).   A representative experiment is shown in Fig. 11.   When the percentage of cell extract in the suspending medium is raised, electron transport reactions are only weakly, but photophosphorylation is strongly affected.   The rate of the latter and the $ATP/2e^-$ ratio decline in parallel.   There are indications of

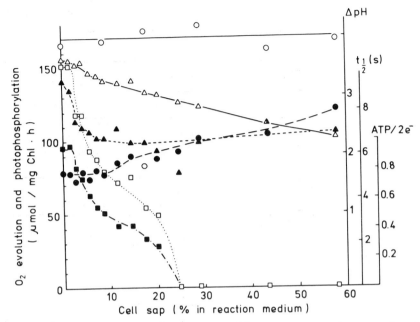

FIGURE 11.   Effects of crude cell extract on photosynthetic activities of thylakoids.   Nonhardened spinach leaves were used to isolate thylakoids in a sorbitol/HEPES medium (9).   For measurements at 20°C, the reaction medium (9) was in part replaced by crude cell sap which was prepared from the same leaf material by centrifuging a leaf homogenate for 6 h at about 150.000 g.   The reaction mixtures were adjusted with 40 mM HEPES to pH 7.6.   Symbols as for Fig. 9.

uncoupling such as enchanced electron transport (under phosphorylating conditions), lowered ΔpH and shortened $t_{1/2}$ of its decay.   But in the range where photophosphorylation steeply declines, these changes are small and seem insufficient to explain the strong inhibition of ATP formation.   Although this pattern of injury varied somewhat with different batches of plant material used, no resemblance to freezing injury in vivo was found.   Thus, a primary damaging effect of vacuolar substances on thylakoids in situ seems unlikely, even though a disturbance of cell compartmentation may take place.   The strict limitation of frost tolerance even of optimally hardened leaves is as more astonishing, as acclimated leaves contain high levels of cryoprotectants.

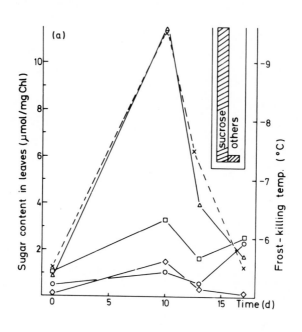

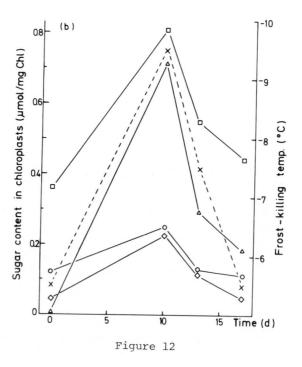

Figure 12

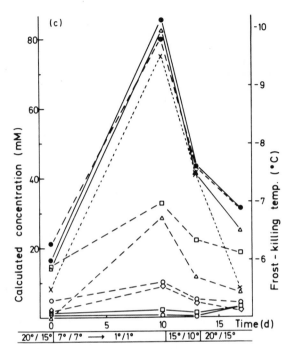

FIGURE 12. Changes of sugar levels in the chloroplasts and extraplastidic space in relation to hardening and de-hardening of spinach leaves. Plants were cold acclimated for 10 days and then de-hardened by altering the growth temperature as given in the figure. Sugar levels were determined enzymatically (2) in the whole leaves and in aqueously isolated intact chloroplasts from plants in nonhardened, hardened, and de-hardened states. Besides sugar levels, the degree of hardiness (frost-killing temp.), --x--, is depicted. Symbols denote: Δ, sucrose; □ , glucose; O, fructose; ◇ , raffinose.

a) Sugar content in the leaf blades. Sucrose is plotted with a factor of 0.1; the inset depicts the sucrose content in relation to the sum of other sugars analyzed in the hardened leaves (10 days).

b) Sugar content in intact isolated chloroplasts. Values are corrected for broken chloroplasts (14-38%) contaminating the preparations.

c) Estimated sugar concentrations in chloroplasts (broken lines) and extrachloroplast space (solid lines); ●, total concentrations. Chloroplast values were calculated assuming an osmotic volume of 25 µl per mg chlorophyll (7). Estimated concentrations in the extraplastidic space are based on the water content of the leaves.

IV. SUGAR ACCUMULATION AND COMPARTMENTATION
RELATED TO COLD-HARDENING

In order to evaluate the role of sugars in cold acclimation, their accumulation in the leaf cell and compartmentation
in the chloroplasts during controlled hardening and
de-hardening of spinach plants was followed. Fig. 12 depicts
the changes in sugar levels occurring in the leaves of one
plant batch. In non-acclimated leaves, the sugar concentration
is low. Most of the glucose and fructose but relatively little
of the sucrose detectable in the leaf reside in the chloroplasts. Cold acclimation leads to a dramatic increase in sugar
levels both in the chloroplasts and in the extraplastidic
space. If the whole leaf is considered (Fig. 12a), sucrose
accumulation is by far the most conspicuous effect. In contrast, glucose, fructose and raffinose substantially contribute, besides sucrose, to the increase in the sugar level of the
chloroplast compartment (Fig. 12b). Returning the plants to
higher temperature largely reversed these changes in parallel
to the loss of frost hardiness. This strongly suggests that
the accumulation of sugars is indeed closely related to
increased frost tolerance. The increase in glucose and fructose levels during de-hardening, which in other experiments was
a transient effect, may be caused by cleavage of sucrose
(cf. ref. 21). In Fig. 12c, estimated concentrations of the
sugars are given. Such estimates may not be very accurate for
chloroplasts, because they rely on an assumed chloroplast
osmotic volume of 25 µl per mg chlorophyll (7), which might
change during acclimation (3). If such volume changes are
neglected, the total sugar concentration after hardening is
about equal inside and outside the chloroplasts. Individual
sugars, however, are not evenly distributed between chloroplasts and external space. The lower portion of sucrose in the
chloroplasts is consistent with the finding (19) that sucrose
synthesis occurs in the cytoplasm. It should be noted, however, that in chloroplasts from hardened cabbage (Brassica
oleracea L.), the most abundant sugar was sucrose (21). Also,
on the basis of calculated concentrations, hardened cabbage
leaves possess higher total sugar levels in the chloroplsts
than in the non-chloroplast space of the cell. However, the
high calculated sugar concentrations for the extraplastidic
space suggest that during hardening of cabbage and spinach
leaves, sugar levels must increase also in the vacuoles to
warrant osmotic equilibrium.
Because of the complex composition of inorganic electrolytes and their largely unknown thermodyamic activities in the
cell, the protective effect of sugars in vivo cannot be
assessed quantitatively at present. As judged from in vitro
experiments (cf. Fig. 9), the sugar levels found in the chloroplasts of acclimated leaves, should exert a significant

protection from adverse effects of inorganic salts, whereas
levels in nonhardened and de-hardened leaves are too small for
an effective protection. It should be noted that sugars are
not the only cryoprotective compounds present in chloroplasts.
Specific low-molecular weight proteins seem to contribute to
frost hardiness (29). Moreover, preliminary experiments (A.
Justenhoven, unpublished) indicate a strong accumulation of
cryoprotective amino acids during hardening. Thus, cold acclimated leaves appear well protected from damaging effects of
rising concentrations of inorganic electrolytes during extracellular ice formation, and such effects are, in fact, not
observed in vivo.

V.   CONCLUSIONS

The present investigation does not disclose a singular, but
several early damaging effects of freezing and thawing on
spinach leaves. By frost treatment within a narrow temperature
range, various vital reactions in the cells are inhibited;
early damage seems to be located at different sites such as
mitochondria, chloroplast envelope and stroma, and thylakoid
membranes. Damage to the photosynthetic system in vivo was
investigated in detail. Obviously, inactivation of CO_2 fixation via the Calvin cycle starts slightly earlier than inhibition of photosynthetic energy-conserving reactions. Inactivation of the thylakoid membranes is characterized primarily by
decreased rates of electron transport; the O_2 evolving apparatus, Protosystems II and I are affected. Photophosphorylation
declines together with electron transport. A primary
uncoupling effect is not observed. We have to conclude that
thylakoid injury in vivo is not caused by toxicity of inorganic
salts arising by ice formation. Thylakoid inactivation by
freezing of leaves resembles the action of amphiphilic substances in vitro, but is not identical with it. Although cell
compartmentation possibly is disturbed in the course of
freezing and thawing, current experiments seem to exclude
thylakoid inactivation by potentially toxic vacuolar solutes.
Thus, we cannot single out a certain type of cryotoxic agent to
be basically responsible for freezing injury of thylakoid membranes in vivo. Rather, the complex composition of the stroma
or other cell compartments may lead to the membrane damage
observed.
As there is no detectable difference in the pattern of
freezing damage in nonhardened and cold acclimated spinach
leaves, hardening obviously does not alter the mechanism of
injury, but only shifts the limit of frost tolerance to lower
temperatures. Increased tolerance to freezing can be well
understood in terms of accumulation of cryoprotective substances such as sugars, certain amino acids and proteins. The

close relationship betwen degree of hardiness and sugar content inside and outside of the chloroplasts shown in this report emphasizes the importance of sugars as cryoprotectants. Hardening of the membrane itself is not unequivocally proven (22). Difficult to understand at present is the sharp limit of frost tolerance found even in acclimated leaves in the presence of high levels of protectants. In addition to solute interaction with the membranes, one has to consider other physical mechanisms of injury. For instance, membrane damage might be aggravated - depending on solute composition - when during freezing a certain critical volume of the unfrozen protoplasts is transgressed, as has been suggested for the lysis of human red cells (18,31), or when the membranes of the cell are rapidly stretched upon thawing (30). Though in vitro experiments exclude purely osmotic changes as sole causes of freezing injury of thylakoid membranes, the very narrow temperature range between beginning damage and fatal injury, as observed in our experiments, suggests that such or similar physical effects might contribute to frost damage in vivo.

ACKNOWLEDGMENTS

We thank Professor K.A. Santarius for critical discussions. The study was supported by the Deutsche Forschungsgemeinschaft.

REFERENCES

1. Aach, H.G., and Heber, U., Z. Pflanzenphysiol. 57:317-328 (1967).
2. Bergmeyer, H.U. (ed.), "Methoden der enzymatischen Analyse" (3rd edition). Verlag Chemie, Weinheim, 1974.
3. Heber, U., Protoplasma 51:284-298 (1959).
4. Heber, U., Biochim. Biophys. Acta 180:302-319 (1969).
5. Heber, U., Tyankova, L., and Santarius, K.A., Biochem. Biophys. Acta 291:23-37 (1973).
6. Heber, U., Volger, H., Overbeck, V., and Santarius, K.A., in "Proteins at Low Temperatures", Advances in Chemistry Series, No. 180 (O. Fennema, ed.), pp. 159-189. Amer. Chem. Soc., Washington, 1979.
7. Heldt, H.W., Werdan, K., Milovancev, M., and Geller, G., Biochim. Biophys. Acta 314:224-241 (1973).
8. Kappen, L., and Ullrich, W.R., Ber. Dtsch. Bot. Ges. 83:265-275 (1970).
9. Klosson, R.J., and Krause, G.H., Planta 151:339-346 (1981).
10. Klosson, R.J., and Krause, G.H., Planta 151:347-352 (1981).
11. Krause, G.H., Biochim. Biophys. Acta 292:715-728 (1973).

12. Krause, G.H., Biochim. Biophys. Acta 333:301-313 (1974).
13. Krause, G.H., Briantais, J.-M., and Vernotte, C., in "Photosynthesis. Proc. Vth Internat. Congr. on Photosynthesis" (G. Akoyunoglou, ed.). Internat. Science Services, Balaban, Israel. (In press), 1981.
14. Krause, G.H., Lorimer, G.H., Heber, U., and Kirk, M., in "Photosynthesis '77. Proc. IVth Internat. Congr. on Photosynthesis" (D.O. Hall, J. Coombs and T.W. Goodwin, eds.), pp. 299-310. The Biochem. Soc., London, 1978.
15. Levitt, L., "Responses of Plants to Environmental Stresses". Vol. 1. (2nd edition). Academic Press, 1980.
16. Lorimer, G.H., Ann. Rev. Plant Physiol. 32:349-383 (1981).
17. Lyons, J.M., Raison, J.K., and Steponkus, P.L., in "Low Temperature Stress in Crop Plants" (J.M. Lyons, D. Graham, and J.K. Raison, eds.), pp. 1-24. Academic Press, New York, 1979.
18. Merryman, H.T., Williams, R.J., and Douglas, M., Cryobiol. 14:287-302 (1977).
19. Robinson, S.P., and Walker, D.A., FEBS Lett. 107:295-299 (1979).
20. Santarius, K.A., Planta 89:23-46 (1969).
21. Santarius, K.A., and Milde, H., Planta 136:163-166 (1977).
22. Santarius, K.A.., Acta Horticulturae 81:9-21 (1978).
23. Santarius, K.A., and Ernst, R., Planta 73:91-108 (1967).
24. Schuldiner, S., Rottenberg, H., and Avron, M., Eur. J. Biochem. 25:64-70 (1972).
25. Thebud, R., and Santarius, K.A., Planta 152:242-247 (1981).
26. Thebud, R., and Santarius, K.A., Plant Physiol. 68:1156-1160 (1981).
27. Ullrich, H., and Heber, U., Planta 57:370-390 (1961).
28. Vernotte, C., Krause, G.H., and Briantais, J.-M. in "Photosynthesis. Proc. Vth Internat. Congr. on Photosynthesis" (G. Akoyunoglou, ed.). Internat. Science Services, Balaban, Israel. (In press), 1981.
29. Volger, H.G., and Heber, U., Biochim. Biophys. Acta 412:335-349 (1975).
30. Wiest, S.C., and Steponkus, P.L., Plant Physiol. 62:699-705 (1978).
31. Williams, R.J., and Shaw, S.K., Cryobiol. 17:530-539 (1980).

A MECHANISM FOR FROST RESISTANCE IN EUCALYPTUS

D. M. Paton

Department of Botany
Australian National University
Canberra, Australia

I. INTRODUCTION

The dramatic effects of frost damage in Eucalyptus were probably first observed as early as 1837 in central Tasmania when severe frosts and associated cold air drainage were considered to be the likely cause of widespread death of large trees growing in valley bottoms and lower slopes (3). Similar devastation occurred at Sukhumi in Southern Russia in 1950 (11) when temperatures approaching -13°C for several days, killed large plantations of E. viminalis and other species. Over the last decade, severe frost damage of Eucalyptus has occurred in Northern California (Hellmers pers. comm.), in Natal (Nixon, pers. comm.), in Brazil (press reports) and near Oatlands in Tasmania (20). As Eucalyptus is at present one of the most widely planted silvicultural crops, damage of this magnitude can have serious economic consequences in areas that are prone to the rare severe frost or cold spell. In some areas however, frost hardy Eucalyptus species are often capable of surviving mild unseasonal frosts that cause severe damage in otherwise winter hardy plants. One such example was observed near Jujuy in the Andes (April 1979) when naturally occurring trees of the semi-evergreen willow, Salix chilensis, were severely damaged by a frost of about -8°C but planted trees of E. viminalis in the same area were unaffected.

One probable reason for the susceptibility of Eucalyptus to leaf temperatures between -10°C and -16°C and lower is that the genus does not exhibit the photoperiodic responses which are involved in the high levels of cold resistance shown by many woody and often deciduous species of the Northern Hemisphere (18). Those Eucalyptus species that can develop levels of resistance approaching -16°C (19) are thus likely to have a mechanism for frost resistance markedly different from that for

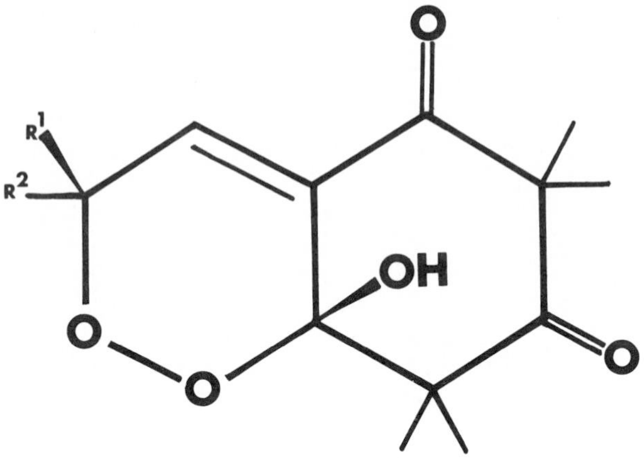

	R^1	R^2
G1	Et	Me
G2	Me	Et
G3	Me	Me

FIG. 1.   The chemical structure of the 3 known forms of G where R_1 and R_2 respectively, are ethyl and methyl groups (G_1), methyl and ethyl (G_2) and both methyl (G_3).

photoperiodically sensitive species.   Indeed, there is good evidence for such a difference as not only are the photo-periodically insensitive processes of hardening and dehardening exceptionally rapid in hardy Eucalyptus species (19) but warm roots are directly involved in the dehardening process of all species so far tested (23).

Another unusual feature of Eucalyptus physiology is the presence of high levels of inhibitory substances in adult leaves of most species (25,26).   One identified component of these inhibitors in E. grandis is G (Fig. 1).   The term G is derived from grandis and for the present at least, this termi-nology allows G to be distinguished from other inhibitory com-pounds such as C in E. citriodora (22).   As promotory effects are included among the various bioassay responses described for G (4,21), G is now considered to be a natural growth regulator in E. grandis.   Nevertheless, the content of G in adult leaves of this species is always high enough to be strongly inhibitory even when diurnal and seasonal variation is taken into account (7).   The adult leaves of all other Eucalyptus species that are capable of rapid hardening, contain inhibitory amounts of non-G compounds.   Conversely, the tropical species E. deglupta which appears incapable of low temperature hardening (19), contains little or no inhibitory compounds in the adult leaves (25).

This is one of several lines of evidence leading to a new concept for the mechanism of frost resistance in E. grandis (20). The concept involves rapid hardening (19) and dehardening (23) both of which possibly involve rapid changes in G content. A direct relationship between endogenous G content and frost resistance in E. grandis is thus critical evidence for evaluation of this concept. The effect of exogenous G on the frost resistance of E. grandis and other species, is equally important. Both approaches are considered in the present study which examines a possible mechanism for frost resistance, not mechanisms of low temperature injury. The proceedings of two recent international seminars also make this distinction as plant cold hardiness is included in the title of one (10) and low temperature stress in the other (12). Nevertheless, determination of levels of resistance or hardiness necessarily involves unequivocal assessment of frost injury.

II. MATERIALS AND METHODS

A. Assessment of Frost Injury

Non-visual assessment of leaf injury was based on the loss of water and possibly other volatile substances when injured leaf tissue finally dies and develops a dry, papery texture (17). The basic assumption in the method is that the percentage ratio of dry weight to wet weight is determined within limits, by the relative amounts of dead and living tissue. A ratio between 30 and 50% indicates that uninjured leaves have a water and volatiles content of between 70 and 50%. A ratio of about 85% in killed plants indicates that the water content of dead tissues was reduced to about 15% in equilibrium with atmospheric moisture. As ratio values between about 40% and 85% agree with subjective visual assessments (17), the ratio provides an objective and quantitative measure of frost injury for individual seedlings. The ratio was determined 2-3 weeks after frosting to allow for changes in either full development of damage or in some cases, repair and recovery from initial damage.

While a radiation frost room was used for all experiments (1), comparison of different experiments and especially different species having different water and volatiles content was possible if the ratio was adjusted to a 10 unit scale separating minimum damage (0) and maximum damage (10) as described previously (19,23). In a few experiments it was convenient to group levels of frost resistance as high, intermediate and low corresponding respectively to damage 0-2, 3-7 and 8-10 on the 10 unit scale (20).

B. Determination of G Content

Extraction, purification and assay of G followed the quantitative methods developed by Osawa (24), and modified by Dhawan et al. (4) and Godden and Paton (7). Extraction time of the cut lamina with methanol (6 x 10 ml) was 24 hr. The extract was evaporated to dryness, taken up in 50 ml water, filtered and the filtrate partitioned using chloroform (2 x 50 ml). The combined chloroform fraction was reduced to 2 ml for purification on silica gel TLC plates. One rather than three plates, was used for separation by the three successive solvents. The first solvent (dichloromethane) and the third solvent (cyclohexane: chloroform: ethylacetate, 4:1:1 v/v) were run in the same direction and at 90° to the second solvent (hexane: ethyl acetate, 2:1 v/v). Use of one plate reduced the 20% loss of G incurred in transferring from one plate to another. The fluorescence-quenching bands that co-chromatographed with authentic G were removed after the third solvent and identified as authentic G by NMR. These bands were eluted with chloroform and used for UV spectrometric assay (245 nm) of the amount of G per g fresh weight of leaf.

C. Membrane Permeability and Electron Transport
in Photosynthesis

The possible effects of G on membrane permeability were examined by following the changes in rubidium (^{86}Rb) uptake by sections of Avena coleoptile, by segments of mung bean hypocotyl, and by discs of red beet tissue. The loss of betacyanin from red beet was also affected by G. These procedures have been developed by Dhawan (unpublished data) and in general, followed those described by Pike and Richardson (28) and by Vianello and Macri (32).
Isolated mesophyll cells of Xanthium strumarium and spinach chloroplasts were used to determine how G affects the photosynthetic apparatus. The methods involved have been described by Sharkey et al. (30).

III. RESULTS

A. Frost Resistance and Ontogenetic Variation in G Content

Provided that roots as well as shoots, were maintained at warm temperatures, dehardening of E. grandis shoots (Fig. 2) was completed within 2-3 days. Such rapid dehardening did not occur if the shoots were warm but the roots remained at low hardening temperatures. This delaying effect of low root temperature on dehardening of warm shoots has been observed in several frost-hardy Eucalyptus species (23). In contrast to

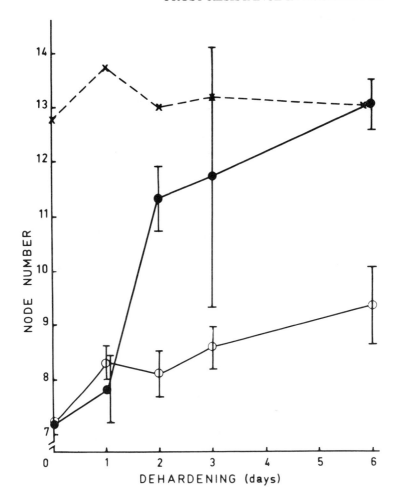

FIG. 2. Relationship in E. grandis seedlings between node position of highest dead leaf pair and days of dehardening before exposure to a frost of -6.5°C. Hardened seedlings were dehardened at glasshouse temperatures with either warm (●) or cold (○) roots. Number of expanded leaves (✗) is given for each treatment. Visual scoring of dead leaves corresponds to damage > 8 on the 10 point scale as described in text. Data from Paton (1981).

these more hardy species, the leaf and stem damage in E. grandis exhibited marked position effects (Fig. 2). At the frost temperature used (-8.5°C), all hardened plants suffered leaf damage at node 7 and below, while leaves at node 8 and above were not affected. In plants with warm roots, the node position separating damaged and undamaged leaves increased to

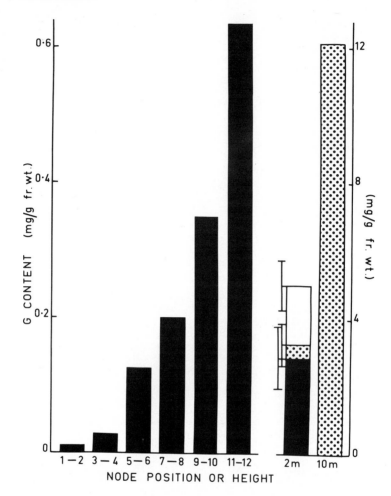

FIG. 3. Variation in G content (mg/g fr. wt.) with increasing ontogenetic age in E. grandis. Data from Paton et al. (1976) for seedling leaves (1-12) and 10 m tree, and from Godden (1980) for 2 m sapling. Diurnal variations in the 2 m sapling averaged to give daily means + SE for extractions made from winter (□), spring/summer (▨) outside grown plants, and from phytotron grown plants (■). All seedlings were glass-house grown plants (■). Data from Paton (1981).

node 11 after two days, and to the apical bud (node 13) after six days of dehardening. In contrast, leaves above node 10 in seedlings with cold roots remained undamaged after 6 days with warm shoot temperatures. Such node effects occurred on lateral shoots as well as on the main axis. The partially damaged plants in Fig. 2 thus appeared as a central cone of dead leaves and twigs with a variable amount of undamaged tissue on the outside determining the size of the cone.

To distinguish the possible roles of ontogenetic age (nodes) and physiological age (maturity of individual leaves) in this cone effect, reciprocal transfers between contrasting glasshouse and bushhouse conditions were made when about seven leaves had developed. When frosted after a further seven leaves had formed, no relationship between leaf history and frost damage was observed. The cone effect in Fig. 2 thus appears related to ontogenetic ranking rather than to different environmental preconditioning during leaf development.

From the ontogenetic changes that occurred in content of the growth regulator G in seedling leaves of E. grandis (Fig. 3), limitation of the height of the damaged cone in hardened plants (Fig. 2) appears to be linked with the rapid increases in G content that occurred above nodes 7-8. Increased G content was thus associated with increased frost resistance. However, different plants and different growing conditions were involved in obtaining the results from Figs. 2 and 3. Thus the ontogenetic increase in frost resistance should not be specifically related to a critical G content greater than about 0.2 mg/g fr. wt.

As there are field records (2) suggesting that the juvenile seedling is one of the most frost sensitive stages in tree growth, the ontogenetic component in frost damage of E. grandis probably extends to other Eucalyptus species as well. Nevertheless, some caution is needed when interpreting this kind of field evidence because seedlings are often exposed to grass temperatures which can be appreciably lower than the screen or air temperatures affecting the higher adult foliage.

The direct relationship between root temperature and rate of dehardening in E. grandis seedlings (Fig. 2) suggests that warm roots may either inactivate a hardening factor from the shoots or supply some dehardening factor to the shoots (23). As roots are also involved in the root-shoot gradient of G content in E. grandis leaves (27), a likely role for warm roots in dehardening is to decrease the G content of leaves. This proposed mechanism explains in part, the low G content of 2 m plants of E. grandis that are fully dehardened during summer months in Canberra (cf Fig. 3). It also explains the observed association between increased frost resistance towards the top of E. grandis seedlings in Fig. 2 and the marked ontogenetic increase in G content in Fig. 3.

B. Seasonal Variation in G Content

In 2 m saplings of E. grandis, maximum G content (5.0 mg/g fr. wt.) occurred in winter grown plants (Fig. 3) when the level of frost resistance was sufficient for the plants to survive -8°C without damage. In spring and summer when damage occurs at about -2.5°C, the G content was much lower (3.0 mg/g fr. wt.) and was not significantly different from that in non-stressed, phytotron grown plants. The standard errors shown for the 2 m plants in Fig. 3 involved diurnal variation in G content (6,7) as well as variation among the same plants sampled on different days. Information of this kind is being collected for adult leaves of larger trees of E. grandis (10 m and higher), which as illustrated in Fig. 3, have the highest G content so far recorded.

As seasonal changes in levels of ABA and other inhibitors occur in several other plant genera (33), it is possible that these may also be related to frost resistance as part of the induction and maintenance of winter dormancy in Northern Hemisphere plants. Similarly in potato tubers, the cold rest period can be correlated with a high content of several components of the inhibitor ß complex (5).

C. Diurnal Variation in G Content

With winter sampling of E. grandis (Fig. 4), G content decreased from a high level of 5.6 mg/g fr. wt. before sunrise (0615 h) to 5.1 mg/g fr. wt. at sunrise (0710 h) and to 4.3 mg/g fr. wt. by 0910 h. After about 12 noon, variation between replicated samples increased markedly although an overall increase did not occur until night fall when the high pre-sunrise level was finally re-established. The G content of adult leaves thus changes during day/night cycles in winter when the balance between rapid dehardening (warm days) and rapid hardening (cold nights) also changed. As the G level remained at pre-sunrise levels for up to 6 hrs in leaves covered with black plastic bags (6) it seems that the winter diurnal decrease in G content was related to light. Once the bags were removed, the G content decreased in a similar way to that observed under natural light conditions.

In contrast to the winter data in Fig. 4, the spring, summer and autumn samplings revealed no decrease in G content during the first four hours of day light when G remained at a relatively constant low level (c. 3.4 mg/g fr. wt.). This level is lower than the minimum (4.3 mg/g fr. wt.) recorded during winter. In Phytotron grown plants, G content was even lower (2.8 mg/g fr. wt., 27) and remained constant under simulated summer conditions.

Restriction of exceptionally high G content in E. grandis to leaves exposed to winter nights, implies that G may be involved in the ability of more hardy specimens of this species to survive radiation frosts (-5°C to -10°C) that occur during the Canberra winter. It is very likely however, that the mechanism for frost resistance in the most hardy Eucalyptus species including E. bicostata (23) involves some modification of this. For example, the various growth regulators in E. bicostata do not include high amounts of G (25). Furthermore, the stomatal conductance values of E. bicostata during winter

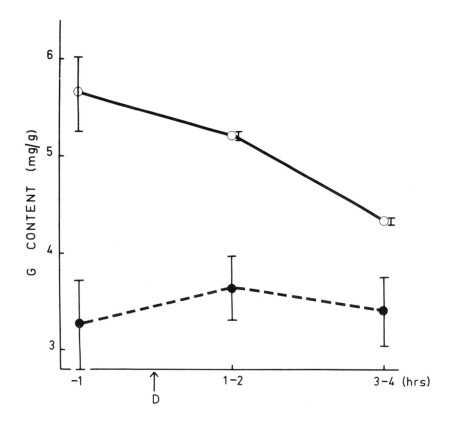

FIG. 4. Variation in G content of adult leaves of E. grandis sampled at various times (hrs) before (-) and after sunrise (D) for plants growing in winter (O) and plants grown in spring, summer and early autumn (●). Means ± SE given for 4-6 replicates at each sampling time.

are substantial when compared with E. grandis (6). For E. grandis, the light dependent decrease in G content in winter grown plants demonstrates that some metabolic activity continues during the Canberra winter when stomatal conductance is consistently low. One such light dependent metabolic activity that could be affected by G is photosynthesis.

D. Effects of Exogenous G on Membrane Permeability and Electron Transport in Photosynthesis

Concentrations of 3 mM G or higher, inhibited CO_2 dependent oxygen evolution during photosynthesis of isolated mesophyll cells of Xanthium strumarium (Fig. 5, 30). About 20% inhibition occurred with 0.5 mM G while 0.1 and 0.3 mM G

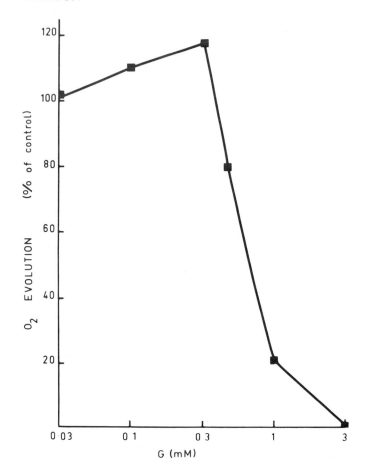

FIG. 5.   Effect of a range of concentrations of G(mM) on
oxygen evolution from isolated cells of <u>Xanthium</u> <u>strumarium</u>.
Oxygen evolution expressed as percentage of control rate which
varied between 121 and 132 μmoles O_2 mg $Chl^{-1}hr^{-1}$.
Data from Sharkey et al. (30).

promoted oxygen evolution. To test whether G affected electron
transport or carbon metabolism, 1 mM parabenzylquinone an electron acceptor that can accept from either PSI or PSII, was
added to the assay medium. In the absence of G, the rate of
oxygen evolution was 260 μmol O_2 mg $chl^{-1}hr^{-1}$ and was
unstable. From this kind of evidence Sharkey et al. (30) consider that G affects photosynthetic electron transport.
    As indicated by the effects of G on oxygen evolution from
Spinach chloroplasts (Table 1, 30), 3 mM G reduced uncoupled
whole chain (water to methyl viologen) electron transport of
spinach chloroplasts by 60% (Table 1). PSII activity (water to

TABLE I. Effects of G on photosynthetic electron transport. Data from Sharkey et al. (30)

Stage in photosynthesis	Electron pathway	Oxygen production	
		EtOH Control	G (3.10^{-3}M)
Whole chain (uncoupled)	water to meth. viol. (+ NH_4Cl)	178	71
Photosystem II (uncoupled)	water to DCIP (+ NH_4Cl)	187	78
Photosystem I (coupled)	ascorb/DCIP to meth. viol. and DCMU (no NH_4Cl)	71	107
Photosystem I (uncoupled)	(+ NH_4Cl)	174	136

DCIP) was also reduced by 60%. PSI activity (ascorbate/DCIP to methyl viologen in the presence of DCMU) was stimulated by G when no uncoupler was added, but when NH_4Cl was present, G caused a slight inhibition. The most obvious effect of G in these two experiments was a reduction in electron transport in a manner that could not be distinguished from the effect of DCMU. Thus G could be acting at the same site between the point of donation of electrons by water and the point at which reduced DCIP donates electrons. Furthermore, G caused some stimulation of PSI electron transport indicating that G can uncouple photophosphorylation from electron transport.

Some preliminary experiments (Dhawan, unpublished data) have indicated that G affects rubidium (^{86}Rb) uptake by Avena, mung bean and red beet tissue (Table 2). This result implies that G may have an effect on membrane transport (21), an effect claimed for ABA (9).

When these membrane and electron transport effects of G are viewed in the light of seasonal (and diurnal) changes in frost resistance and G content, it seems very likely that changes in G content with hardening and dehardening may influence the active transport properties of membranes. This is the mechanism proposed by Palta et al. (15,16) to explain freezing damage of onion bulbs. In Pinus silvestris there appears to be a very similar hardening mechanism which decreases electron transport capacity to a greater extent in 20 yr old trees than

TABLE II. Effects of G on rubidium (^{86}Rb) uptake (cpm). Dark grown coleoptiles (3 replicates) of <u>Avena</u> <u>sativa</u> (cv Victory) provided one 1.5 cm segment (non-apical) from each seedling. Five beet discs were used for each of 3 replicates. Values for counts per minute (cpm) corrected for differences in fresh wt of replicates. Use of mung bean hypocotyl segments gave results similar to those obtained using <u>Avena</u>. Data from Dhawan (unpublished).

| | <u>Avena</u> sections | | Beet discs | |
	2h	4h	2h	4h
Control	11.3	24.8	24.6	26.1
10^{-5}M G	12.4	26.5	22.3*	22.8*
5.10^{-5}M G	5.4*	5.5*	21.6*	22.4*
10^{-4}M G	nil	nil	22.5*	23.3*

* significant differences from control (P 0.05)

in seedlings (13,14). This ontogenetic component of hardening in <u>P. silvestris</u> is comparable to that in <u>E. grandis</u>.

E.   Effects of Exogenous G on Frost Resistance

A series of preliminary experiments involving spraying, watering or soaking with G solutions have all failed to change the frost resistance of <u>E. grandis</u> seedlings. As shown in Fig. 6 however, spraying with G solutions slightly increased the rate of hardening in <u>E. viminalis</u>. It should be noted that this effect of G is small compared with the effect of surfactant alone (31). The fact that similar treatments do not alter the frost resistance of <u>E. grandis</u> seedlings is possibly due to two factors. One is that exogenously applied G may be metabolized in juvenile seedling leaves which contain small amounts of G (4). The other is that endogenous and exogenous G may have different activities as indicated by their non-additive effects in transpiration (6). As endogenous and exogenous ABA have different effects on frost resistance in barley (29), such non-additive effects do not appear to be restricted to G.

One interesting effect observed in experiments involving spraying with G solutions was that prolonged treatment with high concentrations of G protected <u>E. grandis</u> seedlings from low-temperature induced wilting (25). Recovery from such wilting in control plants involved about the same time (1-2 days) as hardening by low temperatures (19). It is possible that the antitranspirant activity of exogenous G (21) reduced

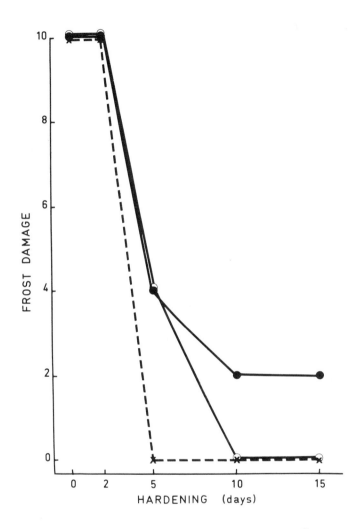

FIG. 6. Effect of duration of hardening at 2°C on level of damage after exposure of E. <u>viminalis</u> seedlings to a frost of -8°C. Seedlings fully dehardened by winter glasshouse conditions (20-25°C) to develop maximum frost damage (killed) with a frost of -6.5°C at time zero. Treatments involved water control (●), surfactant control (O) and repeated spraying with 10^{-3} M G and surfactant (X). Ten experiments have confirmed the surfactant effect but the illustrated G effect was observed 6 times out of 10.

low-temperature induced wilting but the marked between- and within-plant variability has so far prevented a clear demonstration of any relationship of this kind. It is also possible that exogenous G may affect guard cells directly at the same time that it reduces assimilation in the mesophyll by reducing electron transport through photosystem 11 (30).

IV. DISCUSSION

Attempts to obtain clear evidence for a direct causal relationship between G and frost resistance in E. grandis are complicated by the possibility that application of exogenous G does not necessarily ensure specified amounts of G reach the active sites involved in the frost resistance. Perhaps this possibility could be checked in part, if the exogenous G was re-extracted and accurately assayed as is possible in the mung bean antitranspirant bioassay (21). In E. grandis however, the quantitative assay of G is not yet sufficiently sensitive for accurate determination of the low G content in a single seedling leaf (6). The marked between-seedling variability in both G content (6) and frost resistance (17,31) is a further complication suggesting that combining leaves from similar node positions on different seedlings is not a solution to this problem. These and other points can be further examined when radioactive synthetic G becomes available for labelling experiments.

An additional important point that could be clarified by using labelled G is whether diurnal variation in G content consists of metabolic loss of G and subsequent resynthesis or whether some reversible process is involved. Either way, the role of roots in the control of G content appears to be critical for our understanding of the mechanism for frost resistance in E. grandis.

While the available evidence suggests that frost resistance in E. grandis involves a link between high G content and capacity for hardening under low temperatures, it is not known whether this proposed mechanism applies more generally than in this one species. The main difficulty that hinders this kind of generalization is that there has been no systematic survey of Eucalyptus for the high, inhibitory G content present in adult E. grandis.

During low temperature hardening of other herbaceous (29) and woody (34) genera, the balance of endogenous growth regulators often changes because of increased inhibitor content. Winter rape provides a good example where exposure to light and/or cold increases an inhibitor which could possibly be ABA (8). Low temperature also increases ATP and energy charge in winter rape but Kacperska-Palacz (8) suggests that the associated changes in photosynthesis and respiration may involve other than hormonal regulation. As G is the first identified compound that cannot be ruled out as an endogenous regulator of electron transport in photosynthesis (30), such separation of regulatory and metabolic processes does not apply in E. grandis frost resistance. Presumably, G can regulate electron transport in various other processes and especially active transport through membranes. Thus the most likely role for G in frost resistance is regulation of membrane properties to ensure that appropriate membrane functions are maintained during exposure to low temperature.

The photoperiodic insensitivity of the rapid hardening and dehardening in Eucalyptus is a feature of some other woody Angiosperm genera of the Southern Hemisphere (Paton unpublished data). These Gondwana plants appear to have evolved frost mechanisms to survive brief sporadic exposure to frost temperatures between about -8°C and -16°C and in addition, to resume growth within a few days of the recommencement of warm weather. The excellent performance of some Eucalyptus species and especially E. grandis when planted outside Australia, is good demonstration of the high level of adaptability that is associated with a G-mediated mechanism for frost resistance.

ACKNOWLEDGMENT

I am grateful to several colleagues who have contributed to various aspects of this work. The photosynthetic studies involve a collaborative research project with Dr. T. D. Sharkey from the Research School of Biological Science of this University. Figures 2 and 3 are reproduced with permission of the Australian Journal of Botany.

REFERENCES

1. Aston, M.J., and Paton, D.M., Aust. J. Bot. 21:193-199 (1973).
2. Barber, H.N., Evolution, 9:1-14 (1955).
3. Calder, J. E., in "The Hobart Town Courier", Saturday, September 21, 1850.
4. Dhawan, A.K., Paton, D.M., and Willing, R.R., Planta 146:419-422 (1979).
5. Franklin, J., and Hemberg, T., Physiol. Plant. 50:227-232 (1980).
6. Godden, G.F., Honours Thesis, Botany Department, Australian National University (1980).
7. Godden, G.F., and Paton, D.M., unpublished data.

8. Kacperska-Palacz, A., in "Plant Cold Hardiness and Freezing Stress" (P.H. Li and A. Sakai, eds.), pp. 139-152. Academic Press, 1978.

9. Levitt, J., in "Regulation of Cell Membrane Activities in Plants" (E. Marre and O. Ciferri, eds.), pp. 103-106. Elsevier/North-Holland, Amsterdam, 1977.

10. Li, P.H., and Sakai, A., Plant Cold Hardiness and Freezing Stress, Academic Press, New York (1978).

11. Linnard, W., For. Abstr. 30:199-209 (1969).

12. Lyons, J.M., Graham, D., and Raison, J.K., Low Temperature Stress in Crop Plants, Academic Press, New York (1979).

13. Martin, B., Martensson, O., and Öquist, G., Physiol. Plant 43:297-305 (1978).

14. Martin, B., Martensson, O., and Öquist, G., Physiol. Plant 44:102-109 (1978).

15. Palta, J.P., Levitt, J., and Stadelmann, E.J., Plant Physiol. 60:393-397 (1977).

16. Palta, J.P., Levitt, J., and Stadelmann, E.J., Plant Physiol. 60:398-401 (1977).

17. Paton, D.M., Aust. J. Bot. 20:127-139 (1972).

18. Paton, D.M., Aust. J. Bot. 26:633-642 (1978).

19. Paton, D.M., Aust. J. Bot. 28:555-566 (1980).

20. Paton, D.M., Aust. J. Bot. 29:675-688 (1981).

21. Paton, D.M., Dhawan, A.K., and Willing, R.R., Plant Physiol. 66:254-256 (1980).

22. Paton, D.M., and Nishimura, H., unpublished data.

23. Paton, D.M., Slattery, H.D., and Willing, R.R., Ann. Bot. 43:123-124 (1979).

24. Paton, D.M., Osawa, T., Willing, R.R., Mosedale, J., and Crow, W.D., The 9th International Conference on Plant Growth Substances Abstracts 287-289 (1976).

25. Paton, D.M., and Willing, R.R., in "Plant Growth Substances 1973", pp. 126-132. Hirokawa Publishing Co., Tokyo, Japan, 1974.

26. Paton, D.M., Willing, R.R., Nicholls, W., and Pryor, L.D., Aust. J. Bot. 18:175-180 (1970).

27. Paton, D.M., Willing, R.R., and Pryor, L.D., Ann. Bot. 47:835-838 (1981).

28. Pike, C.S., and Richardson, A.E., Plant Physiol. 63:139-141 (1979).

29. Rikin, A., Waldman, M., Richmond, A.E., and Dovrat, A., J. Exp. Bot. 26:175-183 (1975).

30. Sharkey, T.D., Goden, G.F., and Paton, D.M., Plant Physiol. in press (1982).

31. Slattery, H.D., Honours Thesis, Botany Department, Australian National University (1976).

32. Vianello, A., and Macri, F., Planta 143:51-57 (1978).

33. Wright, S.T.C., J. Exp. Bot. 26:161-174 (1975).

34. Weiser, C.J., Science 169:1269-1278 (1970).

A DEGREE GROWTH STAGE (°GS) MODEL AND COLD ACCLIMATION
IN TEMPERATE WOODY PLANTS

L. H. Fuchigami

Department of Horticulture
University of Hawaii
Honolulu, Hawaii

C. J. Weiser and K. Kobayashi

Department of Horticulture
Oregon State University
Corvallis, Oregon

R. Timmis

Weyerhaueser Co.
Tacoma, Washington

L. V. Gusta

Crops Development Center
University of Saskachewan
Saskatoon, Saskachewan Canada

I.   INTRODUCTION

Perennial woody species native to the temperate zone
annually undergo rhythmic growth cycles that are synchronized
with seasonal environmental changes. The timing, rate, and
sequence of changes in bud growth and development are largely
irreversible and induced by environmental signals (31,32,33)
such as changes in temperature and daylength (6,9,38,39,46).
The physiological model (Degree Growth Stage Model) is designed
to numerically describe and predict the annual growth cycle and
hardiness of vegetative buds of such species.

PLANT COLD HARDINESS AND FREEZING STRESS

93

The literature on annual growth cycles and cold hardiness of trees and shrubs is too extensive to discuss in detail. Several good reviews on growth cycles (3,19,22,28,35,45,46,50, 55,57,58,65,71,73,74,75,76) and hardiness (3,4,5,17,56,67,68, 78,79,80) are available.

Attempts to relate growth stages to environmental factors have had limited precision because, to quote Sarvas, "there are no suitable, easily observable phases that can be accurately measured" (60,61). Improved techniques of identifying growth stages help overcome this problem (15,43,44,48,62,63). Progress is also inhibited because the terminology used to identify and describe growth phases is qualitative and often confusing (8,28,46,55,71,74); and because adjacent plant organs may be at different stages of development. The °GS model provides a quantitative means of describing distinct physiological stages of plant growth and development. For simplicity the terms used herein apply only to describing the physiological status of apical (terminal) buds.

In broad terms a living bud may either be growing or dormant (60,61). Growing buds characteristically elongate rapidly during the spring flush of growth, and more slowly as the summer progresses and environmental constraints and/or correlative inhibition by neighboring organs eventually cause the bud to stop growing. When a bud stops growing it is dormant. When all of the buds on a plant stop growing the plant is dormant. Growth cessation in autumn is a prerequisite to cold acclimation in most woody temperate zone species (79), and plants fully deharden when growth resumes in the spring.

A bud may be dormant for several reasons. At some stages of development a dormant bud will rapidly resume growth in response to changes in the physical environment. Dormant buds capable of resuming growth when the environment becomes favorable are dormant quiescent buds. Dormant buds which remain dormant in a favorable environment, but rapidly resume growth if neighboring organs (leaves, buds, etc.) are removed are dormant correlatively inhibited buds.

Dormant buds which remain dormant for prolonged periods even if the environment is favorable for growth and competitive/inhibitory organs are removed are dormant resting buds. Such internally inhibited buds normally require prolonged exposure to inductive chilling temperatures before they can be induced to grow. Figure 1 and 2 schematically illustrates this discussion.

Sarvas (60,61) developed an annual growth model for temperate zone tree species which could be used to mathematically predict the effects of temperature on plant development during the "active growth period" and during the "autumn and winter dormancy" periods. He had difficulty, however, in identifying specific starting and ending points ("point events"), and the intervening periods in plant development ("segment events"). Sarvas' models were well conceived and provided some of the data and the conceptual basis for developing the °GS model. He attempted to relate cold hardiness development to tree growth (60,61), but this portion of his study was speculative, and not supported by experimental data.

Several other mathematical models have been developed to describe the breaking of winter dormancy in temperate plants (1,11,12,29,32,33,37,45,54) and to quantitatively express that portion of the annual growth cycle. All of these models assume that low temperatures (chilling) are needed to break rest, and that warm temperatures are needed to overcome quiescence.

There is a disagreement about the effects of chilling temperatures on breaking rest. Nooden and Weber (46) state that plant responses to temperatures varies widely among species, while Sarvas (60,61) suggests that it does not, and that similar chilling temperatures have the same effect on plants entering rest as in breaking rest (61). It is widely accepted that the onset of winter dormancy, and the duration of chilling (chilling units) required to overcome rest varies with genotype.

Mathematical models such as the Utah chill unit model (37,54) and heat unit models (60) do not accurately predict developmental stages in all environments (2,59), but in some locations they do provide good predictions. These models like Sarvas' do not identify and define physiological starting and ending "points events". They also do not measure physiological segment events and therefore, previous chill unit and heat unit models are not based on physiological processes. This is probably the reason that mathematical models work only in certain climates and with certain plants.

Various terms have been used to describe "point" and "segment" events during the annual developmental cycle of temperate zone woody species. Table 1 summarizes the terminology used in this paper, the sequence of point and segment events, and the corresponding °GS for a typical annual growth cycle. The ensuing discussion describes each event.

Point Events

1. Spring Bud Burst: When visible vegetative growth first appears from the dormant bud. Bud growth is not inhibited by short days, and it is not possible to induce cold acclimation at this stage.

2. Maturity Induction Point: When plants first become responsive to daylength or other environmental factors which promote development of vegetative maturity. In dogwood, growth of buds exposed to short days slows and stops. Growth of buds exposed to long days continues.

3. Vegetative Maturity: When leaf removal no longer stimulates dormant buds to resume growth. Prior to this point leaf removal stimulates correlatively inhibited dormant buds to begin growing within a few days. Whole plants artificially defoliated prior to this point suffer ultimately die-back and do not cold acclimate even in the absence of environmental

TABLE I. Bud growth and dormancy timing and terminology for point events and segment events during the seasonal cycle of development.

Developmental status	Point events	Degree growth stage ($°GS$)	Segment events
Growing	Spring bud break	$0°$	
Growing		$0°-90°$	Rapid growth phase
Growing	Maturity induction point	$90°$	
Growing/inhibited		$90°-180°$	Slow growth/ correlative inhibition phase
Dormant	Vegetative maturity	180	Growth stops/ end of correlative inhibition phase
Dormant		180-270	Deepening rest phase
Dormant	Maximum rest	270	
Dormant		279-315	Decreasing rest phase
Dormant	End of rest	315	
Dormant		315-360	Quiescent phase
Growing	Spring bud break/ end of dormancy	360/0	

stress. Plants defoliated after this point survive prolonged storage and acclimate to freezing temperatures (Fig. 5) (15,43).

4. Maximum Rest: The point at which buds require the longest chilling and/or exposure to the highest concentration of gibberellic acid (72) to resume growth.

5. End of Rest: When the chilling requirement is fully satisfied. Dormant buds will rapidly grow when exposed to favorable environmental conditions, or remain dormant (quiescent) if conditions are unfavorable.

Cold acclimation refers to the natural development of hardiness in the autumn and early winter. Deacclimation refers to the natural loss of hardiness in the late winter and early spring. In dogwood the first stage of acclimation in hardy species is normally induced by short days and warm temperatures after the photoreceptive point ($90°GS$), and the first evidence of acclimation appears at vegetative maturity ($180°GS$) (43),

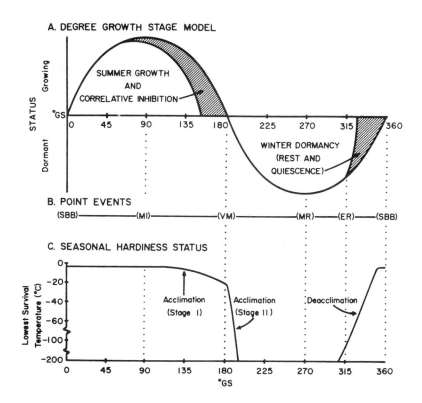

FIG. 1. Diagram of degree growth stage (°GS) model for identifying ontogenetic stages of development (A), point events (B), and hardiness (C) of vegetative buds of temperate woody plants. Point events are indicated: Spring bud break (SBB); Maturity induction point (MI); Vegetative maturity (VM); Maximum rest (MR); End of rest (ER); and End of dormancy/Spring bud break (SBB).

and the second stage by exposure to low temperatures after vegetative maturity (180°GS). Other environmental factors which cause growth cessation and promote development of vegetative maturity may also induce the first stage of cold acclimation. Spring deacclimation is largely dependent upon exposure to warm temperatures. Subsequent exposure to low temperatures may reverse the process (reacclimation) up until growth begins (0°GS).

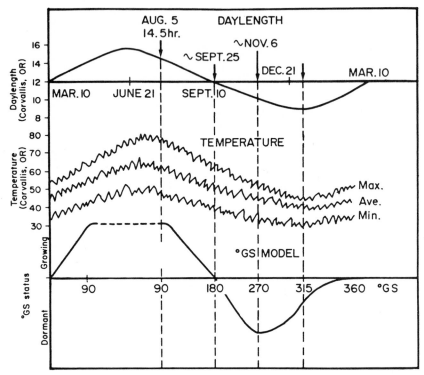

FIG. 2. The relationship of daylength and temperature changes to °GS. Based on studies of red-osier dogwood (Wayland, Massachusetts clone).

II.   THE °GS MODEL

The °GS model is a conceptual numerical cyclical function for quantifying the annual ontogenetic development of vegetative buds of temperature zone woody species (Fig. 1A). The model divides the annual growth cycle into 360 degrees growth stages (°GS) and the cyclical function passes through the five distinct point events described (Fig. 1B). The sine curve illustrates the relative degree of development of the segment events between the point events. The segment events are 1) Rapid growth phase, 2) Slow growth/correlative inhibition phase, 3) Deepening rest phase, 4) Decreasing rest phase and 5) Quiescent phase (Table I).

The amplitude of the sine curve is unimportant, and simply illustrates whether growth is promoted or inhibited. Fig. 1c illustrates the seasonal changes in bud hardiness.

The portion of the sine function from 0 to 180°GS describes the growth period, and the function from 180 to 360°GS describes the dormant period.

Plant growth stages do not occur in a perfect sine function as depicted in Fig. 2. Normally, the curve is skewed, and the extent of skewing depends on environmental influences. Degree growth stages (°GS) represented by this model should not be confused with days of the year (Fig. 2).

One assumption on which this model is based is that under natural conditions the progression of bud growth and development stages are irreversible, e.g. "determination concept" (32). Buds may however reach plateaus of development (transition states) when there is essentially no developmental progress. This can occur when the environment is either non-inductive or inhibitory to development. In contrast cold acclimation and deacclimation are reversible, especially during deacclimation, suggesting that the loss of cold hardiness is a physical rather than a physiological process.

III.   °GS MEASUREMENTS

°GS point events were established by experimentation using a clone of Cornus sericea L. (Wayland, Massachusetts Clone) as the test plant. The tests are relatively simple but time consuming. Faster and simpler techniques are needed. Tests that may someday prove useful for quantifying the °GS are described.

A.   0°GS Spring Bud Burst

0°GS spring bud burst was determined by visually observing terminal vegetative bud elongation, i.e. when the first green leaf tips emerge from the bud scales. The model is based on development of vegetative buds not flower buds. The environmental requirements for vegetative and reproductive bud development are different and vary greatly among genotypes (55). Microscopic observation to determine when stem elongation first begins before visible emergence may be useful for some applications (82). For simplicity we use visual observation.

B.   90°GS Maturity Induction Point

90°GS maturity induction point was determined by subjecting plants to short days and identifying when the least number of days needed to subsequently achieve the 180°GS occurred (Fig. 3). Dogwood plants grown under natural daylength, and at warm temperatures (15°-20°C), from 0°GS were placed in a short photoperiod (8 hr.) at a warm temperature at frequent predetermined intervals. Sub-samples from the short-day treatment were

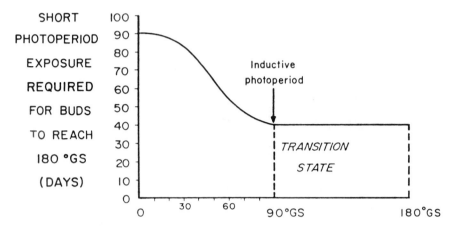

FIG. 3. Diagram illustrating the 90°GS test. Plants growing at 21°C are moved at frequent intervals from a long day environment to a short day environment, and then defoliated. The 180°GS is reached when buds which have become dormant in response to short days fail to resume growth in response to defoliation. The 90°GS is reached when further exposure to short days does not shorten the time required for buds to reach the 180°GS.

defoliated at regular intervals to determine when vegetative maturity occurred. The maturity induction point, 90°GS, was identified by plotting data to determine the earliest sampling date when the least number of subsequent days were required to acquire 180°GS. This test is time consuming and requires controlled environment capabilities and numerous test plants (Fig. 3). Simpler methods are needed.

In red-osier dogwood, about 40 days were required for buds to progress from 90°GS to 180°GS under short days at 21°C (53). Under field conditions in Corvallis, Oregon the Wayland, Massachusetts clone remained in a static transition phase at the 90°GS until the critical shortening photoperiod was reached (Fig. 2). Renquist et al. established by growth chamber and field studies that the critical photoperiod for this clone is about 14.2 hours (53), i.e. development of buds progressed from the 90°GS to the 180°GS in about 40 days at 21°C when plants were exposed to daylengths of 14.2 hours or less.

The maturity induction point is difficult to define and cumbersome to determine. In dogwood its determination is based on the fact that plants are not responsive to photoperiodic influences during the spring flush of growth (32), but later become responsive to photoperiodic stimuli. Many temperate

zone species behave this way, i.e. such plants reach a stage
(90°GS) when they are induced by short photoperiods to proceed
to the next phase of development. Nitsch (45) has described
other types of plants which do not respond in this way
including species that become dormant irrepective to daylength,
and species that do not become dormant. He notes, however,
that plants which eventually become dormant irrespective of
photoperiod often become dormant faster under short days than
long days. This suggests that under natural conditions most
plants do respond to short photoperiods even though photoperiodic effects may be overridden by other environmental
influences such as low temperatures, or be endogenous rhythmic
processes (6,13,14,20,24,41,64).

Preliminary studies indicate that the square wave patterns
generated by an oscilloscope when probes are placed in plant
stems may provide a simpler way for determining the maturity
induction point (66). Square wave patterns observed during
winter dormancy in red-osier dogwood persisted into the spring
flush of growth before assuming a different shape (66).
Further research is needed to determine whether the change in
square wave pattern coincides with the maturity induction point.

C. 0° to 90°GS Rapid Growth Phase

Segment developmental phases between the 0° and 90°GS point
events were not measured experimentally by our group, but
Sarvas's (60) technique for determining period units (p.u.)
during the active period of pollen mother cell meiosis and catkin development is worthy of mention. His major conclusion was
"that the regression of the rate of progress of the active
period on temperature is the same for all the genera, species,
individual trees and the different parts of the active
period." "In all the phases of the active period, the rate of
progress of the active period depends in the same manner on
temperature . . ." Sarvas developed equations based on these
conclusions to predict the following point events (or °GS's)
during the ontogenetic development of temperate tree species.
His equations which are based on observations of plant development, overcome the limitations of models based on heat units
and day-degrees. When Sarvas's equations are adapted to the
°GS model we have:

$$°GS_1 \rightarrow °GS_2 = v(T) (t_2 - t_1) \qquad\qquad (1)$$

where $°GS_1$ and $°GS_2$ are equivalent to Sarvas's period
units, p.u. and $v(T)$ is the rate at which the physiological
processes proceed at temperature T between time t_1 and t_2.

The variable $v(T)$ apparently changes at each °GS as latter
discussion will illustrate. $v(T)$ can be determined experimentally in controlled environment studies. Such studies give
rise to the following relationship:

$$v(T) = v(T)_P \times \frac{hp}{ht} \qquad (2)$$

where $v(T_p)$ is the rate of development at temperature T_p. hp is the time (hours) required for the cycle to pass through a given °GS interval at the base temperature T_p. ht is the time (hours) required for the cycle to pass through a given °GS interval at temperature T.

When equation (1) is adapted to an environment with fluctuating temperatures we have:

$$°GS_1 \rightarrow °GS_2 = \int_{t_1}^{t_2} v\ (T_{(t)})\ dt \qquad (3)$$

in which $T_{(t)}$ is the temperature at a particular time t and v(T) the rate of progress at temperature T.

For practical purposes equation (3) may be written in the form:

$$°GS_1 \rightarrow °GS_2 = \sum_{i=1}^{n} v\ (T_i)\ ti \qquad (4)$$

in which i is the i'th temperature measurement when the time lapse from $°GS_1$ to $°GS_2$ is divided into n equal parts.

Landsberg (31,32,33) used these and similar equations, which he developed independently based on arbitrary "development units" to quantify several readily observed phenomena. He, like Sarvas (58), recognized that the major problem of quantifying growth processes is the identification of accurate developmental stages of plant differentiation or point events, which occur as a result of some stimulus (i.e. daylength). Controlled experiments involving responsiveness to daylength as previously discussed are useful for describing the segment events occurring between the 0°GS and 90°GS point events. As previously noted plants cannot be cold acclimated, either naturally or artificially, during this growth phase (13,70).

D.   180°GS Vegetative Maturity

180°GS vegetative maturity was determined experimentally by defoliation studies (15,16,62,63). Buds are vegetatively mature when no bud regrowth occurs following complete manual defoliation (Fig. 4).

Seibel et al. (61) found that 180°GS coincided with the transition from correlative inhibition (summer dormancy) the onset of rest. This test is simple and straightforward, but it cannot be used to quantify the "segment events" between 90°GS and 180°GS.

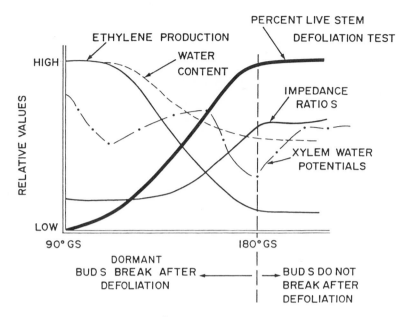

FIG. 4.   Test for 180°GS.   Relationship of several tests to measure 180°GS from 90°GS.   Relative units derived from references 15, 16, 43, 44, 53, 63 and 66.

Other tests (Fig. 4) that have been used with limited success to determine vegetative maturity (180°GS) include ethylene production from nodal stem sections (62), xylem water potential (44), stem water content (44), electrical impedance rations (44,48), square-wave patterns (48,66) and starch accumulation (61). Results tend to verify that the 180°GS is a distict stage of development, but none, as yet, provide as precise a measure of vegetative maturity as the defoliation and regrowth test.

E.   90° to 180°GS Slow Growth and Correlative Inhibition Phase

Segment developmental events between 90° and 180°GS can be characterized by the defoliation test.   In that test (15) plants are completely defoliated manually during the summer and fall, and any resultant regrowth of leaves from previously correlatively inhibited buds is removed daily.   Bud break and stem and bud survival is observed the following spring, and plotted against the initial time of defoliation.   Figure 5 shows the curve that resulted when red-osier dogwood survival in Corvallis, Oregon was studied (15).   The earliest autumn defoliation date which resulted in 100% survival the following spring is the time of the 180°GS.

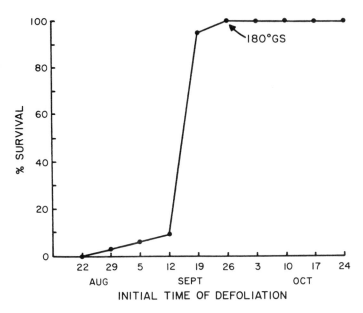

FIG. 5. Percent stem survival test for determining vegetative maturity (180°GS) development. Plants are defoliated in
the fall, and % stem survival is determined the following
spring. The 180°GS is reached when defoliation results in no
stem dieback. Data from Fuchigami et al. (15).

As previously noted a critical shortening daylength had to
be reached in red-osier dogwood (14.2 hrs. for the Wayland, MA
clone) before bud development progressed beyond the 90°GS
(15,53). Temperature effects on the rate of development from
90°GS to 180°GS to within ± 4 days for field grown dogwood
plants (51).
     The equations developed by Sarvas (60,61) are useful for
improving the accuracy of the model in predicting bud development between 90° and 180°GS, i.e. temperature effects established in growth chamber studies (53) were used to compute °GS
using Equation 4. As noted previously, the rate of development
at each °GS is not a constant (27). This increases the complexity of the model, but must be considered to accurately predict developmental events (27).
     In summary this growth phase is characterized in dogwood by
its response to a critical short-day stimulus which triggers
buds to develop beyond the 90°GS (7,15,53). Other environment
variables such as temperature (15,53), mineral nutrition

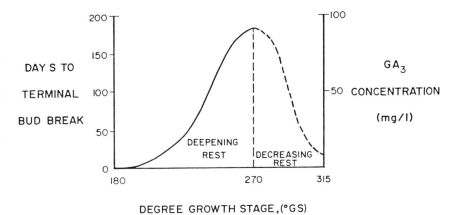

FIG. 6. Regrowth test for 270°GS. Days of exposure to warm temperatures (25°C) and long days required before dormant terminal buds begin growing, or concentrations of GA_3 required to stimulate bud growth.

(16,49,81), and water (81) also influence the rate of development at this stage. Such variables were controlled in this study. Temperature appears to be the most important of these other variables within normal ranges of mineral nutrition and water availability.

F.   270°GS Maximum Rest

270°GS maximum rest can be determined by either of two standard physiological tests (7,15,21,72). The depth of rest can be established by subjecting excised stems to various gibberellic acid concentrations and observing regrowth in a favorable environment (7,20,70). When buds are at the point of maximum rest (270°GS) they require the highest concentration of gibberellic acid (GA_3) to resume growth (7,21,72).

The depth of rest can be established by simply holding intact plants with resting buds in a warm (25°C) environment under long days, and observing how long it takes before buds resume growth. Under such conditions resting buds will eventually begin to grow even though they have been exposed to insufficient chilling. The length of time for growth to resume is proportional to the depth of rest. Maximum rest (270°GS) is the point at which the most time is required before regrowth begins (Fig. 6). In the Wayland, MA dogwood clone almost 200 days were required for regrowth at 270°GS (15,27).

Plants may be triggered to develop maximum rest by defoliating plants prematurely between vegetative maturity (180°GS) and natural maximum rest development (270°GS). Leaves appear

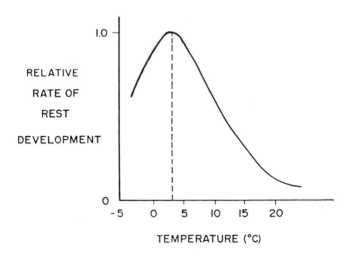

FIG. 7.   Temperature effects on the relative rate of rest development in buds. Relative rates for temperatures below 5°C from Sarvas (57); relative rates for temperatures higher than 5°C from Kobayashi (26).

to possess growth promoters that delay maximum rest development. Under natural conditions low "chilling" temperatures are responsible for stimulating leaf senescence thus reducing the supply of growth promoters (27).

These two tests, or modifications (11), are accurate and useful but both require environmental controls, sufficient material for frequent periodic sampling, and a long time (from 2 to 29 weeks) to evaluate results. There are no other effective chemical, electrical or physical tests for measuring rest.

G.   180 to 270°GS Deepening Rest Phase

The tests described (7,15,21,27,72) above can be used to experimentally quantify the °GS segment events that occur between 180° and 270°GS (Fig. 6). Rest is quantitative, as illustrated by the sigmoid curve that results (Fig. 6) when either concentration of gibberellic acid (21,72) or days to terminal bud break (15,27) is plotted between 180° and 270°GS. The rates (υ(T)) at each °GS may be determined experimentally as a function of temperature. Sarvas (61) reported that the minimum, optimum, and maximum temperatures for promoting deepening rest during this phase of development were -3.5, 3.5 and 10°C, respectively (Fig. 7). Other workers have reported similar results (1,12,27,30) and have suggested that high

temperatures, above 21°C, cause negation of low temperature
effects (12,77).

In the °GS model we used Sarvas's data at temperatures
below 5°C, and expanded the model to 21°C as shown in Fig. 7.
Temperatures above 21°C and below -3.5°C are given a value of
zero, and considered to be ineffective in promoting rest development in buds. It is also assumed that bud development is
irreversible (32). A computer model based on these data and
assumptions has been developed by Kobayashi (27). The model
has predicted maximum rest (270°GS) in red-osier dogwood within
$\pm$ 1 day.

H. 315°GS End of Rest

315°GS end of rest occurs in buds when their chilling
requirement has been satisfied. Experimentally this point may
be determined by placing either excised branches or whole
plants in a favorable warm environment for growth, and
observing the time required for bud break (°0GS). Rest is
ended when further exposure to inductive chilling temperatures
no longer shortens the time for buds to begin growing in a warm
environment favorable for growth. Bud break is normally
observed after some predetermined time (5 to 8 days), or
recorded as the time required for 50% of the buds to break
because it takes awhile for bud growth to become visible
(12,72). In our tests we have arbitrarily determined the end
point by plotting the regressions of the days to bud break
against time of sampling. Two regression equations are produced, one which indicates the rapid loss of rest in response
to chilling temperatures, and the other which indicates the
slow change in bud development during the quiescent period
(Fig. 8). The point where regression lines intersect is considered the end of rest, 315°GS.

I. 270 to 315°GS Decreasing Rest Phase

Regrowth tests described previously to determine the 315°GS
are also used to determine the segment developmental events
(°GS's) between 270° and 315°GS. A reverse sigmoid curve is
obtained when days to bud break or GA concentration is plotted
against °GS during this period (Fig. 9). This indicates that
the rate of decrease in rest ($\upsilon(T)$) is different at each °GS.
This can be verified experimentally in controlled temperature
tests. Sarvas's (59) equations (1,2,3,4) may then be employed
to determine the various °GS.

The low temperature which satisfy the chilling requirement,
and ultimately terminate rest during the 270° to 315°GS phase
are identical to the temperatures required to promote rest
during the 180 to 270°GS phase (Fig. 9). The same procedures
are used to quantify rest status in the deepening rest and the
decreasing rest phases (27).

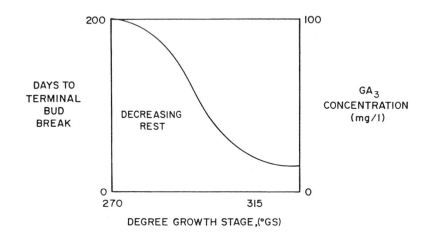

FIG. 8.   Test for 315°GS. Determination of rest completion by locating the point of intersection of two linear regression lines plotted through a growth curve in which days to terminal bud break is plotted against °GS.

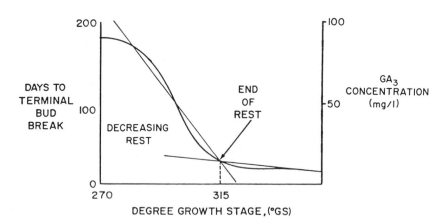

FIG. 9.   Determination of time for terminal bud break under LD-warm temperature environment and concentration of GS_3 required to stimulate bud break.

J.   360°GS

It is the same as the 0°GS (A).

K.   315 to 360°GS Quiescent Phase

The progression of bud development during this phase is promoted by warm temperatures favorable for growth. The tests described by Sarvas to determine υ(T) catkin development during the active period may be used (60), and his adapted equations (1,2,3,4) can then be employed to determine °GS during this period. Spring bud break signals the end of this phase and the beginning of a new growth cycle (0°GS).

IV.   RELATIONSHIP OF COLD ACCLIMATION TO °GS

The general relationship of cold hardiness to °GS has been shown in Fig. 2. Kobayashi, et al. have developed a model to predict hardiness levels based on environmental factors and the degree Growth Stage Model (27). General concepts which will provide the basis for modelling cold acclimation and deacclimation in red-osier dogwood include the following:
Cold acclimation does not begin before growth cessation at 90°GS (15,43), and deacclimation is complete when growth begins at 360°GS (27). In nature the maximum hardiness levels are achieved when plants pass through two or three point events or distinct stages of hardiness as hypothesized by Weiser (78,79) and Tumanov (66,68,69). Landsberg states that before plants can respond to a given stimulus, cells must be in a state in which they are "competent" to act (32). Competence may be either an "all or nothing" state or a quantitative state. For example, in hardiness development, if plants are subjected to cold temperatures before they have aquired the first stage of acclimation induced by short days and warm temperatures (13,15) they will not acclimate as rapidly or to the maximum level (13,20,26,70). Experimental evidence suggests that this period begins at 90°GS and ends at 180°GS (13,15). Similarly, plants exposed to long days (LD) after the first stage of acclimation (between 90° to 180°GS) will not be "competent" to respond to low temperatures, which trigger the second stage of cold acclimation, and may take many days longer to acclimate (6,13,24,26). Apple trees exposed to long days and warm temperatures in a greenhouse were observed to cold acclimate a few degrees in the autumn in spite of the non-inductive environment (24). Some plants may respond to endogenous biological rhythms (6,13,14,20,24,41,64), or have redundant overriding mechanisms for triggering cold acclimation and enhancing the potential for survival (79). In short the sequence of stimuli is important. Plants exposed to an inductive stimuli, such as low temperature

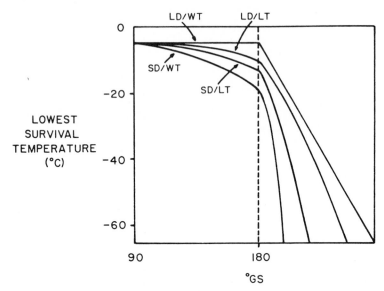

FIG. 10.   Effects of daylength and temperature on cold acclimation before and after 180°GS.   LT is low temperature; WT is warm temperature; SD is short daylength; and LD is long daylength.   Exposure to LT regime after 180°GS.

before they are competent to respond will acclimate, but not as effectively.

Native temperate woody species normally become fully "competent" and achieve the 180°GS in response to shortening photoperiod.   This is not always true of cultivated species and man-made hybrids which are native or adapted to other regions. Such plants are often injured because they fail to achieve 180°GS and the first stage of acclimation (to 20°GS in red-osier dogwood) before lethal freezig conditions occur.

The second stage of hardiness occurs in response to low temperatures after the 180°GS (27,43,83).   After 180°GS the rate of acclimation becomes a function of temperature as shown in Fig. 10, 11.   Frosts are not necessary for the second stage of acclimation as once thought (78,79), but exposure to frost does increase the rate of acclimation at this stage (29,47,67, 68,69,78,79).   Even temperatures as high as 20°C induce plants to acclimate between 180° and 315°GS (Fig. 11), and the upper limit of the acclimation promoting temperature range may extend above 20°C (27).   Studies of red-osier dogwood (27) and Douglas fir (Timmis, unpublished data) suggest that it does in these species.

Equipment limitations have prevented us from investigating the low temperature end of the acclimation promoting temperature range.   Sakai (56) has reported that temperatures between

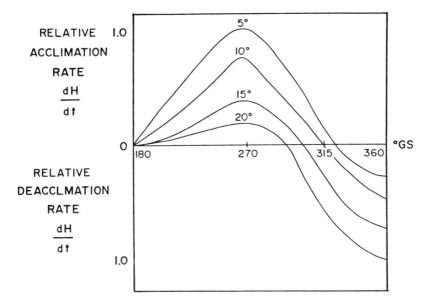

FIG. 11.   Relative rates of cold acclimation and deacclimation in relation to temperature between 180° and 360°GS.   dH/dt
is the rate of hardiness change in relation to temperature.

0 and -5°C were best for acclimating plants at this stage and
states that temperatures below this range were less effective.
    Kobayashi (27) observed that the rate of acclimation
increases between 180°GS and 270°GS.   In addition, he found
that temperatures from 5° to 20°C all caused acclimation in
red-osier dogwood, with effectiveness increasing with
decreasing temperature.   None of the temperatures studied
caused deacclimation between 180° and 270°GS (Fig. 10,11).
This implies that the Landsberg (32) concept of irreversibility
and "determination", may apply to cold acclimation during this
phase of development.   Temperatures higher than 20°C, which we
have not investigated, may cause loss of hardiness (10,18,23,
25,36,51,52).   Maximum acclimation rate occurs at 270°GS
(Maximum Rest) (27).   Between 270° and 315°GS plants continue
to acclimate, but the rate of acclimation decreases from 270°GS
to 360°GS.
    Deacclimation first occurred in dogwood between 270° and
315°GS, and soon after the 315°GS a stage is reached when temperatures as low as 5°C promote deacclimation as shown in
Fig. 11 (27).   The rate of deacclimation increased with
increasing temperatures and the deacclimation rate at each temperature increased from 300°GS to 360°GS.

After 360°GS (0°GS) when the vegetative buds break plants loses hardiness completely. From this growth period to 90°GS the plant can not acclimate under natural conditions and plants continue to grow even at low temperatures and short photoperiods (34). Within a few weeks of rapid growth buds attain "competence" at 90°GS to acclimate once they are exposed to the critical short photoperiod which is genotype specific (79).

V. DISCUSSION AND CONCLUSION

Annual ontogenetic "point" and "segment" events during the annual cycle of bud development are defined and expressed numerically in the degree growth stage (°GS) model. "Point events" were identified at 0°GS, 90°GS, 180°GS, 270°GS, and 315°GS. A sine curve can be drawn through these points from 0° to 360°GS. Sarvas's (60,61) equations were used to quantify point events and segment events. These equations could be refined further to account for environmental factors other than temperature and photoperiod that influences bud development, but they are useful in their present form, and each has a physiological basis that can be experimentally established. Greater detail will ultimately be required to identify °GS between point events. Computers can be used effectively to determine the rate change ($\upsilon(T)$) occurring at each °GS in relationship to temperature (27).

The °GS model is not mechanistic. Hence other physiological processes and environmental factors such as nutrition and moisture were not considered. These were assumed to be important and were controlled during experimentation. Limitations or excesses of these factors affect bud development, and will be considered as sub-routines in future refinements of the model.

The GS model should provide a basis for estimating hardiness levels in plants. Previous attempts to predict hardiness via computer models (40) have failed because the developmental status of plants were undefined, and the models were not based on physiological processes. Kobayashi and others (27) have developed a hardiness model based on the °GS model. This model is quite accurate and illustrates the importance of relating hardiness rates to specific growth stages.

Kobayashi's studies (27) have shown that the rate of acclimation and the effects of temperature on the process changes, between 180° and 360°GS. A bell-shaped curve describing these relationships is generated when the rate of hardiness in response to temperature (dH/dt) is plotted against °GS (Fig. 11). Deacclimtion was not observed until about 300°GS in the decreasing rest phase of bud development. Acclimation has been observed in dogwood after 300°GS, indicating that reacclimation can occur after deacclimation (27). The ability to reacclimate decreases rapidly after 300°GS because of the increased rate of deacclimation at progressively lower temperatures. Similar observations have been reported for other species (19,23,25,29,36,42,51,56).

In addition to its predictive potential the numerical °GS model can help clarify the subjective and complex nomenclature and used to describe developmental events which occur during the annual cycle of temperate woody plants.

ACKNOWLEDGMENTS

The authors acknowledge the technical assistance of F. W. Moeller and the interst, collaboration and contributions of colleagues M. Hotze, P. Nissila, J. Seibel, K. Timmis, and K. Brainerd.

REFERENCES

1. Aron, R.H., Ph.D. Thesis, Oregon State University, Corvallis, OR, 1974.
2. Aron, R.H., HortScience 10(6):559-560 (1975).
3. Burke, M., Gusta, L.V., Quamme, H., Weiser, C.J., and Li, P.H., Ann. Rev. Pl. Phys. 27:507-528 (1976).
4. Chandler, W.H., Lea and Febiger, Philadelphia, Pennsylvania, 1957.
5. Chandler, W.H., Proc. Amer. Soc. Hort. Sci. 64:552-569 (1954).
6. Christersson, L., Physiol. Plant. 44:288-294 (1978).
7. Donoho, C.W., and Walker, D.W., Science 126:1178-1179 (1957).
8. Doorenbos, J., Meted. Landbouwhogesch, Wageningen, 53:1-23 (1953).
9. Downs, R.J., and Borthwick, H.A., Bot. Gaz. 117:310-326 (1956).
10. Edgerton, L.J., Proc. Amer. Soc. Hort. Sci. 64:175-180 (1954).
11. Erez, A., Lavee, S., and Samish, R.M., J. Amer. Soc. Hort. Sci. 96:519-522 (1971).
12. Erez, A., and Lavee, S., I. Amer. Soc. Hort. Sci. 98:221-223 (1971).
13. Fuchigami, L.H., Weiser, C.J., and Evert, D.R., Plant Physiol. 47:98-103 (1971).
14. Fuchigami, L.H., Evert, D.R., and Weiser, C.J., Plant Physiol. 47:164-167 (1971).
15. Fuchigami, L.H., Hotze, M., and Weiser, C.J., J. Am. Soc. Hort. Sci. 102(4)450-452 (1977).

16. Fuchigami, L.H., and Weiser, C.J., J. Am. Soc. Hort. Sci. 106(2):140-143 (1976).
17. Glerum, C., in "Tree Physiology and Yield Improvement" (M.G.R. Cannell and F.T. Last, eds.), pp. 403-420. Academic Press Inc., London, 1976.
18. Hamilton, D.F., J. Amer. Soc. Hort. Sci. 98:221-223 (1973).
19. Hanover, J.W., BioScience 30(11):756-762 (1980).
20. Harrison, L.C., Weiser, C.J., and Burke, M.J., Plant Physiol. 62:894-898 (1978).
21. Hatch, A.H., and Walker, D.R., J. Amer. Soc. Hort. Sci. 94:304-307 (1969).
22. Heslop-Harrison, J., in "Physiological Aspects of Crop Yield", (J.D. Eastin, F.A. Haskins, C.Y. Sullilvan and C.H.M. van Bavel, eds.), pp. 291-325. Amer. Soc. of Agron. Crop Science Society of Amer., Madison, WI, 1969.
23. Howell, G.S., and Weiser, C.J., J. Amer. Soc. Hort. Sci. 95:190-192 (1970).
24. Howell, G.S., and Weiser, C.J., Plant Physiol. 45:1191-1196.
25. Irving, R.M., and Lanphear, F.O., Proc. Amer. Soc. Hort. Sci. 91:699-705 (1967).
26. Irving, R.M., and Lanphear, F.O., Plant Physiol. 42:1191-1196.
27. Kobayashi, K., Ph.D. Thesis, Oregon State University, Corvallis, OR, 1981.
28. Kramer, P.J., and Kozlowski, T.T., "Physiology of Trees", pp. 468-535. McGraw-Hill Book Co. Inc., New York, 1980.
29. Krasavtsev, O.A., Plant Physiol. 16:188-194 (1969).
30. Lamb, R.C., Proc. Amer. Soc. Hort. Sci. 51:313-315 (1948).
31. Landsberg, J.J., Ann. Bot. 38:1013-1023 (1974).
32. Landsberg, J.J., in "Environmental Effects on Crop Physiology", (J.J. Landsberg and C.V. Cutting, eds.), pp. 289-307. Academic Press, Inc., London, 1977.
33. Landsberg, J.J., and Thorpe, M.R., Ann. Bot. 39:689-699 (1975).
34. Lavender, D.P., Sweet, G.B., Zaerr, J.B., and Hermann, R.K., Science 182:838-839 (1973).
35. Levitt, J., "Responses of Plants to Environmental Stresses", Vol. 1. Academic Press, New York, 1980.
36. Litzow, M., and Pellett, H., HortScience 15(1):92-93 (1980).
37. Lombard, P., and Richardson, E.A., in "Modification of the Aerial of the Aerial Environment of Crops", (B.J. Barfield and J.F. Gerber, eds.), pp. 429-440. Am. Soc. of Agric. Eng., Michigan, 1979.
38. Magnesen, S., Medd. Vestland Forstl Forsopssta 14:1-50 (1969).
39. Magneson, S., Medd. Vestland Forstl Forsokssta 14:229-269 (1971).
40. McCarthy, E.F., M.S. Thesis, Rutgers University, New Brunswick, N.J., 1977.
41. McKenzie, J.S., Weiser, C.J., and Burke, M.J., Plant Physiol. 53:783-789 (1974).

42. Meader, E.M., and Blake, M.H., Proc. Amer. Soc. Hort. Sci. 43:91-98 (1943).
43. Nissila, P.C., and Fuchigami, L.H., J. Amer. Soc. Hort. Sci. 103(6):710-711 (1978).
44. Nissila, P.C., and Fuchigami, L.H., J. Amer. Soc. Hort. Sci. 103(6):708-709 (1978).
45. Nitsch, J.P., Proc. Amer. Soc. Hort. Sci. 70:526-544 (1957).
46. Nooden, L.D., and Weber, J.A., in "Dormancy and Developmental Arrest: Experimental Analysis in Plants and Animals", (M.E. Clutter, ed.), pp. 22-268. Academic Press, New York, 1978.
47. Overcash, J.P., and Campbell, J.A., Proc. Amer. Soc. Hort. Sci. 66:87-92 (1955).
48. Parmelee, K., M.S. Thesis, Oregon State University, Corvallis, OR, 1979.
49. Pellett, N.E., J. Amer. Soc. Hort. Sci. 98:82-86 (1973).
50. Perry, T.O., Science 171:29-36 (1971).
51. Proebsting, E.L., Proc. Amer. Soc. Hort. Sci. 83:259-269 (1963).
52. Proebsting, E.L., HortScience 5(5):422-424 (1970).
53. Renquist, A.R., Wensink, R.B., Fuchigami, L.H., Seibel, J.R., Nissik, P.C., and Bates, E.M., J. Amer. Soc. HortSci. 103(6):742-744 (1978).
54. Richardson, E.A., Seeley, S.D., and Walker, D.R., HortScience 9(4):331-332 (1974).
55. Romberger, J.A., U.S. Dept. Agric. Tech. Bull. 1293 (1963).
56. Sakai, A., Plant Physiol. 41:353-359 (1966).
57. Samish, R.M., Ann. Rev. Plant Physiol. 5:183-203 (1954).
58. Samish, R.M., Lavee, S., and Frez, A., Proc. XVII Intl. Hort. Conf. 3:397-408 (1967).
59. Sanders, C.G., HortScience 10(6):559-560 (1975).
60. Sarvas, R., Commun. Inst. for. Fenn. 76.3 (1973).
61. Sarvas, R., Commun. Inst. for. Fenn. 84.1 (1974).
62. Seibel, J.R., and Fuchigami, L.H., J. Amer. Soc. Hort. Sci. 103(6):739-741 (1978).
63. Seibel, J.R., and Fuchigami, L.H., J. Amer. Soc. Hort. Sci. 103(6):737-739 (1978).
64. Siminovitch, D., Gfellar, F., Rheaume, B., in "Cellular Injury and Resistance in Freezing Organisms", (E. Ashahima, ed.). Inst. Low Temp. Sci., Hokkaido University, Sapporo, Japan, 1967.
65. Smith, H., aand Kefford, N.P., Amer. J. Bot. 51:1002-1012 (1964).
66. Timmis, K., Fuchigami, L.H., and Timmis, R., HortScience 16(2):200-202 (1981).
67. Tumanov, I.I., Fed. Proc. 23:793-795 (1963).
68. Tumanov, I.I., Soviet Plant Physiol. 14:440-455 (1967).
69. Tumanov, I.I., Kuzina, G.V., Karnidova, L.D., and Khvalin, N.N., Soviet Plant Physiol. 19:31-39 (1972).
70. Van Hystee, R.B., Weiser, C.J., and Li, P.H., Bot. Gaz. 128:200-205 (1967).

71. Vegis, A., Am. Rev. Plant Physiol. 15:185-224 (1964).
72. Walserr, R.H., Walker, D.R., and Seeley, S.D., J. Amer. Soc. Hort. Sci. 106(1):91-94 (1981).
73. Wareing, P.F., Ann. Rev. Plant Physiol. 7:191-214 (1956).
74. Wareing, P.F., in "The Physiology of Plant Growth and Development", (M.B. Wilkins, ed.), pp. 605-644. McGraw-Hill, England, 1969.
75. Wareing, P.F., Sym. Society Exptl. Biology (D.D. Dacres and M. Balls, eds.), pp. 323-344. Cambridge University Press, 1971.
76. Wareing, P.F., and Phillips, I.D.J., Pergamon Press, 1970.
77. Weinberger, J.H., Proc. Amer. Soc. Hort. Sci. 63:157-162 (1954).
78. Weiser, C.J., HortScience 5(5):403-409 (1970).
79. Weiser, C.J., Science 169:1269-1278 (1970).
80. Weiser, C.J., Quamme, H., Proebsting, E., Burke, M., and Yelenosky, G., in "Modification of the Aerial Environment of Plants", (B.J. Barfield and J.F. Gerber, eds.), pp. 55-84. Am. Soc. Ag. Eng., St. Joseph, MO, 1979.
81. Young, E., and Hanover, J.W., Forest Sci. 24(4):458-467 (1978).
82. Young, L.C.T., Winneberger, J.T., and Bennett, J.P., J. Amer. Soc. Hort. Sci. 99:146-149 (1974).
83. Young, R., HortScience 5(5):411-413 (1970).

MAJOR ACCLIMATION IN LIVING BARK OF SEPT. 16 BLACK LOCUST TREE TRUNK SECTIONS AFTER 5 WEEKS AT 10°C IN THE DARK - EVIDENCE FOR ENDOGENOUS RHYTHMS IN WINTER HARDENING

D. Siminovitch

Chemistry and Biology Research Institute
Research Branch, Agriculture Canada
Ottawa, Ontario, Canada

I.   INTRODUCTION

That external or environmental factors of low temperature and photoperiod are not the sole elements controlling the acclimation of woody plants to winter climates but that some internal factor of seasonal periodicity or annual endogenous rhythm is also implicated in this acclimation has long been appreciated by workers in plant hardiness (3,4,9,12,14,19,24). It has also been realized that even prior to the time that these influences are exerted, there must be a preparatory stage of conditioning in the summer during which high temperature and light intensity promote not only new growth but accumulation of reserves of energy and metabolic substrates both in the leaves and in the living bark (3,4,12,22,24). In late summer and early autumn after slowing down and cessation of growth, these reserves are mobilized for use instead in various chemical synthetic processes associated with hardening (1,5,7,8,11,20). In deciduous trees, events of leaf senescence and autumnal translocation and migration are involved in supplying these nutrients to the living bark cells where they are utilized in the final processes of hardening (1,5,8,18,19,21). Larcher et al. in their review on "Limiting Temperature for Life Function" (3) have tabulated and characterized the sequence of all these events in terms of phases of hardening, namely, predisposition for hardening, prehardening and deep-hardening, the latter corresponding to Tumanov and Krasavtsev's second stage of hardening (25).

There is no dispute now that it is the short photoperiods and cool night temperatures which are the agents triggering the

preparatory events leading to cessation of growth and mobilization of nutrients for the bark cells (3,4,10,12,13,24). The
only dispute concerns the relative contribution which these two
influences make on the preparation of the bark for the final
stages of hardening although the relative contribution must
vary with different species (3,6,10,12,13). But after translocation from the leaves and mobilization processes in the bark
cells are completed, what then drives the final events of
hardening within the bark cells themselves? Low temperature
could have an influence, but it is inconceivable that photoperiod or any light could have an effect through the thick bark
of large branches and trunks of mature trees. After consideration of the reproducibility of timing of autumn hardening of
the locust tree in the Ottawa area over a period of 15 years
despite the incidence in some years of moderately warm conditions, we proposed that "to a considerable extent an endogen-
ously inspired clock mechanism or rhythm located in the bark
cells drives the hardening reaction to completion later in fall
regardless of light or low temperature" (21). Sakai has frequently shown that very low temperatures below freezing are not
required to induce extreme tolerance even to -70°C or lower
(9,10,12). Fortified with this kind of evidence, Sakai liked
to preface his papers or his discussions in these papers with
statements like "Frost hardiness in woody plants shows a
remarkable periodicity throughout the year" (9). Weiser after
considering studies by one of his students which showed that
woody plants acclimate to some extent even when they are subjected to long daylengths and high temperatures in a greenhouse
(2) suggested that there are hardiness rhythms independent of
temperature or photoperiod (24). More convincing evidence of
innate endogenous rhythms is contained in a paper by Smithberg
and Weiser who reported that different clones of dogwood originating from different geographic regions and which had different seasonal patterns of hardening retained, when grown in a
common location, the original timing of their hardening periods
(23). But the most direct evidence is provided in observations
by Schwarz who, after growing young pine trees at constant temperature as high as 5°C with short photoperiod, noted that in
the first year of growth at this temperature the pines
exhibited nearly the same cycle of hardening as trees grown
outdoors (14). We had ourselves, in unpublished studies made a
number of years ago, examined the question of hardiness of
potted young locust trees kept at constant temperature of 10°C
with short photoperiod. The locust trees were placed in a
chamber with this temperature on Sept. 16. By December they
had developed as much hardiness as trees outdoors, that is,
total tolerance to liquid nitrogen. Prompted by these early
results, we addressed ourselves further to the question of the
role of endogenous rhythms in the terminal stages of hardening
in the bark, only this time in order to eliminate the factor of
light or photoperiod, by attempting to harden the bark of

totally severed sections of the trunk at constant temperature in total darkness. Because we used 9 year old trees, light or photoperiod would have had no effect through the thick bark. We used three foot sections of trunk which were placed in cabinets at 10°C without lighting and with constant air flow at 85%-95% humidity.

II. RESULTS

In the first years in which we conducted these trunk section experiments we encountered little success. In order to ensure adequate hardening, we had delayed examination of the bark of the cut sections until December after placing them in the cabinets on Sept. 16 as in the whole tree experiments. By this time, with moisture and rapid air turnover at 98% - 95% R.H., the tissues had become dehydrated and deteriorated. In the fall of 1980 we repeated the experiment placing the sections in the cabinets on Sept. 16 as before but this time we made a preliminary examination of the bark tissues on Oct. 17 after only 5 weeks of storage at which time trees outdoors had hardened considerably. We were rewarded by the results of this early assessment.

Although some desiccation had occurred under the rapid air flow, not only had the bark tissue survived, but the cells had taken on many of the visual characteristics of normally hardened cells. Most of the starch granules that had crowded the protoplasm of the cells on Sept. 16 had nearly disappeared (Fig. 1, A and B) and in neutral red stained cells, the whitish aspect of the nucleus and protoplasmic strands characteristic of the condition of protoplasmic augmentation in hardy cells was easily discernible (Fig. 2). More significantly, when the cells were subjected to plasmolysis stress in strong balanced salt solutions ($NaCl:CaCl_2::9:1$) followed by deplasmolysis in tap water, some survival as tested by neutral red was obtained even at 5 M concentrations (Table 1). On Sept. 16 the cells had not been able to withstand even 1 M completely (Table 1). All indications pointed to some degree of semblance to normal seasonal hardening and of associated chemical processes. Both events were confirmed by actual freezing tests and by preliminary chemical analyses on lyophilized powders.

In Fig. 3 the results of the freezing tests performed on samples of bark taken on Oct. 17 are compared with the results of freezing tests made on bark samples taken on Sept. 16 just prior to storage and again in January following. Freezing tests were conducted by subjecting a series of excised bark samples (3 cm x 2 cm) wrapped in aluminum foil to slow freezing (2°C/hr) at temperatures ranging from -10°C to -70°C in a controlled freezing cabinet followed by immersion in liquid nitrogen. At 10 degree intervals, samples were removed, slowly

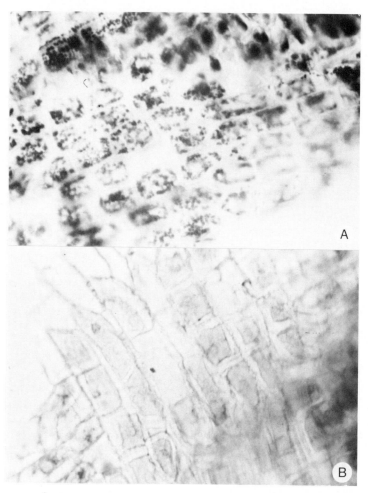

FIG. 1. Depletion of starch in cells of bark of cut trunk
sections of Sept. 16 black locust trees after storage at 10°C
in the dark for 5 weeks. A. Cells with starch granules on
Sept. 16. B. Cells free of starch granules on Oct. 17.

thawed and tested for survival or injury by a combination of
methods which included amino-acid leaching (18), vital staining
with neutral red, turgidity and degree of browning, the latter
occurring in the surface layers of cells of small blocks of
healthy locust bark tissues when immersed and shaken for several weeks in tap water containing 2% sucrose. Although the
maximum hardiness of deep-winter tissues was not reached, it
was evident that in approximately one month a remarkable
increase in hardening had occurred in these bark tissues kept

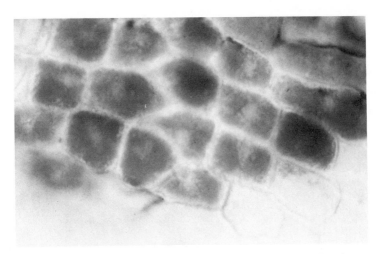

FIG. 2. Appearance of cells of bark of cut sections of Sept. 16 black locust trees after storage at 10°C in the dark for 5 weeks (neutral red stained).

TABLE I. Changes in plasmolysis tolerance of Sept. 16 black locust bark tissue after storage for 5 wks. in the dark at 10°C. (Tolerance measured in percentage survival after plasmolysis in balanced salt solutions ($NaCl$-$CaCl_2$, 9:1) of increasing molarity and deplasmolysis in 0.1M salt solution).

| Tissue | Molarity of plasmolysing salt solution | | | | |
	1M	2M	3M	4M	5M
Sept. 16	90	20	0	0	0
Sept. 16 + 5 wks. (10°C)	100	100	90	45	20
Jan. 5	100	100	100	90	50

at 10°C without light (Fig. 3). Some survival of bark tissues was obtained even in liquid nitrogen as indicated in Fig. 3. Plasmolysed neutral red stained cells of some slices of these tissues taken from bark samples which had been slowly frozen down to -70°C and liquid nitrogen, thawed and then incubated in White's salt solution + 2% sucrose for two weeks are shown in

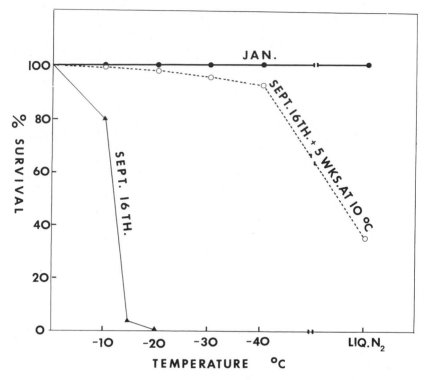

FIG. 3. Freezing tolerance of bark of sections of the trunk of the black locust tree on Sept. 16; on Oct. 17 after storage for 5 weeks at 10°C in the dark; and in January; measured in terms of survival after freezing to various temperatures.

Fig. 4. The values for hardiness of normal trees on Oct. 17 are not shown in Fig. 3. Reference to older data (16,19) would indicate that on Oct. 17 the hardiness of normal trees outdoors would not have reached a maximum either and would not have greatly exceeded that of these logs kept at 10°C. It is unlikely however, due to the desiccating conditions of the constant temperature cabinets used, that any greater hardening could have been obtained by prolonging the period of storage. Until cabinets with better humidifying facilities are used, any speculation about the possibility of obtaining full hardening by extended storage is premature.

The results of tests for starch and soluble sugars on Sept. 16 before storage and on Oct. 17 after the 5 weeks' storage are shown in Figs. 1 and 5. The striking depletion of starch which occurred during the 5 weeks is evident from Fig. 1. This result at 10°C was surprising because dissolution of starch in locust cells is usually noted at lower temperatures (15,17).

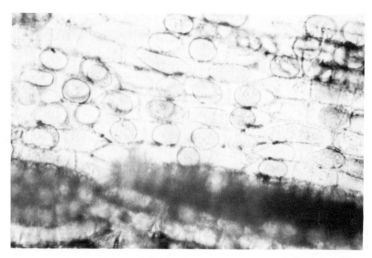

FIG. 4.   Vitally stained and plasmolysed cells taken from
thawed blocks of bark excised from black locust trunk sections
after storage for 5 weeks from Sept. 16 at 10°C and slowly frozen to liquid nitrogen temperatures.

No estimate of soluble reducing and non-reducing sugars was
made by direct analysis. Instead, paper chromatograms were run
on spotted aliquots of 80% alcohol extracts of lyophilized powders of the Sept. and Oct. bark. Soluble sugars were identified   after   72   hours   separation   with   butanol:pyridine:
water::10:3:3 and identification with 1, 3, naphthalenediol in
95% ethyl alcohol + phosphoric acid + HCl. Equal quantities of
a 0.3% sucrose solution were also run as standard. Except for
traces of stachyose and raffinose which appeared at the top of
the chromatogram (Fig. 4), the only sugar which was detected
was sucrose, concentration of which by reference to the standard, would in both barks be about 10% - 12% of dry weight.
Doubtless, in view of previous studies on locust bark, glucose
and fructose were probably present in the bark but at a low
concentration and not detectable at the dilution used for the
chromatogram. Thus, the drastic reduction in starch concentration during storage was not accountable in terms of conversion
to sucrose or to any other sugars. Part of the loss in starch
might have been traced to respiratory activity, but as previous
work has shown, it has always been difficult to completely
account for the fate of some of the starch that is normally
lost to the cells in autumn (15). Regardless, the relative
equality of the sucrose concentration in the extracts of Sept.
and Oct. tissues despite the considerable difference in hardiness adds yet more evidence countering the proposition that
soluble sugars contribute to the hardiness of woody plants.

TABLE II. Changes in moisture and amount of soluble protein and phospholipid in Sept. 16 black locust bark tissue after storage for 5 wks in the dark at 10°C.

	Moisture % f.w.	Soluble protein mg g d.w.	Phospholipid mg (lipid-Px25) 5 gms d.w.
Sept. 16	52.8	31.1	25.0
Sept. 16 + 5 wks (10°C)	36.0	57.0	38.8
Jan. 5	50.0	--	62.5

The changes in percentage moisture, amount of soluble protein and phospholipids during the storage period are presented in Table II. Soluble protein and phospholipids were determined as described in previous publications (19,20). The data in Table II show a loss of moisture, a nearly 100% increase in soluble protein and a 50% increase in phospholipids during this period. No figures for soluble protein in Jan. are given but the increase in soluble protein in the stored tissues is about what would be expected in normal winter tissue. Increases in ribonucleic acids would also be expected to accompany such protein synthesis but no determination of nucleic acids has yet been made. Regardless, the definite increase in both soluble proteins and phospholipids in the bark cells (Table II) clearly indicates that the characterisitic event of protoplasmic and membrane augmentation which has been invariably shown to accompany winter hardening in the block locust had proceeded in good part also in the cells of these bark tissues stored at 10°C in the dark (19,20).

III.   DISCUSSION

The results presented here demonstrate that it is possible in mid-September to achieve a considerable degree of seasonal hardening and associated chemical transformations in the bark of trunk sections of the black locust tree at a constant temperature of 10°C without light even under conditions of complete deprivation of nutrients. Thus, by Sept. 16 in the Ottawa area, bark cells of the trunk have received most of the nutrients and signals from the leaves that may be required for cold hardening. This is to say that the triggering effects of cool night temperatures or shortening day length which are instrumental in promoting cessation of growth, leaf senescence, and autumnal migration of nutrients or metabolic substrates needed for the final processes of hardening in the bark cells, have for the most part been completed on Sept. 16. The bark cells are then already primed at this time without any further influence of photoperiod or of low temperatures or freezing or of supply of nutrients, to independently undergo a major part of seasonal transformations directly associated with the hardening process. To this extent then, these experiments add new convincing evidence that annual periodicities or endogenous rhythms play an important role in the terminal phases of cold hardening in the bark of the black locust tree. It could be argued, in view of reports of hardening of dogwood by desiccation stress and the fact that the trunk sections were somewhat dehydrated, that the partial dehydration during storage contributed to the induction of hardening of the locust sections but the degree of hardening found in dogwood after desiccation stress was much less than that observed here (1). Nor can the possibility be excluded, in the light of the hypothesis we have proposed earlier (17), that the process of dissolution of starch in the cells could in itself have enhanced the hardiness. In any event, efforts must be made to prolong the storage life of the bark tissues by improvement of the humidity conditions in the cabinet not only for the purpose of evaluation of the relative effects of desiccation or starch elimination but to determine if the gap from complete tolerance to −40°C to complete tolerance to liquid nitrogen can be closed by extending storage time. Trees, even under normal conditions outdoors, do not, in the Ottawa area, reach total tolerance to liquid nitrogen by Oct. 17. Also there is no assurance that by Sept. 16 the bark cells in the present experiment had received their full complement of nutrients from the leaves. Bark tissue of trunks intercepted by double girdling in mid-September exhibit some degree of limitation in degree of subsequent hardening (16). Therefore in order to assess the extent of contribution of endogenous rhythms to hardening in these locust bark cells, the initiation of storage of the trunk sections should be delayed as long as possible, short of allowing true hardening to begin. Nevertheless, it is evident from the present experiments that by Sept. 16 the bark cells, without displaying any appreciable hardiness, have already acquired the potential for developing much hardiness at constant moderate temperature without any freezing and without light. There was no question but that many cells in the bark of these log sections, after 5 weeks storage under these conditions, survived even liquid nitrogen so that extreme hardiness was being approached (Fig. 5). Whether the gap can be closed by maintaining the sections for a longer period at 10°C through improvement of humidity conditions or by using higher temperatures of storage needs to be studied. Even as they stand,

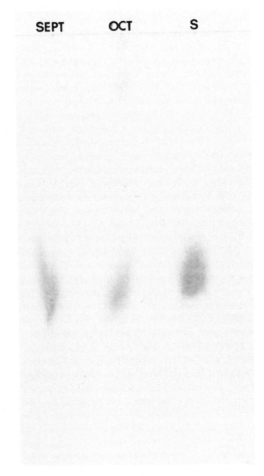

FIG. 5. Chromatogram of soluble sugars of 80% ethyl alcohol extracts of lyophilized powders prepared from the bark of sections of the trunk of the black locust tree on Sept. 16; from bark of trunk sections after storage for 5 weeks from Sept. 16 at 10°C; and of 0.3% sucrose solution.

these experiments with bark of trunk sections, primed as the bark is in mid-September to undergo a large measure of hardening, provide a ready system and springboard for examining the whole question of extreme hardening and the involvement of endogenous rhythms or seasonal biological clocks in this hardening. In a sense therefore, these methods and experiments constitute a sample and convenient tissue culture system for studying the internal factors controlling the final stages of extreme hardening without the complication of sterility or of need for supply of substrates or nutrients.

ACKNOWLEDGMENTS

I wish to thank Mr. B. Rheaume for skilful assistance, also Dr. Yves Cloutier for help in preparing the figures and Dr. I.R. Siddiqui and Z. Collins for the chromatagraphic identification of the soluble sugars.

REFERENCES

1. Chen, P.M., Li, P.H., and Burke, M.J., Plant Physiol. 59:236-239 (1977).
2. Howell, G. Jr., Ph.D. Thesis, Univ. of Minnesota (1969).
3. Larcher, W., Heber, U., and Santarius, K.D., in "Temperature and Life" (H. Precht, J. Christophersen, H. Hensel, and W. Larcher, eds), pp. 195-263. Springer-Verlag, New York, 1973.
4. Levitt, J., "Responses of Plants to Environmental Stresses". Academic Press, New York, 1972.
5. Li, P.H., and Weiser, C.J., Proc. Amer. Soc. Hort. Sci. 91:716-727 (1967).
6. Ormrod, D.P.,mand Layne, R.E.C., HortSci. 9:451-453 (1974).
7. Pomeroy, M.K., Siminovitch, D., and Wightman, F., Can. J. Bot. 48:953-967 (1970).
8. Sakai, A., Low Temp. Sci., Ser. B. (Sapporo) 16:23-24 (1958).
9. Sakai, A., Plant Physiol. 41:353-359 (1966).
10. Sakai, A., and Otsuka, K., Ecology 51:665-671 (1970).
11. Sakai, A., and Weiser, C.J., Ecology 54:118-125 (1973).
12. Sakai, A., Plant and Cell Physiol. 14:1-9 (1973).
13. Scheumann, W., and Bortitz, S., Biologisch. Zentralblatt. 84:489-500 (1965).
14. Schwarz, W., in "Klimaresistenz Photosynthese und Stoffproduktion" (H. Polster, ed.), pp. 55-63. Deut. Akad. Landwirtsch, Berlin, 1968.
15. Siminovitch, D., Wilson, C.M., and Briggs, D.R., Plant Physiol. 28:383-400 (1953).
16. Siminovitch, D., and Briggs, D.R., Plant Physiol. 28:177-200 (1953).
17. Siminovitch, D., and Briggs, D.R., Plant Physiol. 29:331-332 (1954).
18. Siminovitch, D., Therrien, H., Gfeller, G., and Rheaume, B., Can. J. Bot. 42:637-649 (1964).
19. Siminovitch, D., Gfeller, F., and Rheaume, B., in "Cellular Injury and Resistance in Living Organisms" (E. Asahina, ed.), pp. 93-117. Inst. of Low Temp. Sci., Sapporo, Japan, 1967.
20. Siminovitch, D., Rheaume, B., Pomeroy, M.K., and Lepage, M., Cryobiology 5:202-225 (1968).

21. Siminovitch, D., Singh, J., and de la Roche, A.I., Cryobiology 12:144-153 (1975).
22. Siminovitch, D., Cryobiology 18:166-185 (1981).
23. Smithberg, M.H., and Weiser, D.J., Ecology 49:495-505 (1968).
24. Tumanov, I.I., and Krasavtsev, O.A., Soviet Plant Physiol. 6:663-673 (1959).
25. Weiser, C.J., HortScience 5:403-410 (1970).

INFLUENCE OF GROWTH AT COLD-HARDENING TEMPERATURE
ON PROTEIN STRUCTURE AND FUNCTION

N. P. A. Huner, W. G. Hopkins, B. Elfman, and D. B. Hayden

Department of Plant Sciences
University of Western Ontario
London, Canada

M. Griffith

Botany Department
University of British Columbia
Vancouver, Canada

I.   INTRODUCTION

Biological processes remain functional over a relatively narrow range of temperature within which the temperature tolerance of different species varies considerably. The activities of certain enzymes are known to be adapted to the thermal habitat of the species. For example, thermophilic bacteria produce heat stable enzymes (31), and poikilotherms (e.g. fish) contain enzymes which have lower activation energies than the homologous enzymes from homeotherms (birds and mammals) (35). The heat stabilities of enzymes and proteins are also influenced by the adaptations of a given organism to different temperatures. Known cases include urease in cucumber leaves (11), bean leaf Fraction 1 protein (46), fructose bisphosphatase from Nerium oleander (5) and photosystem II of several C_3 and C_4 species (2).

The relationship between proteins and cold-hardiness has been investigated in the past. Siminovitch et al. (47-50) demonstrated a striking parallel between total, soluble protein accumulation and the induction of frost hardiness in cortical cells of black locust trees. Rochat and Therrien subsequently reported that hardy plants have greater ability to incorporate labelled amino acids into water soluble proteins than non-hardy plants (42,43). Bixby and Brown (3) presented evidence that the ribosomal protein structure is altered during the induction of hardiness in black locust seedlings. Since protein synthetic mechanisms do appear to remain active during cold-hardening (8), the ribosomes may be altered to function at lower temperatures.

Heber and co-workers (18,19) demonstrated that chloroplast membranes from winter grown spinach and rye released a protective principle which contained at least two, low molecular weight proteins which protected photophosphorylation and to a lesser extent electron transport reactions against the inactivating effects of freezing.

Roberts observed that changes occur in the ratio of isozymes of several enzymes including invertase (39), peroxidase (40) and phosphatase (18) from wheat leaves during cold acclimation. However, only the changes in invertase appear to be correlated with cold-hardening (39).

Since proteins are the principal component of the dry matter of the protoplasm, Levitt (32) proposed that they were a key to the understanding of freezing injury and in 1962 published his sulfhydryl-disulfide hypothesis. According to this hypothesis for freezing injury, low temperature denatures proteins reversibly, exposing free SH groups. Subsequent freeze-dehydration concentrates the protoplasm which increases the probability of protein aggregation due to intermolecular disulfide bonding. Levitt and Dear (33) suggested that non-hardy plants are those whose membrane proteins irreversibly denature at cold-hardening temperatures and subsequently aggregate due to the formation of intermolecular disulfide bonds. In contrast, hardy plants possess proteins which remain in their native state at low temperature and resist sulfhydryl disulfide interchange thus decreasing the possibility of inactivation due to intermolecular disulfide bond formation. This hypothesis was based on results of freeze dehydration of thiogel and denatured bovine serum albumin (14,34) both being non-plant proteins. Further, indirect support has come from the work of Gaff (12) on drought injury in cabbage and Cothren and Guinn (7) on the effects of low, non-freezing temperature on young cotton plants.

In this report we describe results of experiments which were designed to test both aspects of Levitt's SH hypothesis, that is, low temperature protein denaturation and protein aggregation due to intermolecular disulfide bond formation with respect to both soluble and membrane proteins from plants subjected to growth at cold-hardening (2°-4°C) and non-cold-hardening temperatures (20°-25°C). The purpose is not only to understand the involvement of proteins in injury mechanisms as a result of low temperature stress but also to elucidate possible mechanisms used to overcome the potentially, deleterious effects of growth at low, cold-hardening temperatures on protein structure and function.

II. CHANGES IN THE STRUCTURE AND FUNCTION OF RuBPCase
AND THEIR RELATIONSHIP TO COLD ADAPTATION
AND LEVITT's SH HYPOTHESIS

In recent years, there has been considerable literature published on the effects of cold adaptation in vivo on protein structure and function (21,45). The synthesis of isozymes of differing primary structure can be promoted or repressed by low temperature (1,37,39). Except for increased heat stability of ribulose-1,5-bisphosphate carboxylase-oxygenase (RuBPCase) from Phaseolus acutifolius (46), temperature-induced conformational changes in this major, plant protein are primarily confined to the effects of low temperature in vitro (6,30). Earlier, Huner and Macdowall (22,23) reported that partially purified rye RuBPCase change electrophoretically upon cold-hardening at 4°C for 90 days. This was confirmed upon comparison of the puri- fied enzyme from cold-hardened (RH) and unhardened (RNH) Puma rye when it was shown that the native enzymes differed in their isoelectric points but had the same native molecular weight of 550,000 daltons and the same amino acid composition (24). The kinetics of titration of free SH groups of native RH and RNH RuBPCase with dithiobisnitrobenzoic acid (DTNB) indicated that SH groups were more accessible in RNH than RH RuBPCase. In addition, the denaturation kinetics of RH and RNH RuBPCase in the presence of sodium dodecyl sulfate differed considerably. These results led Huner and Macdowall to conclude that there is an in vivo conformational change in RuBPCase during acclimation of Puma rye to cold-hardening temperatures which results in a decreased exposure of free SH groups even though the total num- ber of free SH per molecule remains constant (24).

Upon investigation of the quaternary structures of RH and RNH RuBPCase, Huner and Macdowall (25) showed that both enzyme forms were made up of eight large subunits (LS) of 55,000 daltons each and eight small subunits (SS) of 14,000 daltons each. However, the presence of a 110,000 dalton peptide was consistently seen in preparations of only RNH RuBPCase in addi- tion to the large and small subunits. This polypeptide (D) has the correct molecular weight for a dimer of the LS. Experi- ments designed to test the structural stability of the two enzyme forms indicated that the LS of RH RuBPCase was less sus- ceptible to aggregation then the LS of RNH RuBPCase in the absence of ß-mercaptoethanol. In addition, Huner and Macdowall showed that the LS of RH RuBPCase was more resistant to aggre- gation due to freezing and thawing than RNH RuBPCase. In all cases, this aggregation phenomenon could be reversed by the addition of ß-mercaptoethanol indicating the involvement of intermolecular sulfhydryl-disulfide interchange as the cause of the aggregation.

The structural differences described above were comple- mented by functional differences between RH and RNH RuBPCase

(26). In addition to being functionally more stable at low temperature, the RH RuBPCase conformer from rye plants grown at 2° to 4°C had a higher apparent affinity for CO_2 between 0° and 5°C than the same enzyme from rye plants grown at moderate temperatures. However, the RNH RuBPCase conformer which was synthesized at 20°C to 25°C had a higher apparent affinity for CO_2 between 20° to 25° than RH RuBPCase. Therefore, growth at cold-hardening temperatures results in a form of RuBPCase which is functionally more efficient at binding CO_2 at low temperatures than the form isolated from plants grown at warm temperatures. Conversely, the enzyme synthesized at warm temperatures was more efficient at binding CO_2 at those temperatures than the conformer synthesized at low temperature. Thus, the structural changes appear to impart significant advantages of low temperature stability and efficiency to this important, photosynthetic enzyme and hence to the whole plant.

According to Levitt's SH hypothesis (32), freezing injury may result from the denaturation of protoplasmic proteins after formation of intermolecular disulfide bonds. This suggests that freeze resistance opposes intermolecular sulfhydryl-disulfide interchange. Since titration with DTNB indicated that SH groups of RH RuBPCase are less accessible than those of RNH RuBPCase, one would predict that RH RuBPCase would have a lower probability of aggregation due to the formation of inter-molecular disulfide bonds. Indeed, this was corroborated by the fact that the LS of the enzyme from cold-hardened rye was less susceptible to aggregation due to intermolecular disulfide bonding induced by the absence of reducing agent or freezing and thawing. Thus, as summarized in Fig. 1, it appears that rye RuBPCase can exist in at least two stable configurations; a high temperature form (HT) which has exposed SH groups and is susceptible to aggregation due to intermolecular disulfide bonding and a low temperature form (LT) which has its SH groups buried to a greater extent than the HT form and is thus not as susceptible to intermolecular disulfide bonding which supports Levitt's SH hypothesis (32). Therefore, altering protein configuration during cold acclimation by burying SH groups would appear to be a unique mechanism for resisting the deleterious effects of freezing and low temperature denaturation.

To determine if a relationship exists between the structure, in terms of exposed SH groups, and the potential of different species to acclimate to cold-hardening temperature, Huner et al. (27) compared the structure and function of RuBPCase from a cold hardy species (Solanum commersonii) and a non-hardy species (Solanum tuberosum), unable to acclimate to low temperature (5), with both species grown under non-acclimating conditions of 20°-25°C. RuBPCase from S. commersonii differed from the enzyme from S. tuberosum in that the native structure of the former has fewer exposed SH groups than that of the latter even though the total number of SH groups titratable with DTNB was the same in the enzyme from both

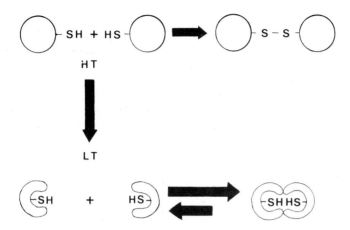

FIG. 1. Schematic model illustrating the configuration change in Puma rye RuBPCase upon growth at cold-hardening temperature and its proposed effect on intermolecular disulfide bonding. HT, high temperature form of RuBPCase; LT, low temperature form of RuBPCase.

species. Furthermore, the enzyme from the cold-hardy species was less sensitive to the absence of reducing agent and freezing and thawing than the enzyme from the non-hardy species. Thus species within a genus which are better able to withstand low temperature may have RuBPCases which are more stable to low temperature as a result of decreased exposure of SH groups than those species less able to withstand low temperatures. RuBPCase from cold-hardy species of potato may be in a low temperature form (LT in Fig. 1) even prior to cold acclimation while the enzyme from non-hardy potato species is in a high temperature form (HT in Fig. 1). However, the results summarized in Table 1 indicate that there is not intergeneric relationship between hardiness and the number of exposed SH groups. Although S. tuberosum is the least hardy of the three species tested, it has fewer exposed SH groups than Secale cereale which is the hardiest of the species tested.

In contrast to the LS of RuBPCase from Pume rye and potato, no differential structural effects were observed for the SS of RuBPCase in the absence of ß-mercaptoethanol or after freezing and thawing (25,27). Thus, it is concluded that the principal locus for change in the structure and function of RuBPCase is associated with the LS. In an attempt to further elucidate these changes in the LS of rye RuBPCase, we have initiated a study into the subunit structure of RH and RNH RuBPCase using the powerful technique of two dimensional slab gel

Table I. Comparison of species hardiness and titratable SH groups in native RuBPCase.

| Species | Killing temp. (°C) | | No. of titratable SH groups | |
	Acclimated	Non-acclimated	Acclimated	Non-acclimated
Secale cereale	-30	-4	25	45
Solanum commersonii	-11	-4	-	10
Solanum tuberosum	-3	-3	-	17

electrophoresis according to the method of O'Farrell (38). This method uses protein separation on the basis of charge in one direction followed by molecular weight separation in a direction perpendicular to the initial charge separation. The gels shown in Fig. 2 and Fig. 3 indicate such separations for purified RNH and RH RuBPCase respectively. These results elegantly depict the major difference in the LS of RNH and RH RuBPCase. In the case of RNH RuBPCase, there are four major types of LS polypeptides of differing isoelectric points and of about 55,000 dalton molecular weight. In contrast, RH RuBPCase has only two types of LS polypeptides indicating that a major change in the microheterogeneity of the large subunit takes place during cold acclimation of Puma rye. The mechanism involved in controlling this alteration in heterogeneity during cold acclimation is presently under investigation.

In addition to the electrophoretic approach, the crystallographic approach is being used to further elucidate the structural differences between RH and RNH RuBPCase (28). Crystals of purified RNH RuBPCase have been grown to a size suitable for X-ray and optical diffraction employing the method of microequilibrium vapour diffusion. To date, RH RuBPCase, does not crystallize under the same conditions as RNH RuBPCase indicating a probable difference in structure.

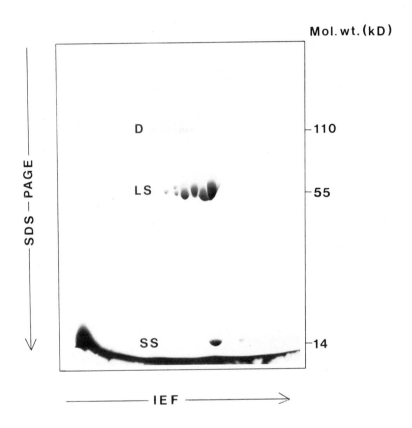

FIG. 2. Result of two dimensional gel electrophoresis of RNH
RuBPCase.   IEF,   isoelectrofocussing;   SDS-PAGE,   SDS   polyacrylamide gel electrophoresis; D, apparent dimer; LS, large
subunit polypeptides; SS, small subunit polypeptides.

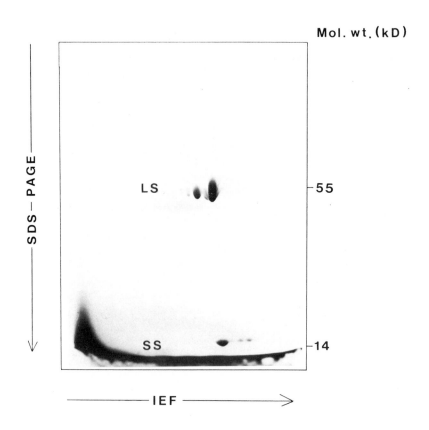

FIG. 3. Result of two dimensional gel electrophoresis of RH RuBPCase. IEF, isoelectrofocussing; SDS-PAGE, SDS polyacrylamide gel elecytrophoresis; LS, large subunit polypeptides; SS, small subunit polypeptides.

III.   STRUCTURAL CHANGES IN THYLAKOID POLYPEPTIDES AND
        THEIR RELATIONSHIP TO COLD ADAPTATION
        AND LEVITT'S SH HYPOTHESIS

Functional studies have shown that thylakoid membranes are sensitive to freezing stress. Heber and Santarius (19) showed that freezing uncoupled photophosphorylation from electron transport, inactivated the light-dependent ATPase and increased proton permeability of washed spinach thylakoids. Alterations in the composition or structure of frozen thylakoid membranes which could account for these functional changes have been investigated. Garber and Steponkus (13) reported that the chloroplast coupling factor (CF_1) was released from spinach thylakoids frozen in vitro. Studies of thylakoids from cold acclimated plants have shown them to be more resistant to the effects of freezing. Senser and Beck (44) reported that thylakoid membranes isolated from winter hardy spruce needles had an increased capacity for cyclic photophosphorylation.

Compositional or structural alterations of thylakoid membranes would seem to be prerequisites for the maintenance of function at low temperatures. However, surveys of the polypeptide complement of thylakoid membranes have failed to show gross compositional changes during cold acclimation of plants (9,29). The results described below are from experiments intended to investigate the possibility that configurational rather than compositional changes in thylakoid proteins are important in adaptation of rye to growth at cold-hardening temperatures (15,16).

Although the chlorophyll a/b ratios, protein/chlorophyll ratios and the general polypeptide patterns of thylakoid proteins from hardened and unhardened Puma rye did not change significantly, a consistent difference was observed in polypeptides of 34 and 35 kD molecular weight. In the absence of ß-mercaptoethanol to RHN thylakoids, the amount of 35 kD polypeptide increased with a concomitant decrease in the 34 kD polypeptide. Furthermore, addition of the oxidizing agent, cupric phenanthroline (CuP), to either RH or RNH preparations eliminated the 35 kD polypeptide and resulted in a concomitant increase in the amount of the 34 kD polypeptide. Similar effects were observed by Moase (36) for the -subunit of isolated CF_1 from wheat. Since the molecular weight of 35 kD is that expected for the γ-subunit of CF_1, we suggest that our observations maybe related to structural changes in the CF_1 of rye. These changes can be summarized as shown in Fig. 4. In the absence of reducing agent or in the presence of an oxidizing agent, the 35 kD polypeptide forms an intramolecular disulfide bond which causes a change in protein configuration such that it becomes more compact and migrates with a lower apparent molecular weight (34 kD). This effect can be reversed by the addition of reducing agent which breaks the disulfide

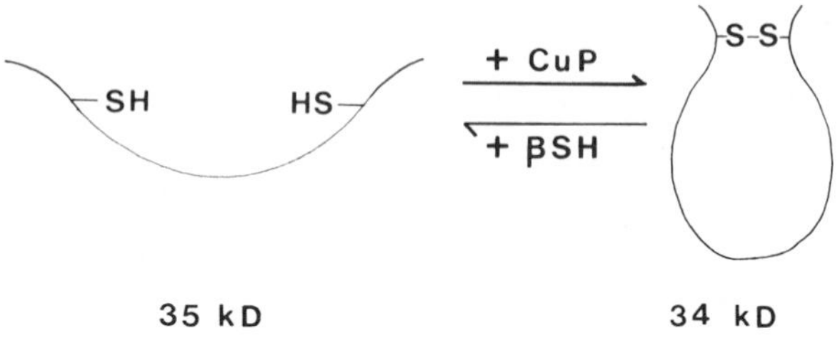

FIG. 4. Schematic model illustrating a possible mechanism for the interconversion of the 35 kD and 34 kD thylakoid polypeptide involving the formation of an intramolecular disulfide bond. CuP, cupric phenanthroline; ßSH, ß-mercaptoethanol.

bond causing the 34 kD polypeptide to unfold and migrate with a slightly higher apparent molecular weight (35 kD). Thus, the RH 35 kD polypeptide appears to be less susceptible to intra-molecular disulfide bond formation than the RNH 35 kD polypep-tide. The importance of this fact can be emphasized by the recent report of Weiss and McCarty (52) who showed that cross-linking SH groups in the γ-subunit of CF_1 uncouples photophosphorylation and enhances proton permeability of the thylakoids. Thus, the results of SDS-PAGE appear to indicate the configurational rather than compositional changes in thylakoid proteins may be important in adaptation to growth at cold-hardening temperatures.

When RNH and RH thylakoids were frozen in vivo, SDS-PAGE revealed no effect on thylakoid membrane polypeptides when com-pared to unfrozen controls. However, freezing thylakoids in vivo always resulted in a greater proportion of the 35 kD poly-peptide relative to the 34 kD polypeptide in RH and RNH prepar-ations. No visible aggregation of membrane polypeptides was observed under freezing conditions except in the presence of ß-mercaptoethanol. It is concluded that intermolecular disul-fide bond formation is not a major freeze denaturation mech-anism in chloroplast thylakoids. Rather, intramolecular disul-fide bonding may be an important mechanism for alteration of membrane proteins at cold-hardening temperatures. Thus, our results indicate that Levitt's SH hypothesis may be consistent with the effects of freezing on soluble proteins (e.g. RuBPCase) but not membrane polypeptides.

IV. EFFECT OF GROWTH AT COLD-HARDENING TEMPERATURES ON THE
STABILITY OF THE THYLAKOID CHLOROPHYLL-PROTEIN COMPLEXES

Chlorophyll-protein complexes can be separated by SDS poly-
acylamide gel electrophoresis (17,51). Higher plant chloroplast thylakoid membranes usually yield three such complexes;
CPI, the P700-chlorophyll a-protein which is associated with
photosystem I, CPII or LHPP, the light harvesting chlorophyll
a/b protein which is associated mainly with photosystem II and
CPIII or FP, which is a detergent-complexed free pigment. The
LHPP' has been shown to be a dimer of LHPP (20). Recently,
Hayden and Hopkins (17) reported the presence of a new pigment-
protein complex (CPIV) associated with photosystem II.
CPI and LHPP can be considered major polypeptide components
of the chloroplast thylakoid membrane since the latter accounts
for almost 50% of the total lamellar protein and, together,
these two complexes account for 50% to 60% of the total chlorophyll in higher plant chloroplasts. The work of Bjorkman et
al. (4) indicates that high temperature (45°C) adaptation is
related to the increased stability of the photosynthetic apparatus associated with photosystem II. Furthermore, Berry and
Bjorkman (2) suggest that acclimation to low temperature (20°C
in this case) primarily involves an increase in the capacity of
temperature-limited   enzymatic   reactions   of   photosynthesis
rather than an increased stability of the photosynthetic apparatus. Because these complexes are major constituents of the
thylakoid membrane and because of their important functional
roles in photosynthesis, a study was initiated to investigate
the structure of these complexes as a function of growth at
cold-hardening temperature (2°-4°C) (10). As shown in Fig. 5,
four pigment-protein complexes (CPI, LHPP', CPIV, LHPP) and
free pigment (FP) were separated in RNH thylakoids by SDS gel
electrophoresis at 23°C essentially by the method of Hayden and
Hopkins (17). The presence of these four pigment-protein complexes is typical for most higher plant chloroplasts. In contrast, only three of the major protein-pigment complexes could
be resolved at 23°C in preparations of RH thylakoids. As shown
in Fig. 6, CPIV does not resolve under these conditions in RH
thylakoid preparations and, in addition, there is significantly
less LHPP' present in RH than in RNH preparations. The stability of CPI also appears to be affected by low growth temperature since there is less of this pigment-protein complex in RH
than in RNH preparations relative to LHPP and is accompanied by
a concomitant increase in the FP. Changes in the fluorescence
induction kinetics at 23°C associated with photosystem II upon
cold-hardening complement the structural changes associated
with CPIV.

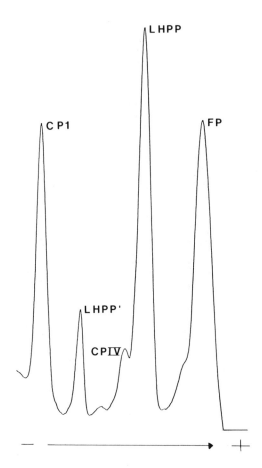

FIG. 5.   Gel scan of chlorophyll-protein complexes from RNH
thylakoid membranes.   Arrow shows direction of migration.   CPI,
P700 chlorophyll a-protein; LHPP', dimer of the light harvesting chlorophyll a/b protein; LHPP, light harvesting chlorophyll a/b protein; FP, free pigment.

However, when the chlorophyll-protein separations were performed at 4°C with lithium dodecyl sulfate, CPIV was resulted
and similar amounts of CPI appeared in both RH and RNH thylakoid preparations.   Thus, CPI and CPIV of RH chloroplasts
appear to be more stable at 4°C than at 23°C.   It appears that
the temperature stability of these complexes may be related to
the temperature regime to which the plant has been adapted.

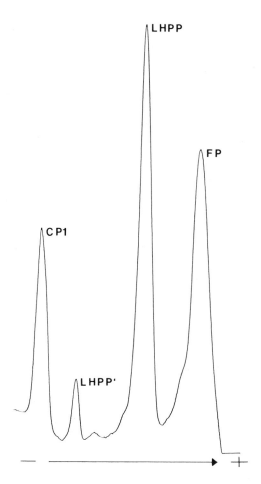

FIG. 6.   Gel scan of chlorophyll-protein complexes from RH
thylakoid membranes.   Arrows shows direction of migration.
CPI, P700 chlorophyll a-protein; LHPP', dimer of the light harvesting chlorophyll a/b protein; LHPP, light harvesting chlorophyll a/b protein; FP, free pigment.

V.   SUMMARY

Plant growth at cold-hardening temperatures has a profound
effect on the structure and function of both soluble and membrane proteins associated with chloroplasts.   The results with
RuBPCase indicate that intermolecular disulfide bonding is
indeed an important mechanism for irreversible denaturation of
this protein due to freezing as predicted by Levitt's SH

hypothesis. Moreover, alteration of protein configuration to reduce the accessibility of free SH groups appears to be a unique mechanism for resisting protein aggregation due to intermolecular disulfide bonds. In contrast, the membrane polypeptides of thylakoids appear to be inherently quite resistant to aggregation due to intermolecular disulfide bonding. However, evidence suggests that intramolecular disulfide bonding may play an important role in the denaturation mechanism of membrane polypeptides. In addition, we have now determined that the chlorophyll-protein complexes of thylakoid membranes undergo subtle but significant changes in structural stability as a result of growth at cold-hardening temperatures.

ACKNOWLEDGMENTS

This work was, in part, supported by grants to N. Huner and to W. Hopkins and D. Hayden by the National Science and Engineering Research Council of Canada. The authors are indebted to R. Johnston, A. Draper and C. Henderson for their capable assistance.

REFERENCES

1. Baldwin, J., and Hochachka, P.W., Biochem. J. 116:883 (1970).
2. Berry, J., and Bjorkman, O., Ann. Rev. Plant Physiol. 31:491 (1980).
3. Bixby, J.A., and Brown, G.N., Plant Physiol. 56:617 (1975).
4. Bjorkman, O., Badger, M., and Armond, P.A., Carnegie Inst. Washington Yearbook 77:262 (1978).
5. Chen, P.M., and Li, P.H., Bot. Gaz. 137:105 (1976).
6. Chollet, R., and Anderson, L.L., Biochim. Biophys. Acta 482:228 (1977).
7. Cothren, J.T., and Guinn, G., Phyton 33:131 (1975).
8. Craker, L.E., Gusta, L.V., and Weiser, C.J., Can. J. Plant Sci. 49:279 (1969).
9. DeYoe, D.R., Ph.D. Thesis, University of Missouri, Columbia (1977).
10. Elfman, B., Huner, N.P.A., Hopkins, W., and Hayden, D., Plant Physiol. Suppl. 67:31 (1981).
11. Feldman, N.L., Dokl.-Botan. Sci. Sect. 166:45 (1966).
12. Gaff, D.F., Aust. J. Biol. Sci. 19:291 (1966).
13. Garber, M.P., and Steponkus, P.L., Plant Physiol. 57:681 (1976).
14. Goodin, R., and Levitt, J., Cryobiology 6:333 (1970).
15. Griffith, M., Ph.D. Thesis, University of Minnesota, St. Paul (1981).

16. Griffith, M., Brown, G.N., and Huner, N.P.A., Plant Physiol. Suppl. 65:153 (1980).
17. Hayden, D.B., and Hopkins, W.G., Can. J. Bot. 55:2525 (1977).
18. Heber, U., Cryobiology 5:188 (1968).
19. Heber, U., and Santarius, K., Plant Physiol. 39:712 (1964).
20. Hiller, R.G., Pilger, D., and Genge, S., Plant Sci. Lett. 1:81 (1973).
21. Hochachka, P.W., and Somero, G.N., in "Strategies of Biochemical Adaptation", pp. 179. W.B. Saunders Co., Toronto, 1973.
22. Huner, N.P.A., and Macdowall, F.D.H., Can. J. Biochem. 54:848 (1976).
23. Huner, N.P.A., and Macdowall, F.D.H., Biochem. Biophys. Res. Comm. 73:411 (1976).
24. Huner, N.P.A., and Macdowall, F.D.H., Can. J. Biochem. 56:1154 (1978).
25. Huner, N.P.A., and Macdowall, F.D.H., Can. J. Biochem. 57:155 (1979).
26. Huner, N.P.A., and Macdowall, F.D.H., Can. J. Biochem. 57:1036 (1979).
27. Huner, N.P.A., Palta, J.P., Li, P.H., and Carter, J.V., Can. J. Biochem. 59:280 (1981).
28. Huner, N.P.A., Bancroft, J.B., Johnston, R., Richardson, J., and Payne, N., Plant Physiol. Suppl. 67:139 (1981).
29. Huner, N.P.A., and Macdowall, D.H., Can. J. Biochem. 54:848 (1976).
30. Kawashima, N., Singh, S., and Wildman, S.G., Biochem. Biophys. Res. Commun. 42:664 (1971).
31. Koffer, H., Mallet, G.E., and Adye, J., Proc. Nat. Acad. Sci. (USA) 43:464 (1957).
32. Levitt, J., J. Theoret. Biol. 3:355 (1962).
33. Levitt, J., and Dear, J., in "Frozen Cell", p. 149. Ciba Foundation Symposium, 1970.
34. Levitt, J., Cryobiology 1:312 (1965).
35. Low, P.S., Bada, J.L., and Somero, G.N., Proc. Nat. Acad. Sci. (USA) 70:430 (1973).
36. Moase, E.H., M.Sc. Thesis, Univ. of British Columbia, Vancouver, 1981.
37. Moon, T.W., and Hochachka, P.W., Biochem. J. 123:695 (1971).
38. O'Farrell, P.H., J. Biol. Chem. 250:4007 (1975).
39. Roberts, D.W.A., Can. J. Bot. 53:1333 (1975).
40. Roberts, D.W.A., Can. J. Bot. 47:263 (1969).
41. Roberts, D.W.A., Enzymologia 39:162 (1969).
42. Rochat, E., and Therrien, H.P., Can. J. Bot. 53:2411 (1975).
43. Rochat, E., and Therrien, H.P., Can. J. Bot. 53:2417 (1975).
44. Senser, M., and Beck, E., Planta 137:195 (1977).
45. Smellie, R.M.S., and Pennock, I.F., in "Biochemical Adaptation to Environmental Change", pp. 1. The Biochemical Society Publications, Colchester, 1977.
46. Sullivan, C.Y., and Kinbacher, E.J., Crop Sci. 7:241 (1967).

47. Siminovitch, D., and Briggs, D.R., Arch. Biochem. Biophys. 23:9 (1949).
48. Siminovitch, D., and Briggs, D.R., Plant Physiol. 28:15 (1953).
49. Siminovitch, D., and Briggs, D.R., Plant Physiol. 28:177 (1953).
50. Siminovitch, D., and Briggs, D.R., Plant Physiol. 29:331 (1954).
51. Thornber, J.P., Ann. Rev. Plant Physiol. 26:127 (1975).
52. Weiss, M.A., and McCarty, R.E., J. Biol. Chem. 252:8007 (1977).

TRANSITION OF LIPID METABOLISM IN RELATION TO FROST HARDINESS IN CHLORELLA ELLIPSOIDEA

S. Hatano and K. Kabata

Department of Food Science and Technology
Faculty of Agriculture, Kyushu University
Fukuoka, Japan

Studies of freezing injury and resistance in plants indicate that the primary site of freezing injury is the cellular membranes (7,12,17) and that the membranes undergo changes in both the total amount and the composition of phospholipids during hardening (3,18,21). However, little attention has been given to any shift of lipid metabolism leading to the membrane changes.

Previous studies have shown that hardened cells of Chlorella ellipsoidea are able to survive slow freezing to -196°C (8) and the amount of total lipids in the cells increases during hardening. A thin-layer chromatogram showed that an unknown non-polar lipid greatly increases with an increase in frost hardiness (5). An electron micrograph of hardened cells showed that a number of lipid bodies appear along plasma membrane during hardening (9).

This paper described the identification of the non-polar lipid, and its possible involvement in the appearance of lipid bodies and in the development of frost hardiness. Furthermore, a distinct shift of lipid synthesis from chloroplasts to a cytoplasmic system during hardening, and the separation of two fatty acid synthetase systems by molecular sieve chromatography are described.

Abbreviation: ACP, acyl carrier protein; CHI, cycloheximide; FFA, free fatty acids.

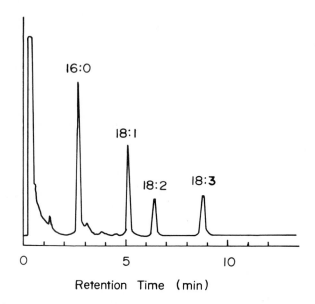

Retention Time (min)

FIG. 1.   Gas-liquid chromatogram of methyl esters of non-polar lipid produced during hardening.   Tentative identification of the methyl esters was achieved by adding authentic fatty acids to the non-polar lipid.   Chain length:number of double bonds; palmitic, 16:0, oleic, 18:1; linoleic, 18:2; linolenic, 18:3.

I.   INTERRELATION AMONG DEVELOPMENT OF FROST HARDINESS, FORMATION OF FREE FATTY ACIDS, AND APPEARANCE OF LIPID BODIES

Chlorella elipsoidea Gerneck (IAM C-27) was grown in synchronous culture at 25°C as described previously (8).   In as much as the cells were hardened at the L_2 stage (an intermediate stage in the ripening phase of the cell cycle) (8), L_2 cells were used here. L_2 cells synchronized at 25°C were directly hardened at 3°C for 48 h.

As reported previously (4), a thin layer chromatogram of non-polar lipids extracted from unhardened and hardened *Chlorella* showed that an unknown non-polar lipid was detected as a band located between triglyceride and stearic acid standards.   Although amounts of phospholipids and glycolipids increased during hardening, an increase in the unknown lipid was most notable in *Chlorella*.

To identify the non-polar lipid, the lipid was eluted from the band and analyzed with an infrared spectrometer.   The infrared spectrum of the unknown lipid showed the carbonyl absorption of free fatty acids at 1,700 kaysers, but did not show the carbonyl stretch band of the unconjugated ester of

triglycerides at 1,740 kaysers.   These results suggested that
the unknown lipid consists of fatty acids.   Therefore, the
lipid was analyzed by gas-liquid chromatography after metanoly-
sis.   Figure 1 shows a gas-liquid chromatogram of the methyl
esters of the lipid.   Palmitic, oleic, linoleic, and linolenic
acids were main components of the non-polar lipid.   The lipid
was also detected when cell lipids were extracted with cold
chloroform-methanol (2:1, v/v) instead of hot methanol by
grinding cells with sea sand at 0°C.   These results indicate
that the non-polar lipid which increased greatly during
hardening mainly consists of the four free fatty acids,
palmitic, oleic, linoleic, and linolenic acids.

It was also noted that triglycerides were not detected in
Chlorella cells.   Yoshida (20) has reported that a decrease in
triglycerides is accompanied by an increase in phospholipids
during hardening of cortical cells from poplar stem.

On the other hand, the morphological changes of Chlorella
during hardening were observed by an electron microscopy.   An
electron micrograph of hardened cells showed that a number of
lipid bodies appear along plasma membrane during hardening
(9).   No lipid body was observed in the unhardened cells and in
the cells hardened in the presence of cycloheximide (CHI).   CHI
has been reported to inhibit completely the development of
frost hardiness (16).   These results indicate that the production of lipid body may be involved in the development of frost
hardiness.

The formation of free fatty acids (FFA) and the appearance
of lipid bodies were remarkable during hardening.   To clarify
further the interrelation among these events and the development of frost hardiness, inhibitory effects of antimetabolites
on them were examined.   The results are summarized in Table I.
The algal cells developed frost hardiness in a similar level in
both the light and the dark in the presence of glucose.   The
FFA formed under both conditions were equivalent in composition
and quantity (Fig. 1).   Cycloheximide, oligomycin or DCMU
inhibited the development of frost hardiness except for DCMU
added in the dark.   All cells which developed a high hardiness
formed the FFA and none of the cells which failed to form the
FFA developed a high hardiness.   The appearance of lipid bodies
was invariably accompanied by the formation of FFA.   These
results suggest that the formation of FFA and lipid bodies are
closely related to the development of frost hardiness.

Significance of the fatty acid formation and lipid body
appearance in the hardening process is still obscure.   To
clarify the significance, the incorporation of ^{14}C into the
FFA was compared with that into total lipids of the cellular
membrane fraction and whole cells.   Chlorella cells were pulse-
labeled with $[^{14}C]NaHCO_3$ for 4 min at zero time and at the
12th h of hardening.   The radioactivies incorporated into
total lipids of whole cells and the membrane fraction (9,000g
pellet) and the FFA were chased at suitable intervals.   As

TABLE I. Interrelation among development of frost hardiness, formation of free fatty acids, and appearance of lipid bodies.

Cells were exposed to cycloheximide (CHI), oligomycin (OGM), and DCMU during hardening in the light and in the dark in the presence of glucose. The viability of cells was determined by the ratio of the increment in packed cell volume in the first 72 h incubation to the value of packed cell volume at zero time, and frost hardiness was estimated by a percentage of the viability of frozen cells to that of unfrozen cells. Fatty acid formation and lipid body appearance were confirmed by TLC and electron microscopy, respectively.

Hardening Condition		Frost	Free	
Light condition	Agent added to culture	hardiness development	fatty acid formation	Lipid body appearance
Control	(unhardened)	(-)[a]	-	-
Light	None	88	+	+
Light	CHI (0.1 mm)	(-)	-	-
Light	OGM (10 μm)	6	+	+
Light	DCMU (1 μm)	(-)	-	-
Dark	None	19	-	-
Dark	Glucose	89	+	+
Dark	Glucose + CHI	(-)	-	-
Dark	Glucose + OGM	(-)	-	-
Dark	Glucose + DCMU	87	+	+

[a]   Packed cell volume of frozen-thawed cells decreased during the first 72 h incubation.

Figure 2 shows, the initial increasing rate in radioactivity of the FFA was significantly lower than that of total lipids of whole cells and the membrane fraction in pulse-labelings at both zero time and the 12th h of hardening. Frost hardiness of the algal cells remarkably increased after 12 h hardening. In pulse-labeling after a 12 h hardening, the incorporations of ^{14}C into total lipids of whole cells and membrane fraction were still actively proceeding. These results suggest that the de novo synthesis of fatty acids is much more active than active insertion of fatty acids into complex lipids, and as a result the surplus fatty acids accumulate in the cells.

Since the cell walls of Chlorella are resistant to enzymic digestion (1), it is difficult to isolate intact lipid bodies from hardened cells. The present studies could not show that the FFA synthesized during hardening are accumulated in lipid bodies. The appearance of lipid bodies, however, was always

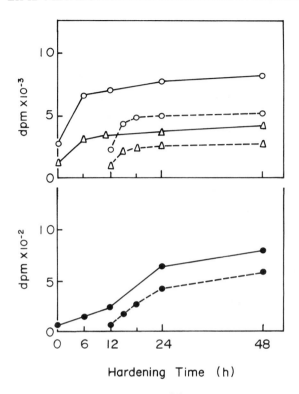

FIG. 2. Incorporation of $[^{14}]$NaHCO$_3$ into total lipids of whole cells and 9,000 g pellet, and into free fatty acids (FFA). Cells were pulse-labeled with $[^{14}C]$NaHCO$_3$ for 4 min at zero time (solid lines) and at the 12th h (broken lines) of hardening. Total lipids, (open circles); 9,000 g pellet, (triangles); FFA, (closed circles).

accompanied by the synthesis of the FFA (Table I). The results suggest a close relation between these two.

Free fatty acids were accumulated during hardening of Chlorella cells (Figs. 1 and 2). Although quantitative changes in phospholipids, triglycerides, glycolipids, and sterols during hardening of some higher plants have been intensively studied (6,19), FFA have not been found to increase in these plants. Pomeroy and Siminovitch (15) indicated that seasonal augmentation of lipid bodies was closely related to the seasonal cycle of frost resistance of black locust. Otsuka (14), however, could not observe any significant changes in quantity and form of lipid bodies during frost hardening of mulberry. In Chlorella, the formations of FFA and lipid bodies were closely related to the development of frost hardiness (Table I); however, an initially increasing rate in the radioactivity incorporated into the FFA was significantly lower than

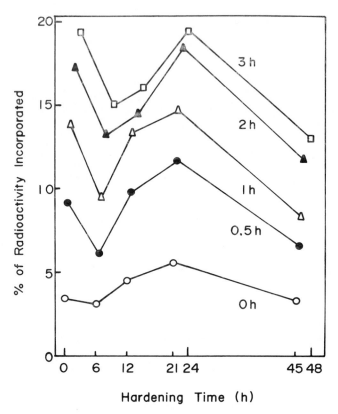

FIG. 3.   Time-dependent changes in the incorporation rate
of [^{14}C] NaHCO$_3$ into total lipids at each chase time. The
incorporation rate was represented by the percentage of the
radioactivity in total lipids at each chase time to the total
radioactivity in the acid-soluble and acid-insoluble fractions
at zero time of chase.   The figures on the abscissa (0, 6, 12,
21, 45) indicate the intervals of pulse labeling during
hardening.   Zero h (open circles), 0.5 h (closed circles), 1 h
(open triangles), 2 h (closed triangles), and 3 h (open square)
represent the intervals of sampling for the chase experiments.

that into total lipids of whole cells and the cellular membrane
fraction (Fig. 2).   From these results, it is inferred that a
metabolic change activating fatty acid synthesis is essential
for the development of frost hardiness in <u>Chlorella</u>, but the
accumulation of FFA and the appearance of lipid bodies <u>per se</u>
are not essential.

II. A DISTINCT SHIFT OF LIPID SYNTHESIS FROM CHLOROPLASTS
TO A CYTOPLASMIC SYSTEM DURING HARDENING

The incorporation of ^{14}C into total lipids of whole cells and membrane fraction was remarkable even at the 12th h of hardening (Fig. 2). In order to study the transition of lipid metabolism during hardening, the cells were pulse-labeled with $[^{14}C]NaHCO_3$ for 4 min at zero time, at the 6th, 12th, 21st, and 45th h from the start of hardening (10). The pulse-labeling was ended by adding unlabeled bicarbonate and the labeled cells were hardened again. Aliquots of the rehardened suspension were removed at intervals of 0, 0.5, 1, 2, 3 h, and other suitable hours to chase radioactivity incorporated into the acid-soluble fraction, the acid-insoluble fraction, and total lipids. As Figure 3 shows, the radioactivity in total lipids at each chase time is expressed as a percentage of the total radioactivity in the acid-soluble and -insoluble fractions at zero time of chase. The incorporation rate at each chase time decreased after a 6-h hardening. In the next 15 h, a distinct increase was noted. These results suggest that two different systems are involved in the lipid synthesis of the cells: an activity-decreasing system and an activity-increasing system.

To clarify further the involvement of two different systems in lipid synthesis, effects of CHI and DCMU on the incorporation rate of ^{14}C into total lipids were examined. When CHI was added to the culture at the start of hardening, CHI completely inhibited the increase in the incorporation rate observed for 15 h from the 6th h of hardening (Figs. 3 and 4). When CHI was added after the end of pulse labeling at various times of hardening, the increase from the 6th h of hardening was noticed. These results indicate that the activity-increasing system involves protein synthesis of cytoplasmic 80S-type ribosomes and, therefore, may be localized in cytoplasm exclusive of chloroplast. DCMU added at the start of hardening markedly lowered the high incorporation rate at zero time of hardening. The results suggest that the major site of the activity-decreasing system is the chloroplast. When DCMU was added after the end of pulse-labeling, the incorporation rate increased until 12 h hardening. The cells which showed the increased incorporation rate developed frost hardiness. When DCMU was added at zero time of hardening, DCMU lowered the high incorporation at zero time and inhibited the development of frost hardiness. As Figure 4 shows, however, CHI did not affect the high incorporation rate but inhibited the hardiness increase. These results suggest that lipid synthesis in the cytoplasmic system is directly involved in the lipid changes during hardening and chloroplastic lipid synthesis is indirect.

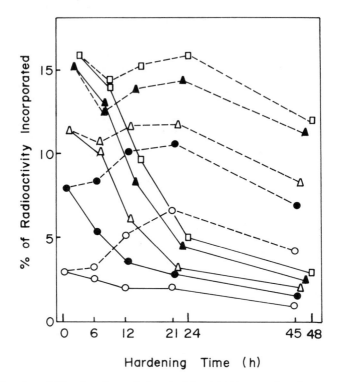

FIG. 4. Effect of cycloheximide (CHI) on the incorporation rate of [14C]NaHCO3 into total lipids. Solid lines, cells exposed to CHI (0.1 mM) during a 48-h hardening except the period of pulse labeling; broken lines, cells exposed to CHI after termination of pulse labeling at different hardening times. The meanings of other symbols are the same as in Figure 3.

The cells were further pulse-labeled with D-[U-14C] glucose in the dark. The incorporation rate of 14C glucose into total lipids increased along with hardiness. The results support the above suggestion.

What can be concluded from the pulse-labeling experiments is that the major site of lipid system in Chlorella shifts from chloroplast to a cytoplasmic system during hardening.

III. SEPARATION OF TWO FATTY ACID SYNTHETASE SYSTEMS IN CHLORELLA

A 0-70% ammonium sulfate fraction of crude extracts obtained from hardened cells was applied to a Bio-Gel A-15m column. Fractions of 4 ml were collected and measured for

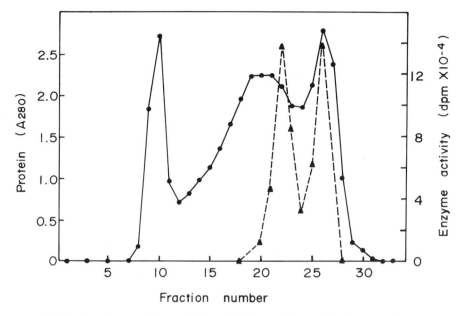

Fraction number

FIG. 5. Separation of two fatty acid synthetase systems on Bio-Gel A-15m. A 0-70% ammonium sulfate fraction obtained from hardened cells was applied to a 2.5 x 23 cm column. Fractions of 4 ml were collected and measured for absorbance at 280 nm (solid line) and fatty acid synthetase activity (broken line). Both the first peak and the second peak activities were depended on acyl carrier protein.

absorbance at 280 nm and fatty acid synthetase activity. The synthetase was measured by the incorporation of $[2-^{14}C]$ malonyl-CoA into chloroform-soluble saponified products. The two fatty acid synthetase systems were distinctly separated (Fig. 5). Both the first peak and the second peak activities were dependent on acyl carrier protein (ACP). The same enzyme fraction that used in Figure 5 except for being frozen at -20°C for 20 days, was put on a Bio-Gel A-5m (Fig. 6). The second peak synthetase lost its activity much more than the first one in freezing storage at -20°C. Experiments will be performed in the near future as to which peak synthetase is localized in cytoplasm exclusive of chloroplast.

Studies of fatty acid synthetase prepared from various organisms have indicated that there are two types of synthetase systems. One type is a multienzyme complex which is independent on ACP for its activity. The other type is a series of discrete enzymes and its activity depends on ACP. Yeast and animals have the multienzyme complex, and bacteria and higher plants have the discrete enzyme system. Euglena has the two type synthetases in a cell. In the etiolated cells of Euglena,

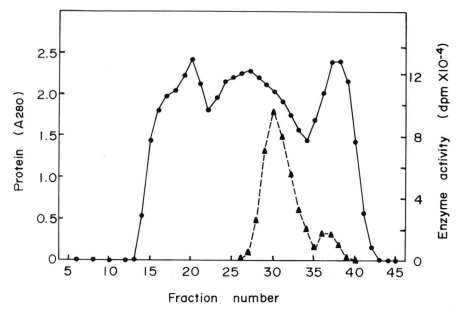

FIG. 6. Separation of two fatty acid synthetase systems of Bio-Gel A-5m. The same enzyme fraction that used in Figure 5 except for being frozen at -20°C for 20 days, was put in a 2.5 x 33 cm column.

a large portion of the total activity stems from the ACP-independent enzyme. Following exposure of the cells to light, the ACP-dependent enzyme system accounts for about 50% of the total activity. In Chlorella, both the first peak and the second peak enzymes were dependent on ACP. They could not synthesize fatty acids in the absence of ACP. The two synthetases in Chlorella were separated in a similar way to that of Euglena on Bio-Gel A-5m. In comparison with the results obtained from Euglena by Ernst-Fonberg (2), the two synthetases in Chlorella definitely differ from those in Euglena in molecular weight and stability to freezing. The molecular weight of multienzyme complex of Euglena was much bigger than that of the first peak enzyme of Chlorella. It is well known that Euglena exhibited characteristics of both plants and animals. We presume that Chlorella has potential for both bacterial and plant characteristics.

It is an interesting question whether or not only algae can have two fatty acid synthetase systems. In spinach leaves, Stumpf and collaborates (13) have recently demonstrated that the chloroplast is the sole site for de novo synthesis of long chain fatty acids by using immunological techniques and gently lysed protoplasts as the source for intact chloroplasts. However, the spinach leaves used by them were not hardened. It is possible that the activities of CO_2 fixation and fatty acid synthesis in the chloroplast are so low at a low temperature and a low light intensity that the chloroplast can not supply enough fatty acids to help develop frost hardiness. If the polar head and the fatty acyl moiety of membrane phospholipids are not translated or modified in the original position, but if de novo synthesized phospholipids are inserted into membrane lipids to make the membranes hardier, active fatty acid synthesis would be necessary for the development of frost hardiness.

De novo synthesis of fatty acid synthetase during hardening should be studied in more detail in Chlorella and leaf cells.

IV. SUMMARY

Chlorella ellipsoidea Gerneck (IAM C-27) was synchronously grown and cells at an intermediate stage in the ripening phase of the cell cycle were hardened at 3°C for 48 h. Palmitic, oleic, linoleic, and linolenic acids increased greatly and lipid bodies appeared during hardening. In pulse labeling with $[^{14}C]NaHCO_3$ for 4 min, the initial incorporation rates of ^{14}C into total lipids of whole cells and cellular membrane fraction were significantly higher than that into free fatty acids. These results suggest that a metabolic change activating fatty acid synthesis is essential for the development of frost hardiness, but the accumulation of free fatty acids and the appearance of lipid bodies per se are not essential.

The cells were pulse-labeled with $[^{14}C]NaHCO_3$ at various times of hardening and the incorporation rate of ^{14}C into total lipids was determined. A high incorporation rate at zero time of hardening decreased after 6 h. In the next 15 h, a distinct increase was noted. This increase occurred prior to the development of frost hardiness. Cycloheximide completely inhibited both the increase and the development, and DCMU remarkably lowered the high incorporation rate at zero time. These results suggest that the major site of lipid synthesis shifts from chloroplasts to a cytoplasmic system during hardening. The two fatty acid synthetase systems were separated by Bio-Gel. These synthetases were dependent on acyl carrier protein.

REFERENCES

1. Atkinson, A.W., Gunning, Jr., B.E.S., and John, P.C.L., Planta (Berl.) 107:1-32 (1972).
2. Ernst-Fonberg, M.L., Biochemistry 12:2449-2455 (1974).
3. Grenier, G., Hope, H.J., and Willemot, C., Plant Physiol. 55:906-912 (1975).

4. Kabata, K., Sadakane, H., Miyachi, M., Nagata, K., Hatano, S., and Watanabe, T., J. Fac. Agric. Kyushu Univ. 23:155-161 (1979).
5. Kabata, K., Sadakane, H., Kurose, M., Kobayakawa, A., Watanabe, T., and Hatano, H., J. Fac. Agric. Kyushu Univ. 25:91-97 (1980).
6. Levitt, J., Responses of Plant to Environmental Stresses, Ed. 2, Vol. 1, Academic Press, New York, 1980.
7. Mazur, P., Ann. Rev. Plant Physiol. 20:419-448 (1969).
8. Hatano, S., Sadakane, H., Tutumi, T., and Watanabe, T., Plant Cell Physiol. 17:451-458 (1976).
9. Hatano, S., in "Plant Cold Hardiness and Freezing Stress" (P.H. Li and A. Sakai, eds.), pp. 175-196. Academic Press, New York, 1978.
10. Hatano, S., Kabata, K., and Sadakane, H., Plant Physiol. 67:216-220 (1981).
11. Hatano, S., Kabata, K., Yoshimoto, M., and Sadakane, H., Plant Physiol. Submitted for publication.
12. Heber, U., Cryobiology 5:188-201 (1968).
13. Ohlrogge, J.B., Kuhn, D.N., and Stumpf, P.K., Proc. Natl. Acad. Sci. USA 76:1194-1198 (1979).
14. Otsuka, K., Low Temp. Sci. Ser. B 30:33-44 (1972).
15. Pomeroy, M.K., and Siminovitch, D., Can. J. Bot. 49:787-795 (1971).
16. Sadakane, H., Kabata, K., Ishibashi, K., Watanabe, T., and Hatano, S., Environ. Exp. Bot. 20:297-305 (1980).
17. Steponkus, P.L., Garber, M.P., Myers, S.P., and Lineberger, R.D., Cryobiology 14:303-321 (1977).
18. Yoshida, S., and Sakai, A., Plant Physiol. 53:509-511 (1974).
19. Yoshida, S., Contribution from Inst. Low Temp. Sci., Ser. B 18:1-14 (1974).
20. Yoshida, S., and Sakai, A., Plant Cell Physiol. 14:353-359 (1973).
21. Yoshida, S., Plant Physiol. 57:710-715 (1976).

ISOENZYMES OF GLUCOSE 6-PHOSPHATE DEHYDROGENASE IN RELATION TO FROST HARDINESS OF <u>CHLORELLA</u> <u>ELLIPSOIDEA</u>

H. Sadakane
S. Hatano

Department of Food Science and Technology
Faculty of Agriculture, Kyushu University
Fukuoka, Japan

A previous study demonstrated that the pentose-phosphate cycle is directly involved in the hardening process of <u>Chlorella</u> (9). Sagisaka (10,11) also showed a distinct metabolic shift of glucose 6-phosphate metabolism from glycolysis to the pentose-phosphate cycle in wintering poplar xylem. Hatano et al. (3) suggested that in <u>Chlorella</u> the activation of lipid synthesis in a cytoplasmic system is essential for the membrane changes associated with frost hardiness. A high activity of the pentose-phosphate cycle is probably necessary for the active lipid synthesis in the cytoplasmic system. Glucose 6-phosphate dehydrogenase (G6P-DH), the first enzyme of the pentose-phosphate cycle, was highly activated during the early phase of hardening by only low temperature (9). G6P-DH seems to be the key enzyme to activate the pentose-phosphate cycle at the early process of hardening.

This paper describes <u>de novo</u> synthesis of G6P-DH isoenzymes during hardening of <u>Chlorella</u> and the differences among the isoenzymes of unhardened and hardened cells in electrophoresis patterns, immunological properties, fatty acids-sensitivity, ATP-sensitivity and the activity of low temperatures.

Abbreviations: G6P-DH, glucose 6-phosphate dehydrogenase (EC 1.1.1.49); 6PG-DH, 6-phosphogluconate dehydrogenase (EC 1.1.1.44).

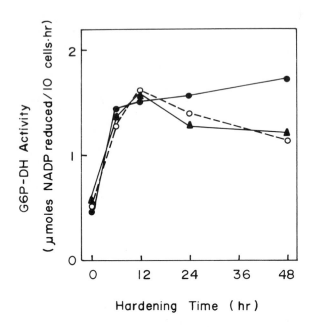

FIGURE 1. Changes in G6P-DH activity during hardening. Open circles, cells hardened in the light; solid circles, cells hardened in the dark without glucose; solid triangles, cells hardened in the dark in the presence of glucose.

I. CHANGES IN THE ACTIVITIES OF G6P-DH AND 6PG-DH
DURING HARDENING

Chlorella ellipsoidea Gerneck (IAM C-27) was grown in synchronous culture at 25°C under 9 to 10 kilolux with 1% CO_2-air at a concentration of 5-15 x 10^9 cells per liter and under a 28-hr light/14-hr dark regime as described previously (4,5). Since the cells were hardened most at the L_2 stage (an intermediate stage in the ripening phase of the cell cycle) (4), L_2 cells were used in this study. L_2 cells were directly hardened at 3°C for 48 hr.

As shown in a previous study (9), G6P-DH was greatly activated during the first 6 hr of hardening time even in the dark without glucose (Fig. 1). Chlorella cells were not hardened without sugars (9). 6-Phosphogluconate dehydrogenase (6PG-DH) activity gradually increased with hardiness increase (9). Since G6P-DH was highly activated during the early phase of hardening by only low temperature, we inferred that this enzyme is the key enzyme to activate the pentose-phosphate cycle at the early process of hardening.

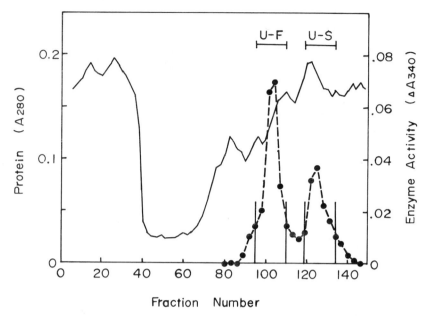

FIGURE 2. Elution pattern of G6P-DH in unhardened cells from DEAE-cellulose column. Solid line, absorbance at 280 mm; Solid circles, G6P-DH activity in ΔA$_{340}$ per min.

II.   SEPARATION OF G6P-DH ISOENZYMES AND THEIR CHARACTERISTICS

The unhardened and dark-hardened (without glucose) cells stored at -20°C were disrupted by Vibrogen Zellmühl (9). After removing cell debris, unbroken cells and nucleic acids, the crude enzyme was precipitated with $(NH_4)_2SO_4$. The precipitate was desalted and passed through DEAE-cellulose column (3.0 x 10 cm). The partially purified G6P-DH was applied to a DEAE-celluose column (2.0 x 50 cm) and eluted with a continuous gradient of phosphate buffer from 0.05 to 0.5 M. Fig. 2 shows the elution pattern of G6P-DH of unhardened cells. The partially purified enzyme was separted into two fractions. The first peak (U-F) was eluted at a phosphate concentration of about 0.3 M and the second peak (U-S) at about 0.4 M. G6P-DH obtained from hardened cells was also separated into two fractions (Fig. 3). The first peak (H-F) showed the same elution pattern as that of U-F, whereas the second peak (H-S) area was larger than that of unhardened cells. Furthermore, H-S had a shoulder. The results suggest that the H-S fraction contains newly synthesized isoenzyme and the new isoenzyme contributes to the activation of G6P-DH during the early phase of hardening.

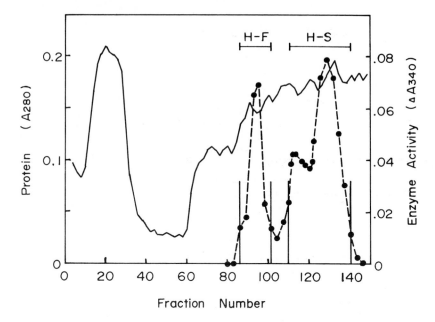

FIGURE 3. Elution pattern of G6P-DH in hardened cells from DEAE-cellulose column. The meanings of the symbols are the same as in Fig. 2.

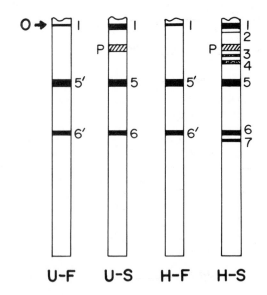

FIGURE 4. Disk electrophoresis patterns of G6P-DH in four fractions. The gels were stained for G6P-DH activity. The origin (top of the small pore gel) is indicated by the letter "O".

We compared U-F, U-S, H-F and H-S for the electrophoretic properties in polyacrylamide gels (Fig. 4). Disk electrophoresis was carried out by the method of Davis (2), using a 3.75% large pore gels and 7.5% small pore gels. The gels were stained for G6P-DH activity (7). The existence of dimer and more aggregated forms of G6P-DH has been reported in blue-green algae, rice plant cells, etc. (6,8,12). In Chlorella also, multiple molecular forms of the enzyme were revealed by electrophoresis in polyacrylamide gels (Fig. 4). Band 1 may be the extensive aggregated form of G6P-DH, bacause it did not penetrate into the gels (Fig. 4). Band 6 and 6' may be monomers (Fig. 4). Band 5 may be a dimer of band 6 and band 5' a dimer of 6'. Although 6 and 6' showed the same electrophoretic migration, the two enzymes differed in DEAE-cellulose affinity (Figs. 2 and 3) and other properties as mentioned later. Band 7 in the H-S fraction is probably a monomer of newly synthesized isoenzyme. The formation of band 2, 3 and 4 may be caused by the various associations of the band 6 and 7 enzymes. Band P had a pink color (Fig. 4). The pink-colored substance is possibly phenazine methosulfate used for staining. The pink color may be adsorbed by band P protein which is abundant in the U-S and H-S fractions on protein staining. It seems likely that the G6P-DH in U-F and H-F are the same enzyme whereas H-S has an isoenzyme which differs in electrophoretic mobility from the isoenzyme in U-S.

To clarify further the differences between the newly synthesized isoenzyme and other isoenzymes, the antigenic properties of the H-S fraction were compared with those of the other fractions. We used U-F and H-S as injection material to make the rabbit antibodies. Samples of U-F and H-S were, respectively, emulsified in Freund complete adjuvant and injected subcutaneously three times into male rabbits at intervals of three weeks. Blood samples were obtained by carotid puncture 1 week later. Enzyme inhibition tests were carried out by preincubation of the four fractions with appropriately diluted antiserums for 5 min. The antibody to U-F markedly inhibited the G6P-DH activity of U-F (Table I). The antibody also suppressed the activity of H-F, but failed to inhibit the activities of U-S and H-S (Table I). The recovery of the activity at the lower dilution may be caused by the excess of antibody. The antibody prepared to H-S was also most inhibitory to the homologous antigen, H-S (Table II). It remarkably inhibited the G6P-DH activity of U-S, but did not inhibit the activities of U-F and H-F (Table II). These results indicate that G6P-DH species in the first and the second fractions do not share common antigenic determinants. The differences of the inhibition between G6P-DH in U-S and that in H-S may be due to the antigenic properties of the newly synthesized isoenzyme.

Pseudomonas has been reported to have two isoenzymes of G6P-DH which differ in sensitivity to inhibition by long chain fatty acids and ATP (1). Hatano et al. (3) reported that the

TABLE I. Effects of antibody to U-F on G6P-DH activity of four fractions.

Dilution rate	Relative G6P-DH activity			
of antibody	U-F	U-S	H-F	H-S
No antibody	100	100	100	100
1 : 3000	42	100	71	100
1 : 1000	21	100	36	100
1 : 500	8	100	13	100
1 : 100	24	100	51	100
1 : 50	86	100	94	100

The values show the percentage of the activity to that in the absence of antibody.

TABLE II. Effects of antibody to H-S on G6P-DH activity of four fractions.

Dilution rate	Relative G6P-DH activity			
of antibody	U-F	U-S	H-F	H-S
No antibody	100	100	100	100
1 : 3000	100	52	100	45
1 : 1000	100	17	98	2
1 : 500	100	25	96	1
1 : 100	100	87	99	21
1 : 50	100	100	100	42

activation of the fatty acid synthesis is closely related to the hardiness increase of Chlorella. We studied, therefore, the fatty acid-sensitivity and ATP-sensitivity of the iso-enzymes of Chlorella. Enzyme inhibition tests were carried out by preincubation of the four fractions with appropriately diluted fatty acids or ATP for 5 min. Oleic acid inhibited the G6P-DH activities of four fractions (Table III). The enzymes in U-S and H-S were more sensitive to oleic acids than those in U-F and H-F, respectively (Table III). Although G6P-DH in U-F and H-F was sensitive to oleic acid at high concentrations, we regarded the enzymes in the first fractions as the fatty acid-insensitive enzymes and those in the second fractions as the

TABLE III. Inhibitory effects of oleic acid on G6P-DH activity
of four fractions.

Concentration of	Relative G6P-DH activity			
oleic acid (M)	U-F	U-S	H-F	H-S
0	100	100	100	100
2×10^{-5}	97	72	100	52
5×10^{-5}	91	71	85	45
2×10^{-4}	79	51	57	16
5×10^{-4}	64	20	49	8
10^{-3}	51	4	40	3

The values show the percentage of activity to that in the
absence of oleic acid.

TABLE IV. Concentrations of fatty acids for half-maximal
inhibition of G6P-DH activity of four fractions.

Fatty acid	Concentrations required for half-maximal inhibition (M)			
	U-F	U-S	H-F	H-S
Miristic acid	10^{-3}	2×10^{-4}	10^{-3}	10^{-4}
Palmitic acid	10^{-3}	5×10^{-4}	10^{-3}	10^{-3}
Stearic acid	5×10^{-4}	10^{-4}	5×10^{-4}	10^{-4}
Oleic acid	10^{-3}	2×10^{-4}	10^{-3}	2×10^{-5}

fatty acid-sensitive enzymes. Table IV shows the concentrations of fatty acids for half-maximal inhibition of G6P-DH
activities of the four fractions. Stearic acid was most inhibitory to G6P-DH in U-F, H-F and U-S whereas oleic acid to the
enzyme in H-S. The enzymes in U-F and H-F exhibited the same
sensitivity to fatty acids. G6P-DH in H-S was more sensitive
to miristic and oleic acids than that of U-S. These results
suggest that the isoenzyme synthesized during hardening was
most closely related to fatty acid synthesis.

TABLE V. Inhibitory effect of ATP on G6P-DH activity of four fractions.

Fraction	Concentrations required for half-maximal inhibition (M)
U-F	2×10^{-3}
U-S	5×10^{-3}
H-F	2×10^{-3}
H-S	5×10^{-3}

TABLE VI. G6P-DH activity of four fractions at low temperatures.

Temperature (°C)	Relative G6P-DH activity			
	U-F	U-S	H-F	H-S
25	100	100	100	100
15	64	60	54	71
10	42	53	41	61
5	26	41	33	48

The values show the percentage of activity to that at 25°C.

The inhibitory effect of ATP on G6P-DH species was studied (Table V). G6P-DH in fatty acid-insensitive fractions was more sensitive to ATP than that in the fatty acid-sensitive fractions. It seems likely that the fatty acid-insensitive G6P-DH is closely associated with the oxidative phosphorylation.

The G6P-DH activities of the four fractions decreased with a decrease in temperature (Table VI). The enzyme activities of the fatty acid-sensitive fractions were more active than those of the fatty acid-insensitive fractions at a low temperature (Table VI). The H-S fraction was most active at a low temperature and kept one-half of the 25°C-measured activity even at 5°C (Table VI).

Fig. 5 shows a schematic hardening process in Chlorella summarizing the results obtained up to date. It appears in Chlorella that almost all ATP and NADPH produced by chloroplasts are used up for carbon-assimilation, and that ATP required for the hardiness is provided by mitochondria and NADP is reduced via the pentose-phosphate cycle. Mitochondria play a principal role in the process of hardening. Mitochondria

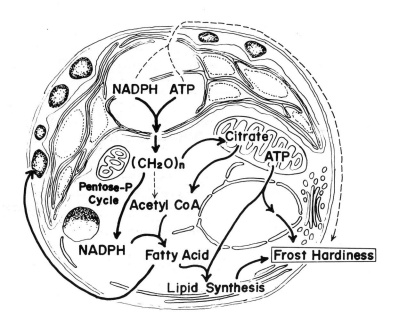

FIGURE 5. Schematic hardening process in Chlorella.

change their membranes into a structure much hardier than the chloroplasts and are essential for the repair of freezing-injured chloroplasts in Chlorella. The isoenzymes of G6P-DH, which is the key enzyme to activate the pentose-phosphate cycle, was synthesized during the early phase of hardening by only low temperature. The activation of G6P-DH was the fastest response in the events which we examined. Therefore, we infer that the synthesis of the G6P-DH isoenzyme is close to a "Trigger" reaction initiating changes in metabolism that lead to hardiness increase. The further studies of the G6P-DH iso-enzyme may serve to solve the initiation process of hardening.

A metabolic change activating fatty acid synthesis followed the activation of the pentose-phosphate cycle. Long chain fatty acids were synthesized more than enough by the fatty acid synthetase localized in cytoplasm exclusive of chloroplasts. The incorporation of ^{14}C into total lipids of membrane fraction was still actively proceeding after a 12 h hardening. It seems to us that de novo synthesized phospholipids are inserted into membrane lipids to make the membrane hardier. Chlorella has two fatty acid synthetase systems, and the major site of lipid synthesis shifts from chloroplasts to a cytoplasmic system during hardening.

III.   SUMMARY

Chlorella ellipsoidea Gerneck (IAM C-27) was synchronously grown at 25°C and cells at an intermediate stage in the ripening phase of the life cycle were hardened at 3°C for 48 h. Glucose 6-phosphate dehydrogenase (G6P-DH) was greatly activated during the first 6 h of hardening time by only low temperature regardless of whether the cells were hardened, whereas 6-phosphogluconate dehydrogenase activity gradually increased with hardiness increase. The partially purified G6P-DH was separated into two fractions by DEAE-cellulose in both the unhardened and hardened cells. The first peak of the hardened cells (H-F) showed the same elution pattern as that of unhardened cells (U-F), whereas the second peak area of the hardened cells was larger than that of unhardened cells. The second peak of hardened cells (H-S) contained a new isoenzyme which differ in electrophoretic mobility from the isoenzyme of the second peak of unhardened cells (U-S). An antibody pre- pared to U-F did not inhibit G6P-DH activities in U-S and H-S and an antibody to H-S did not inhibit those in U-F and H-F. These results indicate that G6P-DH species in the first and second fractions do not share common antigenic determinants. The enzymes in U-F and H-F were fatty acid-insensitive and ATP- sensitive. All tests made suggest that U-F and H-F contain the same isoenzyme. G6P-DH in H-S was more sensitive to oleic acid than that in U-S. The enzyme in H-S was most active at a low temperature and kept one-half of the 25°C-measured activity even at 5°C. It seems likely that the new isoenzyme in H-S is directly involved in the development of frost hardiness.

REFERENCES

1. Cacciapuotic, A.F., and Lessie, T.G., J. Bacteriol. 132:555 (1977).
2. Davis, B.J., Ann. N. Y. Acad. Sci. 121:404 (1964).
3. Hatano, S., Kabata, K., and Sadakane, H., Plant Physiol. 67:216 (1981).
4. Hatano, S., Sadakane, H., Tutumi, M., and Watanabe, T., Plant & Cell Physiol. 17:451 (1976).
5. Hatano, S., Sadakane, H., Tutumi, M., and Watanabe, T., Plant & Cell Physiol. 17:643 (1976).
6. Igaue, I., Tamura, H., Endo, K., and Saito, K., Plant & Cell Physiol. 22:311 (1981).
7. Muto, S., and Uritani, I., Plant & Cell Physiol. 11:767 (1970).
8. Muto, S., and Uritani, I., Agr. Biol. Chem. 35:1459 (1971).
9. Sadakane, H., Kabata, L., Ishibashi, K., Watanabe, T., and Hatano, S., Environ. Exper. Botany 20:297 (1980).

10. Sagisaka, S., Low Temp. Sci. Ser. B 28:43 (1970).
11. Sagisaka, S., Plant Physiol. 50:750 (1972).
12. Schaffer, F., and Steiner, R.Y., Arch. Microbiol. 116:9 (1978).

EFFECT OF LOW TEMPERATURE
ON THE GLUTATHIONE STATUS OF PLANT CELLS

C. L. Guy
J. V. Carter

Laboratory of Plant Hardiness
Department of Horticultural Science and Landscape Architecture
University of Minnesota
St. Paul, Minnesota

I. INTRODUCTION

Glutathione, γ-glutamylcysteinylglycine, is widely distributed throughout both animal and plant kingdoms (14), existing in cells in three different forms (15). Glutathione exists as a tripeptide thiol or reduced form (GSH), an oxidized form, in which a disulfide bond joins two molecules forming a dimer (GSSG), and a mixed disulfide formed by interaction of GSH with protein SH or SS groups (PSSG). In most cells GSSG and PSSG constitute only a small percentage of the total glutathione present. Because of its widespread occurrence in nature, much effort has been devoted to a determination of the function of glutathione in cell metabolism.

It has been suggested that the structural integrity of cellular and organelle membranes is dependent on the ability of the cell to maintain the appropriate GSH status. Both GSH and GSSG are needed to regulate the thiol/disulfide status of membrane proteins through interchange reactions (16). Glutathione is thought to play a role in protection of membrane lipids against peroxidation by free radicals and lipid hydroperoxides (14). It is also closely associated with assembly and disassembly of microtubules (20) and is involved in DNA, RNA and protein synthesis (16).

Recently, GSH was found to be correlated with frost hardiness in _Picea_ _abies_ (8). Reduced glutathione concentration increased as growth temperatures decreased, and reached a maximum during winter, and declined in spring and summer to minimum levels. Reduced glutathione accounted for 95% of the non-protein thiol present in spruce needles with free cysteine accounting for the remainder.

Esterbauer and Grill (8) suggests that high GSH levels are involved in protection of protein SH groups from oxidation during freezing as envisioned by Levitt's sulfhydryl hypothesis (17). Ivanov (13), on the other hand, examined glutathione levels in several citrus species of differing frost hardiness both prior to and after hardening. He found glutathione content to be inversely related to frost hardiness. Other evidence (18,19,22,26) suggested that GSH does not accumulate during cold acclimation and is not related to frost hardiness, but GSH levels were not determined in these four studies.

The purpose of this study was to examine the effect of low temperature exposure on levels of both reduced and oxidized glutathione in a plant that does not acclimate (Solanum tuberosum), a plant that acclimates to moderate levels of hardiness (Hedera helix) and a plant that acclimates to extreme levels of hardiness (Cornus sericea).

II. MATERIALS AND METHODS

A. Plant Culture

Red osier dogwood plants, Cornus sericea L., and english ivy, Hedera helix L. were propagated by stem cuttings from monoclonal sources (4). Rooted cuttings were maintained in the greenhouse with day/night temperatures of 20/15 C and supplemental light to give a 16-hr-long day. Plants were kept in the greenhouse until used for experiments. Potato plants, Solanum tuberosum, were propagated from tuber slices in a soil, sand and peat (3:2:2) mixture in controlled environment chambers under 16-hr-long days and 25/20 C temperatures. Hardening experiments were conducted in controlled environment chambers.

At the beginning of each experiment 20-30 uniform plants of each species were selected and transferred to a growth chamber with 25/20 C day/night temperatures and 16-hr-long days and held for 1 week. At the end of one week tissue samples were taken from 4 different plants of each species, each plant constituting a single experimental unit. All tissue samples were taken at the same time, prior to the beginning of the light period, to avoid possible diurnal fluctuations in physiological factors. Tissue samples from potato and ivy plants consisted of the youngest fully mature leaves while tissue samples of dogwood consisted of bark tissue stripped from the entire stem of the tree. Generally, 5 g fresh weight of leaf tissue and 2-4 g fresh weight of bark tissue was used for each sample. Immediately following the initial sampling, the remaining plants were transferred to another growth chamber with a constant 5 C day/night temperature with a 12 hour photoperiod.

Tissue samples were taken on a weekly basis for the next 4 to 6 weeks.

B. Glutathione Purification

Initial efforts to measure glutathione levels in trichloracetic acid or ethanolic extracts from potato and ivy leaf tissue by traditional methods revealed a high degree of background interference by phenolic and other compounds. The following procedure was developed to remove the interfering compounds and purify glutathione from plant tissue. Fresh tissue was boiled for 1 min in 80% ethanol containing 0.15 mM ethylenediaminetetraacetic acid (EDTA) then homogenized immediately for 1 min. Next, the insoluble residue was collected by vacuum filtration and washed several times with boiling 80% ethanol. The combined washes and filtrate were partitioned 5 times with hexane to remove chlorophyll and pigment. The extract was then degassed by vacuum aspiration and passed through a 10 mm x 140 mm Dowex H^+, 50-4X, 50-100 mesh column equilibrated with 80% ethanol. Glutathione was retained on the column while most of the phenolic and other impurities were eluted in the wash. After washing the column with 50 ml degassed 80% ethanol, glutathione was eluted with degassed 0.4 M citrate buffer pH 4.25. The first 20 ml of buffer was discarded. Glutathione eluted in the following 40 ml of buffer.

C. Measurement of Non-protein Thiol and Glutathione

Total non-protein thiol (NPSH) was determined following titration with 5,5'-dithiobis-(2-nitrobenzoic acid) (DTNB) as described by Ellman (7).

Glutathione was specifically determined by a modification of the cycling enzyme assay of Tietze (25). Total glutathione was determined by reacting 0.4 ml sample with 0.4 ml 0.7 mM DTNB in 1.0 M Tris-HCl buffer pH 8.0, 0.1 ml glutathione reductase EC. 1.6.4.2 (3IU) and 0.1 ml 5.0 mM NADPH and recording the change in absorbance at 412 nm for 3 min with a Beckman dual beam recording spectrophotometer. Glutathione concentrations were determined by comparison with a standard curve prepared with solutions of known glutathione concentration.

GSSG was determined after removal of GSH from the sample. GSH was removed from the reaction mixture by mixing 1.8 ml sample with 0.2 ml 100 mM N-ethylmaleimide (NEM) in Tris-HCl buffer. The mixture was allowed to stand at room temperature for 70 min after which excess NEM was removed by repeated solvent extraction (5 times) with equal volumes of ether. Residual ether was removed by bubbling a nitrogen stream through the solution. GSH was determined by subtraction of GSSG from total glutathione. The difference between NPSH and GSH was also determined by subtraction, the resultant value representing non-glutathione thiol (RSH). Tabulated values

TABLE 1. Absorbance at 412 nm of crude 80% ethanol extracts of
5 gm fresh potato leaf tissue with and without DTNB
added at pH 8.0. Total volume 65 ml.

Sample	−DTNB	+DTNB
1	.544	.720
2	.567	.689
3	.550	.723

TABLE II. Elution of GSH from Dowex 50 − 4x H^+, 140 x 10 mm
column.

Moles Added	Moles Recovered	% Yield
1.30×10^{-5}	1.40×10^{-5}	108
1.30×10^{-5}	1.43×10^{-5}	110
1.30×10^{-5}	1.27×10^{-5}	98

assume RSH to be a monothiol. Glutathione reductase, GSH,
GSSG, DTNB and NEM were all purchased from Sigma.

III.   RESULTS AND DISCUSSION

A.   Glutathione Measurement

Quantitative measurement of non-protein thiol and glutathione present in tissue extracts rely upon colorimetric techniques that depend on the reaction of DTNB with a thiol group
to form an anion that absorbs strongly at 412 nm (7,25).
Initial attempts to measure glutathione in crude ethanolic and
trichloroacetic acid extracts of plant tissue by these methods
revealed a high level of interference (Table 1). Efforts to
dilute the extracts to reduce the interference were not successful since the absorbance generated by the reaction of the
thiols with DTNB went to extinction before the absorbance due
to the interfering compounds did. Control experiments with
authentic glutathione showed that an ion exchange chromato-
graphic procedure for amino acid purification (23) could also
be employed for glutathione purification in that it could be

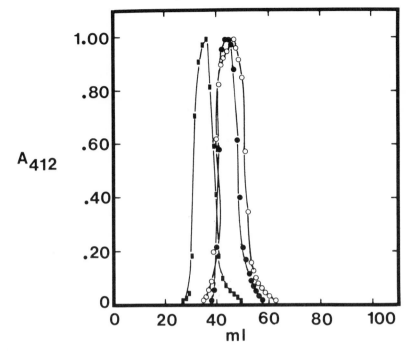

FIGURE 1.   Elution of reduced glutathione (●), oxidized glutathione (O) and cystein (■) from a Dowex ion-exchange column.   Glutathione and cysteine were eluted with 0.4 M citrate buffer pH 4.25 at a rate of 1 ml per minute.

quantitatively bound and recovered from a Dowex cation exchange column (Table 2).   GSH oxidation was minimized by degassing solvents to reduce oxygen concentration, chelating metal ions that catalyze oxidation and maintaining acid pH of solvents. Because our purification method was adapated from amino acid techniques, purified glutathione preparations also contained free amino acids that commonly occur in plant tissues. Cysteine was not separated from GSH or GSSG by our ion-exchange procedure (Fig. 1), and thus, if present, would represent at least part of what we have designated as RSH.   No doubt GSH homologues as well as dipeptides containing cysteine would also co-elute with GSH in our system.

Accurate determination of non-protein thiols present in plant tissues requires rapid denaturation and removal of enzymes that catalyze reactions with glutathione and soluble thiols.   At the same time oxidation of sulfhydryls must be minimized.   Boiling ethanol precipitates proteins and quickly stops enzymatic activity.   EDTA present in the hot ethanol chelates and inhibits oxidation of thiols by Cu^{++} and Fe^{++} cations in the presence of oxygen (14).   Exogenous GSH added to

TABLE III. Quantitative   recovery   of   added   GSH   (6.5   x
10^{-6} mol) prior to homogenation of 5 gm Solanum
tuberosum leaf tissue.

Sample	GSH Added	Moles Recovered	% Yield
1	-	6.2×10^{-7}	-
2	-	7.5×10^{-7}	-
1	+	7.1×10^{-6}	99.7
2	+	7.7×10^{-6}	106.9

TABLE IV. Variability of glutathione assay. 15 gm leaf tissue
was homogenized and divided into 3 equal fractions.

Sample	Moles Recovered	% of Mean
1	1.05×10^{-6}	104
2	1.00×10^{-6}	99
3	0.98×10^{-6}	97

potato leaf tissue prior to homogenation was recovered quantitatively (Table 3) and variability was low (Table 4) by our
analysis procedures.

Generally, GSH represents the majority of non-protein thiol
present in cells (8,10,11,16). Thus, measurement of total
thiol content by Ellman's (7) DTNB titration reflects GSH
status. In the present study, GSH accounted for 50% to 75% of
the total non-protein thiol. As a result, DTNB titration of
crude extracts by Ellman's (7) procedure is not a satisfactory
measure of the GSH status of plant cells. DTNB does not react
specifically with GSH alone and therefore should not be used as
the only method of determination of plant cell GSH content.
Specific assays for glutathione are necessary to avoid ambigu-
ous results.

B. Effect of Growth Temperature on Glutathione Concentrations

At moderate temperatures GSH represented the largest fraction of non-protein thiol in potato and ivy leaves while RSH
was the largest fraction in dogwood bark tissue (Fig. 2).
Levels of GSSG were a small percentage of the total glutathione

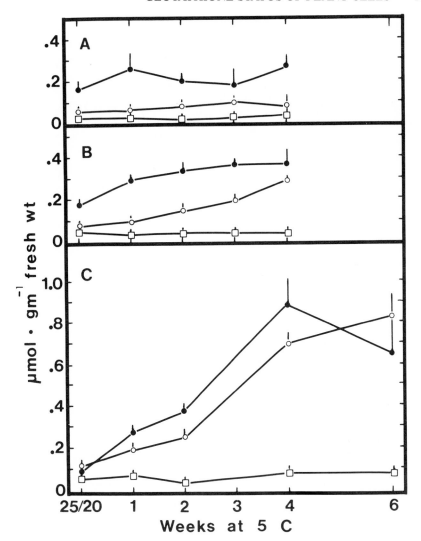

FIGURE 2.   Effect of exposure to a constant 5 C on glutathione levels in potato leaf tissue (A), in ivy leaf tissue
(B), and in dogwood bark tissue (C).   Reduced glutathione (●),
oxidized glutathione (□) and other non-protein thiol (O).

content in all plants, ranging from 10% to 33% of the total.
It remained low at all growth temperatures, suggesting oxidation of GSH was low or that glutathione reductase is efficient
under these conditions.   Normally, GSH/GSSG ratios exceed 10 in
chloroplasts and animal cells (12,16).   But, ratios of less
than 10 have been reported for wheat embryo (9) and corn roots
(3).

When the plants were transferred to low temperatures GSH levels increased with time in all plants. In potato leaves, GSH initially increased, then declined and subsequently increased to its highest level after 4 weeks at 5 C. GSH also increased in ivy leaves and dogwood bark tissue when grown at 5 C. The largest increase in GSH was observed in dogwood while ivy showed an intermediate response between that of dogwood and potato. The increases in GSH content are in agreement with the findings of Esterbauer and Grill (8), but in direct conflict with the conclusions drawn by Levitt et al. (19) that increases in plant tissue SH during cold acclimation are not due to GSH. We find that changes in the levels of GSH and other non-protein thiols during cold acclimation contribute substantially to increases in tissue SH. Since GSH levels increased while GSSG levels remained relatively constant throughout the experiments, the increase of GSH was due to a net increase in total gluta- thione. Whether the accumulation of glutathione is a result of increased synthesis or decreased degradation is not clear. It has been suggested that glutathione may function as a storage form of reduced sulfur in plant cells (21). Therefore, decreased demand for sulfur accompanying the slowdown in growth at low temperatures may lead to the accumulation of sulfur com- pounds such as glutathione. In a subsequent study, dogwood trees grown under natural environmental conditions acclimated to withstand immersion in liquid nitrogen without injury while at the same time showed no accumulation of glutathione. Thus, it appears that increases in GSH are not necessary for stem tissue to fully acclimate.

Also of interest although its implication for cold acclima- tion is not clear, is the increase of other non-protein thiols, RSH, upon low temperature exposure of these plants. Like GSH, RSH reached its highest levels in the bark of dogwood, was lowest in potato leaves and was intermediate in ivy leaves (Fig. 2). Increases in RSH content were more marked and less variable than increases in GSH content in tissues exposed to low temperatures. Identification of the compound(s) making up RSH is presently underway. As stated previously, our prepara- tions could include cysteine. However, cysteine concentrations in cells are usually only a small fraction of the total non- protein thiol content (8,14). Esterbaurer and Grill found cysteine to represent only 5% of the total thiol in spruce needles. These reports cast doubt on the possibility of cysteine being the sole constituent of RSH. It is also possible that RSH may be precursors or breakdown products of GSH which are not substrates for the enzymatic assay of gluta- thione. Another possibility is that homologues of glutathione, such as homoglutathione (1,2) are also present. Although homo- glutathione reacts with glutathione reductase at the same rate as glutathione other homologues may not. Lower rates of reac- tion by glutathione homologues in the glutathione assay would

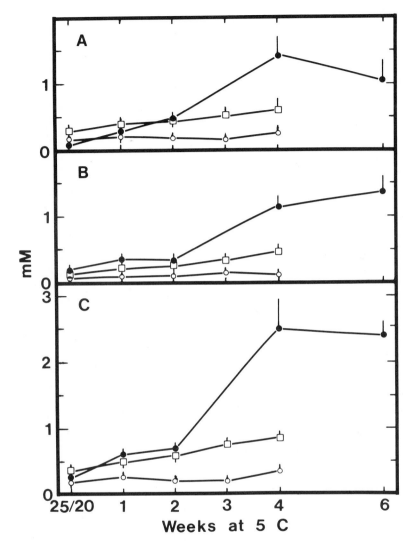

FIGURE 3.   Effect of exposure to a constant 5 C on the concentration of reduced glutathione (A), of other non-protein thiol (B), and total non-protein thiol (GSH+RSH) (C) in potato leaf tissue (O), ivy leaf tissue (□) and dogwood bark tissue (●).   Calculation of concentration is based on total tissue water content.

result in underestimation of the total glutathione content and would overestimate non-glutathione thiol content.

When non-protein thiol concentrations are expressed on a mM basis, the increase of GSH, RSH and NPSH levels in leaf and bark tissue grown at low temperatures becomes readily apparent

(Fig. 3). Providing that most of the non-protein thiols are compartmentalized in the cytoplasm and organelles while the vacuole contains the bulk of the cell's water, the concentrations may be many times the values shown. When RSH and GSH are added together, dogwood bark tissue shows a non-protein thiol concentration of greater than 2.4 mM after 6 weeks growth at 5 C.

In this study, glutathione content is correlated with reported frost hardiness. Potato leaves (Solanum tuberosum), incapable of cold acclimation (5), contained the lowest levels of glutathione of the plants tested. Ivy leaves will acclimate to about -15 C when grown at 5 C for 4 weeks (24) and dogwood stems will acclimate to -25 C after 3 weeks growth at 5 C (4). Ivy leaves contained intermediate levels of GSH and non-protein thiols while these compounds reached the highest concentrations in dogwood bark. Contrary to our findings, Ivanov (13) and de Kok et al. (6) found glutathione not to be associated with frost hardiness in citrus and spinach.

Regardless of whether glutathione is related to plant frost hardiness, our findings suggest growth temperature influences the thiol/disulfide status of plant cells. Because thiols are some of the most reactive compounds present in cells (14) significant changes in the thiol/disulfide equilibrium may influence the oxidation/reduction status of cellular proteins and membranes. Large increases in GSH may alter the rates of, and therefore the equilibrium between assembly and disassembly of microtubules, affect the regulation and conformation of enzymes, and stimulate synthesis of RNA, DNA and protein (16). Grill et al. (11) even suggested high GSH levels may induce a general disruption of fundamental cellular processes. Evidence presented in this study indicates further study of the relationship of glutathione in plant frost hardiness is warranted.

REFERENCES

1. Carnegie, P.R., Biochem. J. 89:459-471 (1963).
2. Carnegie, P.R., Biochem. J. 89:471-478 (1963).
3. Carringer, R.D., Rieck, R.D., Bush, L.P., Weed Sci. 26:167-171 (1978).
4. Chen, H.H., Li, P.H., Plant Physiol. 62:833-835 (1978).
5. Chen, H.H., Li, P.H., Plant Physiol. 65:1146-1148 (1980).
6. de Kok, L.J., de Kan, P.J.L., Tanczos, O.G., Kuiper, P.J.C., Plant Physiol. 67:63 (1981). Abstract only.
7. Ellman, G.L., Arch. Biochem. Biophys. 82:70-77 (1959).
8. Esterbauer, H., Grill, D., Plant Physiol. 61:119-121 (1978).
9. Fahey, R.C., Di Stefano, D.L., Meier, P.G., Bryan, R.N., Plant Physiol. 65: (1980).
10. Grill, D., Esterbauer, H., Phyton (Austria) 15:87-101 (1973).

11. Grill, D., Esterbauer, H., Klosch, U., Environ. Pollut. 19:187-194 (1979).
12. Halliwell, B., Foyer, C.H., Planta 139:9-17 (1978).
13. Ivanov, S.M., C.R. Acad. Sci. U. R. S. S. 277-281 (1939).
14. Jocelyn, P., Biochemistry of the SH Group. Academic Press New York (1972).
15. Kosower, N.S., Kosower, E.M. in "Glutathione" (L. Flohe et al., eds.), Academic Press, New York, 1974.
16. Kosower, N.S., Kosower, E.M., Rev. Cytol. 54:109-160 (1978).
17. Levitt, J., J. Theoret. Biol. 3:355-391 (1962).
18. Levitt, J., Sullivan, C.Y., Johansson, N.O., Pettit, R.M., Plant Physiol. 36:611-616 (1961).
19. Levitt, J., Sullivan, C.Y., Johansson, N.O., Plant Physiol. 37:266-271 (1962).
20. Rebhun, L.I., Miller, M., Schnaitman, T.C., Nath, J., Mellon, M., J. Supramol. Struct. 5:151-171 (1976).
21. Rennenberg, H., Schmitz, K., Bergmann, L., Planta 147:57-62 (1979).
22. Schmuetz, W., Sullivan, C.Y., Levitt, J., Plant Physiol. 36:617-620 (1961).
23. Smith, I.K., Plant Physiol. 55:303-307 (1975).
24. Steponkus, P.L., Lanphear, F.O., Physiol. Plant. 21:777-791 (1968).
25. Tietze, F., Anal. Biochem. 27:502-522 (1969).
26. Waisel, Y., Kohn, H., Levitt, J., Plant Physiol. 37:272-276 (1962).

SORBITOL IN TRACHEAL SAP
OF DORMANT APPLE (Malus domestica Borkh) SHOOTS
AS RELATED TO COLD HARDINESS

S. Ichiki
H. Yamaya

Aomori Apple Experiment Station
Kuroishi, Aomori, Japan

I. INTRODUCTION

Many studies have related seasonal carbohydrate content and temperature to cold hardiness in fruit trees.

More recently, it was reported that levels of sorbitol in the sap of apple shoots were highest at subfreezing temperatures and appeared to be inversely related to temperatures during the autumn, winter, and spring months (8,9). Raese, Williams, and Billingsley (5) also reported that high levels of sorbitol in the sap were significantly related to cold hardiness of holding apple shoot at temperature both above and below freezing. However, relatively little is known about the levels of sorbitol in sap on cold hardiness under field conditions.

Therefore, we examined the relationship of sorbitol levels in the sap to air temperature and cold hardiness of apple trees in Aomori.

II. MATERIALS AND METHODS

Sorbitol in tracheal sap of vigorous current season shoots was investigated during the winters of 1976-77, 1977-78, 1978-79 on Starking Delicious apple trees that were planted at the Aomori Apple Experiment Station. Shoots, approximately 100 cm long were cut from trees before noon at 2 to 5-day intervals for a month. At each sampling 10 shoots were taken

181

Copyright © 1982 by Academic Press, Inc.
All rights of reproduction in any form reserved.
ISBN 0-12-447602-3

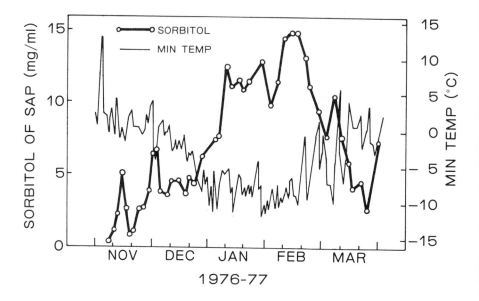

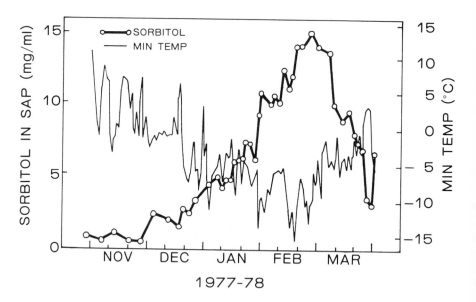

FIGURE 1.   Sorbitol changes in sap from current season shoots of Starking Delicious apple trees, as related to minimum temperatures, during 3 winters.

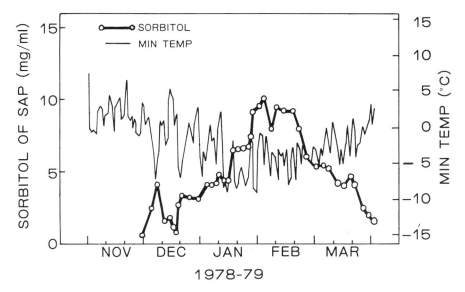

from each of 50 trees. The sap was removed from the wood under vacuum by the method of Bollard (1). The sorbitol in the sap was analyzed by periodate oxidation method by modification of Flood and Priestley (3).

The mid-portion of shoots were used to estimate cold hardiness by the concentration of potassium of the effusate (10), similar to that used by Stayanov (6).

Shoot samples (3 g) were cut into 0.5 cm segments and placed in 20 ml weighing bottles and stored for 17 hours at $0 \pm 1°C$, $-5 \pm 1°C$, $-10 \pm 2°C$, $-15 \pm 2°C$, $-20 \pm 4°C$, $-25 \pm 4°C$, and $-30 \pm 4°C$. After the samples had thawed for two hours at room temperature, and 10 ml deionized water per g of tissue was then added to the 100 ml beakers containing the samples. After 24 hours of diffusion, the shoot samples were removed from the solution and the concentration of potassium was determined by flame photometer (A_t). The sample was boiled for 7 min to kill the tissue, cooled and made to initial volume with deionized water. After an additional 24 hours the concentration of potassium again measured (B_t).

From these potassium determinations, "Index of injury" was calculated using the modified formula by Flint, Boyce, and Beattie (2):

$$It = (R_t - R_o)/(100 - R_o)$$

Where, I_t = Index of injury resulting from exposure to temperature (t)

$R_t = 100 \ A_t/(A_t + B_t)$

$R_o = 100 \ A_o/(A_o + B_o)$

I_t and observation of injury of 3-year old potted Starking Delicious apple trees were compared during the winter of 1977-78 (10). There was no damage in trees with the current season shoot I_t of less than 10. There were varying amounts of damage in the bark and xylem with I_t between 20 and 30. Current season shoots with an I_t greater than 50 were dying (10).

III.   RESULTS AND DISCUSSION

In 3 different winters, sorbitol levels in the sap increased sharply as the minimum temperatures dropped to near 0°C (Fig. 1). Levels of sorbitol in the sap generally increased as the minimum temperatures fell during the dormant season and decreased during warm periods. However, sorbitol values fluctuated sharply with changes in temperature. When minimum temperatures were low before sampling, sorbitol values increased.

The largest amount of sorbitol in the sap occurred after period of the lowest temperature. Twenty to thirty times as much sorbitol was present in the sap in mid-February as occurred in early November. In 1978-79 the levels of sorbitol in the sap were much lower than 1976-77 and 1977-78. A possible explanation for the difference between seasons could be that minimum temperatures were high over a long period during January to February.

Williams and Raese (9) point out that several cold days are necessary for sorbitol in the sap of apple shoots to reach peak levels, while only a single warm day may drastically lower the amount in the sap. A similar pattern exists for cold hardiness of apple trees. Ketchie and Beeman (4) reports the highest correlation between cold resistance and the temperature for the 7 preceding days from September through April. Our data indicate that sorbitol levels in the sap are probably an immediate response to the temperature (Fig. 1).

Figure 1 and 2 show the three sorbitol peak plateaus in 1976-77 and 1977-78. The first peak plateau of sorbitol in the sap is approximately 5 mg·ml^{-1}, the second 12 mg·ml^{-1} and the third 15 mg·ml^{-1}. However, in 1978-79 the most warm season of the 3 winters, the third peak plateau was not evident. It is not apparant that these three sorbitol peak plateaus tend to coincide with 3 stages of cold hardiness in hardy woody plants by the hypothesis of Weiser (7).

Cold hardiness of excised shoots generally followed the minimum temperatures trend in both years. In 1977-78, from early November to late December, a sharp fluctuation in minimum temperature was accompanied by a gradual decrease in t_{10} and t_{20} of the shoots. However, during the cold period in early

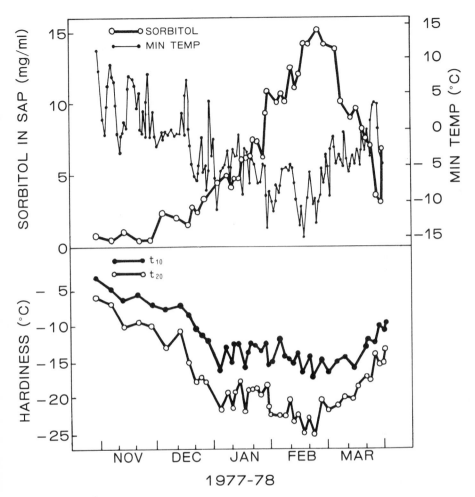

FIGURE 2. Sorbitol status in sap from current season shoots of Starking Delicious apple trees, as related to minimum temperatures and hardiness, expressed as t_{10} and t_{20} which are temperatures required to give an It 10 and It 20 respectively.

January and mid-February, cold hardiness fluctuated with changes in minimum temperatures. Cold hardiness decreased beginning at the latter part of February due to substantially higher temperatures (Fig. 2).

When the values are combined, the correlation between t_{10}, t_{20} values and the corresponding levels of sorbitol in the sap are significant at 1% in both years (Table 1).

Table I. Correlation coefficients (r) of t_{10}, t_{20} values
versus content of sorbitol in the sap from current
season shoots of Starking Delicious apple trees.

| Season | No. of Samples | Correlation coefficient (r) | |
		t_{10} vs. sorbitol (mg/l)	t_{20} vs. sorbitol (mg/l)
1977-78	45	-0.789**	-0.711**
1978-79	42	-0.719**	-0.808**

** Significant at the 1% level.

Cold hardiness, expressed as t_{10}, t_{20} values parallels
the content of sorbitol in sap. A large difference is between
t_{10} and t_{20} within the same year (Fig. 2). The difference
in degrees between t_{10} and t_{20} would indicate irregular
cold hardiness in different tissues of shoot.

Our data indicate that sorbitol in the sap of shoots play
an important role in cold hardiness of apple trees. Although
we have not established the influences of nutrition, soil moisture, crop load and tree vigor on the levels of sorbitol in the
sap during the winter.

IV. SUMMARY

Sorbitol levels in the current season shoots of Starking
Delicious apple trees were compared with minimum temperatures
during 3 winters (1976-79) in order to gain more detailed
information on cold hardiness in relation to temperature conditions in Aomori. Sorbitol in tracheal sap extracted under
vacuum were determined by periodate oxidation method. Cold
hardiness of current season shoots was measured by determining
the percentage of K ions released by artificial freezing. Sorbitol in the sap generally increased with subfreezing temperatures and decreased during warm periods. The first sharp rise
in sorbitol appears to be associated with the onset of low minimum temperatures. However, sorbitol values fluctuated with
changes in daily minimum temperature. In cold winter (1978-79)
the level of sorbitol in the shoot sap was much higher than in
warm (1976-77). There was correlation between cold hardiness
and the levels of sorbitol in the sap during the winters of
1977-78 and 1978-79. Our data indicate that sorbitol plays an
important role in cold hardiness of apple trees.

REFERENCES

1. Bollard, E.G., J. Exp. Bot. 4:363-368 (1953).
2. Flint, H.L., Boyce, B.R., and Beattle, D.J., Can. J. Plant Sci. 47:229-230 (1976).
3. Flood, A.E., and Priestley, C.A., J. Sci. Fd. Agric. 24:945-955 (1973).
4. Ketchie, D.O., and Beeman, C.H., J. Amer. Soc. Hort. Sci. 98:257-261 (1973).
5. Raese, J.T., Williams, M.W., and Billingsley, H.D., J. Amer. Soc. Hort. Sci. 103:796-801 (1978).
6. Stoyanov, S.M., Tidsskrift for planteavl, Sear. Tid. Plant. 77:13-18 (1973).
7. Weiser, C.J., HortScience 5:403-409 (1970).
8. Williams, M.W., and Billingsley, H.D., J. Amer. Soc. Hort. Sci. 98:205-207 (1973).
9. Williams, M.W., Raese, J.T., Physiol. Plant. 30:49-52 (1974).
10. Yamaya, H., and Ichiki, S., Ann. Meeting, Tohoku Soc. Hort. Sci. 35-36 (1980). (In Japanese).

ULTRASTRUCTURAL CHANGES OF PLASMA MEMBRANE
IN CORTICAL PARENCHYMA CELLS
OF MULBERRY TWIG RELATED TO FREEZING TOLERANCE

T. Niki

The Institute of Low Temperature Science
Hokkaido University
Sapporo, Japan

Pomeroy and Siminovitch (6) first demonstrated that seasonal cytological changes were closely related to the seasonal
cycle of freezing tolerance, in particular a seasonal transition of the plasmamembrane from a relative smoothness in the
summer to a highly folded state in the winter. They inferred
that the highly folded state of the plasmamembrane facilitates
the flow of water and, at the same time, alleviates the stress
of contraction and expansion during the freezing-thawing
cycle. They also proposed that the process of the renewal of
the membrane was involved in the mechanism of cold acclimation.

However, in hardy cortical cells of mulberry twig, the
highly folded vesicular state of the plasmamembrane was not
observed in mid-winter. The present studies were undertaken to
clarify the process of presumed changes in the plasmamembrane
associated with freezing tolerance.

I. MATERIALS AND METHODS

A. Plant Materials and Freezing Procedures

One-year-old twigs of 10-year-old mulberry trees grown on
our campus were used. The twigs enclosed in polyethylene bags
were frozen at -5°C for 2 hr, then brought by 5°C increments at
2 hr intervals to -30°C. The twigs were further cooled to
-70°C in 10°C increments at 2 hr intervals.

For artificially hardening of twigs, the twigs enclosed in
polyethylene bags were exposed to 0°C for 13 to 20 days.

The viability of the cells was determined by vital staining
with neutral red and by plasmolysis after thawing.

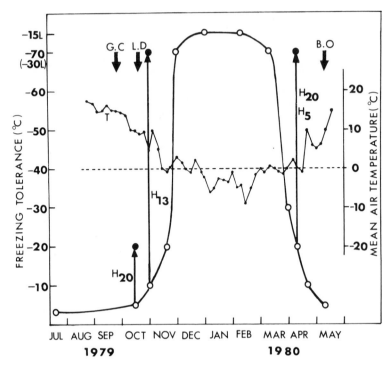

FIGURE 1.   Seasonal changes in freezing tolerance of corti-cal   parenchyma   cells   from   mulberry   tree.   Hardiness   is expressed as the minimum survial temperatures of cells frozen for   16   hr   at   selected   test   temperatures. H_5, H_{13}, H_{20}, hardening at 0°C for 5, 13 and 20 days, respectively. -15L, -30L immersion in liquid nitrogen after prior freezing at -15 or -30°C (7).   T, Mean air temperature; G.C, Growth cessation of twig; L.D, Leaf defoliation; B.O, Bud opening.

B.   Procedures for Electron Microscope

Fixation for electron microscopy was carried out according to the methods of Niki et al. (4).   For the cytochemical studies, ultrathin sections were destained for 30 min in 1% periodic acid.   Sections were rinsed and stained with 1% phos-photungstic acid (PTA) in 10% chromic acid for 50 min.   All steps were carried out at room temperature.

II. RESULTS

A. The Seasonal Cycle of Freezing Tolerance

The seasonal fluctuation in freezing tolerance of cortical parenchyma cells of the mulberry twig is shown in Fig. 1, with the mean air temperature. In twigs collected on July 15, the cortical cells did not survive freezing, even at -5°C. After exposure to 0°C for 20 days, there also was no increase of freezing tolerance. In mid-October, cortical cells of twigs were hardy to -5°C, and freezing tolerance of the cortical cells exposed to 0°C for 20 days increased to -20°C.

Freezing tolerance abruptly increased in late November and reached -70°C, or survived immersion in liquid nitrogen after prefreezing at -30°C. Twigs collected in late October, which were marginally hardy to -20°C, increased their freezing tolerance after exposure to 0°C for 13 days, survived to -70°C. Maximal freezing tolerance was maintained from late December to late January, when twigs survived immersion in liquid nitrogen after prefreezing to -15°C.

Thereafter, freezing tolerance decreased gradually, reaching -5°C in early May immediately before bud opening.

B. Electron Microscope Studies

The previous studies (4) reported that the plasmamembrane of cortical parenchyma cells from twigs collected on October 16 was regular with no conspicuous infolding. However, a number of marked fine structural changes occurred in cells hardened for 20 days (Fig. 2). The most striking of these changes was the transformation in the plasmamembrane. There were marked invaginations which appeared to couple and form vesicles in the peripheral cytoplasm. These vesicles then were engulfed into the vacuole. The cytochemical tests (Fig. 3) showed that these vesicles were similar to the fragments of the plasmamembrane. This suggests that in the first step pieces of the specific cell organelles were detached from those organelles, or that whole organelles were involved, and that the vacuole play an important role in their digestion of these organelles.

Under natural conditions, similar structural features were found in cells collected on November 20, which were hardy to -30°C (4). Although on October 30 the plasmamembrane remained smooth and regular with no conspicuous infoldings, the hardened cells which were collected on October 30 showed the same dramatic changes of ultrastructure as shown in Fig. 2. In cells hardened for 20 days (Fig. 4), microvesicles were generated by the Golgi apparatus. Some were fused with the plasmamembrane (arrows, Fig. 4). By the cytochemical tests shown in Fig. 5, it is clear that these microvesicles were similar to

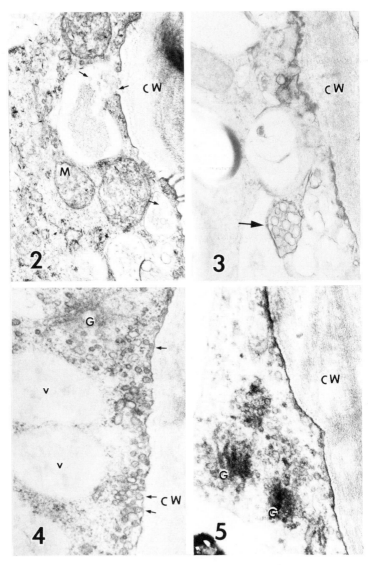

FIGURE 2. Hardened cell from a twig collected on October 16 after exposure to 0°C for 20 days. Invaginations of the plasmamembrane are observed. Vesicles and fibrous materials (arrows) are recognized in the invagination of plasmamembrane. M, Mitochondria; CW, Cell wall. x 17,000.

FIGURE 3. Electron micrograph with PTA staining at the same stage as Fig. 2. The engulfed vesicles (arrow) were positively stained. x 21,250.

plasmamembrane because the PTA was found to be highly specific for the plasmamembrane.

A similar feature was obtained under natural conditions for the cell of November 27 (Fig. 6). Numerous microvesicles were pinched off from the Golgi apparatus. These microvesicles were observed in the peripheral plasmamembrane and fused with the plasmamembrane (Fig. 6, arrows).

In mid-winter, January 29, the nucleus was located in the central portion of the cell, and numerous small vacuoles were dispersed throughout the dense and extensive cytoplasm (Fig. 7). This arrangement is characteristic of extremely hardy cells as reported by Pomeroy and Siminovitch (6). In cortical cells of the mulberry, however, the plasmamembrane was relatively smooth; no irregularities or high degree of folding was observed.

Freezing tolerance gradually decreased. The plasmamembrane from cells of twigs collected on March 10, again exhibited a highly folded structure (Fig. 8) like that shown in Fig. 2. The invaginations of the plasmamembrane contained vesicles and fibrous materials (Fig. 8, arrows). These invaginations were pinched off into the cytoplasm, then were engulfed into the vacuole. This ultrastructure was observed in the cells collected from mid-February to the end of March.

The cells collected on March 30 were hardy to -30°C, and their plasmamembranes became relatively smooth and regular (Fig. 9). Microvesicles appeared near the periphery of the cytoplasm. Thus, microvesicles may be generated by the Golgi apparatus. Thereafter, as freezing tolerance continued to decrease, small vacuoles fused and gradually became a large vacuole.

A diagramatic representation of the sequence of presumed renewal of the plasmamembrane is shown in Fig. 10.

III. DISCUSSION

Pomeroy and Siminovitch (6) inferred that the highly folded condition of membranes in the cells of black locust trees facilitates the flow of water and, at the same time, alleviates

---

FIGURE 4. Hardened cell from a twig collected on October 30 after exposure to 0°C for 13 days. Numerous microvesicles were confined to the peripheral cytoplasm lining the plasmamembrane. Some microvesicles have fused with the plasmamembrane (arrows). V, Vacuole; G, Golgi apparatus. x 17,000.

FIGURE 5. Electron micrograph with PTA staining at the same stage as Fig. 4. The microvesicles generated from Golgi apparatus were positively stained. x 17,000.

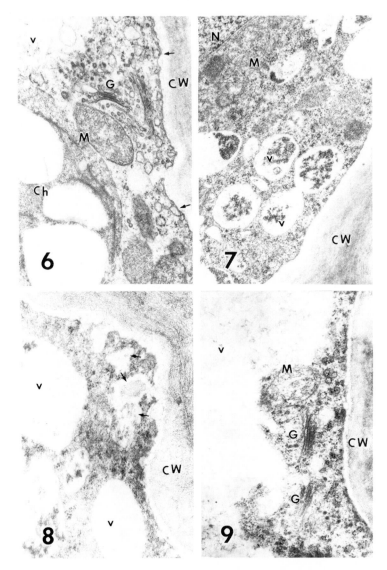

FIGURE 6.   The ultrastructure of a cell from twigs collected on November 27. Many microvesicles were observed in the
peripheral plasmamembrane.   It is clear that some of them were
generated from Golgi apparatus. Some microvesicles (arrows)
have fused with the plasmamembrane. Ch, Chloroplast. x 21,250.

FIGURE 7.   The ultrastructure of a cell from twigs collected on January 29. The plasmamembrane was smooth and regular.   Microvesicles   could   not   be   observed.   N,   Nucleus.
x 15,300.

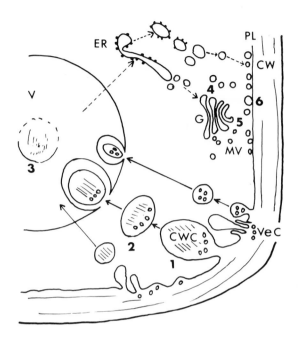

FIGURE 10. Diagramatic representation of the sequence for the presumed renewal of the plasmamembrane. 1, Invagination of vesicles and fibrous materials; 2, Vesicles pinched off; 3, Engulfment of vesicles into vacuole; 4, Generation of microvesicles; 5, Microvesicles appear in the peripheral plasmamembrane; 6, Fusion of microvesicles with the plasmamembrane.

the stress of contraction and expansion during the freezing-thawing cycle. But the plasmamembrane of cortical cells from mulberry twigs has a relatively smooth structure in mid-winter.

In the present report, it became clear that when freezing tolerance of cells was enhanced in autumn, decreased in spring or artificially enhanced by hardening, first, invagination of vesicles and fibrous materials which were detached from the plasmamembrane occurred, next, numerous microvesicles were generated and fused with the plasmamembrane.

FIGURE 8. The ultrastructure of a cell from twigs collected on March 10. The plasmamembrane again showed a highly folded structure. The arrows show inclusion materials (vesicles and fibrous materials). x 15,300.

FIGURE 9. The ultrastructure of a cell from twigs collected on March 30. Microvesicles and the Golgi apparatus near the plasmamembrane were observed. x 15,300.

In the formation of the pollen wall and in the elongation of the pollen tube, microvesicles from the Golgi apparatus and/or the endoplasmic reticulum have been reported to fuse with the plasmamembrane (3). Similar ultrastructural changes have been reported in the formation of the cell plate during cell division in higher plants (7). Consequently, the microvesicles observed in the peripheral cytoplasm probably are involved in the turnover of materials in the plasmamembrane.

However, the relation of those ultrastructural changes to the substance of membrane renewal remains obscure. In the black locust tree, an increase in phospholipids was positively related to increased freezing tolerance (5,8). In poplar twigs, an increase in freezing tolerance always was accompanied by an increase in phospholipid, especially phosphatidyl choline and phosphatidyl ethanol amine (8,10,11), and the phospholipid contents were inversely related to the environmental temperature, showing a remarkable increase at 0°C. However, the triglyceride content decreased sharply at 0°C. The inverse responses of the phospholipid and triglyceride contents to low temperature suggests that there may be interconversion between them, and that the transition of membranes from a phospholipid-depleted to a phospholipid-enriched state is associated with an increase in freezing tolerance.

The incorporation of ^{3}H-labeled acetic acid into cell organelles examined by electronmicroscope-autoradiography has shown that this ^{3}H-labeled substance was first detected in the vacuole, and followed Golgi apparatus and microvesicles during artificial hardening (data not shown). Unfortunately, it was not determined what products came from the labeled acetic acid. But it is proposed that labeled acetic acid incorporated into the vacuole was used as material when microvesicles were generated.

Consequently, it seems that those ultrastructure changes, engulfment of vesicles and fibrous materials, generation of microvesicles, and fusion of microvesicles with the plasmamembrane, were the process of membrane renewal associated with changes of freezing tolerance of cells in summer to winter and vice versa.

ACKNOWLEDGMENTS

The author would like to express his sincere thanks to Professor Dr. A. Sakai, The Institute of Low Temperature Science, Hokkaido University, for many valuable suggestions and discussions.

REFERENCES

1. Mollenhauer, H.H., Whaley, W.G., and Leech, J.H., J. Ultrastruc. Res. 5:193-200 (1961).
2. Moore, J.D., Mollenhauer, H.H., and Bracker, C.B., in "Origin and Continuity of Cell Organelles 2" (J. Reinert and H. Ursprug, eds.), pp. 82-126. Springer-Verlag, Berlin, 1971.
3. Nakamura, S., J. Electron Microsc. 28:275-284 (1979).
4. Niki, T., and Sakai, A., Plant & Cell Physiol. 22:171-183 (1981).
5. Pomeroy, M.K., Siminovitch, D., and Wighman, F., Can. J. Bot. 48:953-967 (1970).
6. Pomeroy, M.K., and Siminovitch, D., Can. J. Bot. 49:787-795 (1971).
7. Sakai, A., Plant Physiol. 40:882-887 (1965).
8. Siminovitch, D., Rheme, B., Pomeroy, K., and Lapage, M., Cryobiology 5:202-225 (1968).
9. Yoshida, S., Contr. Inst. Low Temp. Sci. B 18:1-43 (1974).
10. Yoshida, S., and Sakai, A., Plant & Cell Physiol. 14:353-359 (1973).
11. Yoshida, S., and Sakai, A., Plant Physiol. 53:509-511 (1974).

Part III
Freezing Stress

EXTRAORGAN FREEZING OF PRIMORDIAL SHOOTS
OF WINTER BUDS OF CONIFER

A. Sakai

The Institute of Low Temperature Science
Hokkaido University
Sapporo, Japan

Plants survive subfreezing temperature by several means. Two of the most important are tolerating cell dehydration caused by extracellular freezing or by merely avoiding the freezing in certain tissues or organs by deep-supercooling. The leaves, twigs and bud tissues except for the primordial shoots of very hardy conifers survived freezing as low as -70°C or below, and even the temperature of liquid nitrogen by extracellular freezing (12,13). It is well known that supercooling only confers hardiness to a limited extent and never below the homogeneous nucleation temperature of the tissue solution, about -41°C (3). In excised winter buds of very hardy conifers which belong to Abietoideae of Pinaceae, when cooled very slowly the low temperature exotherm shifted markedly to a lower value and the exotherm became much smaller (11). Also, masses of needle ice were observed mainly beneath the crown of the primordial shoot, increasing with decreasing temperature up to around -30°C (1,11). From these results, it may be considered that most of the water in the primordial shoots gradually migrates out through the crown and freezes as the temperature decreases, which enables primordial shoots to survive at very low temperatures below -50°C. I referred to this type of freezing as "extraorgan freezing" (11).

The degree of resistance to extraorgan freezing in winter coniferous buds is variable depending on species and reaches even the temperature of liquid nitrogen (11). To clarify further the survival mechanism of primordial shoots of winter buds of very hardy conifers, the minimum survival temperature of conifers belong to Abietoideae (2) which are native to different climates, the provenance difference of Picea glauca, and survival mechanism of primordial shoots of conifers other than Pinaceae were investigated.

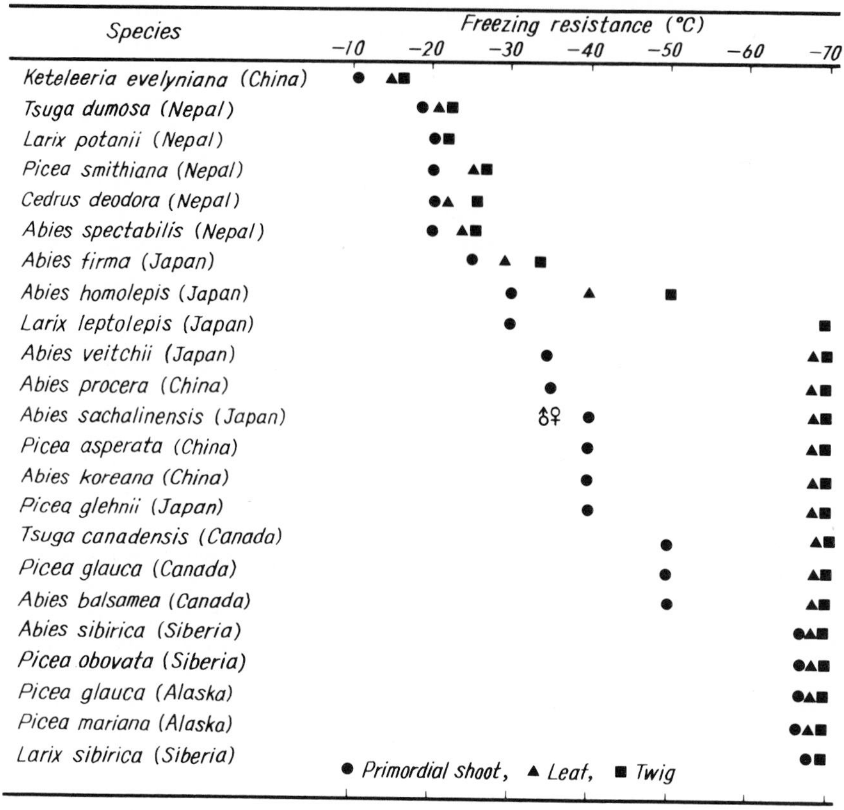

Species	Freezing resistance (°C)						
	−10	−20	−30	−40	−50	−60	−70
Keteleeria evelyniana (China)	● ▲■						
Tsuga dumosa (Nepal)		●▲■					
Larix potanii (Nepal)		●■					
Picea smithiana (Nepal)		● ▲■					
Cedrus deodora (Nepal)		●▲ ■					
Abies spectabilis (Nepal)		● ▲■					
Abies firma (Japan)			● ▲ ■				
Abies homolepis (Japan)			●	▲	■		
Larix leptolepis (Japan)			●				■
Abies veitchii (Japan)				●			▲■
Abies procera (China)				●			▲■
Abies sachalinensis (Japan)				♂♀ ●			▲■
Picea asperata (China)				●			▲■
Abies koreana (China)				●			▲■
Picea glehnii (Japan)				●			▲■
Tsuga canadensis (Canada)					●		▲■
Picea glauca (Canada)					●		▲■
Abies balsamea (Canada)					●		▲■
Abies sibirica (Siberia)							●▲■
Picea obovata (Siberia)							●▲■
Picea glauca (Alaska)							●▲■
Picea mariana (Alaska)							●▲■
Larix sibirica (Siberia)							●■

● *Primordial shoot,* ▲ *Leaf,* ■ *Twig*

FIGURE 1.   Winter hardines of leaf, bud and twig of conifers belonging to Abietoideae.   Winter twigs were cooled in 5°C increments at daily intervals to successively cold temperatures to −70°C. ♂♀: Flower buds.

I.   MATERIALS AND METHODS

One-year twigs of 10 to 15-year-old conifers were mainly collected in the nursery of our Institute (43°03'N), the Tokyo University Forest in Hokkaido (Yamabe about 43°30'N), and the Kyoto University Forest (Kamigamo, Kyoto, 35°01'N) in midwinter.   Spruces wintering in Alaska were sent from Fairbanks by air.   Himalayan fir and hemlock were personally collected twice by author in winter.   These twigs were stored at about −5°C in polyethylene bags containing snow to prevent desiccation for two weeks.   Five twig pieces, 10 cm long, were cut from each twig sample, enclosed in polyethylene bags and cooled in 5°C increments at either 2-hr or daily intervals to successively cold temperatures.   Twigs were remwarmed in air at 0°C.

After thawing, the twig pieces in polyethylene bags were placed at room temperature for at least one month to evaluate their viability. Browning was the criterion for evaluating injury. Freezing tolerance was expressed as the lowest temperature at which little or no injury was sustained.

DTA analysis of excised buds was performed as previously reported (10,11). Exothermic responses were detected with 0.2 mm cu-constantan thermocouples and amplified 40 or 100 times. Ice segregation outside the primordial shoot in frozen buds was observed under a dissecting microscope in a cold room at -10°C. Unfrozen primordial shoots were excised from buds frozen at different temperatures in cold rooms, and the water content was then determined.

II.   RESULTS AND DISCUSSION

Winter twigs of confiers which belong to Abietoideae of Pinaceae were frozen in 5°C increments at daily intervals to -30°C, and then cooled in 10°C steps at daily intervals to successively cold temperatures down to -70°C. The freezing resistance of very hardy confiers of Abietoideae is characterized by the relative susceptibility of the primordial shoots of buds to freezing (Fig. 1). The leaves, twigs and bud tissues other than the primordial shoots tolerated extracellular freezing down to -70°C or below, while the primordial shoots preferentially sustained injury between -30 and -50°C. In the 5 hardiest species the primordial shoot also survived -70°C. As shown in Fig. 1, female and male buds of Abies sachalinensis were slightly less hardy than the vegetative buds in midwinter. In the less hardy conifers, which are marginally hardy to about -20°C, the difference in hardiness between the primordial shoot and the leaf or twig disappears or diminishes. The same trend was observed in other conifers except for the genus Pinus (12). It thus seems difficult to evolve very hardy primordial shoots for conifers. And the minimal survival temperature of primordial shoots of winter buds is among the important factors setting the northern boundaries of the natural ranges of many hardy conifers.

Seasonal hardiness changes in leaf, bud, cortical tissue and xylem of twigs were assessed with Abies sachalinensis native to Hokkaido. By early September, the primordial shoot of bud had developed and in late September the water content was about 72%. The leaf, bud scales and bud axial tissues, and twig tissues were all marginally hardy to -5°C, but the primordial shoots supercooled to -15°C. The same trend was observed in early May immediately before bud opening. In mid-October the hardiness of leaf, primordial shoot and twig tissues was similar. However, during winter the leaf, twig, bud

TABLE I. Freezing resistance of primordial shoots of winter buds of firs cooled at different rates.

Species	Survival temperature of primordial shoots (°C)			Native habitat
	5°C/2h	5°C/day	5°C/day to -20°C, -20°C for 20 days, 10°C/day to -60°C	
Abies spectabilis	-20	-25	—	Temperate (Nepal 3,900m)
A. firma	-23	-30	—	Temperate (Japan)
A. homolepis	-30	-35	—	Cool temperate (Japan)
A. sachalinensis	-30	-40(-45)[a]	-45[a]	Sub-cold (Yamabe, Hokkaido)
A. balsamea	-35	-50	-60,-30LN[b]	Sub-cold (New Brunswick, Canada 47°42'N, 68°14'W, 348m)

[a] Lateral buds along the twig
[b] Immersion in liquid nitrogen from -30°C.
The viability of primordial shoots was determined one month or more after thawing.

TABLE II. Effect of cooling rate upon low exotherm temperature of primordial shoots in excised buds cooled at different rates.

Species	Cooling rate		
	0.11°C/min	0.05°C/min	5°C/day (-5 to -20°C), -0.05°C/min (-20 to -45°C)
Abies homolepis	-24.5 ± 0.5	-29.4 ± 0.2	-33.8 ± 0.6
Abies sachalinensis	-25.4 ± 0.9	-34.4 ± 0.4	no exotherm (killed)

scales and bud axial tissues became much hardier than the primordial shoot.

As shown in Table 1, the freezing resistance of primordial shoots of winter buds of firs differed greatly depending upon the cooling rate. The primordial shoots of Abies balsamea frozen at -20°C for 20 days and then cooled in 10°C increments at daily intervals down to -70°C survived freezing to -60°C at least or immersion in liquid nitrogen after prefreezing to -30°C. In A. sachalinensis lateral buds were much hardier than terminal buds when cooled very slowly.

Differential thermal analysis (DTA) of excised whole buds revealed that the low temperature exotherms derived from freezing of primordial shoots shifted markedly to low values (Table 2). In A. sachalinensis which was cooled to -55°C at cooling rate of 0.05°C/min, no low temperature exotherm was detected on DTA profiles, though these primordial shoots preferentially sustained injury between -35 and -55°C, probably due to freeze-dehydration during slow cooling.

As reported previously (11), in the buds of Abietoideae which were frozen at -10°C for one day, the buds scales and twig axes had frozen, while the primordial shoot remained unfrozen and much needle ice was observed beneath the crown to which the shoots are attached, and outside surrounding the basal area of the primordial shoot (Fig. 2). Ice segregation outside the primordial shoot of coniferous buds increased with decreasing temperatures to about -30°C or below (Table 3). These results indicate that water in the primordial shoot migrates outside mainly through the crown due to a vapour pressure gradient between the supercooled water in the primordial shoots and the ice crystals outside, as observed in cells frozen extracellularly.

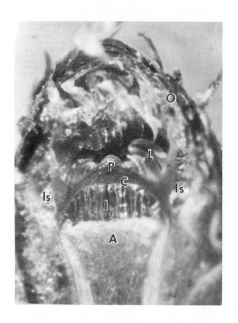

FIGURE 2. A longitudinal section of the winter bud of <u>Larix sibirica</u> held at -20°C for one day. The primordial shoot remained unfrozen. the needle ice labeled (I) is underneath the crown and the ice inside the scales surrounding the basal part of the primordial shoot is labeled (Is). A: Head of axial core pith, C: Crown, P: Primordial shoot, L: Primordial leaves, O: Scale. X35.

TABLE III. Water content of primordial shoot in winter buds frozen at different temperatures.

Freezing temperature (°C)	Water content of primordial shoots	
	Abies homolepis	Abies firma
Unfrozen	57.7	58.8
-10	44.8	45.3
-20	40.1	43.4
-30	37.9	–

Unfrozen primordial shoots were excised from buds frozen at indicated temperatures at cold room.

Excised whole buds enclosed in polyethylene bags were cooled in 5°C increments at daily intervals.

Primordial shots of <u>Abies homolepis</u> and <u>A. firma</u> were killed below -35° and -25°C, respectively.

When winter buds of very hardy spruces were cooled in 5°C increments in 2 day intervals to -40°C, and then at 10°C increments at daily intervals to -70°C, two species native to Hokkaido survived freezing only to -40°C, while <u>Picea glauca</u>

TABLE IV. Freezing resistance of primordial shoots of winter buds of very hardy spruces.

Species	Origin of seed	Collection site	Freezing resistance (°C)		
			Primordial shoot	Leaf	Twig
Picea glenii	Hokkaido	Yamabe	-40	-90[b]	-90[b]
Picea jezoensis	Hokkaido	Yamabe	-40	-90[b]	-90[b]
Picea glauca	Ontario, Canada (44°N, 200m)	Yamabe	-50	-90[b]	-90[b]
	Canadian Rocky, Summit Lake (54°17'N, 122°40'W, 700m)	Yamabe	-60,-40LN	-90[b]	-90[b]
	Bonanza, Inland Alaska (64°51'N,147°44'W,230m)	Yamabe	-70,-30LN	-90[b]	-90[b]
	Fairbanks, Inland Alaska (64°51'N,147°51'W,180m)	Fairbanks[a]	-70,-30LN	-90[b]	-90[b]
Picea mariana	Fairbanks	Yamabe	-70,-30LN	-90[b]	-90[b]
	Fairbanks	Fairbanks[a]	-70,-30LN	-90[b]	-90[b]

a Winter twigs were sent by air from University of Alaska.
b Uninjured at the lowest test temperatures indicated.

Twigs were cooled in 5°C increments at two day intervals. Spruces tested in this experiment were all grown at the same conditions at Tokyo University Forest, Yamabe, Hokkaido.

-30LN: Immersion in liquid nitrogen after prefreezing to -30°C.
The viability of primordial shoots was determined one month and half after thawing.

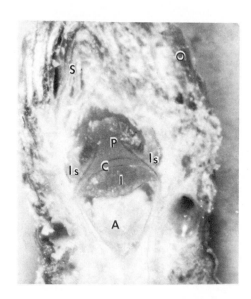

FIGURE 3. A longitudinal section of the winter bud of Picea glauca from Alaska which was held at -20°C for one day. I: Ice segregation beneath the crown, IS: Ice surrounding the basal part of the primordial shoot, C: Crown, A: Head of axial core pith, S,O: Scale. x15.

and Picea mariana which were obtained from Alaska remained alive after freezing as low as -70°C, and even immersion in liquid nitrogen after prefreezing to -30°C (Table 4). Primordial shoots of Picea glauca from different prcvenances which were planted 12 years ago in a uniform environment plot in Tokyo University's Forest in Hokkaido showed freezing tolerance increasing with winter cold of their natural geographic origin (Table 4). As reported previously (11), ice segregation outside primordial shoots of Alaska spruces was also observed (Fig. 3). Thus it appears that the primordial shoots of winter buds of extremely hardy conifers tolerated an intensive freeze-dehydration due to ice segregation. In the genera which belong to Abietoideae and Laricoideae a big pith cavity located between the crown of the primordial shoot and the head of bud axial pith is formed as a result of repeated ice segregation (10).

Some reports (4,6) suggested that the function of flower bud scales was to accomodate ice which was formed from water derived from the flower primordium. This is not the case in primordial shoots of conifers which belong to Abietoideae. However, in many genera belonging to Taxodiaceae, Taxaceae and Cephalotaxceae, which do not have a crown in these primordial shoots, ice segregation outside the primordial shoots was also observed, and in these conifers bud scales acted as an ice sink.

In winter buds of Taxus, Torreya, Cepharotaxus, Cryptomeria, low temperature exotherms due to freezing of primordial shoots were detected on DTA profiles. However, in winter buds of Metasequoia glyptostroboides (Fig. 4), no low temperature exotherm appeared on the DTA profile even when cooled

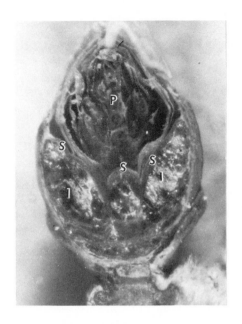

FIGURE 4. A longitudi-
nal section of the win-
ter bud of Metasequoia
glyptostroboides held at
-15°C for one day. I:
Ice, P: Primordial
shoots, S: Scale. In
this bud scales act as
an ice sink. x30.

continuously to -40°C at 0.05°C/min, which is probably due to
rapid water flow to scales from very small primordial shoots,
though buds collected in mid March showed low temperature exotherms around -17°C. Winter bud of Metasequoia cooled below
about -30°C always sustained injury to the primordial shoots.
In the genus Pinus, which does not have a crown associated with
the primordial shoots, low temperature exotherms below -15°C
were not detected. Also, it was difficult to observe ice
segregation outside the primordial shoot, unlike other genera
of Pinaceae, Taxodiaceae and Tacaceae. It thus appears that
primordial shoots of Pinus freeze extracellularly.

Freezing avoidance by supercooling, as observed in xylem
ray parenchyma (3,8) and most of the hydrated seeds (5,14), has
a lower limit around -40°C. Little or no change in minimum
survival temperature was observed by different cooling rates in
xylem (3,8). However, in the freezing avoidance mechanism
where water content decreases substantially during slow
freezing, as observed in primordial shoots of extremely hardy
conifers, there seems to be no low temperature limit, provided
that primordial shoots are cooled very slowly and that they can
resist intensive freeze-dehydration.

Extracellular freezing in hardy cells does not permit ice
penetration into protoplasm when cooled slowly and thus freeze-
dehydration proceeds. The temperature range of freeze-induced
injury varies depending on its ability to withstand freeze-
dehydration of the cell concerned. Neither does extraorgan
freezing permit the penetration of ice into the primordial
shoots, which always results in death. The high ability of

FIGURE 5.   Survival mechanisms of plants to subfreezing temperatures.

Freezing avoidance:
1. Deep-supercooling
2. Freezing point depression
3. Desiccation

Freezing tolerance:
1. Extracellular freezing
2. Extratissue freezing
3. Extraorgan freezing

primordial shoot itself to be supercooled to -12 and -15°C facilitates the ice segregation of primordial shoots. Therefore, the extraorgan freezing does not produce intraorgan freezing in very hardy coniferous buds when cooled extremely slowly. This enables primordial shoots to survive freezing below about -40°C, which is the homogenous nucleation temperature of tissue solution (4), provided that the primordial shoots can withstand intensive freeze-induced dehydration. However, when winter buds of very hardy conifers are cooled continuously at 0.05°C/min or more, low-temperature exotherms derived from intraorgan freezing or primordial shoots appeared on DTA profile during extraorgan freezing.

It thus appears that the rates of cooling and exosmosis of water in the primordial shoots play an important role in determining whether death of primordial shoot is caused by freeze-dehydration or intraorgan freezing.

Ice segregation from primordial shoots through the crown appears to be a phenomenon similar to needle ice near the soil surface (7). Extraorgan freezing is characterized by slow water flow outside the primordial shoots in comparison with very hardy cells frozen extracellularly, probably because the primordial shoot is much larger. In general primordial shoots of very hardy conifers are much smaller than less hardy ones, which permits rapid ice segregation from the primordial shoot due to it's larger specific surface. Especially in Alaskan spruces ice segregation proceeds more rapidly than in other conifers, which is probably due to the very small size of the primordial shoot and/or the morphological properties of the crown (11).

Extraorgan freezing in which substantial decrease in water content occurs during freezing as observed in the primordial shoots of coniferous buds, the flower primordia and some hydrated seeds should be included in freezing tolerance mechanism. Very hardy cells survive extracellular freezing below -70°C, but cannot withstand intracellular freezing even at -10°C. The same is true in the primordial shoots of very hardy conifers. Thus, I presented a new conception "extraorgan freezing" as a freezing tolerance mechanism (10). Survival mechanisms of plants to subfreezing temperatures are indicated in Fig. 5.

REFERENCES

1. Dereuddre, J.   Physiol. Veg. 16:469-489 (1978).
2. Florin, R., Acta Horti. Bergiani. 20(4):121-312 (1960).
3. George, M.F., Burke, M.J., Pellet, H.M., and Johnson, A.G., HortSci. 9:519-522 (1974).
4. George, M.F., Burke, M.J., and Weiser, C.J., Plant Physiol. 29:29-35 (1974).
5. Ishikawa, M., and Sakai, A., Low Temp. Sci. B 36:39-49 (1978).
6. Ishikawa, M., and Sakai, A., Plant and Cell Physiol. 22:953-967 (1981).
7. Outcalt, S.I., in Permafrost Second International Conference, North America Contribution. pp. 229-234 (1970).
8. Quammer, H.A., Weiser, C.J., and Stushnoff, C., Plant Physiol. 53:273-277 (1973.
9. Sakai, A., Plant & Cell Physiol. 19:1439-1446 (1978).
10. Sakai, A., Low Temp. Sci. B37:1-9 (1979).
11. Sakai, A., Plant & Cell Physiol. 20:1381-1390 (1979).
12. Sakai, A., and Okada, S., Silvae Genetica 20:91-97 (1971).
13. Sakai, A., and Weiser, C.J., Ecology 54:118-126 (1973).
14. Stushnoff, C., and Juntila, C., in "Plant Cold Hardiness and Freezing Stress" (P.H. Li and A. Sakai, eds.), pp. 241-247. Academic Press, New York, 1978.

LIQUID WATER DURING SLOW FREEZING BASED ON CELL WATER
RELATIONS AND LIMITED EXPERIMENTAL TESTING

C. Rajashekar

Laboratory of Plant Hardiness
Department of Horticultural Science
University of Minnesota
St. Paul, Minnesota, U.S.A

M. J. Burke

Institute of Food and Agricultural Sciences
Fruit Crops Department
University of Florida
Gainesville, Florida, U.S.A.

I. INTRODUCTION

Olien (10,11) was among the first to make direct determination of frost desiccation during freezing of intact plant tissues and relate this to the winter hardiness of plants. In his
measurements, Olien used a sublimation cell in which ice was
condensed outside the plant tissue and the tissue hydration
level was monitored by weighing the tissue sample. From a
physico-chemical point of view, the sublimation cell is a sound
way of obtaining freezing curves. However, this method does
have a disadvantage in that the ice formation and growth do not
occur in the tissue. An alternative method for obtaining
freezing curves involves the use of pulse nuclear magnetic
resonance (nmr) spectroscopy. This method is described in
detail elsewhere (e.g., see 2). In the pulse nmr experiment,
one attempts to distinguish between liquid water and ice within
a partially frozen tissue. The nmr method has the advantage
that ice forms within the tissue and the disadvantage that the
nmr distinction between liquid water and ice is open to potential error. In any event, pulse nmr methods have been applied
in numerous studies on the freezing of plant tissue (e.g.,

3,5,6,9,12). In this chapter, we will discuss freezing curve data obtained from nmr experiments.

Equilibrium freezing curves obtained by plotting L_T (liquid water at temperature T) vs. T (°C) (temperature) are often found to be hyperbolas (3,5,6,9,12). Because of this observation, Gusta et al. (5) proposed and Stout (13) modified the proposal, with both saying that freezing curves can be plotted in a linear fashion and have the following form:

$$L_T = \frac{(L_o - K)\Delta T_m}{T(°C)} + K \tag{1}$$

L_o is the liquid water at 0°C. K and ΔT_m are constants which relate to the intercept and slope, respectively, when L_T is plotted vs. $1/T(°C)$. Gusta et al. (5) suggested that ΔT_m is the average of the freezing point lowerings of the tissue solutions (ΔT_m = 0.86 osmotic potential in MPascals) and K is the water which cannot be frozen to form crystalline ice. There is reason to believe these suggestions need refinement.

To date the authors know of two experiments designed to compare ΔT_m obtained from equation 1 with melting point lowerings obtained by psycrometry on extracted solutions (9,12). In both cases, ΔT_m obtained from equation 1 is 1.5 times larger than the direct measure of ΔT_m on extracted solutions.

The objective here is to derive an equation for the freezing of plant tissues using the simple considerations of plant water relations and discuss in relation to supercooling and non-supercooling plant tissues.

II. THEORY

In these derivations, we assume that a plant tissue is composed of n independent freezing units which for convenience will be taken as (n-1) cells plus a single extracellular space. Then, if ice in the extracellular space is in equilibrium with the cells, the following equation holds

$$\psi_T(ice) = \psi_{T,i}(cell) = \pi_{T,i} + \tau_{T,i} + P_{T,i} \tag{2}$$

where $\psi_T(ice)$ is the total water potential of ice at temperature T and $\psi_{T,i}(cell)$ is the cell water potential at T for the ith cell. $\pi_{T,i}$ $\tau_{T,i}$ and $P_{T,i}$ are the osmotic potential, matric potential and pressure potential, respectively, which are components of the total cell water potential (8). In using this equation, we are following the norm for describing the water potential of plant cells. In plant tissues it is generally found convenient to describe the water potential of the cell using $\pi_{T,i}$ to cover effects of solutes, $P_{T,i}$ to cover effects of hydrostatic pressure (Pressure potential + barometric pressure), and $\tau_{T,i}$ to cover matric and other effects. We loose no rigor by partitioning the total water potential in this way even though this partitioning is not unique.

The amount of osmotically active solutes in moles within the ith cell ($M_{T,i}$) can be approximated in terms of $\pi_{T,i}$ as follows:

$$M_{T,i} = \frac{\overline{V}}{RT} (\ell_{T,i} \ \pi_{T,i}) \tag{3}$$

where $\overline{V}$ is the molar volume of water, R is the gas constant and $\ell_{T,i}$ is the moles of osmotically active liquid water within the ith cell at T. If osmotically active solutes are not allowed to exchange between cells or the extracellulr space, then as freeze dehydration occurs the following equation can be written for the ith cell:

$$\frac{\ell_{T,i} \ \pi_{T,i}}{T} = \frac{\ell_{o,i} \ \pi_{o,i}}{T_o} \tag{4}$$

T_o is 273°K. This equation states that as osmotically active cell water changes during freezing between 273°K and T ($\ell_{o,i}$ to $\ell_{T,i}$) then cell osmotic potential will change in a corresponding way ($\pi_{o,i}$ to $\pi_{T,i}$). Note that L_T can be written as a summation of liquid water in all cells and the extracellular space

$$L_T = \sum_{i=1}^{n} \ell_{T,i} \tag{5}$$

and also note that equation 4 for each cell and extracellular space can be added and rearranged to give

$$\frac{\frac{L_o}{T_o} \sum_{i=1}^{n} \ell_{o,i} \ \pi_{o,i}}{\sum_{i=1}^{n} \ell_{o,i}} = \frac{1}{T} \sum_{i=1}^{n} \ell_{T,i} \ \pi_{T,i} \tag{6}$$

Substituting $\psi_T(ice) \ -\tau_{T,i} \ -P_{T,i}$ for $\pi_{T,i}$ (equation 2) and rearranging gives

$$\psi_T(ice) = \frac{T \ L_o}{T_o L_T} \frac{\sum_{i=1}^{n} \ell_{o,i} \pi_{o,i}}{\sum_{i=1}^{n} \ell_{o,i}} + \frac{\sum_{i=1}^{n} \ell_{T,i} \tau_{T,i}}{\sum_{i=1}^{n} \ell_{T,i}} + \frac{\sum_{i=1}^{n} \ell_{T,i} P_{T,i}}{\sum_{i=1}^{n} \ell_{T,i}} \tag{7}$$

To simplify this, one can define tissue averages for osmotic potential $\bar{\pi}_o$, matric potential, $\bar{\tau}_T$, and pressure potential, $\bar{P}_T$ as follows

$$\bar{\pi}_o = \frac{\sum\limits_{i=1}^{n} \ell_{o,i}\, \pi_{o,i}}{\sum\limits_{i=1}^{n} \ell_{o,i}} \tag{8}$$

$$\bar{\tau}_T = \frac{\sum\limits_{i=1}^{n} \ell_{T,i}\, \tau_{T,i}}{\sum\limits_{i=1}^{n} \ell_{T,i}}, \qquad \text{and} \tag{9}$$

$$\bar{P}_T = \frac{\sum\limits_{i=1}^{n} \ell_{T,i}\, P_{T,i}}{\sum\limits_{i=1}^{n} \ell_{T,i}} \tag{10}$$

which on substitution gives

$$\psi_T(ice) = \frac{T\; L_o}{T_o\; L_T}\, \bar{\pi}_o + \bar{\tau}_T + \bar{P}_T \tag{11}$$

This is our base equation for relating ψ_T(ice) to L_T. It includes the assumptions that the plant tissue is composed of cells which come to equilibrium with extracellular ice and do not exchange osmotically active solutes.

We would like to modify equation 11 for inclusion of a special component of water, component k. Our reasons for having component k will become apparent on viewing equation 1. We choose component k with the following properties:

a. it is free of osmoticum and/or osmotically inactive such that $\pi_{T,k} = \pi_{o,k} = 0$.

b. it is under no pressure potential with $P_{T,k} = 0$.

c. it comes to equilibrium with ice $[\tau_{T,k} = \psi_T(ice)]$ and its quantity, $\ell_{T,k}$, remains constant i.e., $\ell_{T,k} = K$ for all subfreezing temperatures.

Incorporating component k into equation 7 yields a special case of that equation

$$\psi_T(ice) = T\,\frac{(L_o - K)}{T_o\,(L_T - K)}\, \bar{\pi}_o' + \bar{\tau}_T' + \bar{P}_T' \tag{12}$$

where

$$\bar{\pi}_o' = \frac{\sum_{i=1}^{n} \ell_{o,i} \; \pi_{o,i}}{L_o - K}; \quad i \neq k \tag{13}$$

$$\bar{\tau}_T' = \frac{\sum_{i=1}^{n} \ell_{T,i} \; \tau_{T,i}}{L_T - K}; \quad i \neq k, \text{ and} \tag{14}$$

$$\bar{P}_T' = \frac{\sum_{i=1}^{n} \ell_{T,i} \; P_{T,i}}{L_T - K}; \quad i \neq k \tag{15}$$

If component k is extracellular water then $\bar{\pi}_o'$, $\bar{P}_T'$ and $\bar{\tau}_T'$ are weighted averages for the cells not counting the extracellular space (i.e., not including component k).

The last modification is to evaluate the temperature dependence of ψ_T(ice). The Clapeyron-Clausius equation can be written as follows

$$\frac{d \ln p}{dT} = \frac{\Delta H_e}{RT^2} \tag{16}$$

where p is the vapor pressure of water, and ΔH_e is the heat of evaporation. Since a similar equation can be written for ice and subtracting the equations for ice and supercooled water and integrating between R and 273°K yields

$$\ln \left[\frac{P_{ice}}{P_{liquid}} \right] = \frac{\Delta H_f}{R} \left[\frac{T - 273}{273T} \right] \tag{17}$$

We can write

$$\psi_T(ice) = \frac{RT}{\bar{V}} \ln \left[\frac{P_{ice}}{P_{liquid}} \right] \tag{18}$$

and substituting from equation 17 gives:

$$\psi_T(ice) = \frac{\Delta H_f}{273\bar{V}} \; T(°C) \tag{19}$$

$$= 1.22 \; T(°C)$$

ΔH_f is the heat of fusion of water and ψ_T(ice) is in units of MPascals (10 bars = MPascal). ψ_T(ice) can also be calculated directly from the vapor pressure of ice and supercooled liquid water between 0 and -15°C (7) using:

$$\psi_T(ice) = (1.16 \pm 0.04) \; T(°C) \tag{20}$$

This latter relation will be used. Substituting back into equation 12 and rearranging gives

$$L_T = \frac{T}{T_o} \frac{0.86(L_o - K)\bar{\pi}_o^{'}}{T(°C) - 0.86(\bar{\tau}_T^{'} + \bar{P}_T^{'})} + K \qquad (21)$$

All variables in this equation can be estimated experimently with the exceptions of $\bar{\tau}_T$ and $\bar{P}_T$.

III. APPLICATION OF THEORY TO EXPERIMENT

Freezing curves so far reported fall between two extreme cases shown in Figure 1. One of these cases, Karkhov winter wheat (Triticum aestivum L.) is well described by equation 1 giving a good fit to a line with coefficient of determination, R^2, equal to 0.98. The second case for freezing of hickory twigs clearly is not described by the single line as in equation 1. The low value of the coefficient of determination is consistent with this observation ($R^2=0.32$). The hickory sample is known to have a component of deep supercooled water (4). Therefore a poorness of fit for supercooled samples is generally observed.

A. Freezing Curves which Follow Equation 1

We have derived a relation (equation 21) and observed a similar but different relation (equation 1). The two must be reconciled. As pointed out earlier, all the terms in equation 21 can be obtained experimentally with the exceptions of $\bar{\tau}_T$ and $\bar{P}_T$. $\bar{\pi}_o^{'}$ can be estimated by psycrometry, plasmolysis, etc. $T(°C)$, L_T, L_o and K can be obtained from the freezing curve or they can be replaced by substituting ΔT_m in equation 1 into equation 21 giving the following

$$\bar{\tau}_T^{'} + \bar{P}_T^{'} = 1.16 \left[1 - \frac{\bar{\pi}_o^{'} T}{1.16\Delta T_m T_o} \right] T(°C) \qquad (22)$$

Marcellos and Burke (9) and Palta et al. (12) have measured both $\bar{\pi}_o^{'}$ and ΔT_m. Using their data, the results in Table I are obtained. Note that $\bar{\tau}_T^{'}$ and $\bar{P}_T^{'}$ are negative and quite substantial at low freezing temperatures. Briggs (1) has shown that water can withstand large negative pressures. However there appears to be no evidence for negative turgor pressure in plants at least during water stress (14).

$\bar{\tau}_T^{'}$ and $\bar{P}_T^{'}$ act in such a way as to reduce ice formation. These results suggest that they account for 23 to 46% of the cell water potential at subfreezing temperature.

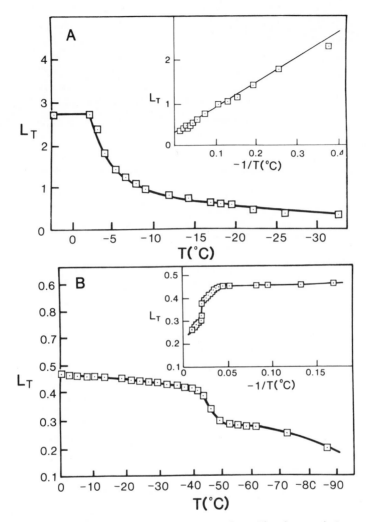

FIGURE 1.   The freezing curves for Kharkov winter wheat
(5), A; and shagbark hickory xylem (1), B. L_T is the liquid
water in grams per gram dry sample.   T(°C) is the temperature
in degrees Celsius.

B.   Freezing Curves Which Do Not Follow Equation 1

    Plant tissues containing a deep supercooled component are
included in this category.   Shagbark hickory (Carya ovata L.)
xylem is an example and it has been studied in some detail
(4).   A freezing curve for shagbark hickory stem is in
Figure 1.   We will restrict our attention to the supercooled
component only in a temperature range where such supercooled

TABLE I. Matric potential, $\bar{\tau}_T'$, and pressure potential $\bar{P}_T'$, calculated from ΔT_m's obtained from the slope of freezing curves and osmotic potentials, $\bar{\pi}_O'$. 1.16 ΔT_m is used here to change the units to MPascals.

Cultivar[1]/ Species	$\bar{\pi}_O'$ (MPascals)	1.16 ΔT_m (Mpascals)	$\bar{\tau}_T' + \bar{P}_T'$ (Mpascals)
Wheat			
Kite	-1.13	-1.69	0.39T (°C)
3488	-1.26	-1.83	0.36T (°C)
Cheyenne 1	-1.55	-2.03	0.27T (°C)
Cheyenne 2	-1.21	-2.21	0.53T (°C)
Onion	-1.07	-1.36	0.25T (°C)
Hickory	-2	---	1.16T (°C)

[1] Wheat leaf results had to be calculated from data of Marcellos and Burke (9); onion epidermis results are from Palta et al. (12); and Hickory xylem results are from George and Burke (4).

water is stable (between 0°C and -40°C). George and Burke (4) have shown that this supercooled component of hickory is in equilibrium with ice; therefore, its freezing behavior should be described by equation 21. Because the fraction does not freeze, $L_O - K = L_T - K$. Substituting this into equation 21 and solving for $\bar{\tau}_T'$ and $\bar{P}_T'$ gives

$$\bar{\tau}_T' + \bar{P}_T' = 1.6\ T(°C) - \frac{\bar{\pi}_O'\ T}{T_O} \tag{23}$$

and values for $\bar{\pi}_O'$ are estimated in Table I. At subfreezing temperatures, $\bar{\tau}_T'$ and $\bar{P}_T'$ must be large and negative. Consequently they account for almost all of the total cell water potential at subfreezing temperatures.

IV. DISCUSSION

A relation has been derived to describe freezing of plant tissues using cell water potential and temperature. This relation has been used to analyze experimental freezing curves and $\bar{\tau}_T'$ (matric potential) plus $\bar{P}_T'$ (pressure potential) are found to be significant parts of the cell water potential.

An interesting question arises. Of the two potentials ($\bar{\tau}_T^{\,\prime}$ and $\bar{P}_T^{\,\prime}$) which is the most important and how does it change with subfreezing temperature. We speculate that $\bar{P}_T^{\,\prime}$ is the dominant term in the following cases:

a. We feel significantly negative $\bar{P}_T^{\,\prime}$ is sufficient to prevent frost dehydration of deep supercooled water in shagbark hickory xylem and supercooled water in other hardwood plants where ice at ambient pressure is in equilibrium with the supercooled component.

b. A large negative $\bar{P}_T^{\,\prime}$ may develop in samples freezing according to equation 1 at small subfreezing temperature (0 to -5°C).

We further speculate that the $\bar{\tau}_T^{\,\prime}$ only becomes dominant at large subfreezing temperatures (below -5°C) where cells freezing according to equation 1 are highly dehydrated. We do not feel $\bar{\tau}_T^{\,\prime}$ to be important for the supercooled water in xylem.

In summary, cell dehydration and collapse occurs during freezing. This collapse includes folding and distortion of the cell wall which must resist this process. To overcome the resistance and flex of the cell wall, a negative hydrostatic pressure will develop in the cell solution and this, we propose, leads to the large negative pressure potential, $\bar{P}_T^{\,\prime}$. The magnitude of the $\bar{P}_T^{\,\prime}$ should depend on the magnitude of the cell wall resistance to folding. For the case of shagbark hickory xylem, the resistance to folding and flexing of the xylem ray parenchyma cell wall structure is great. In the case of leaf cells of the other plants, the cell walls are softer and the resistance to flexing and folding is less.

REFERENCES

1. Briggs, L.J., J. App. Phys. 21:721 (1950).
2. Burke, M.J., George, M.F., and Bryant, R.G., in "Water Relation of Foods" (R.B. Duckworth, ed.). Academic Press, New York (1975).
3. Chen, P.M., Burke, M.J., and Li, P.H., Bot. Gaz. 137:313 (1976).
4. Gorge, M.F., and Burke, M.J., Plant Physiol. 59:319 (1977).
5. Gusta, L.V., Burke, M.J., and Kapoor, A.C., Plant Physiol. 56:707 (1975).
6. Harrison, L.C., Weiser, C.J., and Burke, M.J., Plant Physiol. 62:899 (1978).
7. International Critical Tables of Numerical Data, Physics, Chemistry and Technology, Vol. III. McGraw-Hill, 1928.
8. Kramer, P.J., Knipling, E.B., and Miller, L.N., Science 153:889.
9. Marcellos, H., and Burke, M.J., Aust. J. Plant Physiol. 6:513 (1979).

10. Olien, C.R., Cryobiol. 8:244 (1971).
11. Olien, C.R., U.S. Dept. of Agric. Tech. Bull. No. 1558 (1977).
12. Palta, J.P., Levitt, J., Stadelmann, E.J., and Burke, M.J., Physiol. Plant 41:273 (1977).
13. Stout, D.G., J. Theor. Biol. 88:513 (1981).
14. Tyree, M.J., Can. J. Bot. 54:2738 (1976).

CELL MEMBRANE ALTERATIONS FOLLOWING A SLOW FREEZE-THAW CYCLE: ION LEAKAGE, INJURY AND RECOVERY

J. P. Palta and K. G. Jensen

Department of Botany
University of Iowa
Iowa City, Iowa

P. H. Li

Laboratory of Plant Hardiness
Department of Horticultural Science and Landscape Architecture
University of Minnesota
St. Paul, Minnesota

I.   INTRODUCTION

There is ample evidence in the literature that cell membranes are one of the major targets of freezing injury. In 1914 Osterhout (9) recognized the leakage of ions following freezing injury. Based on this observation Dexter et al. (2) developed an electrical conductivity method that measures the amount of ion efflux from the tissues immediately after thawing. For almost 50 years conductivity method has been used successfully to assess relative freezing injury. It has been thought to measure the percent of cells killed that empty out the vacuolar contents (7). The reason behind is the assumption that freezing injury results in breakdown of membrane semi-permeability or membrane rupture (6,21,22). Recent work of Palta et al. (11,12,13) and Palta and Li (14) have demonstrated that these assumptions are incorrect. They showed that ion efflux following freezing injury was due to specific alterations in the membrane transport properties. Furthermore these alterations were found to be reversible or irreversible depending upon the degree of injury (12,13). This work showed evidence for a repair of the freezing injury and this repair was found to be an active process. We have here attempted to

PLANT COLD HARDINESS AND FREEZING STRESS

present these findings with additional data of direct ultra-structural observations of freeze injured cells.

We recognize, as has been by many earlier workers, that injury varies with freezing and thawing rates by the duration and the degree of freezing stress and the condition of cold acclimation (7). In our work presented here we use relatively slow (approximately 1°C/hr) freezing and thawing rates. Under these conditions ice forms extracellularly (upon nucleation at around -1°C) and results in the collapse of the cell walls (6,7). The primary objective was to study the nature of membrane alterations that are produced by incipient freezing injury.

II.   FREEZING INJURY

A.   Ion Efflux, Tissue Infiltration, and Cell Viability Following Freezing Injury

First visual signs of a freezing injury is the infiltration of the intercellular space with water on thawing. This gives the tissue a soaked appearance. Depending upon the degree of injury a loss of turgor can result in a flaccid tissue (6,13). These observations are usually taken as an indication of membrane rupture. It was implied that because of membrane rupture, cell empties out its vacuolar content and hence the ion leakage. Furthermore it was assumed that because of membrane rupture cells were unable to absorb extracellular water. Experiments with onion bulbs have shown that these assumptions are invalid (11,12,13). Some of these results are shown in Tables 1 and 2.

Onion bulbs were slowly frozen to -11 and -4°C and kept frozen for 6 days. Upon thawing bulb scale tissues were found to be completely or about 50% infiltrated (soaked appearance) in onions frozen to -11 and -4°C, respectively (Table 1). About 2.5 cm^2 pieces of scale were removed and shaken for 1 hr in 25 ml deionized water. Electrical conductivity of this water (effusate) was found to be more than two times compared to controls (Table 1). Inspite of a complete infiltration of the tissues and a two times increase in ion efflux all the cells were found to be alive (able to plasmolyze in a hypertonic solution and exhibited protoplasmic streaming). These results demonstrate that the ion efflux and the infiltration of the tissues following freezing injury is not due to the death of the cell. Since the cells could be plasmolyzed and the membrane semipermeability was still intact in spite of freezing injury.

Recently we have repeated some of these experiments using leaves of various potato species and onion bulb tissue. Besides measuring ion leakage, samples were also observed under

Table I. Effect of freezing stress on ion efflux, infiltration of the tissue, cell viability and the permeability of cell membranes to water. Onion bulbs were transferred directly to freezing chambers maintained at -4 $\pm$ 0.5°C and -11 $\pm$ 0.5°C and kept there for 6 days. Observations were made immediately after thawing. (Source: Ref. 11).

Freezing temp (°C)	Infiltration[a] of scale tissue (%)	Conductivity of effusate (μS/g, %)	Cell viability	Water permeability constant[b] (μm. sec^{-1})
-11	100	152.0	All alive	2.06 $\pm$ 0.26
-4	50	83.2	All alive	1.97 $\pm$ 0.38
+3 (control)	0	66.7	All alive	1.76 $\pm$ 0.14

[a] Estimated visually.
[b] Measured with radiotracer technique using inner epidermal cell layer. For details on method see Palta and Stadelmann (15,16).

Table II. Changes in extent of infiltration and cell survival with time after thawing of previously frozen onion bulb scales. Onions were kept frozen at respective temperatures for 12 days (Source: Ref. 12).

Freezing temp (°C)	Infiltration of the scale tissue (%)[a]			Cell survival (%)[b]		
	Days after thawing			Days after thawing		
	0	7	12	0	7	12
-11	90	100	100	100	40	20
-4	50	0	0	100	100	100
+3 (control)	0	0	0	100	100	100

[a] Visually estimated.
[b] Values are within $\pm$ 5%.

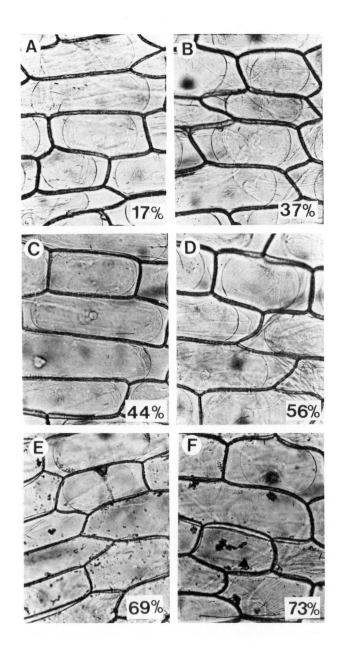

light and electron microscope. In these studies 2 cm^2 onion scale pieces and excised potato leaflets were frozen at the cooling rate of 1°C/hr. Samples were nucleated at -1°C using ice crystals. The details of the procedure were similar to the earlier studies (10,13).

Ion leakage in freeze injured onion tissue was found to be 37, 44, 56, 69 and 73% for the samples frozen to -4, -13, -15, -19 and -21°C respectively (Fig. 1). The control had ion efflux of 17%. Inspite of large ion efflux all epidermal cells frozen down to -15°C could be plasmolyzed just like unfrozen control (Fig. 1). Only less than 10% of the epidermal cells, however showed plasmolysis in -19 and -21°C samples. Samples from the same scale tissues were taken for electron microscopic observations. Results are shown in Figure 2. Compared to control, no ultrastructural changes could be detected in cells frozen to -4, -10 and -13°C (Fig. 2, A-D). However, in many cells frozen to -15°C a swelling of plastids and mitochondria was noticed (Fig. 2, E and F). Furthermore internal membranes were indistinct in many plastids (Fig. 2E, pl). In these samples, in spite of large changes in mitochondria and plastids, the envelope membranes were seen intact. In samples frozen to -19 and -21°C an almost complete loss of compartmentalization was found in almost all the cells (Fig. 2, G-I).

It should be pointed out that onion bulb cells are susceptible to fixation stress. Observations in Figure 2 represent what we noticed in majority of the cells. In general freeze injured cells were found more sensitive to fixation than the control. Examples of fixation stress are shown in Figure 3. Even in some control cells tonoplast could not be seen and membrane vesicles were present (Fig. 3A). In cells frozen to -4 and -7°C membrane vesicles were much more pronounced and tonoplast was absent (Fig. 3, B and D). Furthermore these changes were more frequent in freeze treated cells than control cells. Sometimes a relatively normal cell was found next to a cell with a complete loss of compartmentalization (Fig. 3C). It is important to note that -4 and -7°C stress is nonlethal, and completely reversible in these onion bulb cells, yet these cells were more sensitive to fixation than the control cells.

FIG. 1. Inner epidermal cells of onion (Allium cepa L. cv. Sweet Spanish) bulb scale following a freeze-thaw cycle. Cells were plasmolyzed in 0.8 M mannitol to test membrane intactness. Number in each picture represents the percent ion leakage from the bulb tissue. Freezing temperatures were (A) Unfrozen control; (B) -4°C; (C) -13°C; (D) -15°C; (E) -19°C; (F) -21°C. Excised 2 cm square bulb scale pieces were frozen with a cooling rate of 1°C/hr using the procedure of Palta and Li (14). Samples were thawed slowly over ice.

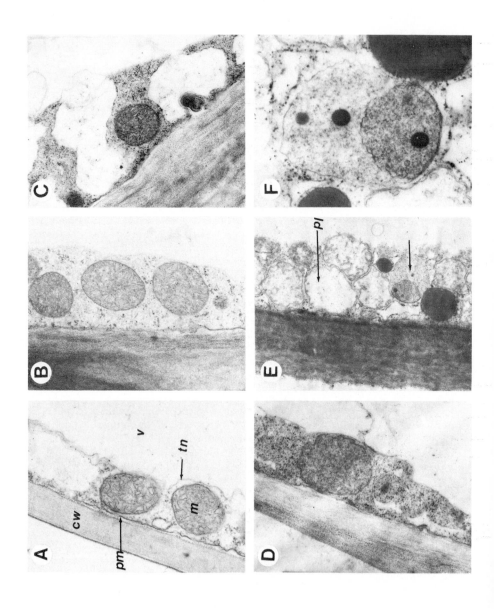

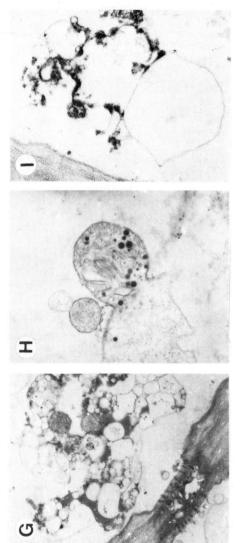

FIG. 2. Onion bulb scale cells from the same experiment as Figure 1 as observed under transmission electron microscope. (A) Control; (B) -4°C; (C) -10°C; (D) -13°C; (E) -15°C; (F) close up of E; (G) and (H) -19°C; (I) -21°C. Samples were fixed in 2% gluteraldehyde and prepared for microscopic observations using standard procedure. cw – cell wall; pm – plasmamembrane; tn – tonoplast; m – mitochondrion; pl – plastid; v – vacuole.

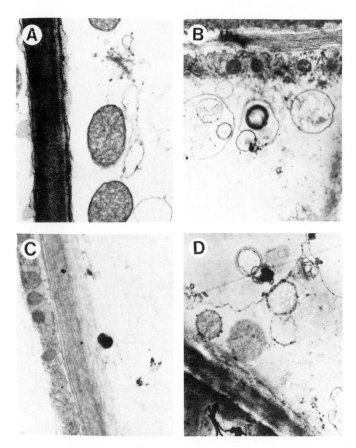

FIG. 3.   Examples of fixation stress in onion bulb scale cells.   Freezing treatments were: (A) Unfrozen control;   (B) -4°C;   (C) and (D) -7°C.

Results with potato leaflets were similar to that of onion bulb tissue.   In general ion efflux and infiltration of the leaf tissues increased with lowering of freezing temperature in Solanum tuberosum and S. commersonii potato species (T_1 data, Table 3).   In these samples the turgidity of leaflet was also observed.   Leaflets of Solanum tuberosum frozen to -3 and -3.5°C showed partial loss of turgor (flaccid).   Similar condition was noted in leaflets of S. commersonii frozen to -8°C (Table 3).   At colder temperatures both the species showed a complete loss of turgor.   Leaflets used in these experiments (Table 3) were also taken for electron microscopic observations (Fig. 4).   Inspite of 51 and 75% ion efflux in samples of S. tuberosum frozen to -2.5 and -3°C no ultrastructural abnormalities could be observed compared to control (Fig. 4 A, B and

Table III. Effect of freezing stress on ion efflux, infiltration and turgidity of the leaflets in two potato species at two time intervals following thawing (T_1-12 hr., T_2=120 hr.). Both the species were grown at 20/15°C, day/night temperatures. Plants were transferred to 5/5°C, day/night regime for one week.

Freezing temp (°C)	Ion efflux (% of leaf tissues)		Infiltration of leaflets (%)		Relative turgidity[a]	
	T_1	T_2	T_1	T_2	T_1	T_2
Solanum tuberosum						
-2.5	51.1	32.2	5	5	A	A
-3.0	75.5	47.4	75	35	B	A
-3.5	80.5	66.7	100	80	B	B
-4.0	83.4	89.6	100	100	B	C
Control	13.7	7.1	0	0	A	A
Solanum commersonii						
-4.0	21.3	12.7	5	0	A	A
-5.0	36.5	15.4	75	10	A	A
-6.0	43.7	13.7	90	25	A	A
-7.0	48.2	56.9	90	100	A	B
-8.0	67.1	82.0	100	100	B	C
Control	11.8	5.8	0	0	A	A

[a] A - same as control, B - partial loss of turgor, C - full loss of turgor.

C). The cell membranes, chloroplasts, mitochondria and nucleus all appeared normal in these cells. In -3.5°C samples ion efflux was 81% and in some cells a breakdown of the cell compartmentalization was noted (Fig. 4 D and E). In this tissue a normal and damaged cell could be seen next to each other (Fig. 4 D and E). Only in -4°C samples a complete breakdown of the cell compartmentalization was noted in all the cells (Fig. 4 F). Even in these cells thalkoid membranes could be seen but the chloroplasts and mitochondria envelope membranes were mostly absent. Some times these envelope membranes were found to be intact when the chloroplasts and mitochondria appeared highly swollen. The damaged cells in -3.5 and -4°C samples showed typical "frost plasmolysis" (Fig. 4 D-F).

Electron microscopic observations were also made on the leaf cells of Solanum acaule which were subjected to a freeze-thaw cycle. Results on this species were basically similar to

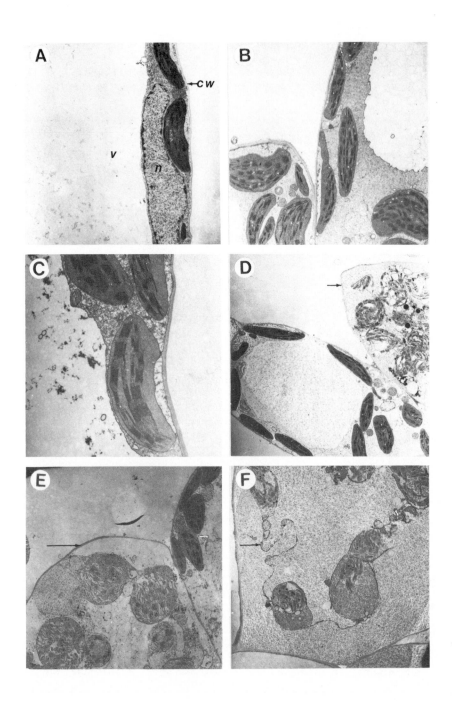

Table IV. Analysis of the efflux; conductivity compared to K^+ content, Ca^{2+} content and pH of the effusate. Measurements on conductivity were made after shaking the sample for about 3 hours and the effusate collected was used for analysis. All the values are average of 4 separate measurements and all (except pH) are expressed as per g of scale tissue (source: Ref. 12).

Freezing temp	Conductivity (µS)	K^+ content (mg/liter)	Ca^{2+} content (mg/liter)	pH
-11	173.0 ± 12.0	50.6 ± 1.6	0.45 ± 0.08	6.45 ± 0.04
-4	119.2 ± 16.1	35.2 ± 5.6	0.40 ± 0.06	5.99 ± 0.11
+3(control)	104.3 ± 13.8	29.6 ± 4.9	0.34 ± 0.04	5.93 ± 0.06

those of S. tuberosum (Fig. 5). In -4°C samples no ultra-structural changes could be observed (Fig. 5 B). In -5°C samples, although some chloroplasts appeared swollen (Fig. 5 D arrow and E), the cell compartmentalization appeared intact. In -6°C samples chloroplasts were highly swollen (Fig. 5 F arrow) and in some cases the chloroplast envelope membrane appeared broken (Fig. 5 G arrow). Even in -7°C samples thalokoid membranes could be seen in intact even though the cell compartmentalization was completely lost in all the cells. Here again highly damaged cells showed "frost plasmolysis" (Fig. 5 H and I).

These ultrastructural observations on onion and potato cells (Fig. 2, 4 and 5) are consistant with our previous data i.e. a large amount of ion leakage can result due to a freeze-thaw cycle without a significant change in the cell ultra-structure. The cell membranes and various other organelle appeared normal in spite of large ion leakage.

FIG. 4. Potato (Solanum tuberosum L., cv. Red Pontiac) mesophyll cells following a freeze-thaw cycle. Third or fourth terminal leaflet from the top of the plant was frozen and thawed according to the procedure previously described (14). Treatments were (A) unfrozen control; (B) -2.5°C; (C) -3.0°C; (D) and (E) -3.5°C; (F) -4°C. Samples were taken and fixed according to procedure used by Palta and Li (14). For ion leakage in these samples refer to Table 2.

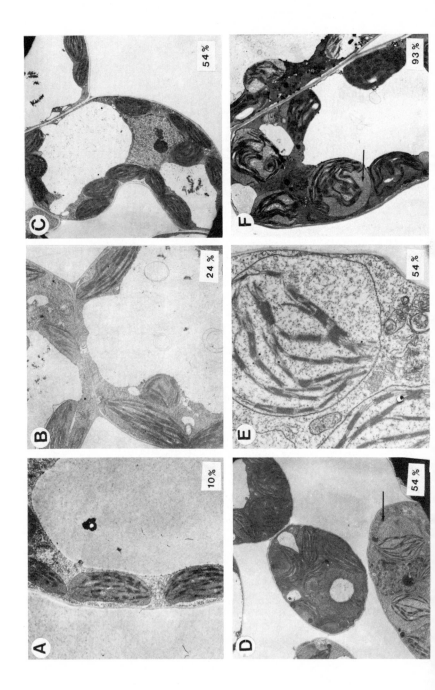

232

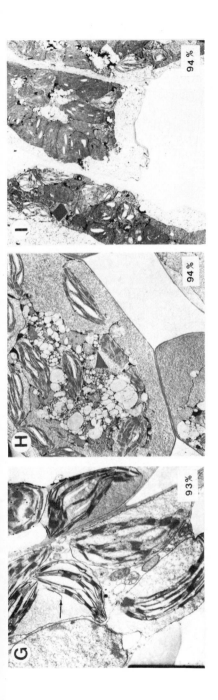

Fig. 5. Potato (<u>Solanum acaule</u> L.) mesophyll cells following a freeze-thaw cycle. Number in each picture represents the percent ion leakage from the leaf tissue. Treatments were: (A) unfrozen control; (B) -4°C; (C), (D) and (E) -5°C; (F) and (G) -6°C; (H) -7°C; (I) -8°C. Sample selection and preparation was same as in Fig. 4.

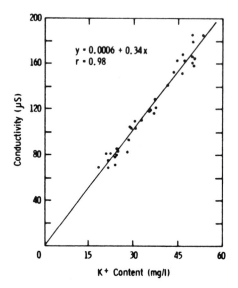

Fig. 6. Relationship between K^+ content and conductivity in freeze-thaw treated onion bulb scale tissue samples (source: Ref. 11).

B. Analysis of Ion Efflux

As pointed out above, in the past, ion efflux was regarded as a result of breakdown of membrane semipermeability and membrane rupture. This was probably the reason for lack of interest on the part of early investigators to analyze the nature of freeze-induced ion efflux. After shaking the freeze injured onion bulb tissues in deionized water, Palta et al. (11,12) collected the effusate and analyzed it for various cations. It was found that K^+ was the main cation present, almost 100 times the concentration of Ca^{++} (Table 4). This was expected since K^+ is the major free cation present in cell sap. By assuming that, corresponding to K^+, the anion present in the effusate was Cl^-, one can calculate the electrical conductance of KCl solution. By the calculation it was found that the amount of K^+ and its corresponding anion present in the effusate accounted for almost all of the conductivity of the effusate.

To determine how well conductivity and K^+ content were related to each other a plot of K^+ content vs conductivity of the effusate was made (Fig. 6). A highly significant correlation was found ($r = 0.98$). The value of the intercept was close to zero confirming the previous calculations (12) that the K^+ along with an anion accounted for all of the conductivity of the effusate.

C. Ion Efflux and Alteration in Cell Membranes

From the results presented above it is safe to conclude that under our experimental conditions ion efflux following freezing injury is not due to a loss of membrane semipermeability or membrane rupture. The freeze injured yet alive cells provide an excellent material for study of the nature of cell membrane alteration that could account for increased ion efflux. One way to study membrane alteration is to measure changes in membrane permeability to water, electrolytes and nonelectrolytes.

Water movement across cell membrane is best explained by the molecular theory of membrane transport advanced by Traeuble (25). According to this theory water molecules are small enough to slip through the imperfections in the hydrocarbon tails (so called "kinks"). Thus a change in water permeability should reflect changes in the lipid portion of the membrane. Palta et al. (11) studied the permeability of freeze injured yet alive onion epidermal cells using a radiotracer technique (15,16). Although there was more than 2-fold increase in ion efflux in -11°C treatment yet no significant change in water permeability could be detected (Table 1). These results indicate that physical status of membrane lipid remain more or less unaltered in spite of freezing injury and increased ion efflux.

The permeability of nonelectrolytes has long been known as a direct function of lipid solubility. In 1933, Collander and Barlund (1) found a direct relationship between permeability constants for nonelectrolytes and the oil/water distribution coefficient of those nonelectrolytes. In recent years this classic work of Collander and Barlund has been confirmed by many workers and most recently by Fineklstein (3). By studying the changes in permeability to nonelctrolytes one should be able to detect the alteration of the membrane lipids. Palta and Li (14) studied the permeability of urea and methyl urea (with oil/water partition coefficient of 15×10^{-5} and 44×10^{-5} respectively) of freeze injured yet alive onion epidermal cells. Compared to unfrozen control cells, no changes in the permeability of these two nonelectrolytes could be detected for freeze injured cells (Fig. 7). These results are in agreement with the results of water permeability measurements (Table 1) and further strengthen the conclusion that the physical status of the membrane lipids remains unaltered by freezing injury.

There is ample evidence in the literature that membrane associated ATPases are responsible for ion transport (19). These ATPases are regarded as the intrinsic membrane proteins that pass entirely through the bilayer (17,18). If inactivated, these intrinsic proteins may serve as channels for passive ion transport. Evidence for the existence of such channels comes, for instance, from the studies with gramicidin A, a simple channel forming molecule (5). Therefore, by

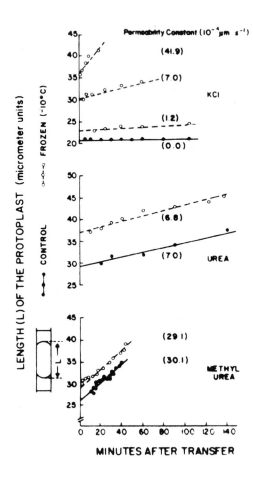

FIG. 7. Membrane permeability of control and freeze-thaw treated onion epidermal cells to urea, methyl urea and KCl. Cells were plasmolyzed in 0.8 M mannitol solution and transferred to equiosmolar solutions of the permeating solutes at zero time. (Source: Ref. 14).

studying the changes in the permeability properties of the membranes to ions, one should be able to detect alterations in the membrane proteins.

Normally plant cells can be kept at a plasmolyzed constant protoplast volume for a long period of time. This means that ordinarily passive permeability of the cell membrane to ions is negligible. As indicated above passive permeability of ion can be dramatically changed if intrinsic proteins are inactivated. Palta and Li (14) studied the passive permeability of freeze injured cells to KCl. Within the experimental time (few hours) no uptake of KCl could be detected in unfrozen control cells (no change in the length of the protoplast, Fig. 7). On the other hand the freeze injured cells were able to take up KCl and as a result protoplast started expanding (Fig. 7). The expansion rate, however, varied in different cells. Since extent of injury to different cells in the same tissue varies, one would expect different permeability values for K^+ from different cells.

A gradual increase in K^+ permeability from nearly 0 in control cells to 41.9 x 10^{-4} $\mu m.s^{-1}$ suggests that membrane intrinsic proteins are altered. These alterations that one would suspect are of the type that render these proteins inactivated. Once inactivated these proteins serve as channels for passive ion transport. Concentration of K^+ in the vacuole is considerably higher than in the extracellular solution. Thus if proteins are altered following freezing injury a passive efflux of ions would result (from vacuole to extracellular solution) in the direction of the concentration gradient. Observed changes in cell surface ATPase activity associated with freezing injury by Jian et al. (8) further strengthen these results.

It cannot, however, be said as to what causes alterations in the properties of the intrinsic membrane proteins. Possible causes could be direct denaturation, lack of energy supply for the functioning of ATPase due to changes in cytoplasm including mitochondria, or due to changes in lipid protein interaction within the membrane. Changes in the phase of the surrounding lipid and reorganization of annular lipids might change the lipid-protein interaction. Denaturation due to aggregation of intrinsic and soluble protein could also be possible. Only further research will help elucidate the mechanism/mechanisms for the observed alterations in membrane transport properties.

III. RECOVERY OF FREEZING INJURY

A. Post-Thaw Injury or Recovery

Study of freeze injured cells during post-thaw period provides another means to look at the mechanism of freezing injury. As discussed above in spite of freezing injury all cells were alive. These observations were made immediately after thawing. Depending upon the degree of injury, however, a recovery or progress in injury was noted during the post-thaw period (12). After a preliminary investigation two temperatures of -4 and -11°C were selected for the study of post-thaw behavior of freeze injured onion bulb cells (12). This was because these two temperatures demonstrated two separate events during the post-thaw behavior, i.e. injury or recovery.

In -11°C treatment the infiltration of the scale tissues increased and the cell survival decreased 7 to 12 days after thawing (Table 2). On the other hand in the bulbs frozen to -4°C the infiltration, that was evident immediately after thawing, disappeared completely after 7 days. In fact onions frozen to -4 were indistinguishable from the controls 7 days after thawing. Similar results were obtained when efflux was

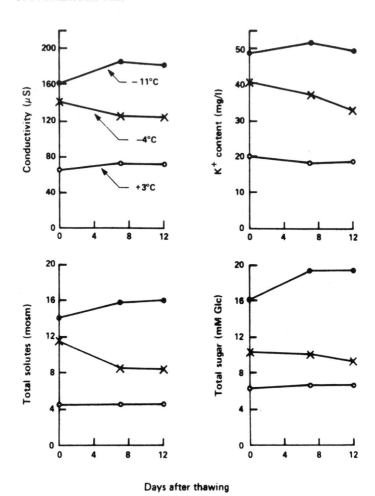

Days after thawing

Fig. 8. Changes in conductivity, K^+ content, total solutes and total sugars during the post thaw period. (Source: Ref. 12).

assayed during the post-thaw period. In bulb tissues frozen to -11°C the ion efflux (conductivity of the effusate) increased whereas it decreased in -4°C during the post-thaw period (Fig. 8). Same was found to be true when the efflux was analyzed for K^+, sugars, and total solutes (Fig. 8).

We also conducted similar experiments with several potato species. Results on two species are given in Table 3. After freezing (1°/hr) to various temperatures excised potato leaflets were examined 12 hr. (T_1) and 120 hr (T_2) after thawing. Depending upon the freezing temperature the infiltration of the tissue and ion efflux either decreased or increased

during the post-thaw period. For instance, at temperatures of
-6 and warmer, in S. commersonii leaflets, the infiltration of
tissues and ion efflux was reduced dramatically at T_2. One
of the consequences of freezing injury is the loss of turgor
(flacid tissue). In the tissue, which suffered higher ion
efflux and increase infiltration during the post-thaw period, a
loss in turgor was found (Table 3).

The recovery of injury must be an active process (i.e. it
requires active transport). Leakage of ions due to freezing
injury result in a decrease in concentration gradient between
vacuole and extracellular solution. The cell therefore loses
turgor and is unable to absorb the extracellular water
resulting in the tissue having a soaked and flaccid appearance. During the recovery an opposite process must take place,
i.e. the cell will have to absorb ions back into the vacuole
which will then lead to uptake of extracellular water (disappearance of infiltration). Since this influx of ions is from
lower concentration (outside) to higher concentration (inside)
it must be considered an active process. The recovery of
injury thus requires the recovery of active transport system.
This conclusion is supported by the results discussed under
freezing injury i.e. freezing injury appears to be due to
alteration of the intrinsic membrane proteins (ATPases involved
in active transport of K^+). The injured cell may have the
ability to repair partially denatured proteins or resynthesize
and reincorporate these proteins into the membrane during
recovery. Recent reports on recovery of solute uptake following a gas shock by cell suspensions of Acer pseudoplatanus
(23,24) supports our view on recovery of freezing injury.

A continued increase in injury after thawing, by same
reasoning, could be due to the inability of recovery of the
active transport process. Since efflux is from higher (vacuole) to lower (extracellular solution) concentration it must be
largely passive diffusion. Again this explanation is supported
by the results discussed under freezing injury i.e. the intrinsic membrane proteins when inactivated serve as an ion channel
leading to an efflux of ions.

B. Secondary Injury: Swelling of the Protoplasm

During the post-thaw period a swelling of the protoplasm
was noticed in irreversibly injured cells (14). It was
explained to be due to the high K^+ content in extracellular
solution. It is believed that when plant cells are left in
K^+ solution for a long time the plasmamembrane becomes highly
permeable to K^+. As K^+ accumulates in the protoplasm then
uptake of water results in a swelling. This gives rise to well
known cap plasmolysis (4,20). Some of the irreversibly injured
cells exhibited cap plasmolysis (14). During the post thaw
period a small but increasing amount of Ca^{2+} efflux was also
observed in irreversibly injured cells. This increased efflux

of Ca^{2+} could be caused by replacing Ca^{2+} in the plasma-membrane with K^+. Such a replacement could weaken the membrane structure as Ca^{2+} is considered important for membrane stability. Thus cap plasmolysis observed in irreversibly injured cells could be due to a replacement of Ca^{2+} with K^+ in the membrane. This could also explain in part why an irreversibly injured cell eventually dies. As indicated above protoplasmic swelling was only observed in irreversibly injured cells.

C. Treatment of Injury

As indicated above, a high amount of K^+ that accumulates in the extracellular space seems to be responsible for secondary injury. Furthermore because of its damaging effect this high K^+ content may in part be responsible for the progress of injury and eventual death of the cell. If that was correct removal of K^+ from extracellular space immediately after thawing should be able to halt the progress of injury. This is exactly what was found by Palta et al. (12). Just by washing the injured onion bulb tissue with distilled water considerably improved the cell survival. We also experimented in similar studies with several potato species. By washing the leaflets with distilled water following thawing we were able to improve survival (visual observation on the improvement of the turgor) by 0.5 to 1°C. This treatment as observed in the onion and the potato was only helpful when injury was not too severe. These observations support the suggestion that increase in injury during the post-thaw period is in part due to accumulation of K^+ in the extracellular space.

IV. SUMMARY

The nature of incipient freezing injury (ion efflux, infiltration of the tissue and loss of turgor), following a slow freeze-thaw cycle, can be best described as follows: 1) Leakage of ions (ion efflux) resulting from a freezing injury is not due to a loss in membrane semipermeability or membrane rupture. Evidence that supports this conclusion is the ability of the injured cell to plasmolyze in a hypertonic solution and no visually ultrastructural changes in cell and organelle membranes. 2) Membrane lipids remain unaltered in spite of freezing injury. Absence of any changes in the membrane permeability to water and nonelectrolytes such as urea and methyl urea, is the basis for this conclusion. 3) Membrane proteins (probably the intrinsic membrane proteins that are ATPases and are responsible for active transport) appear to be the site of membrane alterations. Changes in cell membrane permeability to K^+ and ability of the cell to recover (that requires active process) are the basis for this conclusion.   4) Accumulation of K^+ in the extracellular space explains in part the reasons for the eventual death of irreversibly injured cells.   This conclusion is based on the observation of swelling of protoplasm and the ability to halt the progress of injury by washing out the extracellular solutes.

REFERENCES

1. Collander,   R.,   and Barlund,   H.,   Acta Bot.   Fenn.   11:1-114 (1933).
2. Dexter,   S.T.,   Tottingham,   W.E.,   and   Graber,   L.F.,   Plant Physiol. 7:63-78 (1932).
3. Finkelstein, A., J. Gen. Physiol. 68:127-135 (1976).
4. Hofler, K., Ber Dtsch. Bot. Ges. 46:73-82 (1928).
5. Kolb, H.A., Lauger, P., and Bamberg, E., J. Membrane Biol. 20:133-154 (1975).
6. Levitt,   J.,   "Responses   of   Plants   to   Environmental Stresses." Academic Press, New York, 1972.
7. Levitt,   J.,   "Responses   of   Plants   to   Environmental Stresses," 2nd ed. Vol. 1. Academic Press, New York, 1980.
8. Jian Ling-cheng, J., Long-hua, S., He-zhu, D., and De-lan, S., in "Plant Cold Hardiness and Freezing Stress" (P.H. Li and A. Sakai, eds.), Vol. 2. Academic Press, New York, 1982.
9. Osterhourt, W.J.V., Science 40:488-491 (1914).
10. Palta,   J.P.,   Chen,   H.H.,   and   Li,   P.H.,   Bot.   Gaz. 142:311-315 (1981).
11. Palta,   J.P.,   Levitt,   J.,   and   Stadelmann,   E.J.,   Plant Physiol. 60:393-397 (1977).
12. Palta,   J.P.,   Levitt,   J.,   and   Stadelmann,   E.J.,   Plant Physiol. 60:398-401 (1977).
13. Palta, J.P., Levitt, J., and Stadelmann, E.J., Cryobiol. 14:614-619 (1977).
14. Palta,   J.P.,   and   Li,   P.H.,   Physiol.   Plant.   50:169-175 (1980).
15. Palta,   J.P.,   and   Stadelmann,   E.J.,   J.   Membrane   Biol. 33:231-247 (1977).
16. Palta, J.P., and Stadelmann, E.J., Physiol. Plant. 50:83-90 (1980).
17. Racker, E., Biochem. Soc. Trans. 3:785-802 (1975).
18. Singer, S.J., "Cell Membranes," pp. 35-44. H.P. Publishing Co., New York, 1975.
19. Spanswick, R.N., Lucas, W.T., and Dainty, J., "Plant Membrane Transport: Current Conceptual Issues." Elsevier/ North-Holland Biomedical Press, New York, 1980.
20. Stadelmann, E.J., Methods in Cell Physiol. 2:143-216 (1966).

21. Steponkus, P.L., and Weist, S.C., in "Recent Advances in Plant Cold Hardiness and Freezing Stress," (P.H. Li and A. Sakai, eds.), p. 75-91. Academic Press, New York, 1978.
22. Sukumaran, N.P., and Weiser, C.J., Plant Physiol. 50:564-567 (1972).
23. Thoiron, B., Thoiron, A., LeGuid, J., Luttge, U., and Thellier, M., Physiol. Plant. 46:352-356 (1979).
24. Thoiron, B., Espejo, J., LeGuiel, J., Luttge, U., and Thellier, M., Physiol. Plant. 48:161-167 (1980).
25. Traeuble, H., J. Membrane Biol. 4:193-208 (1971).

CHANGES IN ATPase ACTIVITY
DURING FREEZING INJURY AND COLD HARDENING

Jian Ling-cheng, Sun Long-hua, Dong He-zhu and Sun De-lan

Laboratory of Cytology
Institute of Botany, Academia Sinica
Peking, China

I. INTRODUCTION

That the cellular membranes are inferred to be the primary site of freezing injury has been demonstrated by many experimental results in recent years (8,17,20,21,24,26). However, as to what membrane property is the first to be altered during freezing injury, different investigator's experimental results and concepts are not yet identical. Levitt (17) indicated that the freezing damage of membranes is due to the changes in membrane protein. Lyons and his co-workers (18,20) suggested that the chilling injury may be due to a physical-phase transition of the membrane lipids from a flexible liquid-crystalline to a solid-gel structure. Yoshida's experimental results (28,29) revealed that the regulatory properties of membrane-bound phospholipase D were altered during freezing-thawing, and so the phospholipase D activity reaction was accelerated and resulted in degradation of the membrane phospholipids. Palta and Li (24) proposed that the primary site of sub-lethal freezing injury may be at the active ion-transport system of cell membranes. Steponkus and Wiest (26) indicated that the freezing injury is related to the disruption of intermolecular forces in the membranes. Therefore, it is evident that the problem has yet to be further studied.

In view of the fact that ATPase is a membrane-bound protein, and its associations with membranes and organelles are very extensive (1,7,11), and it has important functions in the energy metabolism, in the absorption and transport of materials (3,6), we made a study of the changes in ATPase activity in the cells of winter wheat seedlings during freezing injury and cold hardening by use of the enzymic cytochemical method. Our experimental results suggested that a close relationship exists between the plasmalemma ATPase activity and the freezing injury and cold hardiness of plants.

II.   MATERIALS AND METHODS

A.   Material

Winter wheat, _Triticum_ _aestivum_ L., cv. Nongke No. 1 was used as the experimental material. The seeds were obtained from Institute of Crop Germplasm Resources, Chinese Academy of Agricultural Sciences.

B.   Freezing Treatment

Wheat seeds were placed on moist filter paper in petri dishes and allowed to germinate for 2 days at 25°C and then hardened for 14 days at 2°C. The young sprouts were subjected to freezing treatment for 12 h or 22 h at -8°C, and then were gradually thawed for 2 h at 3°C. Their survival rates were 54% and 12.8% respectively after the frozen young sprouts were recultured at 25°C.

C.   Cold Hardening

Seedlings at two different growth stages were subjected to cold hardening. One lot of seeds was directly sown in pots with sandy/loam soil. They were held at 15-20°C with 12 h day-length. When the seedlings grew to a 3-leaf stage, they were hardened for 10 days at 2-7°C during the day with 9 h light, and at 2°C in a refrigerator at night. The other lot of seeds was soaked in water for 12 h, then placed on moist filter paper in petri dishes. They were allowed to germinate at 25°C for one day. Then the young sprouts about 1 cm in length were hardened for 20 days at 2°C in a refrigerator in the dark. Seedlings after cold hardening showed that their cold hardiness raised at least 5°C.

D.   Enzymic Cytochemical Procedures

The young leaves about 1 cm in length were sampled from treated materials, and the similar samples were collected from untreated seedlings (control). They were cut into 0.5 x 0.5 mm slices and fixed immediately with 4% formaldehyde-3% glutar-aldehyde in 50 mM sodium-cacodylate buffer (pH 7.2) for 75 min at 22°C. After fixation, samples were washed twice, each time for 45 min, with 50 mM sodium-cacodylate buffer (pH 7.2), and further washed for 1-2 h with 50 mM Tris-maleate buffer (pH 7.2). Then the leaf slices were incubated in a modified Wachstein-Meisel medium (11) containing 3 mM $Pb(NO_3)_2$, 5 mM

MgSO$_4$ and 2 mM ATP in 50 mM Tris-maleate buffer at pH 7.2. The incubation was carried out at two different temperatures, one group at 22°C for 2.5 h, and the other at 5°C for 5 h.

Control slices were incubated in: 1) medium without ATP substrate, and 2) medium with fluoride inhibitor (NaF 0.01 M).

After incubation, leaf slices were washed for 30-60 min 2-3 times with 50 mM sodium-cacodylate buffer (pH 7.2), then post-fixed in 2% OsO$_4$ in 50 mM cacodylate buffer (pH 7.2) over-night at 2-3°C. After being washed with distilled water for 3 h, the slices were stained in 0.5% uranyl acetate for 45 min, then dehydrated in graded ethanol solutions and embedded in Epon 812. Ultrathin sections were obtained with a LKB-8800 ultramicrotome, and examined and photographed with a HU-11A electron microscope and a JEM-100CX electron microscope.

III.   RESULTS

A.   Changes in ATPase Activity during Freezing Injury

When the control (unfrozen-treated seedling) leaf slices were incubated for 2.5 h at 22°C, lead phosphate deposits indicating ATPase activity were localized at the plasmalemma, plasmodesmata, intercellular spaces, nucleoli and chromatin; they were also shown in some vacuoles; but they were not found in the mitochondria, plastids, Golgi bodies and endoplasmic reticulum (Fig. 1). When these samples were incubated in: 1) medium without ATP substrate and 2) medium containing fluoride inhibitor (NaF 0.01 M), there was little or no deposition of reaction products (Fig. 2). It is inferred, therefore, that the presence of lead phosphate deposits as shown in figures is the evidence of ATPase activity.

When winter wheat seedlings were frozen at -8°C for 12 h (54% survival of seedlings), we observed two different situations in changes in ATPase activity: (1) In some young leaf sections, the ATPase activity was markedly decreased at plasmalemma, plasmodesmata and intercellular spaces, but it was shown clearly at the endoplasmic reticulum, Golgi bodies and tono-plasts. Besides, the morphology of all the organelles in these cells was still maintained in perfect state (Fig. 3), and (2) In other leaf sections, the ATPase activity at plasmalemma, plasmodesmata, intercellular spaces, nucleoli and chromatin was almost inactivated, or only maintained very weak activity, but high ATPase activity was found at tonoplasts, Golgi bodies, endoplasmic reticulum and the prolamellar membranes of some proplastids (Fig. 4). Some vesicles were produced in these cells. This indicated that the cellular membrane system had been damaged. When seedlings were subjected to freezing treat-ment at -8°C for 22 h (12.8% survival of seedlings), the changes in ATPase activity in some young leaf sections were

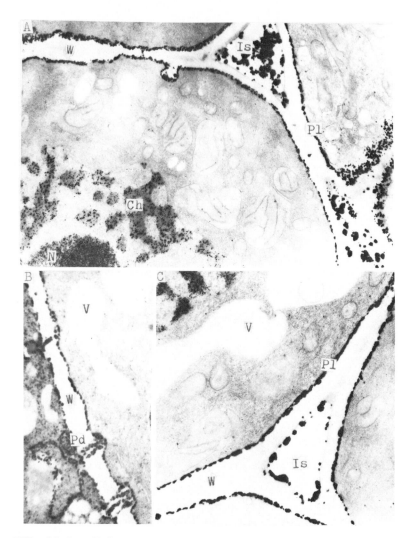

FIG. 1A-C. Unfrozen young leaf cells. They were incubated
for 2.5 h at 22°C. The lead phosphate deposits indicating
ATPase activity were localized at plasmalemma (Pl), plasmodesmata (Pd), intercellular spaces (Is), nucleoli (N) and chromatin (Ch). V: vacuole. W: cell wall. A, x 9350; B, x
15300; C, x 10200.

similar to those in the second case mentioned above (Fig. 5).
In others, the ATPase activity associated with membranes and
organelles was altered and damaged to more severe extent. The
ATPase activity at plasmalemma, plasmodesmata, intercellular
spaces, nucleoli and chromatin was completely lost; it was also

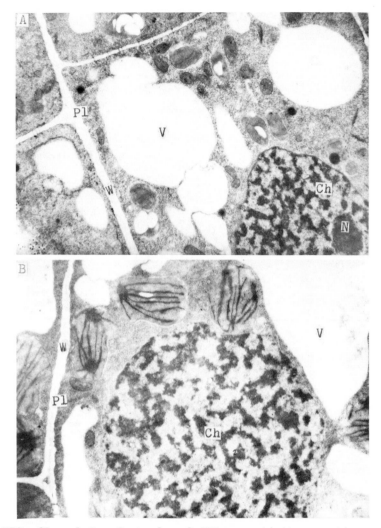

FIG. 2A and B. Controls of ATPase activity reaction. A,
Slices were incubated in medium without ATP for 2.5 h at 22°C.
B, Slices were incubated in medium with 0.01 M NaF for 2.5 h at
22°C. Sections were stained with uranyl acetate and lead
citrate. A, x 6430; B, x 6850.

markedly lowered or wholly inactivated at the tonoplasts and
the prolamellar membranes of some proplastids; the morphology
of the prolamellar membranes in these proplastids was hardly
recognized; and irregular plasmolysis was formed; but heavy
lead deposits indicating ATPase activity were still observed in
the Golgi bodies, endoplasmic reticulum and vacuoles (Fig. 6).

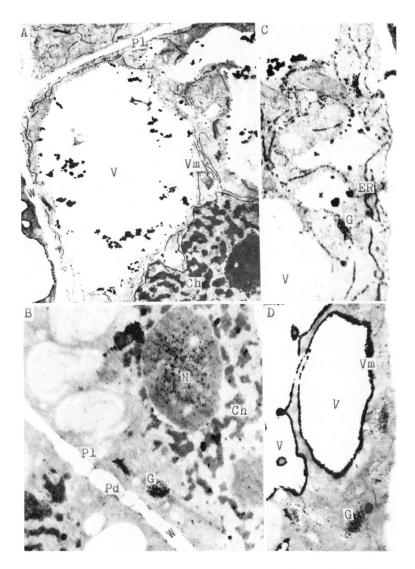

FIG. 3A-C. Young leaf cells frozen for 12 h at -8°C. They were incubated for 2.5 h at 22°C. A, x 5100; B, x 7215; C and D, x 11050. G: Golgi body. ER: endoplasmic reticulum. Vm: tonoplast.

It is evident that during freezing injury the ATPase activity at the cell surface (plasmalemma) was firstly lowered and then completely inactivated. However, the ATPase activity in cytoplasmic organelles (Golgi bodies, ER, proplastids, vacuoles and tonoplasts) was activated.

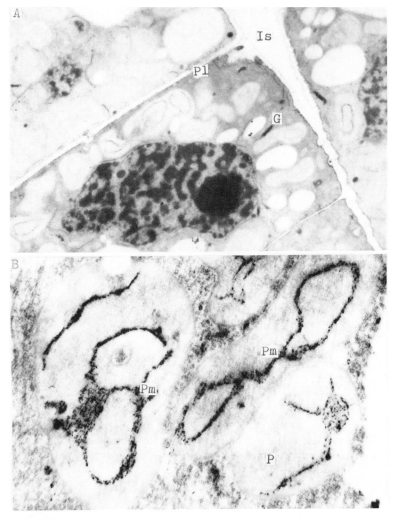

FIG. 4A and B. Young leaf cells frozen for 12 h at -8°C
and incubated for 2.5 h at 22°C. A, x 6000; B, x 24000. P:
proplastids. Pm: prolamellar membranes.

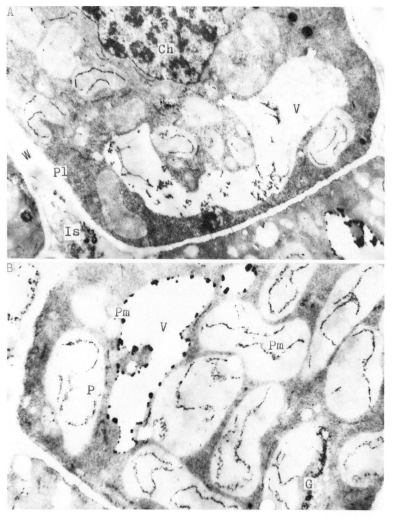

FIG. 5A and B. Young leaf cells frozen for 22 h at -8°C.
They were incubated for 2.5 h at 22°C. A, x 8500; B, x 15000.

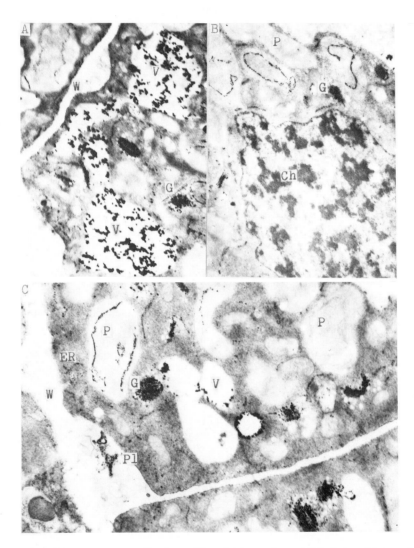

FIG. 6A-C. Young leaf cells frozen for 22 h at -8°C and incubated for 2.5 h at 22°C.   A, x 11000; B, x 10000; C, x 15000.

TABLE I. The difference of ATPase activity between cold hardened and nonhardened winter wheat tissues incubated at 22°C or 5°C ("+" ATPase activity present, "-" ATPase activity absent).

	22°C		5°C	
	nonhardened	hardened	nonhardened	hardened
Plasmalemma	+	+	-	+
Plasmodesmata	+	+	-	+
Nucleoli & chromatin	+	+	+	-
Tonoplast & vacuoles	-	+	-	+

B. Adaptive Changes in ATPase Activity during Cold Hardening

When winter wheat seedlings were subjected to cold hardening under low temperature, the ATPase activity associated with membranes and organelles was altered, and so it was significantly different between cold hardened and non-hardened tissues. These results are summarized in Table 1. From Table 1, it is clear that when tissue slices were incubated at 22°C for 2.5 h, ATPase activity was exhibited at the plasmalemma, plasmodesmata, nucleoli and chromatin in both cold hardened and non-hardened tissues, while that at tonoplasts and in vacuoles was observed only in hardened tissues (Fig. 7 and 8). When tissue slices were incubated at 5°C for 5 h, the differences in ATPase activity between hardened and non-hardened tissues were as follows: (1) ATPase activity at the plasmalemma and plasmodesmata in nonhardened tissues was insignificant or none at all (Fig. 9), while that in cold hardened tissues exhibited high ATPase activity (Fig. 10); (2) high ATPase activity in the nucleoli and chromatin was maintained in nonhardened tissues (Fig. 9), but diminished in cold hardened tissues (Fig. 10); and (3) ATPase activity on the tonoplasts and in vacuoles in hardened tissues was as high as that at 22°C (Fig. 8 and 10), while the tonoplasts and vacuoles in nonhardened tissues did not show any ATPase activity (Fig. 9).

Studies on adaptive changes in some enzyme activities in plants during cold hardening have been reported (5,10,14,15,16). But determinations of the activities of these enzymes were mostly carried out at the optimum temperatures. Only Kacperska-Palacz et al. (14) and Kovacs et al. (15) determined the activity of peroxidase at low temperatures. Their results indicated that the difference in the enzyme activity between cold hardened and nonhardened samples was greater at low temperatures than at the optimum temperature. Our results are in conformity with their observations.

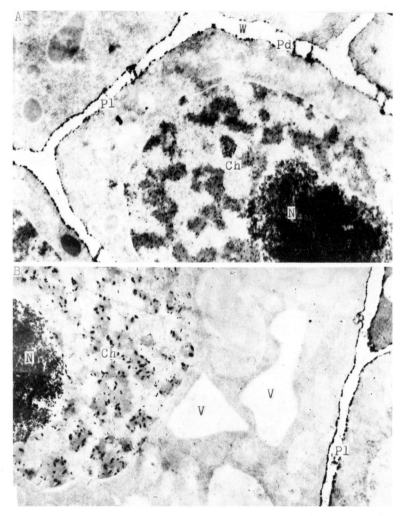

FIG. 7A and B. The young leaf cells of nonhardened wheat seedlings at 3-leaf stage. They were incubated for 2.5 h at 22°C. A, x 15300; B, x 13175.

IV.   DISCUSSION

A.   Plasmalemma ATPase Activity in Relation to Freezing Injury and Cold Hardiness of Plants.

Plasmalemma is a primary sensor and barrier of the cells responding to their environmental factors. Many investigators indicated that the plasmalemma may be the primary site of

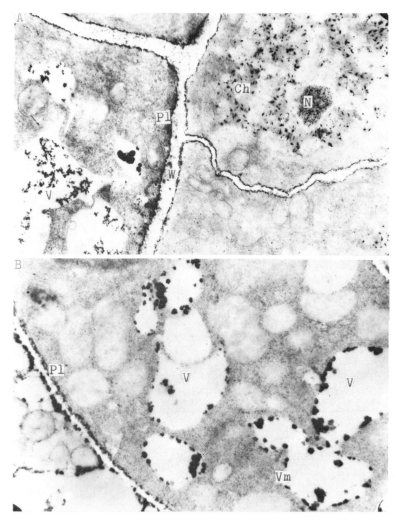

FIG. 8A and B. The young leaf cells of cold hardened wheat seedlings at 3-leaf stage, 2.5 h incubation at 22°C. A, x 10200; B, x 12750.

freezing injury (17,21,24,26). After freezing injury has occurred, one of the primary signs of damage is the loss of membrane semipermeability (17,21). As well known, ATPase activity in cells is, as a rule, chiefly appeared at the plasmalemma (1,7,11), and a close relationship exists between plasmalemma ATPase activity and the active ion-transport function (3,6,25). McMurchie (22) reported that the Arrhenius plots/temperature sensitivity of the ATPase activity in a post-mitochondrial membrane preparation (mainly containing

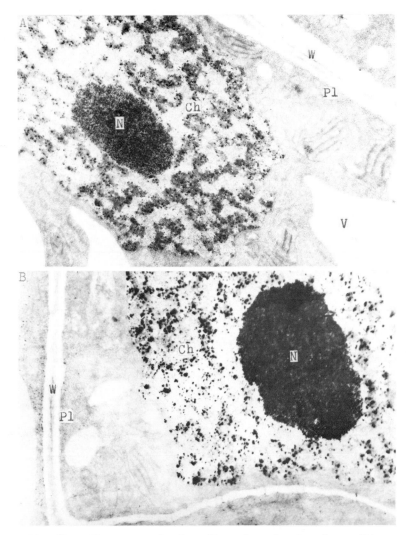

FIG. 9A. The young leaf cells of nonhardened seedlings at 3-leaf stage, 5 h incubation at 5°C, x 10540. Figure 10B. The young leaf cells of non-hardened young sprouts, 5 h incubation at 5°C, x 9350.

plasmalemma) isolated from the tissues of chilling-sensitive tomato and cucumber and from chilling-resistant cauliflower were similar to those observed for mitochondrial membranes (19). Palta and Li (24) suggested that the primary site of sublethal freezing injury may be at the active ion-transport system of cell membranes. As mentioned above, our experimental

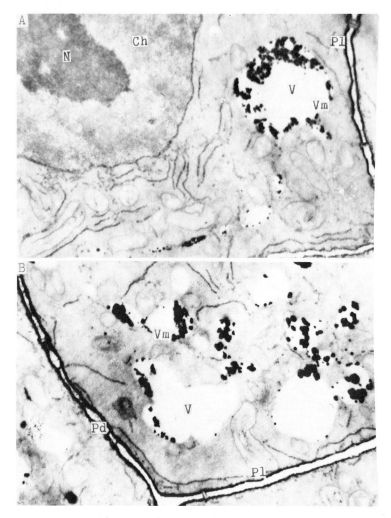

FIG. 10A and B. The young leaf cells of hardened seedlings
at 3-leaf stage were incubated for 5 h at 5°C.   A, x 10115; B,
x 9075.

results indicated that during freezing injury of winter wheat
seedlings, the plasmalemma ATPase activity was firstly lowered
and inactivated.   The change in ATPase activity in cotyledon
cells of tomato and cucumber seedlings during chilling stress
(12,13) is similar to that of wheat mentioned above.   Our
experimental results supported Palta and Li's concept.
       As plasmalemma ATPase activity is very sensitive to cold
stress, cold hardening must involve the increase of the cold-
tolerant characteristic of plasmalemma ATPase activity, so that

it is allowed to survive lower temperature. In fact, we not only found that the plasmalemma ATPase activity was sensitive to freezing injury, but also found that the cold-tolerant characteristic of plasmalemma ATPase activity could be developed during cold hardening. Therefore, it is suggested that a close relationship exists between the plasmalemma ATPase activity and the freezing injury and cold hardiness of plants. The increase of the cold tolerance of plasmalemma ATPase activity does not only enable it to avoid freezing injury, moreover it probably plays an important role in the development of frost resistance of plants. It is possible that if high plasmalemma ATPase activity can be maintained at low temperatures, ion transport will not be depressed, so that plants will function normally during cold hardening process.

B. ATPase Activity Associated with Organelles in Relation
     to Freezing Injury and Cold Hardiness

The ATPase activity in mitochondria, Golgi bodies, endoplasmic reticulum and vacuoles in some plants has been observed by means of the enzymic cytochemical methods (1,7,11), but it is not generally found in mitochondria, endoplasmic reticulum, Golgi bodies and vacuoles in the tillering node cells (11) and young leaf cells of wheat seedlings under normal growth conditions. However, ATPase activity was exhibited at endoplasmic reticulum, Golgi bodies, tonoplasts and the prolamellar membranes of proplastids in the young leaf cells of wheat seedlings during freezing injury. This suggested that the ATPase activity located at endoplasmic reticulum, Golgi bodies, tonoplasts and plastids may be inhibited under normal growth conditions of plants, but it is activated during freezing injury.

As regards the sequence of changes in ATPase activity associated with different organelles during freezing injury, they seem to be as follows. The decrease in ATPase activity located at nucleoli and chromatin was followed by the decrease at plasmalemma, and the activation of ATPase activity at endoplasmic reticulum, Golgi bodies and tonoplasts was induced. These changes are probably recoverable on the basis of the survival percentage of seedlings and cellular structural integrity after freezing. When seedlings were further subjected to freezing injury, the ATPase at the prolamellar membranes of proplastids was activated. This activation may be a sign that plants are going to die at critical temperature. These results are different from the morphological changes in ultrastructure of proplastids (chloroplasts) observed under electron microscope during low temperature stress. When the callus tissues of Cornus stolonifera were subjected to chilling treatment, ultrastructural changes in proplastids were firstly detected, and then rough endoplasmic reticulum, Golgi bodies and tonoplasts were altered (23,27). The membrane alterations of plastids (chloroplasts) in tomato cotyledons preceded other cellular changes too during short periods of chilling (9). During freezing injury of wheat seedlings chloroplast ultra-structures were also altered at first (4). However, the acti-vation of ATPase at mitochondrium membranes was not observed during freezing injury. This result is identical with mito-chondrium ultrastructure which showed more stability than other organelles during low temperature stress (4,23).

When winter wheat seedlings were subjected to cold hardening, insignificant ATPase activity was shown at endo-plasmic reticulum, Golgi bodies, mitochondria and proplastids in hardened tissues as well as nonhardened tissues, while the ATPase at tonoplasts and in vacuoles exhibited its high activ-ity. The latter may play an important role in the accumulation of solutes in vacuoles during cold hardening period. At 5°C incubation, the ATPase activity at nucleoli and chromatin in hardened tissues was lowered or reduced to none. This may be related to the decrease of the physiological activity of nucleoli and the rate of cell mitoses of the seedlings during low temperature hardening in late autumn and early winter (2). It is possible that the ATPase activity plays a regulating role in physiological activity of nucleoli and the rate of cell mitoses. Undoubtedly the physiological activity of nucleoli and the action of cell mitoses require energy. As the ATPase activity is lowered at low temperature, the quantity of energy is decreased, and so the physiological activity of nucleoli and the rate of cell mitoses also slowed down. It is well knwon that the decrease and stop in both cell mitoses and growth is a preceded condition for the development of hardiness in plant.

REFERENCES

1. Bentwood, B.J. and Cronshaw, J., Planta 140:111-120 (1978).
2. Chien (Jian), L.C., and Wu, S.H., Scientia Sinica 15:836-854 (1966).
3. D'Auzac, J., Physiol. Veg. 16:37-65 (1978).
4. Dong, H.Z., Sun, L.H., and Jian, L.C., Acta Botanica Sinica 22:339-342 (1980).
5. Glier, J.H. et al., Cryobiology 14:121-123 (1977).
6. Hall, J. L., in "Ion Transport in Plants" (W.P. Anderson, ed.), pp. 11-24. Academic Press, London, 1973.
7. Hall, J.L., Journal of Microscopy 93:219-225 (1971).
8. Heber, U. Cryobiology 5:188-201 (1968).
9. Ilker, R. et al., in "Low Temperature Stress in Crop Plants" (J.M. Lyons et al., eds.), pp. 97-113, Academic Press, New York, 1979.
10. Jean-Pierre Simon, Plant, Cell and Environment 2:17-21 (1979).
11. Jian, L.C., Dong, H.Z., and Sun, L.H., Acta Biologiae Experimentalis Sinica 12:135-145 (1980).

12. Jian, L.C., Dong, H.Z., and Sun, L.H., Acta Biologiae Experimentalis Sinica 13:437 (1980).
13. Jian, L.C., Dong, H.Z., and Sun, L.H., Acta Botanica Sinica 23:257-261 (1981).
14. Kacperska-Palacz, A. and Uliasz, M., Physiol. Veg. 12:561-570 (1974).
15. Kovacs, I., Fejer, O. and Devay, M., Biochem. Physiol. Pflanz. 173:327-332 (1978).
16. Krasnuk, M., Witham, F.H., and Jung, G.A., Cryobiology 13:225-242 (1976).
17. Levitt, J. "Responses of Plants to Environmental Stresses" Academic Press, New York, 1972.
18. Lyons, J.M., and Asmundson, C.M., J. Amer. Oil. Chem. Soc. 42:1056-1058 (1965).
19. Lyons, J.M., and Raison, J.K., Plant Physiol. 45:386-389 (1970).
20. Lyons, J.M., Ann. Rev. Plant Physiol. 24:445-466 (1973).
21. Lyons, J.M., Raison, J.K., and Steponkus, P.L., in "Low Temperature Stress in Crop Plants" (J.M. Lyons et al., eds.), pp. 1-24. Academic Press, New York, 1979.
22. McMurchie, E.J., in "Low Temperature Stress in Crop Plants" (J.M. Lyons, et al., eds.), pp. 163-176. Academic Press, New York, 1979.
23. Niki, T., Yoshida, S., and Sakai, A., Plant and Cell Physiol. 19:139-148 (1978).
24. Palta, J.P., and Li, P. H., in "Plant Cold Hardiness and Freezing Stress" (P.H. Li and A. Sakai, eds.), pp. 93-115. Acad. Press, New York, 1978.
25. Racker, E., Biochem. Soc. Trans. 3:785-802 (1975).
26. Steponkus, P. L., and Wiest, S.C., in "Plant Cold Hardiness and Freezing Stress" (P.H. Li and A. Sakai, eds.), pp. 75-91. Acad. Press, New York, 1978.
27. Yoshida, S., Niki, T., and Sakai, A., in "Low Temperature Stress in Crop Plants" (J.M. Lyons, et al., eds.), pp. 275-290. Academic Press, New York, 1979.
28. Yoshida, S., in "Plant Cold Hardiness and Freezing Stress" (P.H. Li and A. Sakai, eds.), pp. 117-135. Acad. Press, New York, 1978.
29. Yoshida, S., Plant Physiol. 64:252-256 (1980).

FREEZING-INDUCED MEMBRANE ALTERATIONS:
INJURY OR ADAPTATION?

Elzbieta Sikorska[1] and Alina Kacperska[2]

Institute of Botany
University of Warsaw
Warsaw, Poland

I. INTRODUCTION

Most of the current work on the mechanisms of low temperature action on the plant cell is focused on the question: What
are the primary responses of plant cells to lowered temperature? The molecular ordering of membrane lipids is indicated
as the controlling element in those responses but the role of
certain "master" protein as a sensor of low temperature is also
taken into consideration (10). Recent results of Palta and Li
(11) suggest that the first alteration during the initial
stages of freezing injury concerns the intrinsic membrane proteins that participate in ion transport.

In all reports dealing with the membrane responses to
chilling or freezing temperature, attention is drawn to those
effects that disorder or impair membrane functions and lead to
injury. However, it is well known that many plants can be
hardened against detrimental effects of temperature if first
exposed to a moderate stress. Hence, the following questions
arise:

  (i) - What are the molecular events that allow cell recovery
        and cell remodelling during the post-stress period?
 (ii) - Are the adaptative responses qualitatively different
        from those which are commonly recognized as detri-
        mental effects?

---

[1] Part of the experiment was performed at the Biological
     Research Center, Hungarian Academy of Science, Szeged,
     Hungary, in cooperation with Dr. T. Farkas.
[2] To whom all the correspondence should be sent.

Experiments, discussed below, were undertaken as an attempt to answer these questions.

II. MATERIALS AND METHODS

A. Plant Material and Freezing Procedures

Winter rape plants (Brassica napus, L., var. oleifera L.) were grown in sand supplied with Hogland solution (18). After two weeks of growth under 16 h photoperiod at light intensity of 14 kl and at 20°C (day) or 15°C (night), plants were subjected to hardening procedure, as described earlier (19,22).

Samples for analysis were taken after 5 weeks of hardening at low (>0°C) temperature (the first stage of hardening, 19) and one week later when plants were subjected to subfreezing (-3°C) temperature during nights (the second stage of hardening). Control plants were grown at 20°C (day) or 15°C (night) for 4 weeks ("0" stage of hardening). All the analyses were performed on the 3rd and 4th leaf blades. Leaf blades or leaf discs (14 mm in diameter) were frozen at temperature inducing either small or high injury, the temperature range being chosen according to our previous experiments (19). Samples were kept at the desired temperature for 2 h and thawed for 1 h or otherwise, as specified in the particular experiments. Electrolyte leakage from the frost-thawed leaf discs, stored for 20 h at 5°C in darkness, was taken as a criterion of the tissue injury. It was determined with an electric conductivity method, as described earlier (18).

B. Extraction and Analysis of Phospholipids

Leaf discs (lots of 20 in three replicates) collected at the indicated time and subjected to freezing that caused low ($I_t < 20\%$) or high ($I_t > 60\%$) injuries were homogenized in dry ice-cold chloroform:methanol (1:2, v/v). Lipids were extracted according to the procedure of Allen et al. (1) and separated by two-dimensional thin-layer chromatography (20). The phosphorus content in the spots and in the original extracts was estimated according to the procedure of Ames (2). Quantities of the particular phospholipid fractions were expressed as μg P per 20 discs or as percentage of the total phospholipid content.

C. Phospholipase D Activity Determinations

Two grams (fr.wt.) of a fresh or frost-thawed leaf sample were homogenized in 8 ml of distilled water for two minutes at about 20°C, in duplicates. Homogenate was filtered through 4 layers of cheese-cloth and the filtrate (pH 6.2-6.3) was used

as a source of the enzyme. In most experiments no enzyme acti-
vators were added to the homogenization medium in order to keep
the reaction conditions as close to the native ones as possible. In some experiments, enzyme modifiers such as Ca^{2+}
($CaCl_2$), Mg^{2+} ($MgCl_2$) or ionic detergent, sodium dodecyl
sulphate (SDS), were added to the homogenization medium at the
following respective concentrations: 0.38 $mol \cdot dm^{-3}$,
0.48 $\cdot dm^{-3}$, and 1.5 $mol \cdot dm^{-3}$.

For estimation of phosphatidyl hydrolase activity, the
enzyme extract (0.5 ml), containing about 4 mg of protein (as
checked with the Lowry method, 9) was incubated for 15 min at
27°C with 1.2 μmol phosphatidyl choline (egg lecithin, Merck)
or without exogenous substrate. For estimation of phosphatidyl
transferase activity, 1.2 μmol glycerol were added to 0.5 ml
of the enzyme extract. In some experiments, glycerol was not
added and the reaction was run only in the presence of endogenous glycerol. Other reaction conditions were the same as for
determination of hydrolytic activity.

The reaction was terminated by addition of boiling iso-
propanol (1 ml per 100 μl or 250 μl of the reaction
mixture). The resulting mixture was further subjected to phospholipid analysis. Phospholipids were extracted with the
chloroform:methanol (1:2, v/v) mixture, separated by two-
dimensional thin-layer chromatography and identified according
to Rouser et al. (16). The phosphorus content of the spots was
estimated according to the procedure of Kahovcova and Odavic
(8), all allowing for determination of phosphorus content in a
range of 0.02 μmol to 0.2 μmol in a sample.

The hydrolytic and transferase activities of phospho-
lipase-D were expressed as the amount of liberated phosphatidic
acid (PA) or the amount of formed phosphatidyl glycerol (PG),
respectively, per 1 mg protein and per unit time. In experiments on the effects of freezing and thawing on the enzyme
activity, the activity shown by the homogenized non-frozen
sample was taken as a reference (100%).

III. RESULTS

A. Phospholipid Modification by Freezing

In the previous work (18) we showed that slight frosts were
needed to trigger the second stage of hardening in winter rape
plants. Moreover, a high positive correlation was found
between frost tolerance and total phospholipid content in
leaves collected during the second stage of hardening. Further
studies (19) indicated that pronounced changes in phospholipid
composition took place in the winter rape leaves subjected to
freezing and thawing. We observed a pronounced decrease of
phosphatidyl choline (PC) content, both in slightly and in

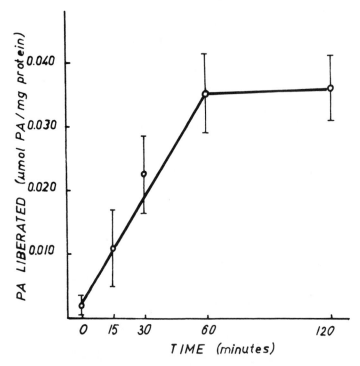

FIG. 1.   Time course of phospholipase D reaction with endogenous substrate (hydrolytic activity).

highly injured leaves.   We also found that the PC decrease resulted in a concomitant increase of phosphatidic acid (PA), but only in the highly injured tissues.   In contrast, in the slightly injured ones, not PA, but phosphatidyl glycerol (PG) and unknown P-containing lipid content increase was observed. On the basis of these results, the possible association of frost hardiness with modified phospholipase D (E.C.3.1.4) activity has been proposed (19).

B.   Frost-Induced Modification of Phospholipase D Activity
in Leaves Hardened to Different Extent

Phospholipase D hydrolytic activity has been observed in many plant tissues of different functions (6).   Nevertheless, its physiological function is still unknown.   The enzyme seems to be one of the structural proteins of membranes (15) but its occurrence in the soluble fraction of the cell was also reported (24).   Hydrolytic activity of the membrane-bound phospholipase D "in vivo", under physiological conditions, was questioned (15).   However, Yoshida (26) was able to demonstrate that activation of the membrane-bound phospholipase D by

Table I. Influence of freezing and thawing of a tissue upon phospholipase D hydrolytic activity (PC added to the reaction medium).

A. The effect of freezing

Time of freezing at -9°C (min.)	The enzyme activity (in percent of the unfrozen control)	Index of injury (%)
1	117	23
30	130	27
60	102	23
120	118	22

B. The effect of tissue storage at 5°C after freezing at -9°C for 2 h.

Period of storage (min.)	The enzyme activity (in percent of the unfrozen control)	Index of injury (%)
0	128	38
15	174	67
30	178	58
60	172	60
120	165	70

freezing and thawing was reponsible for phospholipid degradation in the frost-thawed bark tissues of black locust tree.

Our experiments showed that the amount of phosphatidic acid extracted from the leaves killed in boiling ethanol prior to homogenization, was very low (Fig. 1). However, it increased dramatically within 60 min in homogenates obtained from fresh tissue. This indicates a very low or possibly no hydrolytic acitivity of phospholipase D in the intact cells and activation of the enzyme by a mechanical disruption of the cells. The total activity of the enzyme observed after 1 h was probably both that resulting from liberation of a soluble form of the enzyme from vacuoles (owing to mechanical damage of tonoplast) and that due to activation of the membrane-bound fraction. Thus, in further experiments on effects of freezing and thawing on phospholipase D activity in winter rape leaves, the activity brought about by homogenization of the tissue was taken as a

Table II. Phosphatidyl hydrolase activity in the extracts obtained from leaves showing a small (A) and a high (B) injury (nmol PA/mg protein/15 min $\pm$ S.D.).

A. $I_t < 10\%$

Material	Unhardened stage 0	Hardened	
		stage I	stage II
Unfrozen	239 $\pm$ 30	160 $\pm$ 20	140 $\pm$ 20
Frozen	100 $\pm$ 30	140 $\pm$ 30	139 $\pm$ 10
Inhibition (%)	58	13	1

B. $I_t > 60\%$

Material	stage 0	stage I	stage II
Unfrozen	270 $\pm$ 30	160 $\pm$ 20	140 $\pm$ 20
Frozen	650 $\pm$ 25	240 $\pm$ 40	148 $\pm$ 30
Stimulation (%)	140	50	6

reference (100%). We assumed that the net changes of the activity (above or below the non-frozen control level) would indicate the frost-induced alterations in membranes' functions.

Further experiments showed that freezing of the tissue at $-9°C$ for 2 h caused only a slight increase of phospholipase D activity in comparison to the non-frozen control, when it was determined immediately after thawing (Table I). The length of the freezing period had no effect on the studied activity. However, the activity increased within 15 min of the post-freezing period (Table IB), during which the tissue was kept at 5°C. Simultaneously, about two-fold increase of the injury level was observed (compare data in Table IA and B).

These results indicate that freezing brings about certain alterations in the cell which cause the development of phospholipase D activity at temperatures higher than 0°C. In further experiments on the influence of freezing and thawing on phospholipase D activity, all the determinations were performed after 1 h storage of the previously frozen tissue at 5°C in darkness.

Table III. Phosphatidyl transferase activity in the extracts obtained from leaves of different hardiness level showing a small or a high freezing injury (in nmol PG/mg protein/15 min $\pm$ S.D.).

A.   The reaction run in the presence of 1.2 μmol of glycerol.

Material	Hardiness level		
	0	I	II
Unfrozen	102 ± 15	98 ± 3	78 ± 20
Frozen $I_t < 10\%$	144 ± 25	115 ± 1	77 ± 9
Frozen $I_t > 60\%$	65 ± 8	56 ± 10	52 ± 1

B.   The reaction run in the presence of endogenous glycerol only.

Material	Hardiness level		
	0	I	II
Unfrozen	8 ± 1	48 ± 4	22 ± 2
Frozen $I_t < 10\%$	0	109 ± 3	0
Frozen $I_t > 60\%$	0	50 ± 9	0

Further experiments were aimed at checking whether the enzyme activity in the frost-thawed tissue depended on the degree of injury and on the stage of plant hardening. The results shown in Table II indicate that phosphatidyl hydrolase activity in the highly injured cells was 140% higher than that in the unfrozen tissue. In the tissue injured to a small extent (less than 10%), the opposite effect was observed: the activity was considerably inhibited (by about 60%).

Hardening treatment brought about a marked decrease of the activity in the unfrozen tissue and, simultaneously, it diminished the effects of freezing and thawing: no significant changes in phosphatidyl hydrolase activity were observed in both highly and slightly injured cells when leaves at the second stage of hardening were studied (Table IIA and B). The effect of hardening was already very pronounced in leaves at the first stage of hardening.

Our previous experiments showed that in the slightly injured leaves at the first hardening stage, a decrease of phosphatidyl choline level was concomitant with accumulation of phosphatidyl glycerol and not of phosphatidic acid. Therefore, we examined the phosphatidyl transferase activity in the leaves. Such an activity has previously been observed in experiments "in vitro", performed with phospholipase D extracted from cabbage leaves and peanuts (5,24).

The results presented in Table III indicate that freezing caused pronounced modifications of that activity. In the highly injured tissue there was a decrease of the activity while in the slightly injured one, its increase (by about 40%) was observed, provided that the reaction medium was supplied with exogenous glycerol (Table IIIA). In the presence of endogenous glycerol (Table IIIB), the freezing effects were observed only for leaves at the first hardening stage.

Hardening treatment decreased the transferase activity in unfrozen leaves when reaction was performed in the presence of exogenous glycerol. However, considerably higher (about 6-fold) transferase activity was noted in leaves at the first hardening stage, in comparison to the unfrozen ones, if the reaction was performed in the absence of exogenous glycerol. These apparently inconsistent results may be easily explained if one assumes that glycerol content in the tissue is the reaction-limiting factor in non-hardened and fully hardened leaves, but its content in the leaves at the first hardening stage is sufficient for the reaction to occur "in situ". This is the first indication that frost hardening of the plant tissue may involve activation of phosphatidyl transferase and that glycerol may be a factor of great importance for that process.

The question was subsequently examined whether divalent cations (Ca^{2+} and Mg^{2+}) and anionic detergent, sodium dodecyl sulphate (SDS), factors that are known to affect membrane structure and/or phospholipase D structure and catalytic properties (6,7,13,25), would modify the enzyme activity in winter rape leaves. In that part of the experiments, only phosphatidyl hydrolase activity of phospholipase D was taken into consideration. Data presented in Table IV indicate that the addition of Ca^{2+} or SDS to homogenization medium stimulated the hydrolytic activity of phospholipase D in the extracts from nonhardened leaves, similarly as it was observed for freezing. On the contrary, Mg^{2+} ions markedly inhibited that activity. The first stage of hardening decreased the sensitivity of the tissue to freezing as well as to Ca^{2+} and SDS but was of no effect on magnesium ion action. It can also be noted that the second stage of hardening did not further decrease the effects of Ca^{2+} and of SDS but it decreased tissue sensitivity to Mg^{2+} action. The observations may suggest that the second stage of hardening brought about some qualitative difference in the regulatory properties of the enzyme molecule or in its cellular environment.

Table IV. Comparison of the effects of cations (Ca^{2+} and Mg^{2+}) and of a detergent with freezing and thawing effects on phosphatidyl hydrolase activity in the extracts obtained from leaves of different hardiness level (in nmol PA/mg protein/15 min $\pm$ S.D. or in % of control). The reaction was run in the presence of endogenous PC only.

Treatment	Level of leaf hardiness					
	0		I		II	
		%		%		%
Control	11 ± 9	100	27 ± 7	100	31 ± 4	100
Frozen $I_t > 60\%$	30 ± 5	273	9 ± 5	181	30 ± 5	100
Unfrozen + Ca^{2+}	30 ± 10	273	59 ± 9	219	70 ± 5	226
Unfrozen + SDS	37 ± 20	336	8 ± 10	178	57 ± 10	184
Unfrozen + Mg^2	4 ± 1	37	10 ± 5	37	27 ± 9	87

Table V. The effect of hardening on Ca^{2+} and Mg^{2+} contents in the winter rape leaves (in μmol dm^{-3} $\pm$ S.D.).

Cation	Stage of hardening		
	0	I	II
Ca^{2+}	3.74 ± 0.24	3.70 ± 0.22	4.00 ± 0.30
Mg^{2+}	4.10 ± 0.30	4.70 ± 0.10	5.14 ± 0.26

Analysis of Ca^{2+} and Mg^{2+} content in water extracts of winter rape leaves revealed that freezing did not cause any significant difference in the studied ion levels. However, the content of Mg^{2+} significantly increased during plant hardening, whereas the level of Ca^{2+} remained unchanged (Table V). Thus, one may suppose that hardening process, leading to the increased content of free Mg^{2+} protects the cell against development of phospholipase D activity upon freezing and thawing. The observations are in accordance with the earlier findings of Yoshida (25) for phospholipase D from black locust tree.

IV. DISCUSSION

The results presented here show that freezing and thawing activates phospholipase D in the rape cells and may induce a shift in the enzyme activity from phosphatidyl hydrolase to transferase, depending on the degree of cell membranes' alteration (as indiacted by electrolyte efflux from the tissue). It seems reasonable to propose that either lipid phase transition or separation, or a change of membrane structure from bilayer to hexagonal one, may be the primary reason for development of phospholipase D hydrolytic activity, because the factors which are known to induce similar modifications in artificial phospholipid membranes, e.g. freezing and Ca^{2+} (4,12,17) were now found to promote the hydrolytic activity of the enzyme. Modification of molecular ordering of a lipid domain of the membranes is likely to induce a change in conformation of the enzyme molecule. Such a change was previously proposed by Yoshida (25) in order to explain the opposed effects of Ca^{2+} and Mg^{2+} on phosphatidyl hydrolase activity in the living bark tissues of black locust tree. In that respect, the inhibitory action of Mg^{2+} on phospholipase D hydrolytic activity in less hardy winter rape leaves might be due to its competitive binding to the regulatory subunit of the enzyme, as proposed by Yoshida (25). In more hardy rape leaves (the second stage of hardening) the inhibitory effect of magnesium decreased, this indicates that some qualitative changes in membranes might occur and affect the catalytic properties of the enzyme.

Our studies also showed that an initial modification of membrane properties, which only slightly increased the electrolyte leakage (and did not disturb the ion transport system) allowed promotion of phosphatidyl transferase activity in winter rape leaves. It seems plausible that freezing-induced initiation of lipid separation and the presence of such alcohols as glycerol or ethanol amine (23) are needed to promote that kind of phospholipase D activity. That supposition should be verified in the future. Taking into consideration all the above finds we suppose that changes in a lipid domain of the membranes rather than direct effects of temperature on proteins are primary response of the plant cell to freezing and thawing and possibly to low temperature in general; these changes would lead to development of either phosphatidyl transferase or hydrolase activities. The frost induced modifications of phospholipase D activities may then lead either to injury or to recovery of the cell, depending on the degree of membrane structure alterations (Fig. 2).

In the latter case, the maintenance of a high phosphatidyl choline (or ethanolamine) level would render the cell membranes less susceptible to freezing: phospholipid polar head groups, especially those of phosphatidyl choline and phosphatidyl

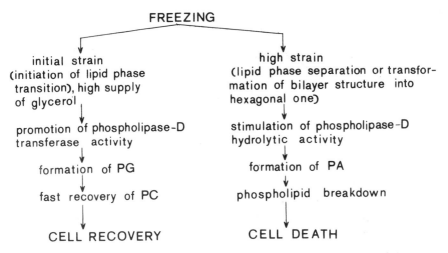

FIG. 2. Proposed scheme of events that may lead either to the cell injury or to the recovery.

ethanolamine, have been shown to play an important role in maintenance of membrane fluidity over a broad range of temperature (3,14). Thus, hardening effect of slight frost during the second stage of hardening of winter rape plants (19) seems to rely not only on activation of a recovery system in the cell (with phosphatidyl transferase taking part in phospholipid turnover) but also on putting into operation mechanism(s) that allow cell structure remodelling.

The reported results are in accordance with our earlier propositions (20) that two phases of plant response to low temperature can be distinguished: (i) the reaction phase, when a transient disturbance in several cell properties may occur and (ii) the restitution phase - when a new structural and functional equilibrium is formed or reestablished. Similar phases were previously observed by Stocker (21) for plants responding to water stress. Searching through literature one can easily find other examples of the reaction and the restitution phase occurring during plant response to other stresses. Thus, the phenomenon may have a general character. Ability of the plant to pass from the reaction phase to the restitution phase may constitute the basic difference between species capable and uncapable for hardening.

REFERENCES

1. Allen, C.F., Good, P., Davis, H.F., Chisum, P., and Fowler, S.D., J. Am. Oil Chem. Soc. 223-230 (1966).
2. Ames, B.N., in "Methods in Enzymology" (S.P. Coolowick, and N.O. Kaplan, eds.), p. 8, 115-118. Academic Press, New York, 1966.
3. Cronan, J.E., Jr., Ann. Rev. Biochem. 47:163-189 (1978).
4. Cullis, P.R., and Verleijl, A.J., Biochim. Biophys. Acta 552:546-551 (1979).
5. Dawson, R.M.C., Biochem. J. 102:205-210 (1967).
6. Heller, M., Advances in Lipid Research 16:267-326 (1978).
7. Heller, M., Mozes, N, Peri (Abramovitz), I., and Mozes, E., Biochim. Biophys. Acta 369:397 (1974).
8. Kahovcova, J., and Odavic, R., J. Chromatogr. 40:90-96 (1969).
9. Lowry, O.H., Rosenbrough, N.J., Farr, A.L. and Randall, R.J., J. Biol. Chem. 193:265-275 (1951).
10. Lyons, J.M., Raison, J.K., and Steponkus, P.L., in "Low Temperature Stress in Crop Plants. The Role of the Membranes" (J.L. Lyons, D. Graham, and J.H. Raison, eds.), pp. 1-24. Academic Press, New York, 1979.
11. Palta, J.P., and Li, P.H., Physiol. Plant. 50:169-175 (1980).
12. Papahadjonoulos, D., Vail, W.J., Newton, C., Nir, S., Jacobson, K., Poste, G., and Lazo, R., Biochim. Biophys. Acta 465:579-598 (1977).
13. Quarles, R.H., and Dawson, R.M., Biochem. J. 122:795 (1969).
14. Reinert, J., and Steim, J.M., Science 168:1580-1582 (1970).
15. Roughan, P.G., and Slack, C.R., Biochim. Biophys. Acta 431:86 (1976).
16. Rouser, G., Fleischer, S., and Yamato, A., Lipids 5:494-496 (1970).
17. Sikorska, E., Ph.D. Thesis, University of Warsaw (1981).
18. Sikorska, E., and Kacperska-Palacz, A., Physiol. Plant 47:144-150 (1979).
19. Sikorska, E., and Kacperska-Palacz, A., Physiol. Plant 48:201-206 (1980).
20. Sobczyk, E.A., and Kacperska-Palacz, A., Acta Physiol. Plant. 2:123-131 (1980).
21. Stocker, O., in "Plant Water Relations in Arid and Semiarid Zones", pp. 63-104. UNESCO, Paris, 1960.
22. Vigh, L., Horvath, I., Horvath, L.I., Dudits, D., and Farkas, T., FEBS Letters 107:291-294 (1979).
23. Yang, S.F., Freer, S., and Benson, A.A., J. Biol. Chem. 242:477-484 (1967).
24. Yoshida, S., Plant Physiol. 64:241-246 (1979).
25. Yoshida, S., Plant Physiol. 64:247-251 (1979).
26. Yoshida, S., Plant Physiol. 64:252-256 (1979).

FLUORESCENCE POLARIZATION STUDIES ON PLASMA MEMBRANE ISOLATED FROM MULBERRY BARK TISSUES

S. Yoshida

The Institute of Low Temperature Science
Hokkaido University
Sapporo, Japan

It is generally accepted that the main targets of freezing injury are cellular membranes, especially plasma membranes (10,11,15,18). It follows therefore that cold acclimation would involve an alteration of the plasma membrane components to meet with the biophysical requirements either to resist freeze-dehydration stress or to resist a thermotropic transition of the lipid bilayers.

There are several approaches to look for the related changes in chemical components of cells, especially membrane lipids during cold acclimtaion (19). This information, however, is based on the analytical data of bulk membrane systems. Therefore, it is not evident whether the observed changes are really related to the changes in plasma membranes.

One of the main reasons for this is laid in the difficulty of the isolation of plasma membrane from those plant cells in sufficiently pure form. Recently the author has succeeded in isolation of plasma membrane vesicles from living bark tissues of mulberry trees in a relatively pure form. This enables us to investigate the chemical and the physical properties of plasma membrane with special reference to cold acclimation and freezing injury of plant cells.

Temperature dependent phase transition of membrane lipids is now a well established phenomena in a wide variety of chilling sensitive plants (9). This phase transition leads to an alteration of membrane function that eventually causes cell injury. However, only limited information is available as to the phase transition of membrane lipids which may be related to frost injury of plant cells (1,4,13,14). The present work was designed to get information about the structural transition of plasma membrane at sub-freezing temperatures, which will be responsible to freezing injury.

I.   MATERIALS AND METHODS

A.   Preparation of Plasma Membrane Vesicles

Living bark tissues of current twigs of mulberry trees (Morus bombycis Koidz. cv. Goroji) sampled in different seasons were the major experimental material. The tissues were ground in a motor-driven mortar and pestle assembly with washed sea sand and Polyclar AT in homogenizing medium. The homogenizing medium consisted of 1.0 M sorbitol, 150 mM Tris-HCl, 15 mM EGTA, 5% PVP and 20 mM ß-mercaptoethanol, pH 7.8. Five ml of the medium, 0.3 g Polyclar AT and 1.4 g of the abrassive were used per 2 g fresh tissues. The brei was passed through two layers of gauze and then one layer of Miracloth, and successively centrifuged at 1,000 g for 7 min, 15,000 g for 20 min and 174,000 g for 30 min. Pellets obtained by centrifugation at 15,000 and 174,000 g were respectively, designated as crude mitochondria and crude microsome. These fractions were washed once with 1.0 M sorbitol-5 mM EDTA-10 mM Tris-MES, pH 7.2 and resuspended in 0.7 M sorbitol-1 mM EDTA-10 mM Tris-MES, pH 7.2 by gentle homogenization in a Teflon homogenizer.

The suspended membrane fractions were homogeneously mixed with Percoll solution in 0.7 M sorbitol-1 mM EDTA-10 mM Tris-MES, pH 7.2 in a final concentration of 16% (v/v). These Percoll-membrane mixtures were centrifuged on Hitachi RP 50.2 (a fixed angle rotor) at 183,000 g for 21 min. After the centrifugation, 1.2 ml aliquots were collected from the top of the tube by using a density gradient fractionator, ISCO 640.

As indicated in Fig. 1, more than 60% of pH 6.5 Mg-ATPase-associated membrane (plasma membrane) were recovered in microsomal fraction, having an activity peak at fraction 9. On the other hand, more than 80% of cyt c oxidase and acid phosphatase activities and more than 70% of IDPase activity were recovered in mitochondrial fraction. The activity peak of pH 6.5 Mg-ATPase in microsomal fraction, however, was overlaped with alkaline ATPase, acid phosphatase, IDPase, and partially with NADH cyt c reductase and cyt c oxidase activities.

The plasma membrane fractions were diluted with 0.7 M sorbitol-1 mM EDTA-10 mM Tris-MES, pH 7.2 and pelleted at 189,000 g for 25 min. To achieve further purification pellets were then suspended in 1 mM EDTA-10 mM Tris-MES, pH 7.2 and subjected to a discontinuous sorbitol gradient of 1.4 and 2.5 M. After this step plasma membrane vesicles were banded at the interface of 1.4/2.5 M sorbitol layers and the acid phosphatase-entrapped membrane vesicles (presumably vesiculated tonoplasts), mitochondria and fragmented chloroplasts were effectively separated from plasma membrane vesicles either at the interface of sample/1.4 M sorbitol layers or at the bottom of the tube. The plasma membrane vesicles were finally

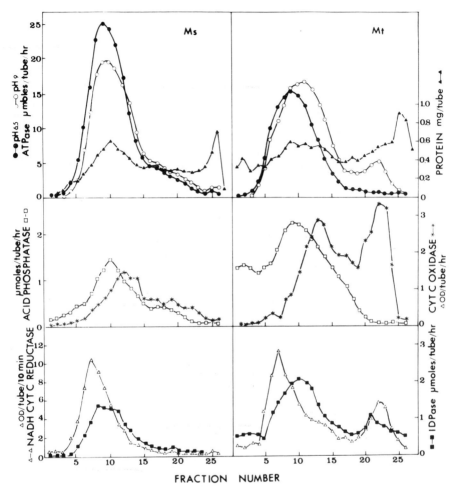

FIGURE 1. Activity profiles of various marker enzymes after Percoll gradient of microsomal (Ms) and mitochondrial (Mt) fractions from mulberry bark tissues. ○ , ● : pH 6.5 and 9.0 ATPases, ▲ ; protein content, □ ; acid phosphatase, ★ ; cyt c oxidase, △ ; NADH cyt c reductase, ■ ; IDPase.

purified on linear sucrose gradient (15–50% w/w) at 96,000 g for 13,5 hr. The plasma membrane thus obtained was free from contaminations by mitochondria, endoplasmic reticulum, acid phosphatase-entrapped vesicles (tonoplast vesicles) and fragmented chloroplasts, however still partially contaminated with Golgi vesicles.

Plasma membrane was identified by phosphotungstate-chromate (PTA-CrO$_3$) stain under electron microscopy and by characterizing the nature of the membrane-bound ATPase. Plasma membrane fraction obtained by the procedure described above was dominantly stained with the PTA-CrO$_3$ and more than 65% purity was estimated by the electron micrograph (data not shown). The membrane-bound ATPase had the optimal pH at 6.5 and slightly stimulated by 50 mM KCl. The ATPase was highly inhibited by vanadate and DCCD which are specific inhibitors for mammal and bacterial plasma membrane ATPases, and also for plant plasma membrane (21), but non-sensitive to azide and oligomycine which are specific inhibitors for mitochondrial ATPase.

B. Fluorescence Polarization Measurement of Membranes and the Liposomes

The fluorescence hydrocarbon 1,6-diphenyl 1,3,5-hexatriene (DPH) was used as a probe for monitoring the thermotropic behavior of lipid bilayer of the membrane and the liposomes. The final concentration of DPH was 10^{-6} M per 100 µg of protein or 100 µg of the extracted lipids. The steady-state fluorescence polarization was measured in Elscient microvicometer mv-1a with a device constructed so that vertically and horizontally polarized emission components could be measured simultaneously. DPH was excited at 375 nm and the emission was detected through a cut-off filter for wave length below 400 nm. The steady-state polarization was expressed as the fluorescence anisotropy, r and as the anisotropy parameter $(r_o/r - 1)^{-1}$, which varies proportionally with the rotation relaxation time of the fluorophore, where r_o is the maximal limiting anisotropy. The value used for r_o was 0.362 for DPH. Within the limitations as reported (7), the anisotropy parameter is related to the relative motional freedom of the DPH without distinguishing specific mechanisms affecting its depolarization motions and provides thereby a comparative index of the fluidity of membrane lipid bilayer. These specific mechanism include alteration of the rate of rotation of DPH owing to the viscous environment, anisotropic rotations and hindered motions owing to structural factors (2,5,6,7). The temperature dependence of the anisotropy parameter was determined over range between 30 and -20°C and the logarithm of the parameter was plotted against 1/T to detect a thermotropic transition by the inflection of the slopes (7).

TABLE I. Phospholipid and free sterol content of plasma membrane vesicles isolated from hardy and tender bark tissues of mulberry trees.

Sample	Hardiness[a] (°C)	Phospholipid Protein μmoles/mg	Sterol Protein μmoles/mg	Sterol Phospholipid μmoles/μmoles
Sept. 21	−3	1.06	0.39	0.36
March 21 (Control)	−70	1.73	0.41	0.23
March 12 (Dehardened)	−10	1.46	0.37	0.25

a   Hardiness is expressed as the lowest temperature at which tissues can survive without injury.

II.   RESULTS

A. Lipid Analysis of Isolated Plasma Membrane Vesicles

Lipid analysis was performed with plasma membrane fractions isolated from mulberry bark tissues at different seasons. As indicated in Table 1, the phospholipid to protein ratio and the free sterols to phospholipid ratio varied significantly with season. As the hardiness increased, plasma membrane became more enriched in phospholipids, however, the content of free sterols per mg protein seemed to be relatively constant. As a result, the relative ratio of sterols to phospholipids was lowered as hardiness increased. Conversely, an artifical dehardening resulted in decrease in phospholipid content and thus resulted in increase of the sterol to phospholipid ratio to some extent.

As can be expected from these results, a marked difference in the effective densities of plasma membrane was observed between tender and hardy cells (data not shown). As the hardiness increased, the effective density was shifted to much lower value, from 1.14 to 1.12.

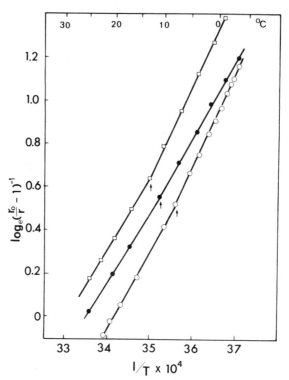

FIGURE 2. Arrhenius plots of the anisotropy parameter of DPH in a sample of plasma membrane vesicles from bark tissues of mulberry. Fluorescence polarization was measured in 0.7 M sorbitol-10 mM Tris-MES-1 mM EDTA (pH 7.2). □ ; September, ● ; October, O ; early March.

B.   Fluorescence Polarization Measurement

To gain insight into the physical nature of the isolated plasma membrane vesicles, steady-state fluorescence polarization of embedded DPH was measured. Fig. 2 shows a marked difference in the anisotropy parameter between different plasma membrane samples above freezing temperatures. The membrane vesicles isolated from actively growing twigs of September shows the highest value in comparison with other samples. In membrane samples isolated from hardy twigs of early March, the smallest value was obtained. The membrane sample from twigs in mid-October showed the intermediate value. Thus, it is likely that the relative fluidity of plasma membrane changes signficantly with the growth stage or the development of cold hardiness. Slight discontinuities were always detected in the Arrhenius plots around 17, 11 and 8°C in the membrane samples

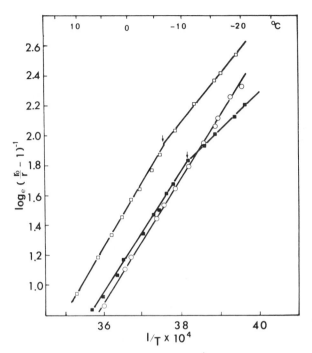

FIGURE 3. Arrhenius plots of the anisotropy parameter of DPH in a sample of plasma membrane vesicles isolated from bark tissues of mulberry trees. Fluorescence polarization was measured in 40% ethylene glycol solution containing 10 mM Tris-MES-1 mM EDTA (pH 7.2). □ ; September, ○ ; early March, ■ ; dehardened sample of early March.

of September, October and early March, respectively. These firstly detected inflections may refer to the initiation of lipid transition.

In order to gain further insight into thermotropic behavior of the plasma membrane vesicles at sub-zero temperatures, the steady-state fluorescence polarization was measured under super-cooled state with presence of 40% ethylene glycol solution. As indicated in Fig. 3, upon further cooling, distinct inflection in the slope was observed around -6°C in plasma membrane vesicles isolated from twigs of September. On the other hand, no inflection was detected in plasma membrane vesicles isolated from hardy twigs of early March, at least within tested temperatures, down to -20°C. After dehardening of these hardy twigs, however, a definite inflection occurred around -11°C. These second inflection temperatures were apparently coincided with the frost killing point of these twig bark cells.

Nearly the same result as obtained in tender twigs of mulberry trees was also confirmed with the plasma membrane

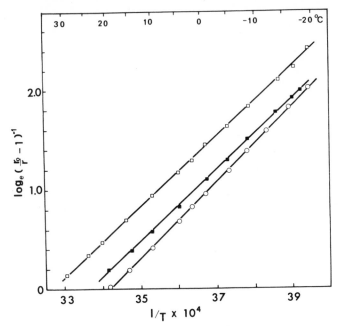

FIGURE 4. Arrhenius plots of the anisotropy parameter of DPH in liposomes prepared from lipid extracts of plasma membrane vesicles isolated from bark tissues of mulberry trees. □ ; September, ○ ; early March, ■ ; dehardened sample of early March.

vesicles of summer growing bark cells of black locust tree. The second inflection was clearly observed around -10°C (data not shown).

To characterize these thermotropic transitions of isolated plasma membrane vesicles in terms of lipid-protein interaction, steady state fluorescence polarization was measured with liposomes which were prepared from these membranes. As shown in Fig. 4, the Arrhenius plots of the anisotropy parameters of the liposomes reveals no inflection between 30 and -20°C, regardless of the twig samples from which the membranes were isolated. The marked difference in the anisotropy parameter, however, was still detected between different liposomes as noted in the corresponding membrane samples (Figs. 2 and 3). This results may indicate that membrane proteins participate in the thermotropic transition of lipid bilayers of those plasma membrane vesicles.

As illustrated in Fig. 5, the fluidity of plasma membrane vesicles from summer mulberry shoots in mid-June was drastically affected by the addition of a local anesthetic, dibucaine, which has a lipophylic molecule with positively

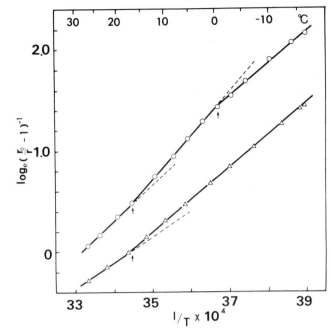

FIGURE 5.   Effects of dibucaine on fluidity and the lower limit of lipid transition.   Plasma membrane vesicles were isolated from newly growing shoots of mulberry trees in mid-July.   O ; control, △ ; with presence of 2 mM dibucaine in the measured system.

charged tertiary amine.   This kind of drug is known to displace the bound Ca^{2+} from membranes and thus affects the membrane functions.   Addition of 2 mM dibucaine resulted in marked increase in the membrane fluidity and the second inflection was depressed or lowered without any effect on the first inflection.   This effect of dibucaine, however, was significantly reversed by the addition of Ca^{2+} (5 mM).   Thus, it would be suggested that the membrane-bound Ca^{2+} may also participate in the regulation of membrane fluidity.

III.   DISCUSSION

In the present study, it was demonstrated that the lipid bilayer of plasma membrane from mulberry bark tissues shifted to a more fluid state as the cold hardiness increased.   Consequently, the fluidity of plasma membrane is likely to be readjusted to some extent to the cold environmental temperatures.

Vigh et al. has examined the fluidity change of plasma membrane of wheat leave protoplasts during cold acclimation by using esr technique (17). They found significant increase in plasma membrane fluidity only in hardy varieties after cold acclimation. However, they could not observe any significant correlation between the membrane fluidity and the increase in unsaturated fatty acids of the bulk lipids extracted from the protoplasts.

Although present author has no comparative data concerning fatty acid composition of plasma membrane lipids from mulberry bark cells, the change in the ratios of sterols to phospholipids and phospholipid to protein in the membrane are predicted to have a prevailing role to determine the difference in the membrane fluidity observed in this study. Nearly the same changes in those ratios have been reported with microsomal fractions from bark cells of black locust trees during dehardening (20).

The current theory explains chilling injury of plant cells in terms of a temperature-induced phase transition of lipid from liquid crystal to gel in cellular membranes. Fey et al. (4) have reported that temperature dependent membrane phase separation occurs in intermediately cold hardy wheat leaves under a freezing temperature below which the leaves sustained injury. Chapman et al. (1) also reported that the increase in fluidity and the lowering in the lower limit of lipid transition of mitochondrial membranes are probably essential feature of the mechanism adapted by Jerusalem artichoke tubers to survive low temperature during winter.

In plasma membrane vesicles from tender mulberry bark tissues, two inflection points were observed in the Arrhenius plots of the anisotropy parameter of embedded DPH. The first inflections, presumably indicating the initiation of thermotropic transition of the membrane lipids, were observed at temperature above zero. The second inflections, indicating lower limit of thermotropic transition, were distinctly detected at subzero temperatures. These second inflections were observed only in plasma membrane isolated from tender bark tissues, ranging from -6 to -10°C, and apparently correlated with the frost killing temperatures of those bark cells. On the other hand, in plasma membrane from hardy bark tissues, no second inflection was detected at sub-zero temperature at least above -20°C, suggesting depression or lowering of the second inflection beyond -20°C.

A great deal of information have been accumulated as to the phase separation of bulk membrane phospholipids related to the cold adaptation of plants (12). Adaptation of some plant species to growth at lower temperatures is accompanied by low phospholipid phase separation temperature. In the present study, however, no discontinuity was noted on the Arrhenius plots of the fluorescence anisotropy parameter of DPH in liposomes prepared either from total lipid extracts or purified phospholipids from corresponding plasma membrane vesicles. Accordingly, the observed inflections in the intact plasma membrane vesicles could not be ascribable to the physical nature of lipids per se. The mechanism including presumed molecular interaction between lipids and proteins in the membranes and/or temperature induced conformational changes in membrane proteins which initiate in the molecular ordering of membrane lipids seem to determine the thermotropic behavior of the membrane.

In the present study, the fluidity of plasma membrane vesicles was drastically affected by a local anesthetic, dibucaine, and in addition the second inflection was markedly depressed or lowered without any effect on the first inflection. These effects of dibucaine were significantly reversed by addition of Ca^{2+}. This kind of local anesthetic has been known to displace the bound Ca^{2+} from membranes and thus affect on the membrane functions (8). Lipid pertubation and distortion of membrane proteins, however, may also be involved in the molecular basis to modify the thermotropic behavior of plasma membrane (3,16).

From these observations, it may be concluded that the compositional alteration in lipids, proteins and presumably change in binding mode of Ca^{2+} in plasma membrane have a profound role to control the physical nature and freezing injury of plant cells.

In the present study, we cannot deduce any information about the effect of freeze-dehydration of cells upon alteration of plasma membrane structure. However, it may be assumed that the thermotropic transition of lipid bilayer or the highly ordered state of lipid bilayer in conjunction with freeze-dehydration produce an irreversible disorganization of membrane structure. Further studies are needed in the future to clarify this point of view.

REFERENCES

1. Chapman, E., Wright, L.C., and Raison, J.K., Plant Physiol. 63:363-366 (1979).
2. Chen, L.A., Dale, R.E., Roth, S., and Brand, L., J. Biol. Chem. 252:2163-2169 (1974).
3. Davio, S.R., and Low, P.S., Biochim. Biophys. Acta 644:157-164 (1981).
4. Fey, R.L., Workman, M., Marcellos, H., and Burke, M.J., Plant Physiol. 63:1220-1222 (1978).
5. Kawato, S., Kinoshita, K., Jr., and Ikegami, A., Biochemistry 16:2318-2329 (1977).
6. Lakowicz, J.R., Prendergast, F.G., and Hogen, D., Biochemistry 18:508-519 (1979).

7. Livingstone, C.J., and Schachter, D., J. Biol. Chem. 255:10902-10908 (1980).

8. Low, P.S., Lloyd, D.H., Stein, T.M., and Rogers, J.A. III, J. Biol. Chem. 254:4119-4125 (1979).

9. Lyons, J.M., Raison, J.K., and Steponkus, P.L., in "Low Temperature Stress in Crop Plants" (J.M. Lyons, D. Graham and J.K. Raison, eds.), pp. 1-24. Academic Press, New York, 1979.

10. Palta, J.P., Levitt, J. and Stadelmann, E.J., Plant Physiol. 60:393-397 (1977).

11. Palta, J.P., and Li, P.H., Physiol. Plant. 50:169-175 (1980).

12. Pike, C.S., and Berry, J.A., Plant Physiol. 66:238-241 (1980).

13. Singh, J., De La Rosche, I.A., and Siminovitch, D., Cryobiol. 14:620-624 (1977).

14. Singh, J., and Miller, R.W., Plant Physiol. 66:349-352 (1980).

15. Steponkus, P.L., and Weist, S.C., in "Low Temperature Stress in Crop Plants" (J.M. Lyons, D. Graham and J.K. Raison, eds.), pp. 231-253. Academic Press, New York, 1979.

16. Ueda, I., Kamaya, H., and Eyring, H., Proc. Natl. Sci. U.S.A. 73:481 (1976).

17. Vigh, L., Horrath, I.H., Dutits, D., and Farkas, T.F., FEBS Letters 107:291-294 (1979).

18. Weist, S.C., and Steponkus, P.L., Plant Physiol. 62:699-705 (1978).

19. Willemot, C., in "Low Temperature Stress in Crop Plants" (J.M. Lyons, D. Graham and J.K. Raison, eds.), pp. 411-430. Academic Press, New York, 1979.

20. Yoshida, S., Plant Physiol. 57:710-715 (1976).

21. Yoshida, S., and Gusta, L.V., in "Plant Cold Hardiness and Freezing Stress" (P.H. Li and A. Sakai, eds.). Academic Press, New York, 1982.

FREEZING RESPONSE OF PLASMA MEMBRANE

Wang Hong-Chun, Mi Rong-Qui, Wang Yu-Qi, and Li Jin-Shu

Environmental Physiology Section
Shanghai Institute of Plant Physiology
Academia Sinica
Shanghai, China

I.   INTRODUCTION

As early as 1912, Maximov (9) concluded that during freezing water molecules moved from the surface of the plasma membrane to the intercellular space, and ice crystals were formed which increased in size. The plasma membrane would be damaged by the force due to ice growth, resulting in injury and death of plant cells. Thereafter the studies of plant cold hardiness in relation to the structure and function of biomembrane have been extensive, and it has been evidenced that the primary freezing injury occurs in the plasma membrane (7,10,11). Freezing injury of permeability of plasma membrane was observed by many researchers. But there is yet no direct evidence on the damage of plasma membrane during freezing. The primary reaction of freezing injury is a central problem of cold hardiness of plants.

Palta and Li (10) proposed a hypothesis about membrane damage during freezing injury, in which the site of primary damage was postulated to be in the tonoplast and the plasma membrane. Taking into account the fact that, although the efflux of the electrolyte and non-electrolyte from the injured cells increased markedly, normally typical plasmolysis could still occur. Moreover, the permeability of plasma membrane to water, urea and methylurea did not change. The following conclusions were reached: the denaturation of membrane protein and the change in lipid-protein interaction of tonoplast and plasma membrane was the primary reaction when the cells were subjected to freezing followed by a slow thawing. ATPases involved in active transport system were inactivated, then electrolytes and non-electrolytes leaked out markedly, followed by change in some other physiological processes and eventually leading to

PLANT COLD HARDINESS AND FREEZING STRESS

285

the breaking down of the membrane system which would lead to the death of the cells.

In this work, the effect of freezing and thawing on membrane bound-enzyme activity and the permeability of plasma membrane were studied using root tips of maize. Attempts were made to elucidate the relation between the activity of membrane-bound-enzyme and the permeability of the membrane during freezing injury.

II. MATERIALS AND METHODS

Plant material used was root tips of maize (Zea mays L. cv. Laorenya) seedlings 48 hours after germination. Low temperature treatments were carried out by using dry ice-alcohol mixture.

Plasma membrane preparation (4) - After freezing treatment, the root tips were ground with cold grinding solution (containing sucrose, 0.25 M; EDTA, 3 mM; Tris-HCl, pH 7.2, 25 mM) to form a homogenate, which was filtered through crude cloth and then centrifuged at 17,000 x g for 10 min. The supernatant was centrifuged at 80,000 x g for 60 min. The sediment was collected and resuspended and centrifuged again.

Determination of the ATPase activity of plasma membrane (2) - The total volume of reaction solution was 1 ml containing 0.5 ml of plasma membrane suspension (protein content 40-80 μg, ml^{-1}), 0.4 ml of reaction solution (final concentration: KCl, 50 mM; NaCl, 50 mM; $MgCl_2$, 1.5 mM; and Tris, pH 7.2, 33 mM) and 0.1 ml ATP solution (final concentration 3 mM). The reaction solution was incubated at 30°C and reaction was stopped with 0.2 ml cold trichloroacetic acid (20%).

Determination of α-ketoglutaric oxidase activity of plasma membrane - The α-ketoglutaric oxidase activity was determined with the oxygen electrode. The volume of the reaction chamber was 2 ml which contained 1.75 ml of reaction solution (final concentration: mannital, 300 mM; phosphate buffer, pH 7.2, 1 mM; $MgCl_2$, 5 mM), 0.2 ml of the plasma membrane sususpension (protein content 2-4 mg ml^{-1}) and 0.05 ml of 100 mM α-ketoglutaric acid. The consumption of oxygen by α-ketoglutaric oxidase was measured at 25°C. The oxidase activity was calculated according to Estabrook (2).

Determination of the permeability of plasma membrane - Root tips after freezing treatment was weighed and put into an equal volume of redistilled water, allowed to undergo vacuum infiltration for 5 min, and then incubated for 30 min at room temperature. Ion leakage was measured with a DDS-11A type conductometer. The material was then heated in boiling-water bath for 5 min. The conductivity was again measured. The ratio of the conductivity measured without and with boiling water was

TABLE I. Influence of the low temperature on the permeability and the TTC reduction of the plasma membrane from the root tips of the maize.[a]

Treatment temperature (°C)	Relative conductivity (%)	TTC reduction (%)
25	14.6 + 2.7	100.0
0	17.0 + 2.7	92.7
-2.5	17.2 + 2.1	90.2
-5	18.3 + 2.9	87.8
-10	97.8 + 2.4	7.3

[a] Two experiments and six measurements.

TABLE II. Effect of the slow thawing on the permeability of the plasma membrane from the root ips of the maize.[a]

Treatment	Relative conductivity (%)
Control	15.1 + 4.1
Trial I[b]	82.5 + 14.9
Trial II[c]	93.1 + 7.3

[a] Five experiments and fifteen measurements.
[b] Freezing at -10°C for 30 min and thawing at 0-3°C for 4 hours.
[c] Freezing at -10°C for 30 min and thawing at 0-3°C for 17 hours.

taken as relative conductivity, which was indicative of the permeability of the plasma membrane.

Other methods were also used to measure the triphenyl tetrazolium chloride (TTC) reduction (12), inorganic phosphorus (1) and protein content (8). Lead phosphate deposits were observed by electron microscopy as an index of ATPase activity (5).

III.   RESULTS

A.   Effect of Low Temperature on the Permeability
of Plasma Membrane and TTC Reduction

With root tips kept at various temperatures for 30 min, the permeability of the plasma membrane and the TTC reduction was determined immediately after the low temperature treatment. Within the range from 25 to -5°C, there was an increase in permeability with only slight decrease in TTC reduction. The value of relative conductivity increased to 97.8% after a 30 min treatment with -10°C and the TTC reduction decreased to 7.3% as of the control (Table I). When the frozen tissue was thawed slowly by placing at 0-3°C, the relative conductivity was still high, being similar to the fast thawing (Table II). The results showed that a temperature as low as -10°C was needed to damage the cells of root tips.

B.   Effect of Freezing Rate on the Activity
of Plasma Membrane Bound-Enzymes

When root tips were frozen at -10°C for 30 min and then subjected to fast thawing by grinding at room temperature for 20 seconds, the ATPase activity of the plasma membrane increased slightly, and the α-ketoglutaric oxidase activity decreased markedly, while the permeability of the plasma membrane increased markedly to 73.2% (Table III). After fast thawing, the amount of lead phosphate deposits observed was nearly the same as the control (Fig. 1, Plate I and II). When root tips were subjected to slow freezing by lowering the temperature at a rate of 2°C/hr from 0°C to -10°C, and kept at -10°C for 60 min and then subjected to fast thawing, the ATPase activity was only slightly higher than the control. But, the permeability increased very markedly as high as 91.4% (Table IV).

C.   Effect of Slow Thawing on the Plasma Membrane
Bound-Enzyme Activity

When frozen root tips were subjected at 0-3°C and allowed to undergo slow thawing for 4 hours, the ATPase activity decreased to about 10% lower than control, but after 17 hours of thawing, the activities of ATPase and α-ketoglutaric oxidase decreased by 40% and 50%, respectively, from those of the control. The decrease in enzyme activity paralleled the increase in the permeability of the plasma membrane (Table V). After slow thawing, the deposits due to ATPase activity of plasma membrane was stil there, being very similar to the control (Fig. 1, Plate I and III).

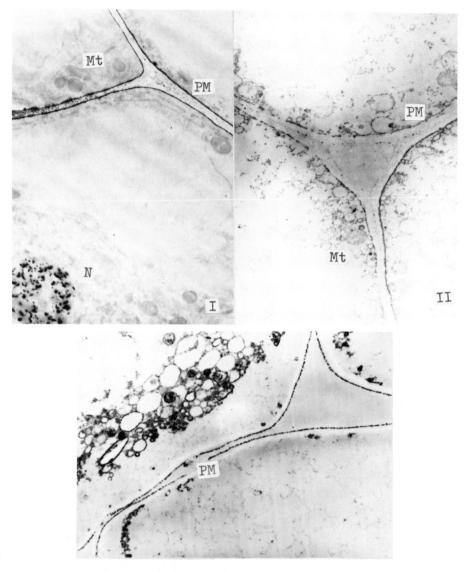

FIGURE 1. Influence of freezing and thawing on the deposits of lead phosphate in the plasma membrane from the root tips of maize.

Plate I. Control. X 5400.

Plate II. The frozen root tips after fast thawing at room temperature for 20 seconds. X 5400.

Plate III. The frozen root tips after slow thawing at 0-3°C for 17 hours. X 5400.

PM=Plasma membrane; Mt=Mitochondria; N=Nucleus

TABLE III. Effect of the fast freezing and the fast thawing on the plasma membrane bound-enzyme activities from the root tips of maize.[a]

| Treatment | ATPase activity | | α-ketoglutaric oxidase activity | | Relative conductivity (%) |
	μmoles Pi mg protein^{-1} hr^{-1}	%	n moles O_2 mg protein^{-1} min^{-1}	%	
Control	7.6 $\pm$ 1.9	100.0	33.5	100.0	10.6 $\pm$ 3.2
Trial[b]	8.4 $\pm$ 3.3	110.5	24.2	72.2	73.2 $\pm$ 24.8

[a] Six experiments and eighteen measurements.
[b] Freezing at -10°C, 30 min.

TABLE IV. Effect of the slow freezing and the fast thawing on
the ATPase activity of plasma membrane from the root
tips of maize.[a]

| Treatment | ATPase activity | | Relative |
	μ moles P_i mg protein^{-1} hr^{-1}	%	conductivity (%)
Control	3.33 ± 0.15	100.0	17.4 ± 4.8
Trial[b]	3.42 ± 0.15	102.9	91.4 ± 1.9

[a] Two experiments and six measurements.
[b] Slow freezing and fast thawing.

D. Effect of Respiration Inhibitor on the Permeability and
        α-ketoglutaric Oxidase Activity of Plasma Membrane

The inactivation of α-ketoglutaric oxidase during the
thawing (Tables III and V) might reduce the supply of energy.
In order to see whether the permeability of plasma membrane
would increase when the supply of energy was curtailed, the
root tips were treated with KCN solutions of different concentration. From Table VI it can be seen that the α-ketoglutaric
oxidase activity decreased with the increase in KCN concentration, but there was not a corresponding change in membrane permeability. It may be concluded that the inactivation of
α-ketoglutaric oxidase is related to freezing, but has no connection with the permeability changes.

IV. DISCUSSION

The inactivation of the ATPase activity is related to the
permeability during the slow thawing. Palta and Li (10) suggested that slow freezing and slow thawing might affect the
inactivation of the active transport system. This might be due
to the denaturation of the intrinsic membrane protein and/or a
change in the lipid-protein interaction. A large passive
efflux of ions and sugars was resulted from the inactivation of
active transport system. They considered that the active
transport system located in the plasma membrane was the primary
site of freezing damage. From our results it is shown that the
ATPase activity of plasma membrane is gradually inactivated
during thawing, while the permeability of the plasma membrane
gradually increased (Table V). The permeability of the plasma

TABLE V. Effect of the slow thawing on the bound-enzyme activity of the plasma membrane from the root tips of maize.[a]

| Treatment | ATPase activity | | α-ketoglutaric oxidase activity | | Relative conductivity (%) |
	μmoles Pi mg protein^{-1} hr^{-1}	%	n moles O$_2$ mg protein^{-1} min^{-1}	%	
Control	3.21 + 0.12	100.0	28.6 + 8.5	100.0	15.1 + 5.0
Trial I[b]	2.86 + 0.54	89.1			90.1 + 5.1
Trial II[c]	1.83 + 0.69	57.0	13.9 + 4.8	48.6	97.0 + 3.3

[a] Four experiments and twelve measurements.

[b] -10°C for 30 min thawing at 0-3°C for 4 hr.

[c] -10°C for 30 min thawing at 0-3°C for 17 hr.

TABLE VI. Influence of the KCN inhibition on the α-ketoglutaric oxidase activity of plasma membrane and of root tips, and the permeability of plasma membrane from the root tips of maize.[a]

KCN conc. (M)	Plasma membrane			Root tips		
	α-ketoglutaric oxidase activity n moles O_2 mg protein^{-1} min^{-1}	%	Relative conductivity (%)	α-ketoglutaric oxidase activity n moles O_2 mg protein^{-1} min^{-1}	%	Relative conductivity (%)
Control	25.1 ± 1.7	100.0	13.9 ± 0.6	234.1 ± 1.6	100.0	11.2 ± 1.9
10^{-5}	19.1 ± 1.4	75.8	11.3 ± 1.2	198.2 ± 16.5	84.7	8.7 ± 1.2
10^{-4}	16.9 ± 1.2	67.1	10.9 ± 1.0	193.1 ± 15.9	82.5	8.1 ± 1.7
10^{-3}	12.1 ± 5.3	47.9	14.0 ± 1.9	109.4 ± 3.9	46.7	11.6 ± 0.9
10^{-2}	8.1 ± 2.2	32.2	16.3 ± 1.2	49.9 ± 0.1	21.3	10.3 ± 0.8

a Two experiments and six measurements.

membrane was negatively correlated with ATPase activity, the correlation coefficient being -0.479, indicating that the inactivation of the ATPase activity is related to the permeability of plasma membrane and the freezing injury.

During fast thawing the ATPase activity was not lower than the control, while the permeability of the plasma membrane increased markedly to as high as 70-90% (Tables III and IV). During the slow thawing for 4 hours and 17 hours, the ATPase activity was 89.1% and 57.0%, respectively, as compared to those of the control, but the permeability of plasma membrane was, however, 90.1% and 97.0%, respectively. Due to the ATPase activity decreased markedly and the permeability increased slighty after different thawing condition, it may conclude that the inactivation of ATPase activity is not intrinsic connection with the permeability of plasma membrane during freezing injury.

As to the relation between the α-ketoglutaric oxidase activity and the freezing injury after fast thawing, the enzyme activity decreased markedly (Table III). Its decrease was more marked during slow thawing than the control after 17 hours (Table V). The increase in peremability of the plasma membrane were 73.2% and 97.0%, respectively (Tables III and V). The results indicate that the inactivation of the oxidase activity of the plasma membrane may reduce the energy supply, hence the active transport of ions, so that a large passive efflux of ions results in. But from the results of KCN inhibition experiment, that the inhibition of α-ketoglutaric oxidase activity had no effect on the permeability of the plasma membrane. It may thus be concluded that the effect of freezing injury in denaturating membrane protein and inducing change in the lipid-protein interaction will result in the inactivation of the α-ketoglutaric oxidase, which may be unrelated to the damage of permeability.

As to the cause of the permeability change of the plasma membrane during the freezing injury upon slow thawing, the inactivation of the ATPase activity was related to the increase in permeability of the plasma membrane, while there was none during fast thawing. In the mean time, the inactivation of the α-ketoglutaric oxidase of the plasma membrane was not connected with the permeability increase. The mechanism of the damage of the plasma membrane in its permeability properties deserves further research. Wilson (13) and Yoshida (14) reported a marked change in the membrane lipid composition during the short freezing time. Kuiper (6) observed different effects of various membrane lipid on the penetration of ions through membrane. Graziani (3) suggested that membrane lipid may have regulatory action on both active and passive transport of ions. According to our results we suggest that the efflux of electrolytes and non-electrolytes due to the damage of the plasma membrane played an important role during freezing injury. In addition to the change in membrane bound-enzyme activity, the changes in membrane lipid composition and lipid-protein interaction may also be responsible for the freezing damage of the cells.

REFERENCES

1. Ames, B.N., in "Methods in Enzymology", VIII, pp. 115, 1966.
2. Estabrook, R.W., in "Methods in Enzymology" (R.W. Estabrook, and M.W. Pullman, eds.), X, pp. 41-47. Academic Press, New York, 1967.
3. Graziani, Y.A., and Livne, A., J. Membrane Biol. 7, 275-284 (1972).
4. Hodges, T.K., Leonard, R.T., Bracker, C.E., and Keenan, T.W., Proc. Nat. Acad. Sci. USA, 69, 3307-3311 (1972).
5. Jian, Ling-Cheng, Dong, Hezhu, Sun, Long-Hua, Acta Biologiae Exp. Sinica, 13, 135-146 (1980).
6. Kuiper, P.J.C., Plant Physiol. 44, 968-972 (1969).
7. Levitt, J., in "Plant Cold Hardiness and Freezing Stress" (P.H. Li, and A. Sakai, eds.), pp. 3-15. Academic Press, New York, 1978.
8. Lowry, O.H., Rosebrough, N.J., Farr, A.L., and Randall, R.J., J. Biol. Chem. 193, 265-275 (1951).
9. Maximov, N.A., Ber. Der. Beutsch. Bot. Gesell. 30, 293, 504 (1912).
10. Palta, J.P., and Li, P.H., in "Plant Cold Hardiness and Freezing Stress" (P.H. Li, and A. Sakai, eds.), pp. 93-115. Academic Press, New York, 1978.
11. Siminovitch, D., Rheaume, B., and Sacher, R., in "Molecular Mechanism of Temperature Adaptation" (C.L. Prosser, ed.), pp. 3. Amer. D. C. Assoc. Adv. Sci., Washington, 1967.
12. Steponkus, P.L., and Lanphear, F.O., Plant Physiol. 42, 1423-1426 (1967).
13. Wilson, R.F., and Rinne, R.W., Plant Physiol. 57, 270-273 (1976).
14. Yoshida, S., in "Plant Cold Hardiness and Freezing Stress" (P.H. Li, and A. Sakai, eds.), pp. 117-135. Academic Press, New York, 1978.

SURFACE LABELLING AND ISOLATION OF PLASMA MEMBRANE
FROM SUSPENSION CULTURE OF BROME GRASS

S. Yoshida

The Institute of Low Temperature Science
Hokkaido University
Sapporo, Japan

L. V. Gusta

Crop Development Center
University of Saskatchewan
Saskatoon, Canada

I.   INTRODUCTION

The plasma membrane in plant cells has been implicated in such important roles as cellulose biosynthesis (13), active ion transports (8), phytochrome responses (11), mediation of auxin responses (1) and host-pathogen responses (14).   In addition to these function, plasma membrane has been suggested to have a profound role to control freezing injury and cold acclimation of plant cells (9,12,17).

To elucidate these roles of the membranes, it is imperative to establish a method to isolate plasma membrane from plant cells in a relatively large amount and also in a relatively pure form.   One of the major problems encountered by investigators attempting to isolate plasma membrane from plant cells involves uncertain membrane identification because of a lack of suitable markers.   The pH 6.5-KCl stimulate ATPase and PTA-CrO_3 stain (8) are generally used for identification of plasma membrane, however, the reliability of these methods has been controversial (11).

In the present study, we examined the utility of iodine-surface labelling mediated by lactoperoxidase to monitor plasma membrane in susbsequent fractionation by using suspension culture of brome grass as the experimental material.

II. MATERIALS AND METHODS

A. Materials

A cell suspension culture of brome grass (<u>Bromus inermis</u> Leyss cv. Manch) was grown in a defined medium of B-5 containing 0.1 mg/l 2,4-D with subculturing every one week. The three day old cultures were mainly used for the experimental materials.

B. Iodination of Intact Plant Cells

Iodination of intact cultured cells was performed according to Hendriks (6) with a slight modification. In brief, five grams of washed and blotted cells were suspended in 10 ml of 50 mM potassium phosphate buffer containing 5 mM $MgSO_4$, pH 6.8. 100 µCi of carrier free $Na^{125}I$, 125 µg of lactoperoxidase and 10 µl of 0.03% H_2O_2 were added into the cell suspension. Hydrogen perioxide solution was supplemented every five min during incubation. After incubation at 25 C for 20 min, cold KI solution was added in a final concentration of 2 mM and the labelled cells were washed at least 6 times with the buffer used for incubation. The labelled cells were combined with the same amount of non-labelled cells and disrupted in a cell homogenizer MSK with 30 gram of glas beads for 20 seconds. The homogenizing medium was consisted of 0.5 M sorbitol, 50 mM Tris-MES and 3 mM EDTA, pH 7.3. The crude homogenate was passed through two layers of Miracloth and subsequently subjected to a differential centrifugation.

C. Protoplast Isolation and its Iodination

Protoplasts were prepared from 3-day old culture. Cell wall digestion was performed in mixture of 2% cellulose Onozuka R-10, 1% pectinase and 1% hemicellulose in 0.8 M glucose, 44.1 mg $CaCl_2$, 29.2 mg MES and 4.9 mg Na_2HPO_4 per 100 ml, pH adjusted to 5.7. Incubation was carried out at first at 4 C for 14 hr and then followed at 26 C for 2.5 hr.

The released protoplasts were filtered through Nylon mesh (45 mu) and repeatedly washed in 0.45 M sorbitol-5 mM $MgSO_4$-20 mM potassium phosphate buffer, pH 6.8.

The iodination procedure for the isolated protoplasts was similar to that for intact cultured cells as described above except for addition of 0.45 M sorbitol as the osmotic stabilizer. The labelled protoplasts were homogenized in a glas homogenizer with tightly fit Teflon plunger.

D. Percoll Gradient

A particulated cell fraction obtained by a differential centrifugation was resuspended in 0.5 M sorbitol - 10 mM Tris-MES-1 mM EDTA-25 mM KCl, pH 7.3 and added by Percoll solution in a final concentration of 16% (w/w). The Percoll mixture was centrifuged at 35,000 rpm for 30 min on fixed-angle rotor of Beckman Type 60. This process resulted in a spontaneous generation of a density gradient in situ due to the heterogeneity of the PVP-coated silica gel particles and thus membranes were separated isopycnically on the generated gradients. Enzyme assays were done as reported before (18).

III. RESULTS AND DISCUSSION

As shown in Table 1, the iodination reaction in intact cells was strictly dependent on the presence of lactoperoxidase. The incorporation of radio-activity into particulated fractions was negligible in the absence of lactoperoxidase. The highest total and specific radio activity was found to be incorporated into the 37,000 g pellet. Low radio activity was also incorporated into the 80,000 g supernatant, however, the specific activity was quite low compared with those in particulated fraction. These results rule out the possibility of participation of the endogenous peroxidase system in the iodination reaction.

Table II indicates the coincidence of membrane maker enzymes and radio activity. More than 80% of mitochondria were sedimented by low centrifugal forces below 3,900 g, however, less than 27% of the total radio activity was recovered in these particulated fractions. Both ATPases and NADH cyt c reductase were broadly distributed over several fractions. Coincidence between pH 6.5 ATPase and radio activity was relatively poor. As described later, this was due to the existence of two kinds of pH 6.5 ATPase associated membranes in brome grass culture.

The 37,000 g fraction was subjected to a Percoll gradient to achieve further separation of membranes. As shown in Fig. 1, two peaks of pH 6.5 ATPase activity were observed at fraction 14 and 25, respectively. The pH 6.5 ATPase at fraction 14 was co-fractionated with the labelled membrane, however, no coincidence was evident between the pH 6.5 ATPase associated membrane at fraction 25 and the incorportated radio activity. Most of mitochondria were banded nearly at the bottom of the tube at fractions 25-27 and the radio activity recovered in these fractions was very low.

Significant level of pH 9 ATPase, however, was overlapped with the labelled membrane. Endoplasmic reticulum were banded

TABLE I. Lactoperoxidase requirement for the ^{125}I-labelling reaction. Reaction mixture contained 125 µg lactoperoxidase (when added) 100 µCi Na ^{125}I and 5 g of suspension culture in 10 ml of 50 mM potassium phosphate – 5 mM $MgSO_4$ (pH 6.8). Reaction was started by addition of 10 µl of 0.03% H_2O_2. H_2O_2 was sequentially added by 5 min. Incubation was carried out at 25 C for 20 min.

Fractions	Control (- lactoperoxidase)			Experimental (+ lactoperoxidase)		
	Protein	total activity	Specific activity	Protein	Total activity	Specific activity
	mg	cpm x 10^4	cpm x 10^4/ mg protein	Protein	cpm x 10^4	cpm x 10^4/ mg protein
400-1900.g pellet	6.5	0.92	0.14	7.1	28.55	4.01
1900-37.000.g pellet	12.6	1.23	0.09	13.1	80.06	6.11
37.000-80.000.g pellet	4.2	0.41	0.09	5.4	15.07	2.79
80.000.g Sup	62.8	1.40	0.02	59.2	5.21	0.09

TABLE II. Coincidence of marker enzymes with ^{125}I surface labelling of membranes. Reaction mixture contained 125 µg lactoperoxidase, 200 µCi $Na^{125}I$ and 5 g of suspension culture in 10 ml of 50 mM potassium phosphate – 5 mM $MgSO_4$ (pH 6.8). Reaction was started by addition of 10 µl of 0.03% H_2O_2. H_2O_2 was sequentially added every 5 min. Incubation was carried out at 25 C for 25 min.

Fractions	Protein	Total activity		Specific activity	ATPase pH 6.5 pH 9	Cytc oxidase	NADH Cyt C reductase
	mg		cpm x 10^4	cpm x 10^4/ mg protein	umoles Pi/hr	ΔOD_{550}/ min	ΔOD_{550}/ min
400–1900·g pellet	12.4	193.92		15.63	48.5 99.1	123.2	28.0
1900–3900·g pellet	8.0	149.36		18.67	39.3 72.8	70.4	22.4
3900–37000·g pellet	11.4	387.36		33.97	58.2 88.7	31.2	31.2
37000·g Sup	76.0	510.41		6.71	n.d n.d	3.0	21.3

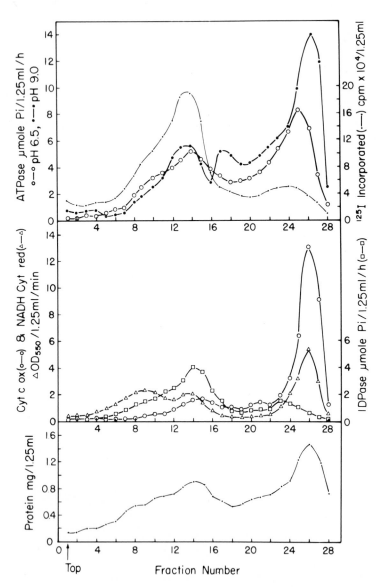

FIGURE 1.   Distribution profiles of marker enzymes and radio activity on Percoll gradient of 37,000 g pellet from iodinated intact cells.

at fraction 7-10 and well separated from the labelled membrane. Major peak of latent IDPase activity also occurred around the peak of the radio activity.

The labelled membrane fractions were collected and pelleted by centrifugation. When this pellet was subjected to washing with a hypotonic buffer solution containing 1 mM EDTA and subsequent washing with 0.5 M KCl solution, a remarkable reduction in the pH 9 ATPase activity in the pellet was observed without any loss of the radio actviity into the supernatant. This fact implies that the pH 9 ATPase occurred at the labelled membrane fractions would have been derived from entrapped or adsorbed solubilized ATPase from mitochondria. The KCl-washed labelled membrane was designated as plasma membrane-enriched fraction.

According to Quail and Browning (10), it is observed autoradiographically that a non-plasma membrane, extracellular components becomes heavily labelled upon lactoperoxidase-mediated iodination of excised tissues segments of Cucumita seedlings. Upon subsequent fractionation, this component behaves as a preferentially labelled and discrete membrane fraction other than plasma membrane.

To overcome this possibility, we tried the lactoperoxidase-mediated iodination by using protoplasts of brome grass culture. As revealed in Fig. 2, when the 15,000 g pellet from iodinated protoplast homogenate was subjected to a Percoll gradient centrifugation, nearly the same profile of marker enzymes and radio activity as observed in the labelled intact cells (Fig. 2). Two kinds of pH 6.5 ATPase associated membranes are also distinctly demonstrated and the membrane with lower density was specifically labelled with iodine.

From both experiments using intact cells and its protoplasts, plasma membrane of brome grass cells is considered to be specifically labelled with iodine. Thus, the labelled membrane occurred at fractoin 14 on Percoll gradient represents the iodinated plasma membrane.

The ATPase bound on plasma membrane of brome grass cells had optimum pH at 6.5 as reported for several kinds of plant species. The KCl-stimulation at the optimum pH was very low but it was relatively high at lower pH than the optimum.

The ATPase was stimulated by Co^{2+}, Mg^{2+} and Mn^{2+} in this order and the specificity for Mg^{2+} was relatively low.

The substrate specificity of the enzyme revealed to be relatively low compared with plasma membrane ATPase as reported for other plant species. ATP and UTP were hydrolyzed in a similar level, however, the activity for CTP and ITP was reasonably low. The high activity for IDP may suggest that the plasma membrane fraction is contaminated with Golgi vesicles and not due to the ATPase activity per se.

The plasma membrane ATPase from brome grass culture was non-sensitive to azyde as a potent inhibitor for mitrochondrial ATPase (4), suggesting no contamination with mitochondrial membrane. However, the enzyme was very sensitive to vanadate, as reported to be specific inhibitor for mammal (2) and bacterial plasma membrane (3) ATPases (Fig. 3). Fifty percent inhibition was achieved by addition of 15 μM vanadate. On the other

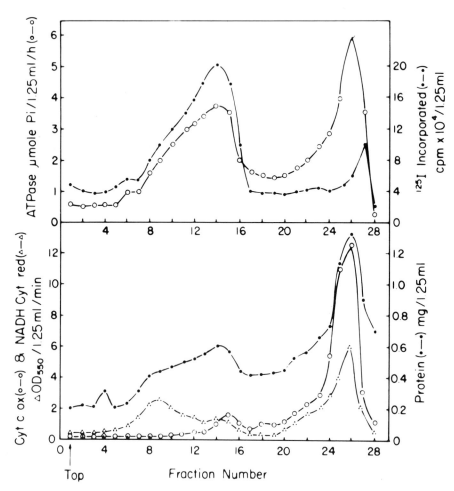

FIGURE 2. Distribution profiles of marker enzymes and radio activity on Percoll gradient of 15,000 g pellets from iodinated protoplasts.

hand, the ATPase associated with heavy membrane was less sensitive to vanadate. Very recently, it has been reported that plasma memebrane ATPase from various kinds of plant species is specifically inhibited by vanadate (5,7,15). On the other hand, vanadate does not inhibit tonoplast ATPase (16).

As already mentioned, there were at least two kinds of pH 6.5 ATPase associated membranes in brome grass cultured cells. The pH 6.5 ATPase bound membrane with higher density other than plasma membrane was not identified in the present study. However, it was easily distingushed from plasma membrane due to the differences in the density of Percoll gradient, sensitivity

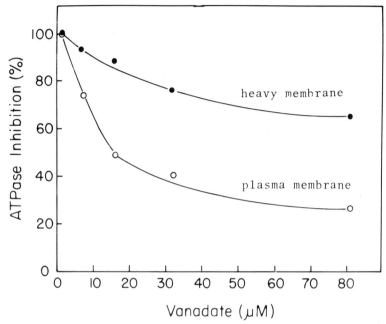

FIGURE   3.   Vanadate   sensitivities   of   membrane   bound
ATPases.   ; plasma membrane vesicles,   ; heavy membrane.

to a low pH treatment and vanadate sensitivity of the ATPase.
Plasma membrane was very sensitive to low pH and easily aggregated below pH 5, on the other hand, the other membrane with
higher density was observed to be stable under such a low pH
condition.

From these facts, it is concluded that the vanadate-
sensitive ATPase and lactoperoxidase-mediated iodination could
be used as convenient and reliable marker for plant plasma membrane. This would make it easy to monitor the membrane during
fractionation processes.

REFERENCES

1. Beffagna, N., Cocucci, S., and Marre, E., Plant Sci. Lett.
   8:91-98 (1977).
2. Bond, G.H., and Hudgins, P.M., Biochem. Biophys. Acta
   600:781-790 (1980).
3. Bowman, B.J., and Slaymann, C.W., J. Biol. Chem.
   253:2928-2934 (1979).
4. Delhex, J., Dufair, J., Thines, J., and Coffean, A., A.
   Eur. J. Biochem. 79:319-328 (1978).

5. DuPont, F.M., Burke, L.O., and Spanswick, R.M., Plant Physiol. 67:59-63 (1981).

6. Henriks, T., Plant Sci. Lett. 7:347-357 (1976).

7. Hendrix, D.L., Rust, M., and Pierce, W.S., Plant Physiol. 67:Supplement p. 66 (1981).

8. Hodges, T.K., Leonard, R.T., Brucker, C.E., and Keena, T.W., Proc. Natl. Acad. Sci. U.S.A. 60:3307-3311 (1972)

9. Palta, J.P., Levitt, J., and Stadelmann, E.J., Plant Physiol. 60:393-397 (1977).

10. Quail, P.H., and Browning, A., Plant Physiol. 59:754-766 (1977).

11. Quail, P.H., Ann. Rev. Plant Physiol. 30:425-484 (1979).

12. Rajasheker, C., Gusta, L.V., and Burke, M.J., in "Low Temperature Stress in Crop Plants" (J.M. Lyons, D.Graham and J.K. Raison, eds.), pp. 252-274. Academic Press, New York, 1979.

13. Shore, G., and MacLachlin, G.A., J. Cell Biol. 64:557-571 (1975).

14. Strobel, G.A., and Hess, W.M., Proc. Natl. Acad. Sci. U.S.A. 71:1413-1417 (1974).

15. Taiz, L., Jacobs, M., Gepstein, A., and Mettler, I.J., Plant Physiol. 67:Supplement p. 8 (1981).

16. Wagner, G.J., and Mulready, P., Plant Physiol. 67:Supplement p. 8 (1981).

17. Weist, S.C., Steponkus, P.L., Plant Physiol. 62:699-705 (1978).

18. Yoshida, S., Plant Physiol. 57:710-715 (1976).

WINTER HARDINESS LIMITATIONS AND PHYSIOGRAPHY OF WOODY TIMBERLINE FLORA

M. R. Becwar

National Seed Storage Laboratory
Colorado State University
Fort Collins, Colorado

M. J. Burke

Institute of Food and Agricultural Science
Fruit Crops Department
University of Florida
Gainesville, Florida

I.   INTRODUCTION

Timberline is a prominent physiographic feature setting the upper forest limit in the mountains of Colorado and similar mountain regions elsewhere. In such areas, it is a line of dwarfed, battered, and distorted trees which abruptly marks the uppermost edge of a healthy and thriving mountainous forest. Timberline in Rocky Mountain National Park is at 3340 m elevation and only rarely deviates from this elevation. This uniformity of elevation is of central interest here, and looking at Figure 1, one can see the uniformity. As will be shown later, this timberline is near the same elevation regardless of factors such as exposure direction, length of growing season, water availability, etc. Figure 1 also shows timberline to be an abrupt boundary. Trees often reach full size only 200 m below timberline. The most important tree species in the Colorado timberline are Engelmann spruce (Picea engelmannii Parry), and subalpine fir (Abies lasiocarpa (Hook) Nutt.); other species are present at much lower levels. However, regardless of the species present, timberline occurs at this fixed elevation.

In spite of the scenic nature of the timberline ecotone, timberline is a region of severe and harsh conditions.

PLANT COLD HARDINESS AND FREEZING STRESS

307

Obviously, the harsh environment causes timberline. Temperature has been suggested as a primary factor; however, the mechanism of its limitation on tree growth and how this might
explain a uniform elevation are unclear. Good reviews on the
subject are available (6,20,21,25). Examples of temperature
causes include shortness of growing season and summer frosts
which retard development of structures for prevention of winter
desiccation (19,21,24). The fact that timberline elevation in
the temperate zone often coincides with the elevation of the
10°C mean July temperature isotherm was noted in many early
studies (6,23). In addition to temperature stress, timberline
has high winds, sand and snow blasting, deep and variable snow
conditions, poor soil conditions, atmospheric rarity, high UV
and visible radiation, lightning, fires, steep rocky terrain,
rock slides, avalanches, etc.

That trees stop growing in such an environment is not a
surprise. However, it is quite interesting that a complex
environment has led to the observed invariant elevation of this
timberline (Fig. 1). We report here an investigation of timberline in the Rocky Mountains of North America, with particular emphasis on timberline in the limited region of north central Colorado in and near Rocky Mountain National Park. Our
purpose is not to discuss why a timberline occurs; the environment is severe enough to account for that. Our purpose is to
question what aspect of this severe environment is responsible
for the observed elevational uniformity of the Colorado
timberline.

II. TIMBERLINE ELEVATIONS

In the temperate zone of North America, timberline in the
Rocky Mountains occurs as far south as 35°N latitude (6).
There the elevation of timberline occurs at 3500 m elevation.
Sixteen hundred kilometers to the north, in Glacier National
Park (48°N), timberline occurs at 1900 m elevation and it continues to decrease in elevation as one moves further north. We
have attempted to better quantitate timberline elevations using
U.S. Geological Survey topographical maps (U.S.G.S., Denver,
Colorado) of six regions of timberline in North America between
35 and 65°N latitude (Table I). Two procedures of analysis of
the maps were used. In the first, elevational contours were
traced and the proportion of closed forest for each individual
contour was calculated. Those data were used in developing
transition curves from closed forest to open alpine zone
(Figure 2A). In the second procedure, the elevation of timberline was recorded at 1.6 km (one mile) intervals along the line
forming the upper limit of the closed forest signified on the
topographical maps, and the mean timberline elevation was calculated (Table I).

Two important features of timberline in these regions are shown in Figure 2A. First, the elevation of timberline changes with latitude. The elevation of timberline decreases about 125 m per degree latitude in moving from south to north. Second, the transition curves from closed forest to open alpine zone are very steep in southern regions, Arizona and Colorado, and more gradual in northern regions, Montana and Alaska. In our investigation of these and other timberlines of western North America, this uniformity of timberline elevation brought out by the steepness of the transition curve is unique to Colorado and neighboring areas. In Rocky Mountain National Park, 80% of the transition occurs over a 200 m change in elevation. In contrast, for Glacier National Park, 80% of the transition occurs over a 550 m change in elevation (Fig. 2B). The results of mean timberline elevation calculations also show the pronounced uniformity of timberline in the southern regions of the Rocky Mountains (Table I). Therefore, the elevational uniformity of timberline is most pronounced in the region we have chosen for study, Rocky Mountain National Park of north central Colorado.

This uniformity of elevation holds when comparing different exposure directions in Rocky Mountain National Park. Analysis of 177 km of timberline in this region revealed the following dependence on exposure:

3340 $\pm$ 110 m, all exposures included
3330 $\pm$ 90 m, north exposures included
3330 $\pm$ 120 m, east exposures included
3370 $\pm$ 90 m, south exposures included
3350 $\pm$ 140 m, west exposures included

FIGURE 1. Timberline in the northern central Colorado Rocky Mountains is an abrupt transition from coniferous forest to open alpine tundra. Approximately 23 km of timberline are shown near Berthoud Pass (extreme right). The slope has north-west exposure and the mean timberline elevation is 3440 m. This figure explains why Mills (13) compared timberline to the ocean's edge: "it is as strange and as abrupt a boundary as the crooked and irregular shoreline of the sea,...like the ocean's edge, timberline has miles that are as straight and level as a die; but in places it sweeps outward around a penin-sula and follows the crooked line of an invading canyon." The mean elevation was measured using U.S. Geological Survey topo-graphical maps which had forest cover indicated. The standard deviation for 113 data points on the above timberline was 70 m. Photograph by D. Flaherty, Washington State University, Pullman, Washington.

Figure 1

TABLE I. Regions of timberline analyzed on U.S. Geological Survey topographical maps. See text for details of the analysis.

Location (latitude)	Topographical maps[1]	Timberline analyzed[2] (km)	Mean timberline elevation $\pm$ S.E. (m)
San Francisco Mountains, Arizona (35°N)	Humphreys Peak	19	3490 $\pm$ 120
Rocky Mountain National Park, Colorado (40°N)	Estes Park, Trail Ridge, Fall River Pass, Pingree Park, Comanche Peak, Clark Peak	177	3340 $\pm$ 110
Grand Teton National Park, Wyoming, (44°N)	Granite Basin, Mount Bannon, Grand Teton	148	2630 $\pm$ 210
Yellowstone National Park, Wyoming (45°N)	Sunlight Peak	314	2880 $\pm$ 170
Glacier National Park, Montana (48°N)	Rising Sun, Many Glacier, Logan Pass, Lake Sherburne, Mount Cannon, Ahern Pass	307	1870 $\pm$ 220
Denali National Park, Alaska (65°N) formerly McKinley National Park	Mt. McKinley, Talkeetna, Talkeetna Mountains	--	Not determined[3]

[1] U.S. Geological Survey maps 1:24,000 scale, except those for Yellowstone National Park (1:62,500) and Denali National Park (1:250,000).

[2] The distance of timberline analyzed was estimated by tracing the distance along the major contour closest to the mean timberline elevation for each region.

[3] Only the timberline transition curve (Fig. 2A) was determined for this region.

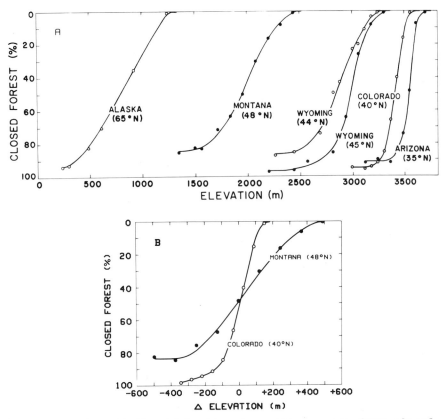

FIGURE 2.   A) The timberline transition curve from closed
forest to open alpine zone of six mountainous regions in the
western United States.   The open and closed circles are percent
closed forest values calculated at various elevations on contour lines of U.S.G.S. topographical maps of each region.   Note
that the timberline transition occurs at higher elevations and
is very abrupt at southern latitudes, whereas the transition at
northern latitudes occurs at lower elevations and is more gradual.   B) Comparison of transition curves for Rocky Mountain
National Park (O) and Glacier National Park (●).   Elevational
uniformity of timberline can be judged from steepness of the
transition curves and is observed primarily in southern regions
of the Rocky Mountains.

There data are essentially identical.   It seems unlikely to us
that a group of environmental factors would cause timberline to
be at 3330 m on east-facing slopes and that a different group
of factors would cause it to be at a nearly identical elevation
on north-facing slopes.

Uniformity of elevation over a large region independent of exposure direction suggests that one group of environmental factors must be limiting the survival of trees at timberline in this region. In addition, the intensity of that group of factors must be almost solely dependent on elevation and independent of exposure direction.

III. TEMPERATURE AND THE TIMBERLINE ENVIRONMENT

Midwinter minimum temperatures are of particular interest to us as potential causes in limiting the timberline elevation, and we will therefore emphasize them here. The timberline environment is severe in many ways, and winter minimum temperatures do not initially stand out as potentially important stresses. Barry (3) has reported on the timberline environment of Niwot Ridge of the north central Colorado Rocky Mountains, and the environment is severe indeed. Among the most impressive things about timberline in this region of Colorado is the high wind velocity. Barry reports that the average annual wind velocity is 47 km/hr, with wind gusts of over 120 km/hr on 30% of midwinter days and 160 km/hr on 10% of midwinter days. The mean annual temperature is low, -3.8°C, and the growing season is short, 50 days. Not all things in the timberline environment are that stressful, however. The precipitation at timberline is 100 cm/yr and increases at higher elevations. There is also snow cover which provides winter protection for low growing plants. The lowest winter temperature extreme over 18 years at 3050 m (300 m below timberline) was only -37°C. The extreme low temperatures in January over a 10-year period (1967-1976) at U.S. Weather Bureau stations located near timberline on Berthoud Pass, Colorado (3440 m) and at Climax, Colorado (3440 m) were only -36 and -34°C, respectively (22). Forested regions in Canada and Alaska are routinely subjected to temperatures well below -40°C and these forests survive. In mountainous regions, extreme minimum temperatures normally drop with a 5 to 10°C lapse rate (i.e., temperature drop per 1000 m elevation gain) (2,21). Based on a comparison to latitudinal changes in temperature (14), such elevational temperature changes in mountainous regions make a 1000 m elevation gain in north central Colorado equivalent to a 3 to 6° latitudinal change (400 to 800 km) in central North America.

We attempt here to obtain a rough approximation of timberline minimum temperatures using lapse rate data and U.S. Weather Bureau data (22) from stations located below timberline. We made the approximations as follows: Regions of interest were in the Rocky Mountains having both Engelmann spruce and subalpine fir at timberline (12). At least five weather stations had to be within 1600 m of the local timberline elevation. A lapse rate of 5°C/1000 m was used to correct

TABLE II. Winter temperature data corrected to the local timberline elevation (1) of mountainous regions in the
Rocky Mountains. Values in parentheses are the number of local weather stations used in calculation of
the predicted timberline temperatures for each location. See text for details of the calculations.

| Location | Elevation of timberline (m) | Predicted temperature at timberline (°C) | |
		Average Jan. minimum	Extreme Jan. minimum
New Mexico, north central (8)	3660	-33 ± 4	-41 ± 5
Colorado, north central (11)	3440	-37 ± 3	-40 ± 3
Wyoming, south central (6)	3290	-33 ± 3	-43 ± 2
Wyoming, west central (6)	3110	-37 ± 3	-42 ± 5
Wyoming, northwest (9)	3080	-38 ± 4	-43 ± 3
Idaho, south central (7)	2960	-35 ± 4	-40 ± 5
Montana, southwest (10)	2900	-36 ± 3	-41 ± 2
Montana, northwest (5)	2350	-38 ± 2	-47 ± 2

the ten-year average of the lowest minimum temperatures
occurring in January (1967-1976) and the ten-year extreme minimum temperature in January at each weather station. The data
were corrected to the regional elevation of timberline. The
corrected data were then used to calculate the means given in
Table II. This approximation suggests that temperatures near
-40°C occur at timberline over a wide region of the Rocky Mountains. In general, temperature extremes would be less severe
on the slopes below timberline and more severe above it. All
of this suggests that there is a very narrow banding pattern of

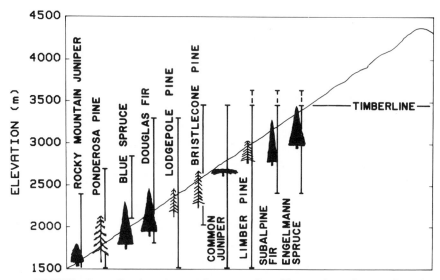

FIGURE 3.   Dominant native woody flora in the Colorado
Rocky Mountains.   In this region, 40° north latitude, timberline occurs at 3340 m.   Solid vertical bars by each species
indicate its elevational distribution as an upright tree.
Dashed extensions of bars indicate that the species extends to
higher elevations as a compact, low growing shrub.

winter hardiness zones on the slopes of mountains.   Based on
the -40°C approximation of timberline temperature extremes and
a comparison to latitudinal hardiness zones timberline in Rocky
Mountain National Park would fall in hardiness zone 3 on level
terrain (14).   Hardiness zone 3 follows the U.S.A.-Canada bor-
der in the central and eastern portions of North America, and
one should note that it is this hardiness zone that marks the
northern limit of the Eastern Deciduous Forest.

IV.   FROST HARDINESS OF WOODY FLORA AT TIMBERLINE

In reviewing literature on frost resistance of timberline
flora, Tranquillini (21) cites several reports on levels of
midwinter frost hardiness near -40°C in several species of
fully cold acclimated woody plants.   Sakai and coworkers
(16,17) have observed frost damage near -40°C in several
species of fully cold acclimated woody plants.   Schwarz (18)
reported the midwinter frost hardiness of needles of Pinus
cembra, the dominant tree at timberline in the Austrian Alps,
to be -43°C.   These observations prompted Becwar et al. (4) to

undertake a thorough investigation into the midwinter frost hardiness of the woody flora at timberline in Colorado. The elevational distribution of the dominant members of that flora are depicted in Figure 3. By far the most apparent members of the timberline flora are Engelmann spruce and subalpine fir (10).

Plant tissues survive winter temperature extremes by two primary mechanisms that involve either tolerating or avoiding ice formation (11). Plant cells that survive by tolerance of ice do so by dehydration in winter. The freezing of water occurs outside the cell, extracellularly, without injury to the living cellular constituents. Such plant tissue in midwinter is extremely hardy and can even survive immersion in liquid nitrogen, $-196°C$, without injury (15). Plant cells that survive by avoidance of ice do so by deep supercooling of tissue solutions (5). Such plants are limited in winter hardiness to about $-40°C$. A supercooled solution is a solution that remains liquid below its equilibrium freezing point and is in a meta-stable state. Such solutions are said to be deep supercooled when they remain liquid to the low temperature limit for super-cooling, the homogeneous nucleation temperature for the solution. The homogeneous nucleation temperature for in situ plant solutions studied previously is about $-40°C$, and this property of solutions sets the lower limit on winter hardiness in super-cooling plants (7). Another characteristic of deep super-cooling systems is that while they freeze near $-40°C$, they thaw near $0°C$. The freezing of such systems can be observed as an exotherm during cooling at a constant rate by differential thermal analysis. In terms of ecological significance, most Eastern Deciduous Forest trees have tissues which survive low temperatures by deep supercooling. They freeze and are injured near $-40°C$, and their northern distribution limit closely parallels plant hardiness zone 3 and the $-40°C$ average annual minimum temperature isotherm for North America (8). In con-trast, boreal forest trees of more northern latitudes, where $-40°C$ temperatures occur with regularity, survive freezing by tolerating ice, a mechanism which does not impart low tempera-ture limits in winter hardiness.

FIGURE 4. Differential Thermal Analysis (DTA) freezing profiles of stems of species native to the Colorado Rocky Moun-tains with low temperature exotherms that indicate freezing of deep supercooled tissue water (4). A) gymnosperms; B) angio-sperms. DTA profiles are plotted as the relative temperature difference between the sample and the reference on the ordinate vs. the reference temperature on the abscissa. The large exo-therm in the Engelmann spruce DTA profile indicates approxi-mately a 3°C temperature difference between sample and refer-ence. Solid circles on the temperature scale indicate the visual injury temperature as determined in separate viability experiments on stems collected at the same time.

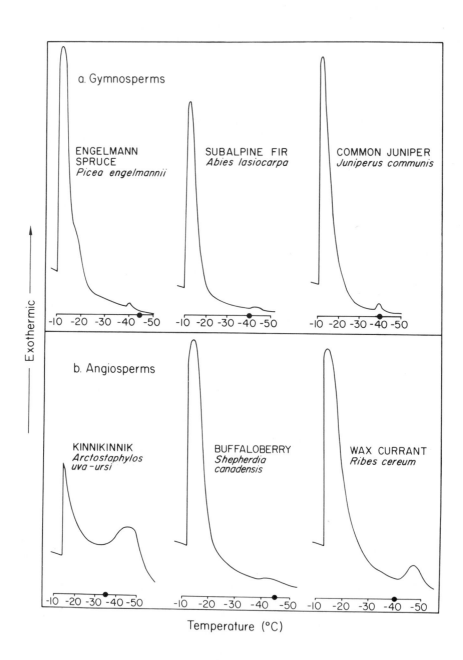

Figure 4

Figure 4 shows the results of differential thermal analysis studies on tree and shrub species native to Colorado and having an elevational limit at or just below timberline. Peaks in the freezing profiles indicate the freezing of water. All species have a large peak near -10°C which is not injurious to the tissue in midwinter. The presence of freezing peaks near -40°C indicates the freezing of deep supercooled tissue water. Such peaks were found in the stem and leaf tissue of numerous conifers studied and in the stem tissue of the woody shrubs studied. Note the close association of the injury temperature (solid circle) with the low temperature freezing peak in all species (Fig. 4). The results of winter hadiness studies (4) are summarized in Table III.

V.   DISCUSSION

To these writers it seems unreasonable that several different groups of environmental conditions could lead to such a uniform elevational limit for timberline as 3340 $\pm$ 110 m in Rocky Mountain National Park of Colorado. It seems unreasonable that the dominant limiting factor at 3340 m in one region of the park is wind, in another at 3340 m it is UV radiation, and in yet another at 3340 m it is temperature. It seems reasonable that one group of factors becomes limiting for plant survival at timberline and that the intensity of these factors depends primarily on elevation. This single group of limiting factors would have to be the same on north, south, east, and west slopes. They would have to lead to elevational uniformity of timberline and an abrupt transition in the forest.

Taking as axiomatic that a single group of environmental factors is involved, one can argue that timberline results from one of two general causes. The first is that some environmental factor has an abrupt transition at timberline which makes plant survival above timberline impossible. So far, there has been little evidence for such an abrupt environmental transition on passing through the timberline elevation. The second general possibility for the cause of timberline is that the environment on passing through timberline changes gradually and the plants respond to one of the gradually worsening environmental factors in a cataclysmic manner. That is, a small change in a single environmental factor causes severe plant stress damage. Stress physiology studies show several stresses which sometimes lead to cataclysmic injury to plants when a slight stress intensification occurs (11). Among the best examples are low temperature stresses. For example, in winter, Engelmann spruce twigs survive prolonged exposure at -35°C and are injured after brief exposure to -45°C (4). Wheat leaves may survive prolonged exposure (days) at -15°C, but are killed after brief exposure to -17°C (9). Other stresses such as

TABLE III. Winter hardiness characteristics and upper elevational limits of gymnosperms and woody angiosperm shrubs native to the Colorado Rocky Mountains. The exotherm temperature (low temperature exotherm) and stem injury temperature are given. Also listed in parenthesis for the gymnosperms, under stem injury temperature, is the midpoint temperature in the transition from low to high relative electrolyte loss from the stem tissue.

Species	Exotherm[1] temp (°C)	Stem injury temp (°C)	Upper elevational limit (m)
Gymnosperms (conifers):			
Engelmann spruce, Picea engelmannii (Parry) Engelm.	-39 (S, L)	-45 (-43)	3660
Subalpine fir, Abies lasiocarpa (Hook.) Nutt	-39 (S, L)	-40 (-38)	3660
Limber pine, Pinus flexilis James	-46 (L)	-45 (-38)	3660
Common juniper, Juniperus communis L.	-38 (S)	-40 (-33)	3440
Lodgepole pine, Pinus contorta Dougl.	not found	none[2] (none[2])	3350
Rocky Mountain juniper, Juniperus scopulorum Sarg.	-41 (S)	-45 (-45)	3350
Douglas fir, Pseudotsuga menziesii (Mirb.) Franco	-40 (S, L)	-40 (-40)	3350
White fir, Abies concolor (Gordon) Coopes	-39 (S)	not determined	3050
Blue spruce, Picea pungens Engelm.	not found	none[2] (-50)	2900
Ponderosa pine, Pinus ponderosa Laws.	not found	-45 (-45)	2830

Angiosperms (woody shrubs):

	not found	none[2]	
Willow, Salix L. spp.	-37	-35	4270
Kinnikinnik, Arctostaphylos uva-ursi L.	-39	-40	3570
Raspberry, Rubus L. spp.	-40	-45	3570
Buffaloberry, Shepherdia canadensis (L.). Nutt.	-40	-40	3510
Wax currant, Ribes cereum Dougl.	-39	-40	3470
Mountain maple, Acer glabrum Torr.	-39	-35	3200
Snowberry, Symphoricarpos Duhamel, spp.	-34	-35	3200
Sagebrush, Artemisia tridentata Nutt.	-38	-40	3080
Mountain Mahogany, Cercocarpus montanus Raf.	-40	-40	3050
Ninebark, Physocarpus monogynus (Torr.) Coult.	-38	-35	3050
Snakeweed, Gutierrezia sarothrae (Pursh) Britt. & Rusby	-38	-40	3050
Bitterbrush, Purshia tridentata (Pursh) DC.	-39	-40	2740
Skunkbrush, Rhus trilobata Nutt.	-40	-40	2740
Wild grape, Vitis riparia Michx.	-40	-40	2130

1 Exotherm temperatures for stem (S) and/or leaf (L) tissue in gymnosperms and stem tissue only in angiosperms.

2 Survived the lowest test temperature, -60°C.

wind, water, light intensity, low CO_2 concentration, salt, mineral deficiency, mechanical damage, etc., generally lead to injury of plants in such a way that gradual increases in stress intensity or duration lead to gradual changes in plant growth (11). It is unlikely that a small increase in wind speed or a small decrease in soil water potential will lead to death of an otherwise healthy plant. However, a small change in the temperature (e.g. from -35 to -45°C) of an Engelmann spruce tree can lead to injury of a previously healthy tree.

Our results show that in Rocky Mountain National Park of north central Colorado the timberline transition is abrupt and at an extremely uniform elevation. This suggests that one group of limiting factors causes the cessation of tree growth in this region. The sharpness of this timberline transition suggests that an abrupt plant stress response to a gradually worsening environment at higher elevations may be the cause. We predict that temperatures approaching -40°C occur at timberline and because of temperature lapse with elevation, the low temperature stress factor will depend strongly on elevation. By analysis of the midwinter frost hardiness of the woody timberline flora, we find that most members of the flora suffer frost injury at -40°C and furthermore, that -40°C is the low temperature hardiness limit for this timberline flora.

REFERENCES

1. Arno, S.F., "Interpreting the Timberline". National Park Service, U.S. Dept. of Interior, San Francisco, Calif. (1967).
2. Baker, F.S., Ecol. Monogr. 14:223-254 (1944).
3. Barry, R.G., Arct. Alp. Res. 5:89-110 (1973).
4. Becwar, M.R., Rajashekar, C., Hansen-Bristow, K.J., and Burke, M.J., Plant Physiol. 68:111-114 (1981).
5. Burke, M.J., Gusta, L.V., Quamme, H.A., Weiser, C.J., and Li, P.H., Annu. Rev. Plant Physiol. 27:507-528 (1976).
6. Daubenmire, R., Butler Univ. Bot. Stud. 11:119-136 (1954).
7. George, M.F., and Burke, M.J., Curr. Adv. Plant Sci. 8:349-360 (1976).
8. George, M.F., Burke, M.J., Pellet, H.M., and Johnson, A.G., HortScience 9:519-522 (1974).
9. Gusta, L.V., Burke, M.J., and Kapoor, A.C., Plant Physiol. 56:707-709 (1975).
10. Harrington, H.D., "Manual of the Plants of Colorado". Swallow Press Inc., Chicago, Ill., 1964.
11. Levitt, J., "Responses of Plants to Environmental Stresses", Vols. I and II. Academic Press, New York, N.Y., 1980.

12. Little, E.L., "Atlas of United States Trees", Vol. 1. U.S. Dept. Agriculture, Forest Service, Misc. Publ. No. 1146, Washington, D.C., 1971.
13. Mills, E., "The Adventures of a Nature Guide". Doubleday Publ., New York, N.Y., 1920.
14. National Arboretum, Agricultural Research Service, "Plant Hardiness Zone Map". U.S. Dept. Agriculture, Misc. Publ. 814, Washington, D.C., 1965.
15. Sakai, A., Nature 185:392-394, 1960.
16. Sakai, A., and Okada, S., Silvae Genetica 20:91-97 (1971).
17. Sakai, A., and Weiser, C.J., Ecology 54:118-126 (1973).
18. Schwarz, W., Flora 159:258-275 (1970).
19. Tranquillini, W., Planta 49:612-661 (1957).
20. Tranquillini, W., Annu. Rev. Plant Physiol. 15:345-362 (1964).
21. Tranquillini, W., "Physiological Ecology of the Alpine Timberline". Springer-Verlag, New York, N.Y., 1979.
22. United States Department of Commerce, "Climatological Data". National Oceanic and Atmospheric Administration, Environmental Data Service, Asheville, N.C., 1967-1976.
23. Wardle, P., Ecology 49:483-495 (1968).
24. Wardle, P., New Zealand J. Bot. 9:371-402 (1971).
25. Wardle, P., in "Arctic and Alpine Environments" (J.D. Ives and R.G. Barry, eds.), pp. 371-402. Methuen and Co. Ltd., London, 1974.

CHARACTERISTICS OF FREEZING AVOIDANCE IN COMPARISON WITH FREEZING TOLERANCE: A DEMONSTRATION OF EXTRAORGAN FREEZING

M. Ishikawa and A. Sakai

The Institute of Low Temperature Science
Hokkaido University
Sapporo, Japan

Freezing avoidance[1] by supercooling as a strategy against freezing stress has been shown to be employed by overwintering xylem ray parenchyma cells of many temperate trees (5,13,17,23), flower buds of Rhododendron (4,6,8,10,15), Vaccinium (19), Cornus (26), and Prunus (1,17,21,22), leaf buds of some conifers (2,25,27,28,29), and hydrated seeds (3,9,13).

In contrast to deep supercooling in xylem, natural occurrence and characteristics of freezing avoidance by deep supercooling in flower buds and seeds have not been clarified sufficiently. And also inadequate is the inquiry into questions such as why some tissues avoid freezing while other tissues tolerate freezing even within an individual plant and what are the differences between tissues that avoid and tissues that tolerate freezing.

Regarding these situations the present report deals with (1) a survey for natural occurrence of freezing avoidance in flower buds and seeds, (2) some characteristics of freezing avoidance in flower buds and seeds, (3) where the difference between freezing avoidance and tolerance lies, (4) classification of freezing avoidance and tolerance, to have better understanding of the nature of freezing events in plants.

As plant materials, flower buds of 62 woody plant species and seeds of 32 woody and herbaceous plant species occurring in Japan, mainly in Hokkaido and cultivated in the Botanical Garden of Hokkaido University, were collected during late

---

[1] In this report, by freezing avoidance the authors refer to "freezing avoidance by (deep) supercooling" and not to "avoidance by being in a desiccated state (with no freezable water)" among the ones suggested by Levitt (16).

autumn to early spring, and used for the survey for the types of resistance with DTA[2]. Most of the species used are listed in Table I and Table II.

For the study on the characteristics of freezing avoidance, overwintering flower buds of Rhododendron japonicum (8,10) and Cornus officinalis, and hydrated seeds of lettuce (Lactuca sativa cv. Great Lakes), Oenothera sp. and Vitis coignetiae (11) were mainly used. Hydrated seeds of these species were obtained by imbibition in water at 0°C for 24 to 200 hours.

Methods used in the present study were microscopic observation of frozen plant materials and their freezing process, DTA, determination of water content, and cold hardiness evaluation.

DTA was conducted according to the methods described elsewhere (10). Exotherm responses were amplified 40 to 100 times and recorded together with the temperature of the sample. Programmed cooling rates of 2 to 240°C/hr were used.

The water content was expressed as the percentage of fresh weight (fr. wt.) or dry weight (dr. wt.). Dry weight of flower buds was determined after being heated at 70 to 80°C for 24 hrs and that of seeds at 105°C for 24 hrs. To know the water content of each part of a frozen flower bud, buds with 15 cm of twigs in polyethylene bags which were saturated with a little snow were cooled at 5°C decrements daily from 0 to -15°C. At each temperature buds were separated into parts and the water content was determined in the manner described elsewhere (10).

Microscopic observations of flower buds and seeds cooled at daily 5°C decrements to -10°C in polyethylene bags were conducted in a cold room held at -10°C to determine the localization of ice. Microscopic observation of the freezing process was also made with petal, ovule, and receptacle tissues of florets mounted in silicone oil, and ice-inoculated at -1 to -2°C.

Cold hardiness of seeds was determined by cooling them from 0 to -15°C at 5°C decrements per hour (5°C/hr), 4 hrs (1.3°C/hr), or 12 hrs (0.4°C/hr) and seeds were kept at the desired temperature for 2 hrs then rewarmed at 0°C. Some seeds were cooled to temperatures of -5 to -30°C at various cooling rates prior to immersion in liquid nitrogen. After 15 min storage in liquid nitrogen, they were kept at -10°C for an hour and then rewarmed at 0°C to eliminate the possibility of survival by rapid cooling and rewarming (30). Survival was expressed as the percentage of germinated seeds per that of untreated control. Cold hardiness of flower buds was also measured in the same manner but at the cooling rate of 2°C/hr. To ascertain that the exotherm temperatures are correlated with the killing temperatures of the florets and seeds, samples were taken from the freezer just before or after the emergence of the exotherm and rewarmed. Flower buds were incubated at room

---

[2]   Abbreviations: DTA, differential thermal analysis; HTE, high temperature exotherm; LTE, low temperature exotherm.

Table I. Results of DTA studies of woody plant flower buds.

Family	Species	Range of LTE[a](°C)
Caprifoliaceae	Viburnum furcatum	ND[b]
	Sambucus racemosa var. miquelii	ND
Oleaceae	Syringa vulgaris	ND
	Forsythia suspensa	−25 −30
	F. koreana	−24 −28
Ericaceae	Vaccinium smallii	−12 −20
	V. vitis-idaea	−15 −29
	Gaultheria miqueliana	ND
	G. adenothrix	ND
	Andromeda polifolia	−18 −22
	Pieris japonica	−10 −23
	Lyonia ovalifolia	?
	Parapyrola asiatica	−18 −19
	Chamaedaphne calyculata	−21 −26
	Harrimanella stelleriana	?
	Cassiope lycopodioides	−18 −25
	Arcterica nana	−18 −21
	Loiseleuria procumbens	−15 −22
	Phyllodoce sp.	−12 −21
	Tsusiophyllum tanakae	−20 −24
	Rhododendron japonicum	−20 −36
	R. dilatatum	−16 −25
	R. ripense	−11 −13
	R. obtusum	−19 −29
	R. tschonoskii	−20 −24
	R. brachycarpum	−22 −28
	R. dauricum	−27 −31
	R. keiskei	−11 −17
	Menziesia multiflora	−21 −24
	M. multiflora var. purpurea	−14 −24
	M. pentandra	−18 −28
	Ledum palsture subsp. diversipilosum	−22 −25
Cornaceae	Cornus stolonifera	ND
	C. officinalis	−17 −32
	C. kousa	−19 −30
	C. florida	−21 −30
Hippocastanaceae	Aesculus turbinata	ND
Cercidiphyllaceae	Cercidiphyllum japonicum	−19 −24
Ulmaceae	Ulmus davidiana var. japonica	−25 −33
	U. pumila	−21 −27
Betulaceae	Betula platyphylla var. japonica	ND
Salicaceae	Populus sp.	ND

[a] Low temperature exotherm. Cooling rate was 2 to 9°C/hr.
[b] LTE were not detected even when amplified 100 times.

temperature for a week to rate the injury by browning. Seeds were incubated at room temperature for a week to three months to count the number of germinated seeds or for 6 hrs in distilled water to rate the injury by the conductivity method as described elsewhere (11).

I. SURVEY FOR THE NATURAL OCCURRENCE OF FREEZING AVOIDANCE

    Table I shows the results of DTA of flower buds of woody plant species. Since the date of collection of each species varies from November to March and the cooling rate which will be shown to affect the supercooling ability was between 2 and 9°C/hr, the range of LTE shown may not necessarily represent the maximum hardiness. But at the given cooling rates, the florets of the buds were killed just after the emergence of exotherms and alive before their emergence.
    In flower buds of Viburnum furcatum, Sambucus racemosa var. miquelii, Syringa vulgaris, Gaultheria miqueliana, G. adenothrix, Cornus stolonifera, Aesculus turbinata, Betula platyphylla var. japonica, and Populus sp., exotherms (LTE) were not detected even when amplified 100 times. On the other hand, flower buds of 2 species in Forsythia, 2 species in Ulmus, Cercidiphyllum japonicum, and 43 species which belong to 13 genera in Ericaceae (23 species of which are listed in Table I and the remaining 20, which belong to Rhododendron are omitted) were newly found to yield LTE.
    In Cornus flower buds 3 species exhibited LTE while C. stolonifera did not. This seems to be similar to the case of Prunus flower buds in which some very hardy species did not show LTE while other less hardy species did show LTE (1,22). Some taxonomists classify Cornus species in different ways; C. officinalis Sieb. et Zucc. to be Macrocarpium officinale Nakai, C. kousa Buerger, ex Hance to be Benthamidia japonica Hara, and C. florida Linn. to be B. florida Spach. So it is still a question whether the case of Cornus is an exception to the usual case, i.e., species in a genus sharing the same strategy. The species in Ericaceae seem to share the same strategy (freezing avoidance) at the family level although there were some exceptions. They seem to suggest a possible involvement of genealogical background in the strategy against freezing stress (avoidance or tolerance) in flower buds which will be further discussed later.
    Table II shows the results of DTA with hydrated seeds and seeds surrounded by the fruits or arils of temperate species. As far as the species surveyed and the cooling rate of 5°C/hr were concerned, seeds with water content above 20% exhibited LTE and seeds were killed after the emergence of LTE. When cooled more slowly, seeds, especially smaller ones increased their hardiness which will be shown later.

TABLE II. Results of DTA studies of seeds of woody and herbaceous plants.

Family	Species	Range of LTE[a](°C)		Water content (% fr. wt.)
Woody plants				
Caprifoliaceae	Viburnum dilatatum*[b]	-23	-27	32.9
	V. wrightii*	-20	-22	34.8
Styracaceae	Styrax japonica	-22	-28	37.7
Ericaceae	Vaccinium oldhami	-14	-22	60.1
	V. bracteatum	-9	-20	35.1
Araliaceae	Kalopanax pictus*	-24	-30	----
Vitaceae	Vitis coignetiae*	-15	-19	28.9
Celastraceae	Celastrus orbiculatus*	-22	-24	34.9
	Euonymus alatus*	-19	-23	45.0
Aquifoliaceae	Ilex crenata	-19	-21	45.2
	I. pedunculosa	-13	-16	32.2
Coriariaceae	Coriaria japonica*	-14	-24	----
Rutaceae	Phellodendron amurense*	-21	-25	----
Rosaceae	Rhaphiolepsis indica	-8	-13	44.0
	Rosa rugosa*	-11	-25	----
	R. multiflora*	-16	-27	----
Saxifragaceae	Hydrangea paniculata	-18	-23	53.5
Theaceae	Eurya japonica	-14	-22	32.0
	E. emarginata	-15	-21	41.4
Berberidaceae	Berberis sp.*	-17	-24	45.0
Magnoliaceae	Magnolia obovata*	-14	-15	30.1
Herbaceous plants				
Compositae	Adenocaulon himalaicum	-15	-17	49.1
	Carpesium abrotanoides	-16	-21	51.2
Gentianaceae	Tripterospermum japonicum*	-17	-22	54.0
Onagraceae	Oenothera sp.	-20	-26	30.9
Buxaceae	Pachysandra terminalis*	-15	-18	----
Rosaceae	Geum japonicum	-12	-22	51.5
Berberidaceae	Diphylleia cymosa*	-18	-21	----
Amaranthaceae	Achyranthes japonica	-14	-18	48.7
Polygonaceae	Polygonum filiforme	-15	-18	36.0
Liliaceae	Smilax china	-18	-22	38.3
	Maianthemum dilatatum*	-15	-18	49.7

a Low temperature exotherm. Cooling rate was about 5°C/hr.
b Seeds of asterisked species were within their fruits or arils when used for DTA.

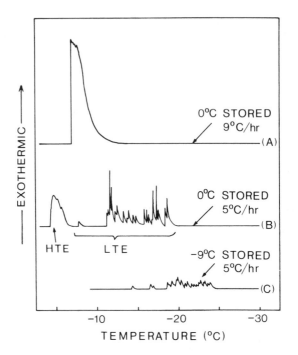

FIG. 1.   DTA profiles of florets of C. officinalis flower buds of early April cooled at different rates. (A): Scales and florets froze together as indicated by the large exotherm when cooled at 9°C/hr. (B): Scales froze first as shown by HTE and small spikes (LTE) between -8 and -18°C indicated lethal freezing of florets. (C): Buds were precooled to -9°C in two days (scales had already frozen), then cooled further at 5°C/hr without thawing.

II.   CHARACTERISTICS OF FREEZING AVOIDANCE IN FLOWER BUDS

One characteristic of freezing avoidance in flower buds is that most of them are cooling rate dependent, especially those with smaller florets. An example is C. officinalis in early April (Fig. 1). A flower bud of C. officinalis has 20 to 25 florets surrounded by four involucral scales (Fig. 2A). Each floret supercooled to -18 to -23°C as indicated by small spikes (LTE) in Fig. 1C when precooled to -9°C at daily decrements of about 5°C. When the bud was cooled at 5°C/hr, LTE appeared at higher temperatures (Fig. 1B). When the cooling rate was 9°C/hr, the florets froze together with the scales and other tissues and no spikes appeared on the DTA profile (Fig. 1A).

During the cooling to -9°C at about 5°C daily decrements, ice accumulated in the scales (Fig. 3C), water content of the

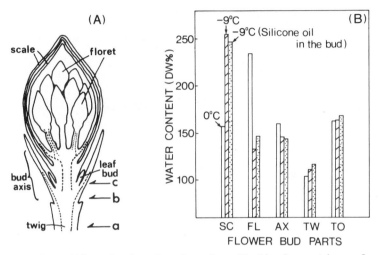

FIG. 2.   (A):   A sketch of a longitudinal section of a C. officinalis flower bud.   Broken lines indicate vascular tissues.   a, b, c, show where the bud was cut from the twig for DTA (See text).   (B):   Migration of water from the florets to the scales in C. officinalis flower buds of early April during the cooling from 0 to -9°C in two days. ▭ , water content of flower bud parts at 0°C (intact); ▨ , water content at -9°C (intact); ▨ , water content at -9°C (silicone oil in the buds).

scales increased and that of florets decreased while the other parts of the flower buds showed little change in their water content (Fig. 2B).   It seems that the water in the florets migrates to the scales where it freezes while dehydrated florets increase their supercooling ability during the slow cooling.   Similar water migration from florets to scales was also observed in several Rhododendron flower buds cooled slowly from 0 to -15°C, but in some cases of Rhododendron a decrease in the water content of the flower bud axis was also observed (8,10).

The question that arises is that through where does the water migrate from the florets to the scales.   To investigate this, silicone oil was inserted with a syringe into the space between the florets and the scales or florets were wrapped with parafilm without damaging the buds to prevent direct migration of water as vapour from the surface of the florets to the scales (Fig. 2B).   Even in these cases during cooling to -9°C the floret water content decreased and that of the scales increased to almost the same degree as in intact buds.   This indicates that water in the floret may migrate by way of

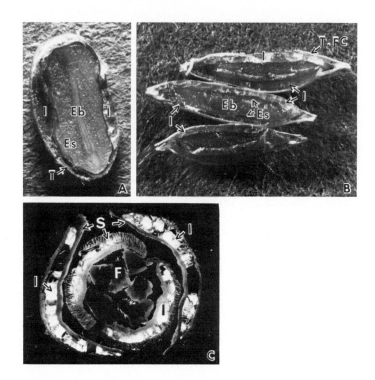

FIG. 3. Extraorgan freezing of frozen seeds of Celastrus orbiculatus (A) and lettuce (B) cooled to -10°C. (C): A cross section of a frozen flower bud of C. officinalis cooled to -9°C photographed with polarized light. (A), X8.5; (B), x11.9; (C), x8.5. (Eb), embryo; (Es), endosperm; (F), floret; (FC), fruit coat; (I), ice; (S), scale; (T), testa.

vascular systems to the scales rather than through the surface of florets to the scales.

Involvement of vascular systems in the freezing avoidance in C. officinalis flower buds was also suggested in another experiment. At the cooling rate of 2 to 3°C/hr florets in a winter bud excised at line a (Fig. 2A) normally supercooled (LTE: -19 to -23°C). On the other hand, in the buds cut at line b LTE were shifted to higher temperature (LTE: -5 to -20°C), and when excised at line c, they did not supercool and froze together with other tissues (LTE: -6 to -7°C).

Microscopic observation of the freezing process of floret tissues of Rhododendron species and C. officinalis was made to check whether or not freezing avoiding tissues have the ability to freeze extracellularly. Petal, receptacle, and ovule tissues mounted in silicone oil were cooled at 10°C/hr and ice-inoculated around -1°C which causes extracellular freezing in

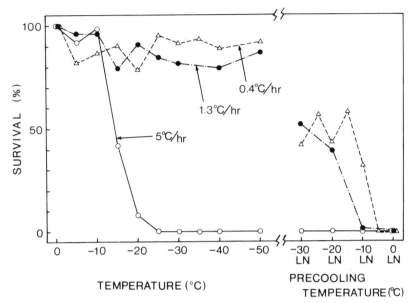

FIG. 4. Effect of cooling rate on the survival temperature of hydrated lettuce seeds with water content of 48.7% (fr. wt.). Some seeds were precooled to -10 to -30°C at various cooling rates prior to immersion in liquid nitrogen. Corresponding LTE (Mean ± SD) of seeds with the same water content was -14.9 ± 1.7°C when cooled at 5°C/hr.

usual overwintering tissues. Then most of the cells in these tissues flashed one after another around -2°C causing intracellular freezing and were killed.

III. CHARACTERISTICS OF FREEZING AVOIDANCE IN SEEDS

Supercooling ability of some seeds was also cooling rate dependent as shown in Fig. 4 and Table III. Cold hardiness of water-imbibed lettuce seeds (water content: 48.7% fr. wt.) was considerably affected by cooling rate. When cooled at 5°C/hr, 50% killing temperature was -14 to -15°C, which coincided with mean LTE (-14.9 ± 1.7°C; cooling rate, 5°C/hr) of the equally hydrated seeds. When the cooling rate was slower than 1.3°C/hr, LTE were scarcely detected on DTA profile of the lettuce seeds and they survived -50°C. Moreover nearly 40% of the seeds withstood the temperature of liquid nitrogen after being precooled to -20°C at 1.3°C/hr. When the seeds were precooled to -10°C at 0.4°C/hr, over 30% of the seeds survived

Table III. Effect of storage at subzero temperatures on the exotherm temperature of Vitis coignetiae seeds with water content of 36% (fr. wt.).

Treatment	LTE[a] (Mean $\pm$ SD)
0°C stored (control)	-20.6 $\pm$ 1.2 (n = 23)
-5°C for 3 days	-22.5 $\pm$ 1.1 (n = 21)
-5°C for 2 days and -10°C for 2 days	-26.8 $\pm$ 2.3 (n = 14)

[a]   Low temperature exotherm, cooling rate was about 5°C/hr.

immersion in liquid nitrogen. The seeming survival in liquid nitrogen was rather low, but this was because a considerable number of the seeds were mechanically cracked by the plunge into liquid nitrogen.

Survival in liquid nitrogen at slow cooling rates was more clearly demonstrated in seeds of Oenothera sp. which are smaller (ca. 1.6 x 0.9 x 0.6 mm) than lettuce seeds (ca. 4.0 x 1.0 x 0.5 mm) (11). Nearly 100% of Oenothera seeds (water content 27 to 31% fr. wt.) survived immersion in liquid nitrogen after being precooled to -15°C at 1°C/hr, while at the rate of 5°C/hr the 50% killing point was about -23°C.

In the case of grape (Vitis coignetiae) seeds which are larger (ca. 5.6 x 3.9 x 3.0 mm) than lettuce, there was no marked change in the exotherm temperature at the cooling rates between 0.4 and 5°C/hr. But the storage of the seeds at -5 to -10°C for a few days (Table III) shifted exotherms to a lower temperature.

Microscopic observation of the seeds at -10°C revealed that ice segregated from the embryos and endosperms during slow cooling or storage at -5 to -10°C. Fig. 3A,B shows the case of Celastrus and lettuce seeds, respectively. In lettuce ice was formed between the integument and the endosperm layer and in Celastrus ice accumulated mainly between the endosperm and the testa. In the case of grape and Oenothera seeds, more ice was formed outside the testa than in the space between the testa and the endosperm. Ice seemed to be formed in a particular space depending on species and consequently the dehydrated embryos and endosperms seem to increase their hardiness as far as they tolerate the dehydrated state. Ice segregation in slowly cooled seeds might be a reverse phenomenon of water absorption.

Junttila and Stushnoff (13) indicated the necessity of the integrity of the endosperm sac in supercooling of lettuce seeds (cooling rate: 20°C/hr). This holds true even in the foregoing cases of ice segregation induced by slower cooling rates (0.4, 1.3°C/hr). When the endosperm sac was stuck with a needle, ice was mainly formed between the endosperm sac and the embryo upon cooling to -5°C. Because of the small size of the seeds and a great amount of storage substances in the embryo cells, we were unsuccessful in determining whether ice occurred within the embryo and whether the embryo cells had the ability to freeze extracellularly or caused intracellular freezing when they were artificially ice-seeded. When ice was formed inside the endosperm sac in a needle-stuck lettuce seed, the radicle was most susceptible and injured during cooling to -5°C. Most of the seeds recovered the injury by developing adventitious roots, but growth of the seedlings was considerably delayed. The other parts of the embryo were hardier than the radicle, but they scarcely tolerated -15°C while in intact seeds they survive -196°C following slow precooling to -10°C. The endosperm sac which consists of two cell layers of endosperm and cuticular layer on the surface, seems to protect the embryo in a manner that it works as a barrier which blocks ice propagation from outside and is permeable to water or vapour from inside or outside.

IV.   CLASSIFICATION AND COMPARISON OF FREEZING AVOIDANCE AND TOLERANCE

In summarizing the characteristics of freezing avoidance in florets and seeds it seems that conventionally used "supercooling" or "deep supercooling" is not sufficient to describe some cases of freezing avoidance in florets and seeds especially at naturally occurring cooling rates. Table IV shows some modification in the classification of freezing events in plants based on the new findings and speculations.

Freezing avoidance by supercooling, whose concept was established by Levitt (16) based on the results of early investigators who used rapid cooling rates (mostly over 5°C/hr), can be categorized into two types at slower cooling rates which occur in nature. One type, which can be termed "extraorgan freezing", accompanies ice segregation from supercooled organ to some specific space outside the organ, resulting in dehydration of the organ. This type is cooling rate dependent; the slower the rate, the higher the ability to suprcool (8,10). The other type is a real "supercooling" which does not accompany water migration and is unaffected by slow cooling. Supercooling in overwintering xylem ray parenchyma cells in temperate woody species may belong to this type (20).

The former type (extraorgan freezing) can be further divided into three types according to the cause of survival or death. The first type is that the organ is fully dehydrated when cooled slowly, so that there is no freezable water within

Table IV. Classification of freezing events in plants and their characteristics

Levitt's classifica-tion	Types of freezing (Strategies)	Theoretical limit around -40°C	Barrier against ice	Ability to freeze extra-cellularly	Speed of water migration	Effect of slow cooling	Ice localization
Freezing tolerance	Extracellular freezing	–	Cell level	+			
	Extratissue freezing	–	Cell level	+			
Freezing avoidance by supercooling	Extraorgan		Organ level?	–?			
	(1) tolerant of fully dehydrated state	+/–					
	(2) intolerant of fully dehydrated state	–					
	(3) with partial dehydration	+					
	(Deep) Supercooling	+	?	–?			

the organ and the organ is desiccation tolerant. Then the organ can tolerate even immersion in liquid nitrogen after precooling. Water imbibed seeds of lettuce and Oenothera sp. (11) and very hardy conifer primordial shoots native to Alaska (27,28) belong to this type. The second type is that the organ can be fully dehydrated during slow cooling but is not desiccation tolerant and killed by extreme dehydration. Very hardy Prunus flower buds which are reported (22) to show LTE, but to tolerate -80°C after prefreezing for several days may belong to the first or second type. Thirdly, even though the organ is dehydrated partially during slow cooling, freezable water remains in the organ and the organ is killed by the break of supercooling. Most of the seeds surveyed in this report including grape and Celastrus, seeds of Ligustrum reported by Gazeau (3), flower buds of Cornus officinalis and Rhododendron japonicum (10) may belong to this type. Leaf buds of less hardy conifers (27) may belong to the second or third type. The third and second type are very difficult to distinguish, because the size of exotherms gets too small to detect when the organs are cooled very slowly. Theoretical limit to the survival temperature around -40°C which corresponds to the homogeneous nucleatin temperature of pure water in a finely dispersed state as indicated in the supercooling of xylem tissues (5) exists in the third type of extraorgan freezing. In this sense, the third type is close to "supercooling". It may be termed extraorgan freezing with respect to ice localization while it could be supercooling with respect to the state of water remaining in the tissue.

Supercooling in hydrated lettuce seeds reported by Junttila and Stushnoff (13) does exist at cooling rates above 5°C/hr, but at slower cooling rates which usually occur in nature lettuce seeds undergo first type of extraorgan freezing rather than supercooling. Supercooling and the third type of extraorgan freezing seem to occur in larger seeds in natural conditions and may work as hardiness mechanism in desiccation intolerant seeds.

Focusing on the ice localization, there is "extratissue freezing" as a derivative of freezing tolerance by extracellular freezing. In this case most of the ice accumulated in a particular space outside the tissue, and a little in the intercellular space within the tissue. For example, ice was formed between the veins and spongy tissue in "frostblase" of Buxus leaves (7), and in the vascular bundle ring region in frozen root of table beet (31).

Looking through the new classification, there is a gradient in the rate of dehydration or water migration from the cell, tissue or organ to some space; rapid in extracellular freezing, slow in extraorgan freezing and little migration in "supercooling". There is also a gradient in the effect of slow freezing in enhancing the cold hardiness; greater in extraorgan freezing, smaller in extracellular freezing and in "supercooling". Particularly the first type of extraorgan freezing, which behaves like supercooling when cooled rapidly yet it can survive -196°C when cooled slowly.

In comparing extracellular freezing and extraorgan freezing, extraorgan freezing is similar to extracellular freezing in a sense that the tissues are more or less dehydrated provided that they are cooled slowly. But the difference between the two seems to lie in the fact that in extra-organ freezing the cells in the intact organ have a barrier against ice propagation at the organ level or outside the organ and do not seem to have one within the organ or at the cellular level. This is indicated by the intracellular freezing of intact floret tissues when they were ice-inoculated artificially at -1 to -2°C and the inability of radicles of needle-stuck lettuce seeds to withstand -5°C. But we do not deny the possibility that an organ has the ability to freeze extra-organly and the cells in the organ have the ability to freeze extracellularly at the same time.

In extracellular freezing the plasma membrane works as a barrier, while in extraorgan freezing the endosperm cuticular layer or testa in the case of seeds, and vascular systems in the case of florets seem to be involved in the prevention of ice propagation and at the same time in the water translocation from the supercooled organ to the ice sink. One possible reason for the inability of floret tissues to undergo extra-cellular freezing is that there is no large intercellular space for extracellular ice formation within the tissues. Another possibility is that the plasma membrane or cell wall is somehow different from those of tissues which freeze extracellularly. Histological and cytological studies in this respect are needed.

As for terminology, the boundary between freezing avoidance and tolerance has become obscure by the emergence of extraorgan freezing which lies between the two strategies. And it is a matter of question which extraorgan freezing belongs to, freezing avoidance or tolerance. The three types of extraorgan freezing could be freezing tolerance when focusing on freeze dehydration at naturally occurring cooling rates while the third type of extraorgan freezing could be freezing avoidance when focusing on the cause of death (break of supercooling). However, the senior author (M.I.) considers it better to recognize actual plant tissue strategies against freezing stress with terms "extracellular freezing", "extraorgan freezing" and (deep) "supercooling" rather than "freezing avoidance and tolerance" since these terms better imply kinds of stresses imposed on cells, tissues, or organs (12).

Taking the foregoing characteristics of extraorgan freezing into consideration, the cooling rates between 2 and 9°C/hr used to survery for the strategies of florets might be insufficient to judge extracellular freezing, extraorgan freezing, or super-cooling. There is a possibility that flower buds in which LTE were not detected might undergo extraorgan freezing and would supercool if they were cooled more rapidly. Exotherm yielding flower buds might tolerate immersion in liquid nitrogen if cooled very slowly. However, it is considered that at these cooling rates flower buds which yielded LTE are extraorgan freezing or supercooling and that those which did not yield exotherms are closer to extracellular, extratissue freezing, or to extraorgan freezing which accompanies high rate of water translocation. It is conceivable that very hardy species of Prunus (1,22) and Cornus in which LTE were not detected and which tolerated -50 to -80°C might have tiny florets which undergo extraorgan freezing rather than extracellular freezing. If it were true, the surveyed species in each genus (Prunus, Cornus) would share the same strategy, i.e., extraorgan freezing.

Implicated genealogical background behind the strategy against freezing stress might arise from the possible involvement of vascular tissue or morphological feature of the flower buds in the strategies against freezing. Another possibility is that the size or developmental stage of florets in the surveyed winter buds is very similar in one genus or family and that a genus which has large florets might yield LTE while one with tiny florets might be without LTE. Further studies in this direction are also required.

ACKNOWLEDGMENT

We wish to thank M. F. Yoshie of the Inst. of Low Temp. Sci. for kindly providing some of the seeds used in the experiments.

REFERENCES

1. Burke, M.J. and Stushnoff, C., in "Stress Physiology in Crop Plants" (H. Mussel and R.C. Staples, eds.), pp. 197-226. John Wiley & Sons, New York, 1979.
2. Dereuddre, J., Physiol. Veg. 16:469 (1978).
3. Gazeau, C. and Dereuddre, J., C. R. Acad. Sc. Paris Ser. D 290:1443 (1980).
4. George, M.F., Burke, M.J., and Weiser, C.J., Plant Physiol. 54:29 (1974).
5. George, M.F., and Burke, M.J., Curr. Adv. Plant Sci. 8:349 (1976).
6. Graham, P.R., and Mullin, R., J. Amer. Soc. Hort. Sci. 101:7 (1976).
7. Hatakeyama, I. and Kato, J., Planta (Berl.) 65:259 (1965).
8. Ishikawa, M., Master's Thesis, The Inst. of Low Temp. Sci., Hokkaido Univ. (1979).

9. Ishikawa, M. and Sakai, A., Low Temp. Sci. Ser. B 36:39 (1978).
10. Ishikawa, M. and Sakai, A., Plant & Cell Physiol. 22:953 (1981).
11. Ishikawa, M. and Sakai, A., Cryobiology 19:Submitted (1982).
12. Ishikawa, M., Doctor's Thesis, The Inst. of Low Temp. Sci., Hokkaido Univ. (1982).
13. Junttila, O. and Stushnoff, C., Nature 269:325 (1977).
14. Kaku, S. and Iwaya, M., in "Plant Cold Hardiness and Freezing Stress" (P.H. Li and A. Sakai, eds.), pp. 227-239. Academic Press, New York, 1978.
15. Kaku, S., Iwaya, M., and Kunishige, M., Plant & Cell Physiol. 21:1205 (1980).
16. Levitt, J., "Responses of Plants to Environmental Stresses". 2nd edition. Academic Press, New York, 1980.
17. Proebsting, E.L., and Sakai, A., HortScience 14:597 (1979).
18. Quamme, H.A., Stushnoff, C., and Weiser, C.J., J. Amer. Soc. Hort. Sci. 97:608 (1972).
19. Quamme, H.A., Stushnoff, C., and Weiser, C.J., HortScience 7:500 (1972).
20. Quamme, H.A., Weiser, C.J. and Stushnoff, C., Plant Physiol. 51:273 (1973).
21. Quamme, H.A., J. Amer. Soc. Hort. Sci. 99:315 (1974).
22. Rajashekar, C., and Burke, M.J., in "Plant Cold Hardiness and Freezing Stress" (P.H. Li and A. Sakai, eds.), pp. 213-225. Academic Press, New York, 1978.
23. Sakai, A., Plant Physiol. 40:882 (1965).
24. Sakai, A., Low Temp. Sci. Ser. B 36:1 (1978).
25. Sakai, A., Plant & Cell Physiol. 19:1439 (1978).
26. Sakai, A., HortScience 14:69 (1979).
27. Sakai, A., Plant & Cell Physiol. 20:1381 (1979).
28. Sakai, A., Low Temp Sci. Ser. B 37:1 (1979).
29. Sakai, A., and Ishikawa, M., J. Jap. For. Soc. 61:15 (1980).
30. Sakai, A., and Yoshida, S., Plant Physiol. 42:1695 (1967).
31. Terumoto, I., Low Temp. Sci. Ser. B 18:39 (1960).

TEMPERATURE EFFECTS ON ACCLIMATION AND DEACCLIMATION OF SUPERCOOLING IN APPLE XYLEM

Sung Gak Hong
Department of Forestry
Kon-Kuk University
Seoul, Korea

Edward Sucoff
Forest Resources Department
University of Minnesota
St. Paul, Minnesota

I. INTRODUCTION

The hardiness of apple (<u>Malus</u> sp.) twigs increases from above -10°C in summer to below -30°C in midwinter (6,9,13,16). This increase in hardiness may occur in stages (6,9,16) with temperature among the controlling factors. Unfortunately, the numerous papers on the influence of temperature either examined only bark tissues (e.g. 6) or did not distinguish whether xylem or bark tissues were being evaluated (e.g. 16).

Since xylem parenchyma of apple acclimate by deep supercooling (12,13) while cambium, phloem, and bark cortex acclimate through dehydration and other mechanisms, the physiology and environmental control of acclimation must be considered separately for bark and xylem. To date, little is known about how environment controls acclimation of xylem (5), although Quamme et al. (13) presented an annual cycle of hardiness.

The present study was designed to examine in more detail the annual cycle of deep supercooling in apple xylem and to explore by laboratory treatments the influence of temperature and metabolic poisons on parts of that cycle.

341

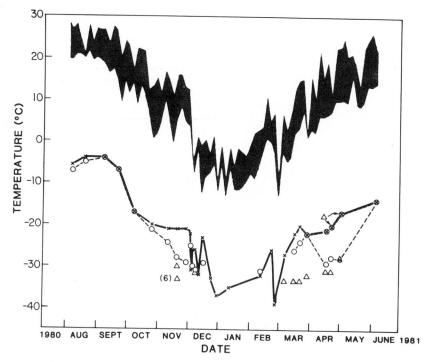

FIGURE 1. Seasonal change in the start of the LTE of EM 26 apple (solid line with X). The dashed line (with 0) shows the seasonal change of the start of the LTE after freshly-collected twigs were exposed to -5C for 3h. The triangles show the start of the exotherm after 3h at -5C followed by 18h or 6 days (6) at -15C. In April and May, the lower circles and triangles are for twig portions with closed buds. The shaded area gives the range of air temperatures.

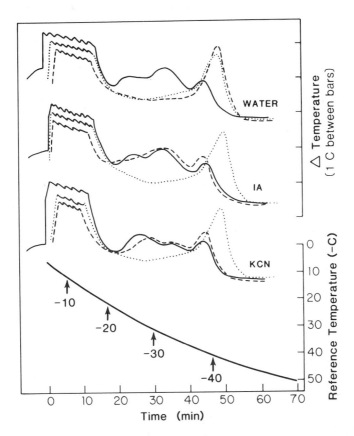

FIGURE 2. The effects of infiltration with water, KCN, and iodoacetate (IA) on the LTE of apple xylem given no -5C treatment (solid line), exposed to -5C for 3h after infiltration (dashed line), or exposed to -5C for 3h before infiltration (dotted line). All twigs were collected December 25, and stored 15 days at 5C before the experiment. Further details are in Table 5.

Table I. Effect of date of collection and temperature treatments on the LTE of apple xylem. Twigs were kept at temperature of collection through placement in the DTA. In DTA they were cooled directly (fresh), collection at -5C for 3h (-5C), or held at -5C for 3h and then cooled to -15C for about 18h (-15C).

Date collected 1980-81	Range of temp during 4 days before collection	Temp at collection	Temp at LTE Start			Temp at LTE Midpoint		
			Fresh	-5C	-5C and -15C	Fresh	-5C	-5C and -15C
6/3[a]	9 to 24	22	-4	-4	--[b]	-10	-10	--
8/7	21 to 28	23	-6	-7	--	-19	-19	--
8/20	20 to 28	27	-4	-5	--	-16	-16	--
9/8	20 to 27	27	-4	-4	--	-20	-20	--
9/8	20 to 27	27	-12	-13	--	-21	-26	--[c]
9/22	11 to 27	20	-7	-7	--	-28	-28	--
10/8	12 to 24	17	-17	-17	--	-30	-32	--
10/26	3 to 19	6	-20	-21	--	-33	-36	--
11/9	-2 to 17	6	-21	-24	--	-32	-37	--
11/20	5 to 20	16	-21	-28	-30	-33	-39	-39
11/29	5 to 13	6	-21	-29	--	-32	-39	--
12/3	-1 to 12	3	-22	-25[d]	--	-30	-40	--
12/4[k]	-9 to 12	-9	-31	--[e]	--	-41	--	--
12/5	-10 to 5	5	-26	-30[f]	-31	-40	-40	-41
12/8	-9 to 5	5	-26	-31	--	-38	-41	--
12/13	-15 to 6 (1.5d)[g]	-11	-32	--	--	-42	--	--
12/15	-15 to 0	0	-23	-29	--	-35	-39	--
12/23	-10 to 2 (0.5d)	-2	-33	--	--	-42	--	--
12/29	-16 to -4 (4d)	-10	-37	--	--	-42	--	--

Date	Temp range							
1/10	-12 to 1 (1d)	-7	-35	--	--	-43	--	--
2/10	-4 to 3 (0.5d)	-2	-32	-31	--	-43	-43	--
2/21	-6 to 5	1	-26	--	--	-41	--	--
2/26	-16 to 7 (2d)	-10	-39	--	--	-42	--	--
3/6	-5 to 9 (0.5d)	-1	-27	--	-33	-42	--	-43
3/16	-1 to 21	9	-22	-26	-33	-34	-38	-43
3/21	2 to 19	11	-20	-24	-33	-30	-35	-43
3/28	-1 to 13	5	-22	-22	-32	-33	-41	-42
4/16	5 to 20	13	-21	-29	-31	-31	-41	-43
4/21	6 to 19	19	-20[h]	-20	-20	-31	-32	-30
4/21	6 to 19	19	-21[i]	-28	-31	-32	-39	-41
5/1	6 to 27	25	-17[h]	-17	-17	-26	-26	-26
5/1	6 to 19	25	-17[i]	-28	-28	-27	-37	-36
6/3[j]	9 to 24	22	-14	-14	--	-20	-21	--

a   Current year twig collected in 1981.
b   -- means data not collected.
c   This twig without leaves was stored 10 days at 5C before DTA analysis.
d   This sample still on the tree was encased in snow.
e   When LTE started at -30C or below, the -5C treatment was often omitted since previous results (5) showed that it had no effect.
f   -5C for 48h.
g   (1.5d) means sample was continuously below 0C for the 1.5 days before collection.
h   Portions of twig with open buds.
i   Lower portions of twig, buds closed.
j   The second-year wood excised from the now two-year old twig in June 1981.
k   Leaves turned from green to brown.

345

II.   MATERIALS AND METHODS

Periodically between August 7, 1980, and June 3, 1981, current year branches, about 50 cm long, were removed from 2 m tall trees of EM 7 and EM 26 apple (<u>Malus</u> <u>pumila</u> Mill.), Kon Kuk University, Clonal Orchard, Seoul, Korea. Following collection, the twigs were immediately analyzed or stored dark and moist, at 5, -5, or -15C, for later use.

The temperatures at which the xylem froze were determined by a differential thermal analysis (DTA) system similar to one previously described (3), except that the cooling rate changed continuously from about 1.3°C/min at 0°C, to 0.7°C/min at -40°C. Unless otherwise noted, the sample was transferred to the DTA apparatus and cooling was started at the same temperature at which the sample had been collected or stored.

For each experiment matching stem pieces 30 to 40 mm long and 4 to 7 mm in diameter were cut from a single branch. The stem piece, without bark, was placed in a small container along with one junction of a differential thermocouple. The second junction was placed in a similar container filled with paraffin in which a third independent thermocouple was embedded. Both containers were inserted into wells in an aluminum block.

The DTA produced typical freezing curves (Figure 2) with a large first exotherm representing extracellular water and a low temperature exotherm (LTE) representing xylem parenchyma and pith. As previous work (3,4) suggests that pith and ray parenchyma behave similarly, the term xylem will be applied to both. The start and midpoint of the LTE were used to evaluate freezing and hardiness as previous work showed a direct correspondence between the percent of water frozen and the percent of ray cells killed (4). The estimate of the midpoint is only approximate because of temperature lags in heat dissipation.

Twig sections collected throughout the year were given various combinations of the following temperature treatments. (a) In the near 0C treatment (NZT), twigs were exposed to -5C or -3C for 3h just before or during DTA. (b) In the -15C temperature treatment, twigs were exposed to -15C for a specified time, usually 18h; the twigs were exposed just before or during DTA. (c) In the prolonged storage treatments, twigs were stored in snow-filled bags at -5 or -15C usually for 30 to 180 days. (d) In the warm temperature treatments, twigs were exposed to 5C to 36C for specified periods.

Many of these laboratory experiments involved vacuum infiltration of matching twigs with water and with 0.02M KCN or 0.02M iodoacetate.

III.   RESULTS

A.   Seasonal Pattern of the LTE

The LTE began at -4C in June and continued to start at high temperatures into September (Table 1, Figure 1). By October 8, the start/midpoint of the LTE dropped to -17/-30C and this hardiness was maintained into mid-December with fluctuations to as low as -31/-41C. From late December through February the start/midpoint fluctuated between its minimum of -39/-42C and higher temperatures. Dehardening began in March and reached -14/-21C by June in the now second-year wood.

B.   Seasonal Response of the LTE to NZT and 18h of -15C

As shown in Table 1 and Figure 1, the response of the xylem to the NZT varied with both the date of collection and the hardiness of untreated twigs. The start of the LTE was not affected by the NZT before October 26, although the midpoint shifted earlier. Between October 26 and late November the response to the NZT increased. From then till March the NZT lowered the start of the LTE to -29C. The NZT had no effect on tissue when the LTE already started near -30C. During March the xylem had limited sensitivity to the NZT, while during April the NZT shifted the LTE to -28C for twig sections with closed buds but had no effect on twig sections with expanding buds.

The effect on the LTE of the NZT followed by 18h of -15C was examined during natural hardening and dehardening (Table 1, Figure 1). Compared to the NZT alone, this treatment had little effect on November 20, December 5, and during April; but during March it lowered the LTE 7 to 10C.

C.   Prolonged Storage and Subsequent Hardening at -15C

The effects of -5C on the LTE changed with duration of storage; 3h at -5C increased hardiness or had no effect (Table 1), but 4 to 180 days at -5C dehardened the xylem (Tables 2,3,4). In contrast, storage at -15C for 6 to 40 days increased hardiness (Tables 1 and 2). After prolonged storage at -5C the xylem hardened more rapidly on exposure to -15C. For the December 5 collection, 18h at -15C shifted the start of the LTE of freshly collected twigs to -31C (Table 1) and the start for twigs stored 120 days to -37C (Table 3). The effects of storage on the LTE varied with collection date. Twigs collected early, October 26, did not deharden during 120 days at -5C but maintained the NZT effect (Tables 1 and 2). They also only hardened to -26C after a 30 day exposure to -15C.

TABLE II. Effect of storage at -5 or -15C on the starting temperature (C) of the LTE of apple xylem.

Date collected	Freshly collected	Days at -5C		At -15C for 90 to 180 days	Days at -15C after 120 to 180 days at -5C			
		0.8	180	180 days	0.1	0.8	4	30
10/26	-20	-22	--	--	--	--	--	-26
11/29	-21	-29	-22	--	-25	-30	-36	-37[a]
12/4	-31	--	-26	-40	--	-37	--	--
12/23	-33	--	-20	--	--	-29	-32	--
2/26	-39	--	-22	-39	--	--	--	--
4/16	-21	-29	-25[b]	-31[c]	--	--	--	--

[a] Not a matching twig.
[b] Only 3 or 20 days at -5C.
[c] Only 13 days at -15C.

D.   Rapid Dehardening and Rehardening

The twigs dehardened and rehardened in response to short term changes in natural air temperature. For example, at noon December 3, when the air was 3C the LTE started at -22C (Table 1). That afternoon a snow storm encased the twig for 5h at 0C $\pm$ 1C and the LTE started at -25C. By midnight after 7h of air temperatures decreasing to -9C, the LTE started at -31C. By noon, when the air was 5C the LTE again started at -25C.

Laboratory experiments on twigs which had been warmed after prolonged storage gave similar results. Dehardening occurred after 4h at 5C (Expt. B, Table 3), or after 3h at 0C (Expts. C and D, Table 4). Warmer temperatures and longer exposures produced greater dehardening and increased the time required to reharden (Table 4). Rehardening by exposure to -15C also began within 3h (Table 3). In responsive cells hardiness increased with duration at -15C (Tables 2 and 4). In contrast, exposure to -5C for up to 23h had no effect on the LTE of previously warmed twigs (Table 3).

E.   Anti-metabolic Agents

Infiltration with 0.02M KCN or 0.02M iodoacetate prior to the NZT prevented the downward shift in the LTE. Similar infiltration after the NZT did not reverse the shift (Table 5, Fig. 2). In contrast, infiltration with 0.02M KCN, 0.2M KCN or

TABLE III. Effect of storage condition, 0.02M KCN and low temperature treatment on subsequent response to -5C and -15C. After storage all samples were infiltrated with water or 0.02M KCN or 0.02M iodoacetate for 1h at 2C. Low temperature treatments followed infiltration.

Expt/ collection date	Storage	Low temp treatment				LTE temp, Water infiltrated start	midpoint
A 12/5	-5C (120d)	None				-28	-41
		-5C (3h)	-15C (18h)			-37	-44
		-5C (3h)	-15C (18h)	+ 20C (23h)	-5C (23h)	-25	-40
		-5C (3h)	-15C (18h)	+ 20C (23h)	-15C(23h)	-37	-44
B 4/24	-5C (4d)	0C (1h)				-25	-36
		-5C (3h)	-15C (66h)			-32	-41
		-5C (3h)	-15C (66h)	+ 20C (4h)		-21	-29
		-5C (3h)	-15C (66h)	+ 20 (4h)	-5C (3h)	-22	-32
		-5C (3h)	-15C (66h)	+ 20 (4h)	-15C(3h)	-32	-41

a Differences in LTE were 2C or less between the water, KCN, and iodoacetate infiltrated twigs.

TABLE IV. Dehardening and rehardening of apple xylem. Experiment A: EM 26 apple collected on 12/20/80 was stored 120 days at -5C. Following warm temperature pretreatments, twigs were infiltrated with water or 0.02M KCN or 0.02M iodoacetate for 1h at 25C before subsequent cold treatments. Experiment B: EM 7 apple twigs collected 12/5/80 were stored 180 days at -5C and infiltrated for 3h at 2C before warm temperature treatment. In Experiments C (EM 7 apple) and D (EM 26 apple), twigs were collected December 8 and stored 180 days at -15C. Results below are for water. Results with antimetabolites were almost identical.

| Experiment | Pre-Infiltration Warm Treatment | | Temp (C) of LTE at start/midpoint with these post-infiltration treatments: | | | |
	Temp (C)	Duration (h)	none	-5C 18h	-15Ca 20h	-15C (days)
A	none	--	--	-24/-37	-34/-45	--
	5	360	-21/-31b	-24/-39	-34/-45	-37/-44 (2d)
	25	24	--	--	-32/-42	--
	25	48	--	--	-27/-42	--
	25	96	-19/-28	--	-21/-39	-31/-41 (4d)
B	5	4	-20/-28	--	-29/-41	-31/-41 (3d)
	23	4	-20/-26	--	-27/-36	-28/-40 (3d)
	31	4	-19/-26	--	-24/-37	-26/-39 (3d)
	36	4	-17/-25	--	-21/-35	-25/-39 (3d)

C	--	0	-36/-43	--	--	--
	2	3^c	-24/-42	--	-34/-43	--
D	--	0	-38/-45	--	--	--
	2	3^c	-31/-43	--	-35/-44	--

a Experiment A was -11C
b Twigs kept at 5C for 3h.
c Includes period during infiltration.

TABLE V. Effects of infiltration with 0.02M KCN (12/25 and 12/30) or 0.2M KCN (12/29) on the LTE of apple xylem exposed to hardening temperatures. Twigs were stored at various temperatures for various durations (days) before and after infiltration.

Collection Date	Storage before Infiltration (I)	Duration Infiltration at 20 or 25C	Storage after Infiltration	Temp of LTE with Infiltration in			
				Water		KCN	
				Start	Midpoint	Start	Midpoint
12/25	5C (15d)	(20C) (1h)		-21	-34	-21	-36
	5C (15d)	(20C) (1h)	-5C (3h)	-27	-41	-21	-37
	5C (15d) -5C (3h)	(20C) (2h)	0C (3-5h) -5(0.5h)	-27	-41	-29	-41
12/30	-5C (120d) 5C (15d)	(25C) (1h)	5C (3h)	-21	-31	-21	-29
	-5C (120d) 5C (15d)	(25C) (1h)	-11C (3h)	-34	-45	-34	-44
12/29	-5C (180d) -15C (30d) 5C (2d)	(25C) (1h)	5C (1h)	-24	-35	-24	-35
	-5C (180d) -15C (30d) 5C (2d)	(25C) (1h)	-15C (40h)	-34	--	-34	--

0.02M iodoacetate did not prevent the downward shift achieved by -11 to -15C (Tables 3,4,5).

IV. DISCUSSION

The results were synthesized into a tentative proposal describing the acclimation and deacclimation of apple xylem and pith.

Newly formed xylem parenchyma supercool (June 3, Table 1), but the freezing is at high temperatures (-4/-10C for the start/midpoint of the LTE). Initial hardening begins between cell formation and early August when many cells are freezing below -19C (the LTE midpoint). A major increase in cold hardiness occurs before October 8; by then the LTE starts at -17C. This initial hardening did not result from low air temperatures as minimum air temperatures from September 8 to October 8 ranged from 8 to 16C (Figure 1). Furthermore, the LTE midpoint did not shift in September 8 twigs held at 5C for 10 days (Table 1).

During the next step of acclimation (October 26 to mid-December), the LTE started near -20C when the air was warm but the freezing point of the parenchyma became progressively responsive to temperatures near 0C. By November 29, 3h at -5C could shift the start of the LTE to -29C, the maximum hardiness achievable by the NZT (5). In nature, from late November into mid-December the LTE fluctuated between -30C in freezing nights and -22 to -25C in the warmer days. The progressive ability to respond to -5C was probably initiated by the cool nights of late October and early November. Independent evidence that low positive temperatures make cells responsive to the NZT is given by the LTE shift in the September 8 collection stored 10 days at 5C (Table 1). Through early December, colder temperatures, 18h at -15C, lowered the LTE only slightly more than the NZT.

A further step in acclimation hardened the parenchyma to -35 to -40C. The probable environmental stimulus was temperatures below -5C as these low LTE's were in twigs continuously frozen for 1 to 4 days and collected at -7C or lower (Table 1). Indirect evidence (Tables 2,3,4; Figure 1) suggests that as the season progresses, briefer exposures to these low temperatures are required to lower the LTE to near -40C. On October 26, 30 days of -15C (after -5C storage) produced little hardening (Table 2), while on November 20, 6 days at -15C dropped the LTE to -33C (Figure 1). The responses of stored twigs (Tables 3 and 4) suggest that in nature twigs acclimate and deacclimate according to current air temperatures. For both field and laboratory twigs, these rapid changes in the start of the LTE involved a relatively small fraction of cells, since the midpoint of the LTE remained almost constant.

Natural dehardening began in March coincident with warm temperatures (Figure 1). During March, the LTE responded little to the NZT but was lowered to -33C by 18h at -15C. In April, dehardening continued, and the ability to reharden in response to -5 or -15C was lost when buds swelled (Table 1, Figure 1). At the same time, xylem adjacent to closed buds, even if on the same branch with open buds, hardened to near -30C after the NZT or 18h of -15C (Table 1, Figure 1). The LTE of the totally dehardened twig started at -14C indicating that hardening was not totally reversible.

While all xylem and pith cells pass through these various steps of acclimation and deacclimation, changes in the shape of the LTE with time indicate that all cells are not acclimating in synchrony.

The steps just proposed for the acclimation and deacclimation of apple xylem partially resemble those described previously for bark cortex, cambium, and phloem (6,7,13,16). The timing is almost the same; initial acclimation does not require cool temperatures (Figure 1, Table 1); freezing temperatures are needed to reach maximum hardiness; and during spring, rapid hardening and dehardening can occur in response to fluctuating air temperatures. While these similarities between bark and xylem are logical in terms of plant survival, they are physiologically surprising since only xylem avoids freezing by deep supercooling. Parallel studies of bark and xylem are needed.

The physiologic mechanisms by which xylem parenchyma gain and lose the ability to deep supercool remain unresolved and controversial (2-5,8,10,11,15,16). This study suggests that at least three discrete physiologic steps are involved. Initial acclimation is irreversible and does not require cool temperature. It may be under photoperiodic control as is initial acclimation in bark (6), although unlike bark, the initial changes in xylem are irreversible.

A second acclimation process is the shifting of the start of the LTE from -20 to -30C by temperatures near 0C. This shift requires normal, living cells as it is prevented by KCN or iodoacetate. However, the shift is not reversed by these antimetabolites (Figure 2, Table 5). The sensitivity of cells to temperatures near 0C changes seasonally, increasing during autumn and disappearing by bud swell.

The third process is the response of the cells to -11 or -15C. These or similar temperatures below -5C were needed to harden xylem to -40C. Unlike the response to the NZT, hardening occurred with -15C even if twigs were infiltrated with KCN or iodoacetate (Tables 3,4, and 5). This suggests that hardening in response to -15C does not require the metabolic activity which is related to the antimetabolites and supports the previous proposals that deep supercooling to -40C involves physical changes in the cell wall (2). Additional work is needed, however, to document that the KCN or iodoacetate entered and was active in the cells. Kantser (8) reported that the lignin content of apple shoots increased after exposure to -15C for 7 days. He also found that lignin content increased from about 25% to 27% between November and February and then decreased when spring growth began. Lignification would stiffen the wall and aid parenchyma to withstand the hydrostatic tensions (-40MPa) postulated by George and Burke (2).

Krasavtsev (10) related hardening to the control of water-flux through a complex barrier composed of cell wall, plasma-lemma, and outer cytoplasm. This barrier was altered during hardening (10). In contrast, other work found no or little dehydration as cells hardened in response to temperatures near zero (5) or 90 days of storage at -15C (current study). Both the current and an earlier study (5) noted that twigs collected at maximum hardiness in midwinter had less deep supercooled water than those collected earlier or later. A quantitative study to detect the timing of the dehydration in relation to hardening is still needed.

Our results indicate that apple xylem attains maximum supercooling using different mechanisms than the florets of Prunus pennsylvanica (1) or the primordial shoots of conifer buds (14). These buds lost large amounts of water as they acclimated to -40C or lower. As indicated, no dehydration was associated with the shift in hardiness following NZT, and sometimes, only modest dehydration was observed in the shift to maximum hardiness.

V. SUMMARY

The xylem of apple deep supercools in stages. The first stage, which is irreversible, shifts the start of the LTE to -15 to -20C and occurs in midsummer and early fall while the air is still warm. A second stage involves gaining the ability to harden from near -20C to -30C in response to a brief exposure to temperatures near 0C. This ability is gained when nights become cool; the shift from -20 to -30C requires normal living cells. The shift from -30C to maximum hardiness results from temperatures below -5C and tentative evidence suggests it is not inhibited by antimetabolites. Dehardening coincides with warm temperatures but the ability to reharden is maintained until the buds swell.

REFERENCES

1. Burke, M.J., and Stushnoff, C., in "Stress Physiology in Crop Plants" (H. Mussel and R. Staples, eds.), p. 197. John Wiley and Sons, New York, 1979.
2. George, M.F., and Burke, M.J., Plant Physiol. 59:319(1977).
3. Hong, S.G., and Sucoff, E., Plant Physiol. 66:40 (1980).
4. Hong, S.G., Sucoff, E., and Lee-Stadelmann, O.Y., Bot. Gaz. 141:464 (1980).
5. Hong, S.G., and Sucoff, E., Plant Physiol. (in press) (1982).
6. Howell, G.S., and Weiser, C.J., Plant Physiol. 45:390 (1970).
7. Howell, G.S., and Weiser, C.J., J. Amer. Soc. Hort. Sci. 95:190 (1970).
8. Kantser, A.N., Fiziol. Biokhim. Jul't. Rast. 4:92 (1972).
9. Krasavtsev, O.A., Soviet Plant Physiol. 16:228 (1969).
10. Krasavtsev, O.A., Soviet Plant Physiol. 26:415 (1979).
11. Levitt, J., Responses of Plants to Environmental Stresses I, 2nd Ed. Academic Press, New York, 1980.
12. Quamme, H.A., Weiser, C.J., and Stushnoff, C., Plant Physiol. 51:273 (1973).
13. Quamme, H., Stushnoff, C.J., and Weiser, C.J., J. Amer. Soc. Hort. Sci. 97:608 (1972).
14. Sakai, A., Plant and Cell Physiol. 20:1381 (1979).
15. Rasmussen, D.H., Macauley, M.N., and Mackenzie, A.P., Cryobiol. 12:328 (1975).
16. Tumanov, I.I., and Krasavtsev, O.A., Soviet Plant Physiol. 6:663 (1959).

SUPERCOOLING ABILITY AND COLD HARDINESS OF RHODODENDRON
FLOWER BUDS WITH REFERENCE TO WINTER WATER RELATIONS

S. Kaku and M. Iwaya

Biological Laboratory
College of General Education,
Kyushu University
Fukuoka, Japan

K. B. Jeon

Department of Biology
Jeonbug National University
Jeonju, Korea

I. INTRODUCTION

Deep supercooling in Rhododendron flower buds has become a
matter of current interest in its physiological and ecological
implications as a frost avoidance mechanism. However, there
has been little work done on the susceptibility of flower buds
among the various organs and tissues in the genus Rhododendron. Recently, we determined the killing temperatures of
various organs or tissues for 23 native Rhododendron taxa
having different potentialities for northern distributions, and
the flower bud was characterized as the most sensitive organ in
Rhododendron (4). Thus, it is conceivable that the degree of
hardiness in flower buds might become one of the very important
limiting factors for cold adaptation and the distribution of
certain species to northern or high altitudes.

It has been pointed out that the freezing resistance of
Rhododendron flower buds is determined in a great measure by
the supercooling ability of florets and this ability to avoid
freezing is strongly affected by water content and the cooling
rate (2,3,5). But, further studies on the relationship between
the supercooling ability and winter water relations are necessary to understand the full eco-physiological significance of
this deep supercooling in flower buds.

PLANT COLD HARDINESS AND FREEZING STRESS

The purpose of this paper was to explore the relationship between the supercooling ability in flower buds and water content and killing temperature (KT) during cold acclimation and deacclimation using R. x akebono and R. kiusianum, both originated from the warm temperate Kyushu region.

II. SEASONAL CHANGES OF THE SUPERCOOLING ABILITY, KILLING TEMPERATURE AND WATER CONTENT IN FLOWER BUDS.

Freezing resistance of Rhododendron flower buds is exclusively dependent on the supercooling ability of florets. This ability is strongly affected by many factors such as water content in florets, the cooling rate, the season, etc. (1,2,3,5). Graham and Mullin (2) have surmised that cold hardiness of azalea florets would depend on the rate of water removal, and the water removal response in hardy species is more rapid and to a greater degree than in less hardy ones. Recently, Ishikawa (3) has demonstrated that deep supercooling in flower buds in some hardy deciduous Rhododendron species resulted from marked water migration from florets to scales in response to freezing. However, it is uncertain whether or not such rapid water migration within bud tissue during the freezing process generally occur in many other Rhododendron species, and whether or not the variance of water content within bud fluctuate daily and is seasonal or not. Moreover, it is well known that woody plants often decrease in water content during cold acclimation and plants with lower water content during winter are generally more hardy (7,8,9).

The relationship between the supercooling ability of flower buds and water content and KTs during cold acclimation and deacclimation was studied with R. x akebono and R. kiusianum. R. x akebono is regarded as the hybrid of R. scabrum and R. ripense and was one of the most popular cultivars planted in the Kyushu mainland. There are two kinds of clones in R. kiusianum. One is named the "flowering clone" as its blossoming continues partially throughout fall to spring, and the other is termed the "dormant clone" because it does not flower till late spring. The natural range for this species is restricted to mountain regions about 1000 m above sea level on Kyushu mainland. The species is also used for gardening as a dwarf azalea.

The determination of the freezing process and the exotherm temperature distribution (ETD) in florets in excised whole flower buds was carried out with differential thermal analysis (DTA). The apparatus for DTA and the KT determination were described previously (5). The cooling rate was 1°C/hr by using a programatic regulator attached to a freezer throughout the experiment.

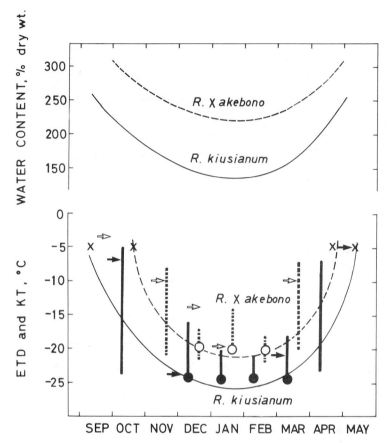

Fig. 1.   Changes of ETD, KT and water content of florets in flower buds for R. x akebono (- - - -) and R. kiusianum (dormant clone) (——) during cold acclimation and deacclimation.   X: Continuous freezing from bud scale to floret; vertical bars (| ¦): the range of ETD; o ● :   high frequency of ETD; → :   KT.

Figure 1 shows the seasonal changes in freezing patterns, ETDs, KTs and water content of florets. The general trends of these changes were basically the same in both species.   The beginning of acclimation in R. kiusianum was one month earlier in autumn, and that of deacclimation was one month later in spring as compared with those of R. x akebono.   Flower buds in non-acclimated plants showed a single large exotherm resulting from the continuous freezing from scales to florets.   In the early stage of acclimation, separation of the freezing in scales and florets and the wide distribution of multiple floret exotherms occurred, and the water content in florets clearly decreased.   In mid-winter, the multiple floret exotherms concentrated in a narrow range around -20°C for R. x akebono and

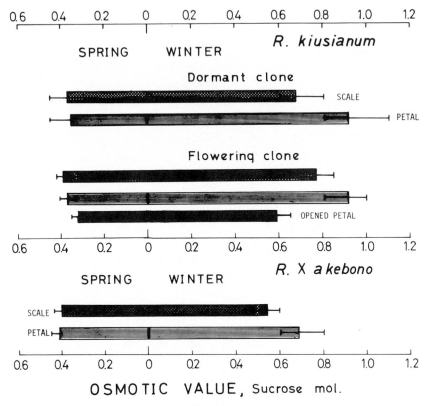

Fig. 2.   Osmotic pressures of bud tissues in flower bud (scale ▨, petal ▤, opened petal ▥) for R. kiusianum and R. x akebono in winter and spring.   Horizontal lines at the top of bars indicate the concentration range in which 5 measurements were made.

-25°C for R. kiusianum and subsequently remained at almost the same level until spring.

The decrease of floret water content resulting from the artificial desiccation of flower buds also enhanced the supercooling ability of florets.   The ETDs in non-acclimated R. x akebono clearly lowered, correlating with the water content decrease in florets within a certain extent (about 300 to 200% in dry wt. basis), but there was a maximum supercooling ability in florets regardless of the marked decrease of water contents in florets as low as about 150 to 100% (6).

The fully acclimated period of flower buds for R. kiusianum was two months longer than those of R. x akebono and the supercooling ability and KTs in the former were lower by about 5°C and the water content of florets was less by about 100% as compared with the latter (Fig. 1).

The osmotic pressures of bud tissues in winter and spring for R. x akebono and R. kiusianum were determined by the incipient plasmolysis using sucrose solution as a plasmolyticum. The osmotic pressure of flower bud tissues in winter in R. kiusianum were markedly higher than in those of R. x akebono, but the specific difference was fairly diminished in spring. In winter, osmotic pressures of petals in florets were extremely high and were clearly higher than those of scales in both species, while the difference between both tissues disappeared in spring (Fig. 2). Therefore, the high supercooling ability in florets during midwinter is quite conceivable from such high osmotic pressures in winter florets. Thus, it seems that the overwintering flower buds in R. kiusianum have a native adaptability as a high mountain species to a cold environment.

KTs of flower buds in the early stage of acclimation were higher than the range of high frequency in ETDs but they coincided with each other in midwinter. Namely, the lowering of KTs was preceded by the enhancement of the supercooling ability in florets at the early stage of acclimation but they corresponded with each other in midwinter.

Fig. 3 shows the difference of viability of flower buds for R. x akebono in November and December exposed at -10°C for one day. In mid-November, the viability of flower buds gradually decreased correlating with the length of exposure time at -10°C, while those in December were almost the same. This suggests that the supercooling ability of flower buds in December was clearly superior to that of November.

The trend of seasonal changes in freezing patterns, ETDs and KTs of flower buds for R. kiusianum was basically the same in the dormant clones and the flowering ones. However, the occurrence of multiple floret exotherms accompanying cold acclimation and the re-widening of ETD and the occurrence of continuous freezing resulting from cold deacclimation in spring began more earlier in the flowering clones than in those of the dormant ones. This means that the acclimation and deacclimation in flower buds occur earlier in the flowering clones rather than in the dormant ones. It may be caused by a different intensity of acclimation in both clones as described later (Fig. 4).

In conclusion, the changes in freezing pattern, the concentration of ETDs to lower subzero temperatures, and a marked decrease of water content in florets are regarded as the symptoms of cold acclimation in flower buds, and vice-versa in spring buds during deacclimation. The decrease of floret water content for both seasonal (acclimational) dependent and artificial dehydrations enhance the supercooling ability of florets according to different extents in each species.

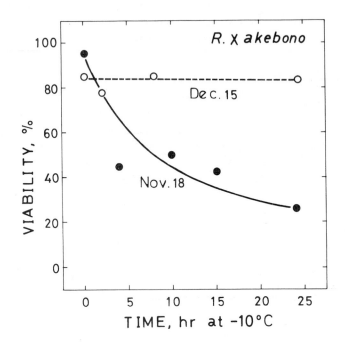

Fig. 3. Viability of flower buds for R. x akebono in November and December exposed to -10°C for a day. KTs were -10°C in November and -15°C in December.

III. INTENSITY OF COLD ACCLIMATION IN WINTER FLOWER BUDS AND CHANGES IN WATER CONTENT WITHIN BUD TISSUES IN RESPONSE TO FREEZING

In flower buds for R. x akebono, the symptoms of deacclimation were indicated by the re-widening of ETD and the occurrence of continuous freezing, and the increase of floret water content appeared as an effect of the dehardening treatment (+15°C) for three weeks, while few effects were observed in the dehardening for one week. In R. kiusianum, the effect of four weeks dehardening treatment markedly manifested itself in the flowering clones and did not in the dormant ones. The water content increase in florets for the flowering clones was about four times as much as in the dormant ones (Fig. 4). These results show that the intensity of acclimation for the dormant clones in R. kiusianum is the highest among the species examined and that of the flowering ones is less intensive than in R. x akebono.

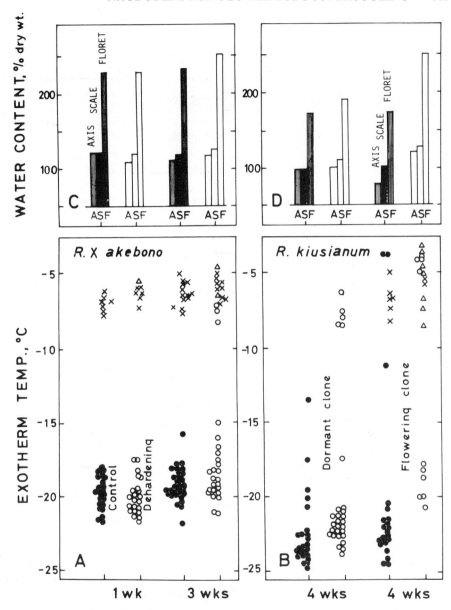

Fig. 4. Effect of dehardening treatment on ETD of floret (A,B) and water content (C,D) of flower bud tissues (axis ▭, scale ▨▨▨, floret ▥) for R. x akebono (A,C) and R. kiusianum (B,D). Closed or shaded symbols indicate the control stored at +3°C, and opened symbols indicate the dehardening treatment at +15°C. △ : Continuous freezing from bud scale to floret; X: exotherm temperature of bud scale; o ● : exotherm temperature of florets.

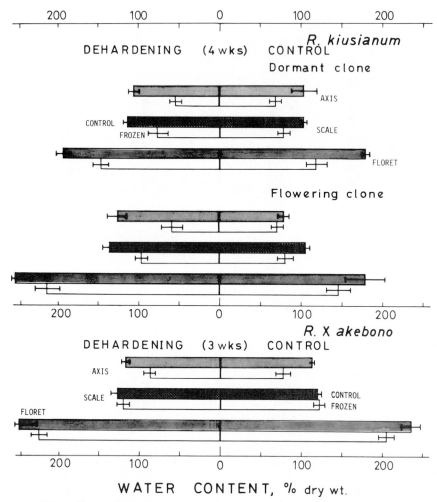

Fig. 5. Changes of water contents within bud tissues (axis ▬, scale ▨, floret ▤) in response to freezing in acclimated and deacclimated flower buds for R. kiusianum and R. x akebono. Shaded bars show the control bud tissues and blank bars show frozen ones. Horizontal lines at the top of bars indicate standard deviation in average 8 samples examined.

It has been suggested that the increased supercooling ability of florets in some hardy deciduous Rhododendron species result from the rapid water migration from florets to scales and other tissues during the early stage of the freezing process (3). The changes of water content within bud tissues in response to freezing were compared with acclimated buds and the dehardened ones (Fig. 5).

The comparison of water content changes in bud tissues in response to freezing were made with the same buds used for DTA and with the control buds which were kept at +3°C for the same period of DTA. Each bud tissue from a frozen bud was excised as quickly as possible with tweezers, cooled to -30°C, on a dissecting plate which was also cooled to -30°C. The dissection was carried out in a cooling chamber cooled to -30°C. After freezing, a considerable decrease of water content was observed in most bud tissues except for the scales of R. x akebono. Although the most remarkable decrease of floret water content in frozen buds was seen in the dormant clones of R. kiusianum, there was no significant difference between acclimated and deacclimated buds. The water content of florets in deacclimated buds was much more than in those of acclimated ones, even after freezing, and it resulted in the occurrence of exotherm temperatures at a higher subzero temperature. Specifically, there was no significant increase of scale water content, and the water migration from florets to scales did not intensify, even in acclimated flower buds. Thus, it can be concluded that deep supercooling in florets might result not necessarily from rapid water migration from florets to scales during the freezing process, but from low water content in situ of acclimated or artificially dehydrated buds.

IV. SUMMARY

The relationship between the supercooling ability, the water content, and the KT of flower buds were studied during cold acclimation and deacclimation of R. x akebono and R. kiusianum. The occurrence of multiple floret exotherms and their shift to lower subzero temperatures and the marked decrease of floret water content were the symptoms of cold acclimation in winter buds, and vice-versa in spring buds during deacclimation. Although the lowering of KTs was preceded by the enhancement of the supercooling ability in florets in the early stage of acclimation, they coincided with each other in midwinter. The fully acclimated period for R. kiusianum was two months longer than that of R. x akebono and the supercooling ability for the former was about -25°C and about -20°C in the latter. The water content decrease in florets for both seasonal (acclimational) dependent and artificial dehydrations enhanced the supercooling ability in florets. The intensity of acclimation in winter buds was also studied by artificial dehardening, and the dormant clones for R. kiusianum showed the highest intensity. Although the changes of water content within bud tissues in response to freezing were examined in acclimated and dehardened buds, no remarkable decrease of water content in florets nor a significant increase of scale water content could be observed even in acclimated buds.

Thus, it is conceivable that deep supercooling in florets might result not necessarily from the rapid water migration from floret to scale in response to freezing, but from the low water content in situ of cold acclimated or dehydrated flower buds.

REFERENCES

1. George, M.F., Burke, M.J. and Weiser, C.J., Plant Physiol. 54:29-35 (1974).
2. Graham, P.R. and Mullin, R., J. Amer. Soc. Hort. Sci. 101:7-10 (1976).
3. Ishikawa, M., Master Thesis, The Inst. Low Temp. Sci., Hokkaido Univ. (1979).
4. Iwaya, M. and Kaku, S., The 13th International Botanical Congress, Sydney. Abstracts p. 92 (1981).
5. Kaku, S., Iwaya, M. and Kunishige, M., Plant and Cell Physiol. 21:1205-1216 (1980).
6. Kaku, S., Iwaya, M. and Jeon, K., Plant and Cell Physiol. 22:1561-1569 (1981).
7. Lumis, G.P., Mecklenburg, R.A. and Sink, K.C., J. Amer. Soc. Hort. Sci. 97:124-127 (1972).
8. McKenzie, J.S., Weiser, C.J. and Li, P.H., J. Amer. Soc. Hort. Sci. 99:223-228 (1974).
9. Parsons, L.R., Plant Physiol. 62:64-70 (1978).

FREEZING AVOIDANCE BY SUPERCOOLING OF TISSUE WATER
IN VEGETATIVE AND REPRODUCTIVE STRUCTURES
OF Juniperus virginiana

Milon F. George

School of Forestry, Fisheries and Wildlife
University of Missouri
Columbia, Missouri

I. INTRODUCTION

Freezing avoidance by supercooling of tissue water is an important mechanism of freezing survival in woody plants of temperate regions (3-5). It has been shown repeatedly that hardy vegetative and reproductive tissues can, in many cases supercool to temperatures in the range -40 to -47°C before they freeze and are injured (8). The biophysics of supercooling has been investigated in detail in the xylem of Malus pumila (15,20), Carya ovata (7,15) and Quercus velutina (6) and in floral primordia of various Rhododendron and Prunus species (10,13,21). The ecological implications of supercooling on the latitudinal and alpine distributions of woody angiosperms have also been examined in some detail (1,2,11,16,21,25).

Although supercooling is now well established as a survival mechanism in woody angiosperms, only a few reports have shown that supercooling is a survival mechanism in conifers. For example, low temperature freezing peaks have been reported by Sakai (23) between -20 and -30°C in winter buds of Abies homolepis, A. veitchii, A. sachalinensis, A. balsamea, A. firma, Pseudotsuga menziesii and Larix leptolepis. He has also described a possible anatomical mechanism to explain how the primordial shoots remain supercooled while other bud tissues contain ice (24). Becwar and Burke 912) and Becwar et al. (2) have recently documented freezing avoidance by supercooling to near -40°C in twigs of Abies concolor, A. lasiocarpa, Juniperus communis, J. scopulorum, Picea englemannii, Pinus flexilis and Pseudotsuga menziesii. They propose that supercooling is a factor contributing to the absence of these species above timberline in the Colorado Rocky Mountains of the western United

States. Similarly, George et al. (12) measured low temperature freezing in twigs of J. virginiana, a conifer common throughout much of the eastern deciduous forest of the United States and southeastern Canada. The potential consequence of supercooling on the northern distribution of J. virginiana was also noted. Additionally, George and Carrasquilla (9) observed freezing and injury near -40°C in hardy male cones of J. virginiana.

The above reports reflect much of the current knowledge of supercooling in coniferous species. The work reported below is a summary and extension of freezing studies conducted on J. virginiana by George et al. (12) and George and Carrasquilla (9). Supercooling is documented in hardy twigs and hardy male cones, low temperature freezing in nonliving periderm is discussed and depression of freezing to below -50°C in twigs and cones is noted.

II. MATERIALS AND METHODS

A. Plant Materials

Twigs and male cones were collected from native J. virginiana trees located near Ashland, Missouri during 1979 and 1980. Samples were transferred from the field to the Tree Physiology Laboratory at the University of Missouri, Columbia wrapped in wet paper towels and sealed in polyethylene bags. Prior to testing, samples were stored either at +2°C in polyethylene bags (as above) or at -10°C packed in ice in polyethylene bags.

B. Freezing and Thawing of Tissue Water

Freezing and/or thawing of tissue water was monitored by differential thermal analysis (DTA). Briefly, the DTA method involves cooling a sample and reference at the same rate, but in separate chambers (19). The sample is enclosed in a small aluminum foil cup, attached to a temperature sensor and placed into one chamber. The reference is constructed by attaching an empty aluminum foil cup to an identical temperature sensor. This arrangement is then placed into the second chamber. Freezing (or thawing) events occurring in the sample cause warming (or cooling) of the sample with respect to the reference. The resulting temperature difference is displayed on a strip chart recorder. Ambient temperature of the reference is also recorded.

A DTA system similar to that described by George et al. (10) has been constructed in the Tree Physiology Laboratory. However, there are two major diffferences in the DTA used here and that described earlier. First, the freezing unit is a Tenney model TJR environmental chamber equipped with a linear

ramp generator and temperature controller. Secondly, the differential temperature sensors are Yellow Springs Instruments
Co., Inc. series 400 (probe no. 423) thermistors instead of the
copper-constantan thermocouples previously used. A schematic
representation of the DTA system is shown in Figure 1. The
rather unsophisticated signal conditioning network uses Precision Monolithics Incorporated ultra-low offset voltage OP-07C
operational amplifers and can achieve sensitivities of 1°C/100
millivolts and detect temperature differences on the order of
0.001°C. In part, the ability to achieve such high sensitivities is related to the long term input offset voltage stability (0.2 microvolts/month) and low average input offset voltage
drift (0.5 microvolts/°C) of the OP-07C operational amplifier.

Although the signal conditioning network is simple, it has
a unique feature. That feature is an analog output which
essentially gives the ratio of the resistance of the reference
thermistor and the sample thermistor (the ratio is modified by
the parallel resistor, R). This feature is important due to
the inherent nonlinearity of the negative temperature coefficient thermistors. The ratiometric measurement of reference
and sample resistance aids in maintenance of the sensitivity
(2.5°C/100 millivolts in experiments reported here) within
approximately $\pm$ 5% over the range -20 to -70°C. Generally,
differential thermistor bridge networks measure the difference
in resistance between the sample and reference. This type of
measurement requires a somewhat more complex arrangement of
thermistors to stabilize the output (14). For analysis of peak
areas and temperatures, the slight nonlinearity of the ratiometric network was eliminated by using the system equation
(Fig. 1) for calculating the temperature difference (Δ)
between sample and reference.

Most DTA runs described below were conducted by cooling or
thawing samples in a continuous fashion from room temperature
to below -45°C or from below -45°C to slightly above 0°C in the
case of thawing. In one case (see below) male cones were held
at -12°C for an extended period to measure the resistance of
supercooled water to desiccation from extracellular ice.
Unless otherwise noted, the cooling rates were between 50 and
60°C/hr.

C. Viability Testing

Twig samples with needles, twig samples with distinct periderm and/or male cones were wrapped in wet cheesecloth with a
24 gauge copper-constantan thermocouple attached. The samples
were then placed into insulated flasks and cooled in a low temperature freezer at a rate not exceeding 25°C/hr. The wet
cheesecloth protruded from the flasks into the freezer and
served to nucleate the samples slightly below 0°C. Sample temperature was monitored on a multipoint recorder. After cooling
to selected temperatures, flasks were removed and then thawed

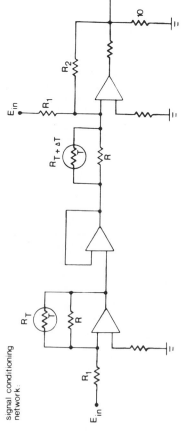

THERMISTOR dta

FIG. 1. Schematic representation of the thermistor DTA system used to measure freezing and thawing of tissue water.

signal conditioning network:

equation for ΔT:

$$\Delta T = -\cfrac{b}{\ln\left[\cfrac{(E_{out}/a + 1)(1 + R_c e^{-b/T}) - 1}{R_c}\right]} - T$$

where

E_{in} = reference voltage

$a \doteq E_{in} \cdot (R_2/R_1)$ for $R_2 \gg 10$

b, c = thermistor constants

$R_T = c e^{b/T}$

dta chamber:

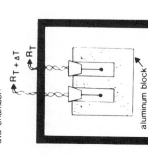

slowly at +2°C. When thawing was complete, the samples were wrapped in wet paper towels and placed in polyethylene bags. The samples were then incubated at room temperature for 5 to 11 days before being evaluated microscopically for browning from freezing damage. In experiments reported here, the incubation period was not found to be critical. This is a routine viability test that has been described previously (17,18).

III. RESULTS AND DISCUSSION

A. Supercooling of Water and Injury in Hardy Twigs and Male Cones

Winter hardy twigs with needles, twigs with distinct periderm (the periderm, other bark tissues and cambium were removed for DTA) and male cones were observed by DTA to have freezing peaks slightly below -40°C (Fig. 2). Integration of the area under the high and low temperature peaks of other samples indicated that less than approximately 2% of total observable freezing occurred near -40°C in twigs with needles and male cones (9,12). Preliminary analysis gave a heat of fusion estimate of approximately 100 cal/g of water freezing at low temperature in twigs with needles (12). Thawing curves (not shown) showed no melting at low temperature in either twigs with needles or male cones. As in past reports on other species, these results support the concept of spontaneous crystallization of a supercooled fraction of tissue water near its expected homogeneous nucleation temperature (3-5,7).

Viability tests revealed browning injury in hardy twigs with needles and hardy male cones near the temperatures where freezing was observed in DTA experiments (twigs with distinct periderm were not tested). In twigs with needles, both needle tissue and xylem tissue displayed browning injury near -40°C. In male cones, the axis upon which the pollen sacs are borne appeared to be the most severely damaged. Damage to specific cell types was not ascertained in either plant part. The results of viability testing are again consistent with the numerous past reports of injury associated with low temperature freezing in woody plants (8).

Although detailed studies on acclimation and deacclimation are not reported here, a few results of freezing tests on non-hardy tissues are worth noting. It was found that low temperature freezing was absent or much reduced in magnitude in male cones during late August, 1979 and early April, 1980. At these times, male cones were severely damaged by freezing to only -10°C. New spring shoot growth behaved in the same fashion. However, in past seasons shoot growth collected in late May, 1980 freezing peaks were still observed near -40°C in both needles (xylem removed) and xylem (needles and bark tissues

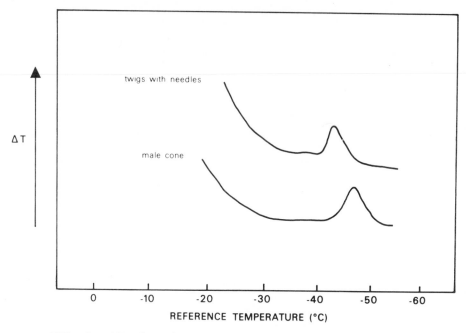

FIG. 2.   DTA freezing curves for twig sections with needles and a male cone of J. virginiana collected on November 29, 1979 near Ashland, Missouri (high temperature freezing is not shown). The male cone was tested on November 30, 1979 and the twig sections were tested on December 7, 1979. Samples were stored at +2°C, 100% RH before testing. Viability testing revealed that browning occurred between -40 and -45°C.

removed) despite deacclimation of the tissues to -20°C as determined by viability testing. In these twig tissues, super-cooling of water seems to be unrelated to freezing resistance. This result suggests that certain anatomical features necessary for supercooling are established during the first year of growth in these tissues. These features then persist regard-less of the winter resistance of the tissue. In fact, dis-tilled water saturated J. virginiana heartwood exhibits freezing near -38°C in an analogous fashion to living twigs (M.F. George, unpublished data).

B. Supercooling of Water in Nonliving Periderm

During investigations of low temperature freezing in whole twigs with distinct periderm, it was observed that storage of hardy samples for 2 to 5 days at +2°C, 100% relative humidity (RH) resulted in the appearance of an additional freezing peak at low temperature. The peak occurred near -38°C. By removal of successive twig tissues, it was determined that this freezing occurs in the outer nonliving region of the periderm. The peak was absent from samples stored at -10°C for periods greater than 7 days. Similar results were obtained in twigs of J. communis (M.F. George, unpublished data).

Because of the ease with which thin periderm strips can be peeled from the living twigs, it was surmised that this freezing might be observed visually. DTA analysis of thin strips showed distant freezing near -40°C and generally little or no freezing at high temperature. Preliminary experiments, however, in which periderm strips were mounted in immersion oil and cooled on a Bausch and Lomb LP2890 microscope placed in a low temperature freezer, have revealed no observable freezing in the nonliving periderm cell lumens. It may be postulated that the fraction of supercooled water in the periderm is located in cell wall pores and that crystallization could not be observed. Perhaps crystallization simply could not be observed with the microscope used for the experiment. These postulates can only be resolved by future research. In any case, this is another example of freezing of supercooled water near -40°C that is unrelated to injury in the tissues involved. It also supports the proposed anatomical nature of supercooling (6,7,10,20,24).

C. Influence of Storage at Moderate Subfreezing Temperatures on Supercooling in Twigs and Male Cones

The influence of long term storage at -10°C, followed by short periods at +2°C, 100% RH on twigs with distinct periderm is shown in Figure 3. Twigs collected February 26, 1979 and stored at -10°C to May 22, 1979 displayed a low temperature peak centered at -52.7°C. During warming, DTA revealed a small melting peak near -7.3°C as well as a large melting peak at higher temperature (not shown). Samples removed from -10°C on May 21, 1979 and stored 2 days at +2°C, 100% RH displayed a low temperature peak centered at -49.2°C and a small melting peak at -5.9°C. Samples removed from -10°C on May 21, 1979 and stored 8 days at +2°C, 100% RH displayed a low temperature peak at -41.2°C and a small melting peak at -3.4°C. It cannot be stated categorically that the freezing and melting peaks are due to the same fraction of water; however, the relationship between depression of the homogeneous ice nucleation temperature and concentration in ideal solutions supports that conclusion. Using Rasmussen and MacKenzie's data (22), homogeneous ice nucleation temperatures of -50.9, -48.5 and -44.0°C would be predicted for NaCl solutions having melting points of -7.3, -5.9 and -3.4°C, respectively. It therefore appears that a fraction of cellular water is concentrating due to slow extracellular freezing during long term storage at -10°C. This freezing behavior is contrary to that reported for storage of

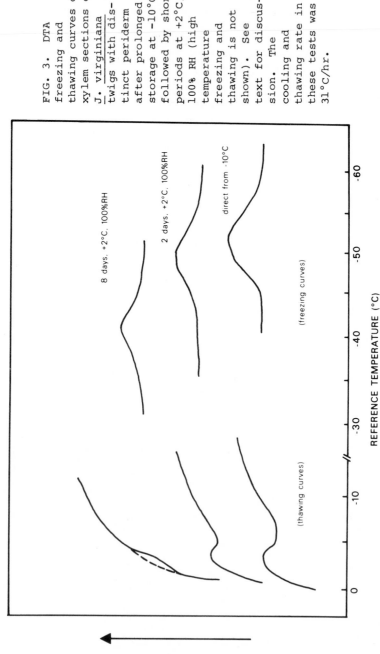

FIG. 3. DTA freezing and thawing curves of xylem sections of _J. virginiana_ twigs with distinct periderm after prolonged storage at -10°C followed by short periods at +2°C, 100% RH (high temperature freezing and thawing is not shown). See text for discussion. The cooling and thawing rate in these tests was 31°C/hr.

C. ovata and Q. velutina twigs at subfreezing temperatures
above -20°C (6,7). In these species, the freezing behavior of
supercooled water is unchanged after storage (i.e. the supercooled water is resistant to desiccation from extracellular ice
at moderate subfreezing temperatures).

It is tempting to believe that the supercooled fraction of
water is located in xylem parenchyma cells of the J. virginiana
twigs. A preliminary viability test conducted April 2, 1979 on
samples stored at -10°C from February 23, 1979 to March 29,
1979 and then at +2°C, 100% RH to April 2, however, revealed no
discernible browning injury to xylem tissues. In these twigs
the low temperature peak was centered at -48.6°C.

Similar experiments were conducted on male cones collected
January 28, 1980 and stored at -12°C for 29 days beginning on
January 31, 1980. In these experiments, samples used for viability testing and DTA were prepared prior to freezing in
January 31. After the storage period at -12°C all samples were
cooled directly to lower temperatures. When storage at -12°C
was completed, samples were not observed to be damaged severely
until -60°C. A small anomalous peak was observed by DTA at
-37.1°C, but no freezing below that temperature was detected
(the DTA cooling rate was 25°C/hr). This result is similar to
that reported by Rajashekar and Burke (21) for flower buds of
hardy Prunus species exposed to moderate subfreezing temperatures for short periods. They showed that low temperature
freezing peaks move to lower temperatures and become smaller
after preconditioning. Sakai (24) has similarly determined
that slow cooling of A. sachalinensis and A. balsamea buds
causes the low temperature freezing peaks to disappear. As in
these resports, the supercooled water in male cones of J.
virginiana is not resistant to desiccation at moderate subfreezing temperature and freezing resistance appears to
increase during storage.

IV. SUMMARY AND CONCLUSION

In this report, freezing of supercooled tissue water at
slightly below -40°C is documented in hardy twigs with needles,
hardy twigs with distinct periderm and hardy male cones of J.
virginiana. This freezing is associated with browning injury
in twigs with needles and in male cones. The supercooled fraction of water appears to freeze slowly and extracellularly in
twigs with distinct periderm and also in male cones during prolonged storage near -10°C. Freezing avoidance by supercooling
of tissue water, therefore, appears to be an important survival
mechanism in twigs and male cones of J. virginiana. However,
during long term exposure to moderate subfreezing temperatures,
supercooling may become less important than freezing tolerance
in these same tissues.

In addition to the supercooling recorded in living tissues of hardy twigs and hardy male cones, it is also noted that supercooling to near -40°C occurs in deacclimated needles and xylem of one year old twigs and in nonliving periderm. Although the supercooling is not associated with freezing resistance in these tissues, it does emphasize the structural nature of freezing avoidance in plant parts. It also points out that every DTA measurement of freezing of supercooled tissue water near its homogenous nucleation temperature is not a priori a lethal event.

ACKNOWLEDGMENT

The author wishes to thank Richard P. Guyette, John E. Roberts and Desmond E. Smith for their technical assistance.

REFERENCES

1. Becwar, M.R., and Burke, M., in "Proceedings of the Intermountain Nurserymen's Association Meeting" (USDA, Forest Service), pp. 70-73. Mt. Sopris Nursery, Aspen, Colorado, 1979.
2. Becwar, M.R., Rajashekar, C., Hansen Bristow, K.J., and Burke, M.J., Plant Physiol. 68:111-114 (1981).
3. Burke, M.J., in "Comparative Mechanisms of Cold Adaptation" (L.S. Underwood, L.L. Tieszen, A.B. Callahan and G.E. Fork, eds.), pp. 259-281. Academic Press, New York, 1979.
4. Burke, M.J., and Stushnoff, C., in "Stress Physiology in Crop Plants" (H. Mussell and R.C. Staples, eds.), pp. 197-225. John Wiley & Sons, New York, 1979.
5. Burke, M.J., Gusta, L.V., Quamme, H.A., Weiser, C.J., and Li, P.H., Ann. Rev. Plant Physiol. 27:507-528 (1976).
6. Carrasquilla, M.L., and George, M.F., in "Proceedings of the Central Hardwood Forest Conference III" (H.E. Garrett and G.S. Cox, eds.), pp. 159-166. University of Missouri, Columbia, Missouri, 1980.
7. George, M.F., and Burke, M.J., Plant Physiol. 59:319-325 (1977).
8. George, M.F., and Burke, M.J., in "Commentaries in Plant Science" (H. Smith, ed). Vol. 2. Pergamon Press, Inc., New York, 1980.
9. George, M.F., and Carrasquilla, M.L., Annual Meeting of the American Society of Plant Physiologists. Abs. 576, 1979.
10. George, M.F., Burke, M.J. and Weiser, C.J., Plant Physiol. 54:29-35 (1974).
11. George, M.F., Hong, S.G., and Burke, M.J., Ecology 58:674-680 (1977).

12. George, M.F., Smith, D.E., and Guyette, R.P., Transactions of the Missouri Academy of Science 13:173 (1979).
13. Graham, P.R., and Mullin, R., J. Amer. Soc. Hort. Sci. 101(1):3-7 (1976).
14. Haruff, R.W., Electronic Design 3:88-90 (February 1, 1980).
15. Hong, S.G., and Sucoff, E., Plant Physiol. 66:40-45 (1980).
16. Kaku, S., and Iwaya, M., Oikos 33:402-411 (1979).
17. Palta, J.P., Levitt, J., and Stadelmann, E., Cryobiology 15:249-255 (1978).
18. Pellett, H., Can. J. Plant Sci. 51:193-195 (1971).
19. Pope, M.I., and Judd, M.D., "Differential Thermal Analysis". Heyden & Son, Inc., New Jersey, 1977.
20. Quamme, H., Weiser, C.J., and Stushnoff, C., Plant Physiol. 51:273-277 (1973).
21. Rajashekar, C., and Burke, M.J. in "Plant Cold Hardiness and Freezing Stress: Mechanisms and Crop Implications" (P.H. LI and A. Sakai, eds.), pp. 213-225. Academic Press, New York, 1978.
22. Rasmussen, D.H., and MacKenzie, A.P., in "Water Structure at the Water-Polymer Interface" (H.H.G. Jellinek, ed.), pp. 126-145. Pelnum Press, New York, 1972.
23. Sakai, A., Plant and Cell Physiol. 19(8):1439-1446 (1979).
24. Sakai, A., Plant and Cell Physiol. 20(7):1381-1390 (1980).
25. Sakai, A., and Hakoda, N., J. Amer. Soc. Hort. Sci. 104(1):53-57 (1979).

FROST INJURY IN SUPERCOOLED LEAVES
OF TUBER-BEARING S OLANUM SPECIES

C. Rajashekar, P. H. Li and J. V. Carter

Laboratory of Plant Hardiness
Department of Horticultural Science & L.A.
University of Minnesota
St. Paul, Minnesota

I. INTRODUCTION

Plants seldom freeze at the freezing points of tissue water, but rather supercool to varying degrees (2,3,11,14,17, 18,28). The extent of plant supercooling in nature is primarily dependent on the ice-nucleating ability of the plant tissue and its immediate environment. In the absence of an external source of ice nucleation, the effectiveness of the ice-nucleating ability of plant tissue plays a major role in controlling the plant supercooling. Marcellos and Single (17) investigated the ice nucleating ability of many herbaceous plants. They found considerable supercooling in all the species studied and noted that the average freezing temperature for spring wheat was as low as -14.4°C. Kaku (10) found that the higher ice nucleation activity of plants was associated with the structural integrity of plant tissue and that it occurred in distinct localized regions. For example, the highest ice nucleating ability of leaves of Veronica sp. was observed in the mesophyll tissue while in leaves of Buxus sp. it was associated with the vascular tissue. Ability of plants to initiate ice, not only depends on the nature of the tissue but also on its maturity.

There have been few attempts to locate the source of ice nucleation in plants under field conditions. Marcellos and Single (16) found that wheat plants can supercool below -10°C. Air samples collected from the field did not contain significant amounts of ice nuclei which were active above -5°C. From the point of view of atmospheric ice nucleation, many studies have focussed on the nature and origin of ice forming nuclei (15,27). The sources of atmospheric ice nuclei are rare particularly at temperatures above -15°C (23). However recent studies (20,21,23) indicate that biogenic source such as decaying plant matter can be a good source of ice nucleation, notably certain bacteria which can initiate ice at temperatures warmer than -3°C.

At present we do not understand the role of plant super-cooling in herbaceous tissues well enough to know whether it is always beneficial, as a means of frost avoidance, or if it is injurious due to subsequent ice formation. Maintaining plants in supercooled state for a limited time causes no apparent injury, however, symptoms similar to those in chilling injury occur if the exposure time is increased (13). Lindow et al. (12) have shown that some chilling sensitive plants could tol-erate subzero temperatures by supercooling, promoted by ridding the plants of bacterial source of ice nucleation. However, if the supercooled plants freeze, they are likely to be injured at the temperature of ice formation as has been shown in several crop plants (1,25,28). It is important to note that with the increasing supercooling, plants risk greater danger of injury. The importance of ice initiation temperature in relation to plant injury has long been recognized (1,25). Earlier workers (1,6) have emphasized the need to avoid excessive supercooling before freezing to avoid lethal intracellular ice formation. Siminovitch and Scarth (25) suggested that nonhardy plant tis-sues do not tolerate ice initiation even just below the freezing points of their tissue water. They found intra-cellular injury at ice initiation temperature as high as -2°C in the nonhardy plant tissues and concluded that supercooling and intracellular freezing is often the main cause of frost injury in leaves of cabbage. The injurious intracellular ice formation in plants is generally avoided by inducing extra-cellular freezing. This is achieved by circumventing super-cooling through inoculation of plant tissue with ice just below its freezing point. The extent of supercooling that is not injurious following freezing in different plant species is not known. No information is available on the factors that may control this safe limit of supercooling in plants.

In the present investigation we have studied the ice nucle-ating ability of leaves of various tuber-bearing Solanum potatoes, and the relationship between the freezing of super-cooled leaf and its injury. We will also discuss some aspects of preventing leaf supercooling through the use of artificial nucleants.

II. SUPERCOOLING IN SOLANUM LEAVES

The ice nucleating characteristics of leaves were studied in Solanum tuberosum, S. acaule and four Solanum crosses namely, 379483 = [ajh[a] x stn x phu] x [acl x phu x adg], 37729-1 x 700882 x ajh (ajh x stn) x (adg x ajh), 76977-6 x 377748-4 = [(adg x adg) x Juz] x (ajn x stn), and Nevada x 376977-8 = (Phu x Phu) x [(adg x adg) x Juz] referred to here as 483, 29-1, 77-6 and Nevada respectively. The seeds of these crosses were provided by Dr. N. Estrada, Bogota, Colombia. The first planting of these crosses was by the seeds and the later plantings were made using tubers. Solanum tuberosum (cv. Red Pontiac) was also grown from tubers while stem/leaf cuttings were used for S. acaule. The plants were grown at 20°/18°C (day/night) in growth chambers. Cold acclimation was induced by holding 8-10 week old plants at constant 5°C for at least 4 weeks.

Ice nucleation activity of leaves was determined by the freezing drop method as described by Vali (26). Leaf discs of 1.5 mm in diameter from mature leaves were suspended in droplets (0.01 cm^3) of double distilled water. The droplets were placed on a metal plate and cooled at a constant rate of 3.5°C/hr in a closed chamber. Table 1 summarizes the ice nucleation activity of leaves of Solanum. The threshold temperature for freezing in nonhardened leaves was always lower than -6.9°C. The lowest threshold temperature was -8.1°C in the case of 483. The mean freezing temperature, which is the mid-point of the temperature range over which the freezing events occurred, for the Solanums was below -9°C. From these results, it can be concluded that the leaves did not contain ice nuclei active at warm temperatures and thus can inherently supercool to a great deal. The ice nuclei concentration for leaves of unhardened Solanums is presented in Figure 1. The ice nuclei concentration increased roughly in log-linear fashion with decreasing temperature except in the case of 29-1 in which the curve reached a plateau between -8.5° and -10.5°C. The ice nuclei concentration was also determined in the leaf homogenates. The threshold and mean freezing temperatures for 2% (w/v in distilled water) leaf homogenate are shown in Table 1. The threshold temperature of leaf homogenates for all the Solanums was lower than that for intact leaf discs except in the case of 483. The difference in threshold temperatures between the intact leaf and its homogenate was as much as 4°C for S. acaule. Thus, structural intactness appears to be important for the higher ice nucleation activity in these

---

[a] Acl - S. acaule, adg = S. andigena, ajh = S. ajanhuiri, Juz = S. juzepczukii, phu = S. phureja, stn = S. stenotomum.

TABLE I. The threshold and mean freezing temperatures for the intact leaves and 2% leaf homogenates with 3.5°C/hr cooling rate.

Solanums	Threshold Temp (°C)			Mean Freezing Temp (°C)[a]		
	Unhardened		Hardened	Unhardened		Hardened
	Intact[b]	Homogenate	Homogenate[c]	Intact	Homogenate	Homogenate[c]
S. tuberosum	-7.2	-8.1	-8.4	-10.1	-10.8	-12.0
S. acaule	-6.9	-11.0	-	-9.2	-14.3	-
483	-8.1	-6.9	-10.2	-11.3	-11.3	-13.6
29-1	-7.5	-8.7	-11.3	-10.3	-10.7	-14.9
77-6	-7.8	-10.2	-9.7	10.5	-12.7	-12.0
Nevada	-7.6	-8.7	-9.6	10.7	-11.6	-12.2

a   The mid-point of temperature range of the freezing distribution.
b   Small leaf discs (1.5 mm diameter) suspended in droplets (0.01 cm³).
c   2% leaf homogenate were filtered through Whatman #1 paper.

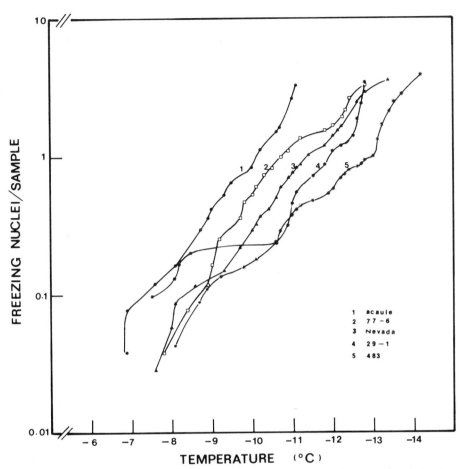

FIGURE 1. Freezing nuclei concentration of intact unhardened leaf samples of Solanums. Leaf discs (1.5 mm in diameter) were suspended in droplets (0.01 cm³) of distilled water. The cooling rate was 3.5°C/hr.

leaves. Similar results were reported by Kaku (9,10) in leaves of Veronica and Buxus species. By homogenizing the leaves, the ice nucleation activity moves to lower temperatures and further loss of activity was noticed when the homogenate was filtered through Whatman #1 filter paper. In the case of hardened leaves, only filtered homogenates were used to determine the ice nucleus concentration (Table 1).

The extent of supercooling in the case of S. tuberosum and 483 was determined by holding the samples at subzero temperatures for longer period of time. Leaf discs (2.0 cm in diameter) were exposed to each test temperature for 6 hours to monitor their freezing. No freezing event was observed till -5°C

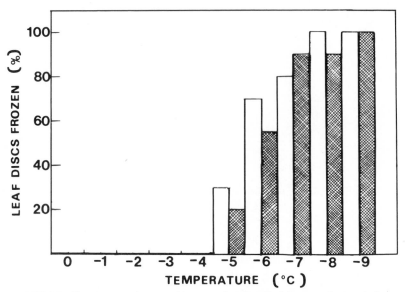

FIGURE 2. Freezing patterns in leaf discs of S. tuberosum (open bars) and 483 (hatched bars). Large leaf discs (2 cm) were held for 6 hrs at each testing temperatures until freezing began (-5°C). Subsequently the samples were cooled at 1°C/hr.

(Fig. 2) in both S. tuberosum and 483. With the initiation of freezing at -5°C, the samples were cooled at 1°C/hr and the remaining freezing events were recorded. All the leaf discs were frozen by -8°C in S. tuberosum and -9°C in 483. Similar experiments were conducted using the whole plants with a constant rate of cooling. Potted plants of S. tuberosum and 483 were cooled in a cold chamber at 2°C/hr. The ice initiation temperature, as measured by a copper-constantan thermocouple inserted in the stem, was -6.5°C for S. tuberosum and -6.8°C for 483. These values of supercooling for the whole plant are not too much different from the threshold temperatures of leaves determined using the freezing drop method. Similar values for the supercooling of S. tuberosum and S. acaule have been reported by Hudson and Idle (7).

III. FROST INJURY IN SUPERCOOLED PLANTS

So far we have discussed the inherent ability of leaves of Solanums to supercool to varying degrees. As mentioned earlier, the supercooling in plant tissue has often been thought undesirable and there are a number of reports indicating the harmful effects of initiating freezing in

TABLE II. The frost hardiness and the lethal ice initiation temperature of leaves of nonhardened and hardened Solanums.

Solanums	Frost hardiness ($^\circ$C)[a]		Lethal ice initiation temperature ($^\circ$C)[b]
	Unhardened	Hardened	
S. tuberosum	-3.5	-3.6	-1.5
S. acaule	-5.5	-8.5	-2.0 (-2.5)[c]
483	-4.5	-6.5	-1.5
29-1	-4.5	-6.5	-1.5
77-6	-5.5	-5.5	-1.2
Nevada	-5.5	-6.5	-1.3

[a]   Hardiness was determined in the absence of supercooling by inoculating the leaves at -1°C with ice.
[b]   Temperature at which leaf injury occurred as a result of ice formation in the supercooled leaves.
[c]   Hardened leaves.

supercooled tissues (6,25). In many cases indirect evidence to this effect exists where plant tissues have been observed to be injured at the ice formation temperatures which are generally below -5°C (7,8,28). The only means of low temperature tolerance in plants appears to be through extracellular freezing. Plants can survive extracellular freezing to much lower temperature if the supercooling and subsequent intracellular injury can be avoided (6). So we have investigated the extracellular freezing tolerance and the extent of supercooling prior to ice formation that is not injurious in the leaves of Solanums.

The freezing tolerance of the leaves of Solanums was determined using the conventional method of frost hardiness evaluation wherein the supercooling of leaves was precluded. Leaves, wrapped in moist cheese cloth, were inoculated with ice at -1°C and held for 30 min to ensure their freezing. The samples were subsequently cooled at 2°C/hr. After reaching the test temperature, the samples were slowly thawed on ice bath and the injury was determined based on visual ratings as well as electrolyte leakage of the stressed tissue. The results of the freezing tolerance of nonhardened and hardened leaves of Solanum are summarized in Table 2. The electrolyte leakage data for the leaves of S. tuberosum are given in Figure 3. With the exception of S. tuberosum, all the Solanums studied were moderately frost resistant. The unhardened leaves of frost resistant Solanums survived between -4.5° and 5.5°C.

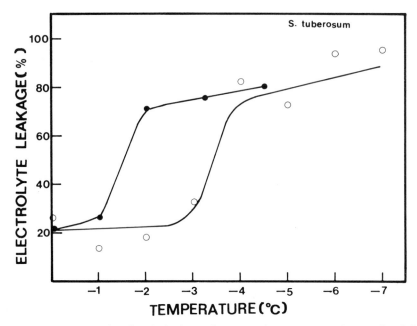

FIGURE 3. The leaf injury in S. tuberosum as determined by the electrolyte leakage. The frost hardiness was determined by inoculating the leaves (O) with ice at -1°C and subsequently cooled at 2°C/hr. Leaves (●) were supercooled and ice initiated at each test temperature with ice.

Leaves of only two Solanums, namely S. tuberosum and 77-6 failed to harden in response to the acclimating conditions. The largest increase in frost tolerance due to hardening was in the leaves of S. acaule which survived -8.5°C when fully hardened. However in other frost resistant Solanums the increase was 1 to 2°C due to the hardening process. These results indicate the freezing tolerance of the leaves in the absence of any supercooling. The freezing tolerance determined by this method provides information on the extracellular freezing tolerance and such methods may not exactly reflect what plants would experience under natural conditions. Discrepancies between the freezing tolerance measured in the laboratory, and that observed under field conditions have been a major concern in potato crop (19). The frost tolerance under field conditions is usually much less than that estimated by the laboratory methods. Richardson and Weiser (19) compiled data on the frost tolerance, primarily from the field observations, of leaves of various Solanums. It is significant to note that the majority of observations indicate that S. acaule, S. multidissectum and S. chomatophilum, considered frost resistant on the basis of laboratory frost hardiness tests, were in

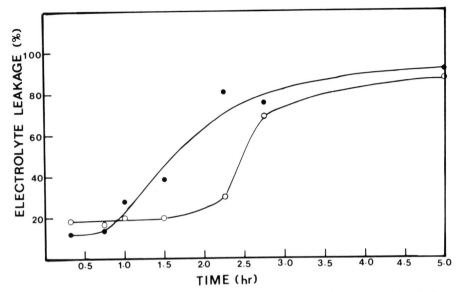

FIGURE 4. Electrolyte leakage from leaf discs (2 cm) of S. tuberosum (●) and 483 (O) frozen in 2.5 ml distilled water. Samples were inoculated at -1°C and held at -1.5°C.

fact frost killed above -3°C. In view of the considerable supercooling exhibited by the leaves of Solanums, it seems important to recognize the role played by plant supercooling in contributing to frost injury. Although the supercooling of plants under natural conditions prior to ice formation remains a possibility for higher killing temperatures as observed in the field, there are other factors, such as time of low temperature exposure and ice encasement, which could influence the killing temperatures. In a separate experiment we examined whether or not the time factor and ice encasement have any effect on the frost tolerance of leaves of S. tuberosum and 483. Leaf discs (2.0 cm in diameter) were submerged in distilled water and inoculated at -1°C and held at -1.5°C for various lengths of time. In similar tests, but subjecting them to 30 min at each test temperature and without ice encasement, the leaves of S. tuberosum and 483 survived -3.5° and -4.5°C respectively (Table 1). Injury to leaves with ice encasement at -1.5°C was observed after 1.5 hr in the case of S. tuberosum and after 2.5 hr in the case of 483 (Figure 4).

The extent of supercooling in the leaves of Solanums that do not result in injury subsequent to ice formation was determined as described below. The whole leaves, placed in a clean, dry test tube could be generally supercooled between -5° and -7°C. The supercooled leaves were inoculated at various test temperatures with crushed ice and held at the test temperature for approximately 30 min to make sure that the leaf was fully frozen. The leaves were then thawed to measure the electrolyte leakage (Fig. 3). The temperature of the leaves was monitored by copper-constantan thermocouple. The ice initiation temperatures in supercooled leaves that resulted in injury are presented in Table 2. The leaves of all Solanums except S. acaule were injured if ice was intiated at temperatures below -1.5°C. In the leaves of S. acaule the injury occurred if ice was initiated below -2°C. The lethal ice initiation temperature for nonhardened and hardened leaves were similar with the exception of S. acaule in which the hardened leaves survived slightly lower ice initiation temperature. The leaves of Solanums were injured if ice was initiated in supercooled leaves at or below -2°C, while the same leaves when inoculated at -1°C could survive extracellular freezing to much lower temperatures. For instance the leaves of S. acaule if inoculated at -1°C, could tolerate freezing to -5.5°C while if they were inoculated below -2°C, injury occurred at the temperature of ice formation. The injury in the supercooled plant tissues has been attributed to the intracellular ice formation (1,25). Slight supercooling has been known to cause injury in potato and other herbaceous tissues (13,25). Gusta and Fowler (5) reported that if crowns of winter wheat were allowed to supercool to -3°C and subsequently freeze, it resulted in injury. Ice formation at temperature as high as -2°C was found injurious in many plant tissues (25). Hudson and Idle (7) found that both S. acaule and S. tuberosum supercooled to -6°C and subsequently froze resulting in lethal injury. The results indicate that the leaves of the frost resistant Solanums including S. acaule can only tolerate ice formation at low degree of supercooling . This obviously increases the risks of frost injury at higher temperature than can be tolerated by extracellular freezing.

The results from the experiments using excised leaves described above were consistent with those using whole plants. The plants were maintained in cold chambers at -2° or -3°C for various lengths of time until they froze. Within one hour of ice formation in the stem, plants were thawed and visually rated for injury. The ice formation in the plants was not uniform, but often occurred in isolated patches on the leaf blades. Generally older leaves froze readily while younger leaves remained completely unfrozen. Kaku (9) has made similar observations. However more uniform ice formation was achieved by spraying the plants with water before transferring to cold chambers. All the Solanums including S. acaule was injured by ice formation at -3°C. In the event of extensive ice formation, the whole plant tops were killed. However in many cases new shoots were found to grow from the base of the dead top plants. Thus it is advantageous to induce extracellular freezing and avoid supercooling in the leaves of frost resistant Solanums to achieve a greater degree of low temperature protection. In the case of plants which have little or no tolerance to extracellular freezing supercooling may prove to be an important tool in providing low temperature protection. The relative merits of these approaches depend on the degree of tolerance to extracellular freezing and the extent of plant supercooling that can be sustained under field conditions.

IV. PREVENTION OF SUPERCOOLING

With the observations that the leaves of frost resistant Solanums cannot tolerate ice formation at some considerable supercooling we have attempted to use some external source of ice nucleation to prevent plant supercooling. The following summarizes the studies on the ice nucleation activity of some artificial nucleants and their application to prevent super-cooling in the leaves of Solanums.

The main objective of this investigation was to bind arti-ficial nucleants that can initiate ice in the leaves of Solanums at relatively warm temperatures. Efficiency of ice nucleation materials such as acetoacetanilide (Sigma Chemical Co.), fluorophlogopite (-200 + 325, Mykroy Ceramics) phthalic anhydride (Matheson Coleman & Bell), metaldehyde (ICN Pharma-ceuticals), 2,4-6 trichloroaniline (Eastman Kodak) and soil particles were tested by measuring the ice nucleus content as described earlier. Freshly ground materials were used to pre-pare 1 to 2% suspension in double distilled water. Droplets of uniform volume (0.005 to 0.020 cm^3) prepared from well agi-tated suspension were cooled at $2°C/hr$. In the case of soil, the sample was a mixture of soil, sand and peat (2:1:1 v/v) from pots containing actively growing potato plants. One gram of soil was mixed with 100 ml of distilled water for an hour on a mechanical stirrer and subsequently filtered through Whatman #1 filter paper. The clean soil extract was used as the test material. Figure 5 shows the ice nuclei concentration per ml of the suspension for various nucleants. Fluorophlogopite and acetoacetanilide were found to have the most effective ice nucleation activity. The threshold temperatures for ice initiation in the case of fluorophlogopite and acetoacetanilide were $-0.7°$ and $-0.8°C$ respectively. The threshold temperature for the nucleants ranged from $-0.7°$ to $-5.4°C$ and the poorest activity was in phthalic anhydride. Although fluorophlogopite and acetoacetanilide had similar threshold temperatures, from the ice nuclei concentration as a function of temperature, it is evident that fluorophlogopite, a fluorine mica, is more active than acetoacetanilide. A sharp log-linear increase of ice nucleus concentration with decreasing temperature was observed for fluorophlogopite. The distribution of the freezing events was over a narrow temperature range and all the drops were frozen at temperature as warm as $-2.4°C$. On the

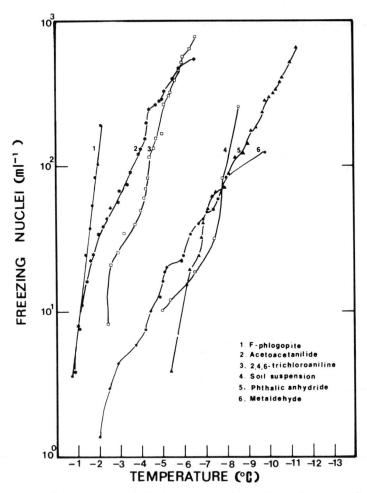

FIGURE 5.   Ice nuclei concentration of various nucleants.
1-2% of aqueous suspensions were used.   Soil extract was
obtained by passing 1% soil suspension through Whatman #1 filter paper.

other hand, the temperature range of freezing distribution for
acetoacetanilide is rather broader and the freezing was not
completed until -7.3°C. The high ice nucelation activity of
fluorophlogopite has been explained on the basis of the chemical and structural characterisitcs of the mica in which fluorine atoms play a key role in ice initiation at temperatures
close to 0°C (24).

Metaldehyde and 2,4-6 trichloroaniline have moderate ice
nucleation activity, with threshold temperature of -2.0°C and
-2.4°C respectively. These values are slightly lower than

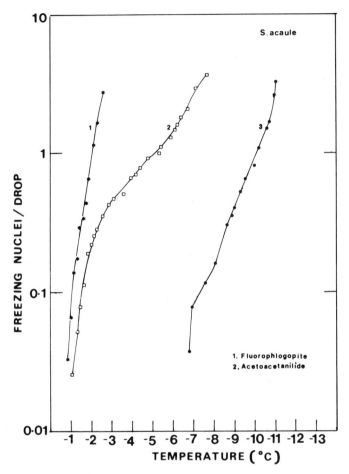

FIGURE 6. Ice nuclei concentration using leaf discs of S. acaule. The leaf discs were suspended in 1% fluorophlogopite (1), 1% acetoacetanilide (2) and distilled water (3). Ice nuclei concentration is expressed per droplet containing the leaf disc.

those found by Fukuta (4). Soil extract used in this study was not an effective source of ice nucleation; no ice nuclei active above -5°C were detected. The high ice nucleation of soil is known to be dependent on the type and amount of organic matter present in the soil (22). It is not unusual to find the inorganic components of soils to have significant ice nucleation activity at temperatures as low as -15°C, while decaying plant materials can be active at -4°C (23).

TABLE III. Frost hardiness of the leaves of Solanums.

| Solanums | Frost Hardiness (°C)[a] | |
	Unhardened	Hardened
S. tuberosum	-1.0	-
S. acaule	-3.5	-6.0
483	-1.0	-5.0
29-1	-2.0	-6.0
77-6	-2.5	-4.5
Nevada	-2.0	-5.0

[a] Supercooling was prevented by inoculating the leaves with 1% fluorophologopite at about -1°C.

To have ice nucleation activity at higher temperature, it was essential to use the freshly ground materials and to prepare the suspension just prior to use. Aging of the suspension for more than 2 days substantially lowered the threshold temperatures.

With the finding that fluorophlogopite and acetoacetanilide can initiate ice in distilled water at temperatures slightly above -1°C, the obvious step was to find if these nucleants are effective in the presence of leaf tissue. The results of such an experiment with S. acaule is presented in Figure 6. Leaf discs (1.5 mm in diameter) were suspended in droplets of 1% suspension of fluorophlogopite and acetoacetanilide. Figure 6 shows the ice nucleus concentration per drop containing the leaf sample. There appears to be no significant effect of leaf tissue on the ice nucleation activity of the nucleants. Similar experiments were done using whole leaves immersed in the suspension of these nucleants. Ice can be initiated by holding samples for approximately 30 min at -1°C. Although both nucleants were effective in ice initiation at -1°C, acetoacetanilide was found to be toxic to leaves. Leaves suspended in 1% suspension of acetoacetanilide for more than 18 hours appeared water soaked and resulted in large electrolyte leakage. No such adverse effects were observed in the case of fluorophlogopite. Therefore in the subsequent studies fluorophlogopite was used to find the effect of preventing supercooling on the leaf injury of Solanums. One or two small drops of freshly prepared 1% fluorophlogopite suspension was deposited at the base of leaf blade. Ice encasement with the suspension caused injury at relatively higher temperatures. It was often necessary to hold the sample for 30 min just above -1°C to make sure that the leaves freeze. After ice formation, the samples were cooled at a constant rate of 2°C/hr. The

results on frost tolerance using 1% fluorophlogopite are summarized in Table 3. The nonhardened leaves are injured at temperatures much higher than normally expected from the extracellular freezing tolerance using ice as nucleant. Leaves of
S. tuberosum and 483 show injury at -1°C while those of S.
acaule tolerate -3.5°C. In nonhardened leaves use of fluorophlogopite to prevent supercooling resulted in little or no
additional increase in the low temperature protection compared
to the lethal ice initiation temperatures. On the other hand,
hardened leaves were found to survive significantly lower temperatures by ice nucleation with fluorophlogopite. The
hardened leaves of 29-1, for instance, survived -6°C by ice
nucleation with fluorophlogopite. This was comparable to the
survival level obtained using ice as nucleant. Thus considerable low temperature protection can be achieved in the hardened
leaves of Solanums by using fluorophlogopite as a nucleant. It
is important, however, to note that even in the hardened leaves
the leaf injury occurred at slightly higher temperatures using
fluorophlogopite than those using ice as nucleant. Although
fluorophlogopite is an effective ice nucleating agent at temperature as high as -1°C, it is not clear as to what causes the
difference in the killing temperatures obtained by using these
two different nucleants. Further work is under progress to
understand these aspects and the ice nucleation phenomenon of
fluorophlogopite in plant tissues in relation to injury.

REFERENCES

1. Asahina, E., The Institute of Low Temperature Science
   10:83-126 (1956).
2. Cary, J.W., and Mayland, H.F., Agron. J. 62:715-719 (1970).
3. Chandler, W.H., Amer. Soc. Hort. Sci. 64:552-572 (1954).
4. Fukuta, N., J. Atmos. Sci. 23:191-196 (1966).
5. Gusta, L.V., and Fowler, D.B., Can. J. Plant Sci.
   57:213-219 (1977).
6. Hudson, M.A., and Brustkern, P., Planta (Berl.) 66:135-155
   (1965).
7. Hudson, M.A., and Idle, D.B., Planta 57:718-730 (1962).
8. Hutcheson, C.E., and Wiltbank, W.J., HortSci. 7:27-28
   (1972).
9. Kaku, S., Plant and Cell Physiol. 12:147-155 (1971).
10. Kaku, S., Plant and Cell Physiol. 14:1035-1038 (1973).
11. Kitaura, K., in "Cellular Injury and Resistance in Freezing
    Organisms" (E. Asahina, ed.), pp. 143-156. Hokkaido Uni-
    versity, Sapporo, Japan, 1967.
12. Lindow, S.E., Arny, D.C., Upper, C.D., and Barchet, W.R.,
    in "Plant Cold Hardiness and Freezing Stress" (P.H. Li and
    A. Sakai, eds.), pp. 249-263. Academic Press, New York,
    1978.

13. Lindstrom, O.M., Ph.D. Thesis, University of Minnesota, St. Paul, 83 pp. (1981).
14. Lucas, J.W., Plant Physiol. 29:245-250 (1954).
15. Maki, L.R., and Willoughby, K.J., J. Appl. Meteorol. 17:1049-1053 (1978).
16. Marcellos, H., and Single, W.V., Agric. Meteorol. 16:125-129 (1976).
17. Marcellos, H., and Single, W.V., Crybiology 16:74-77 (1979).
18. Modlibowska, I., J. Hort. Sci. 37:249-261 (1962).
19. Richardson, D.G., and Weiser, C.J., HortSci. 7:19-22 (1972).
20. Schnell, R.C., Biogenic Sources of Atmospheric Ice Nuclei, Dept. of Atmos. Resources, University of Wyoming Report No. AR111 p. 45 (1974).
21. Schnell, R.C., J. Atmos. Sci. 34:1299-1305 (1977).
22. Schnell, R.C., and Vali, G., Nature 236:163-164 (1972).
23. Schnell, R.C., and Vali, G., J. Atmos. Sci. 33:1554-1564 (1976).
24. Shen, J.H., Klier, K., and Zettlemoyer, A.C., J. Atmos. Sci. 34:957-960 (1977).
25. Siminovitch, D., and Scarth, G.W., Can. J. Res., C 16:467-481 (1938).
26. Vali, G., J. Atmos. Sci. 28:402-409 (1971).
27. Vali, G., Christensen, M., Fresh, R.W., Galyan, E.L., Maki, L.R., and Schnell, R.C., J. Atmos. Sci. 33:1565-1570 (1976).
28. Yelenosky, G., and Horanic, G., Cryobiology 5:281-283 (1969).

POPULATION DYNAMICS OF EPIPHYTIC ICE NUCLEATION ACTIVE BACTERIA
ON FROST SENSITIVE PLANTS AND FROST CONTROL
BY MEANS OF ANTAGONISTIC BACTERIA

S. E. Lindow

Department of Plant Pathology
University of California
Berkeley, California

ABSTRACT

Seasonal variations in the populations of ice nucleation active (INA) strains of Pseudomonas syringae, Erwinia herbicola, and Pseudomonas fluorescens were measured on several frost sensitive California crop plants including pear, potato, tomato, almond, navel orange, and avocado to determine the role of these bacteria in inciting frost damage to the plants on which they reside and to determine optimum times for alteration of epiphytic mircobial antagonists. Maximum populations of INA bacteria varied from less than approximately 100 cells/g fresh weight of leaf tissue on citrus and avocado to in excess of 10^6 cells/g fresh weight on almond. Populations of INA bacteria on most crops increased approximately 1000-fold during the 6 week period extending from shortly after germination of annual plants or bud break of deciduous perennial plants, coinciding with the period of maximum frost hazard. Populations of INA bacteria on citrus and avocado increased over 100-fold from October until February before again declining during spring and summer months. The nucleation frequency of INA bacteria on all plants was similar, i.e., approximately 100 to 5,000 cells per ice nucleus active at -5 C. The nucleation frequency of INA bacteria on plants was found to increase rapidly with decreasing temperature from -2 C to -8 C. Non-INA bacteria applied to pear at 20% bloom reduced increases in populations of INA bacteria to about 1% that on untreated plants. Frost injury at -3 C to pear treated with antagonistic bacteria was decreased significantly compared to untreated trees; frost injury to pear was correlated significantly with numbers of ice nuclei on leaves active at -5 C.

I. INTRODUCTION

Increased attention has recently focused on factors influencing frost damage in frost sensitive plants. Since frost sensitive plants are injured only if ice forms within their tissues (2,4,30,31,33), this increased attention has been focused on factors which influence ice formation. Considerable work has also recently been done concerning heterogeneous ice nucleation in plant tissues. The liquid-solid phase transition of water at temperatures above -10 to -20 C requires the presence of a nonaqueous catalyst. Heterogeneous ice nuclei are thus required for formation of ice at the warm temperatures of -2 to -6 C associated with frost injury to frost-tender plants (2,32).

The ability of many frost sensitive plants to supercool, thus avoiding frost injury has been recognized for some time (4,25,30,31). Modibowska has shown that flowers of small fruit trees can supercool to -2 C before ice formation occurs (32). Extensive supercooling has been reported for lemon, grapefruit and other Citrus species (7,44-48). Potato leaves have been reported to supercool to at least -2.5 C (40) to -6 C before ice formation occurred. Similarly, tomato and wheat leaves have been observed to supercool to -4.5 C to -5.0 C, respectively (37-39). Several recent reports also indicate variability in the degree of supercooling of a large number of different plant species which ranged from -2 C to -14 C (8-12,29). It is therefore obvious that frost sensitive plants have an innate ability to supercool and avoid damaging ice formation above certain critical temperatures.

Since ice formation and subsequent frost damage most often occur at temperatures warmer than the determined temperature limits of supercooling, efficient heterogeneous ice nuclei must limit supercooling under field conditions in most cases. These heterogeneous ice nuclei have been associated with the surface of plant parts (25). Kaku (8-12) has shown that ice nuclei are not uniformly distributed on a given leaf and that these nuclei vary in quantity with maturity of leaves and also vary among plant species. Similar results ave been presented by other workers (6,29). Atmospheric ice nuclei have long been assumed to be the sources of ice nuclei on plants. However, Marcellos and Single have presented evidence that ice nucleation on leaves by air-borne particles is unlikely (28).

Maki and co-workers have found that cells of the bacteria Pseudomonas syringae van Hall and Pseudomonas fluorescens Migula are efficient ice nuclei, catalyzing ice formation at temperatures warmer than -2 C (26,27). Vali and Schnell and co-workers have discussed the importance of biogenic ice nuclei that arise from leaves presumably due to the activity of these bacteria (34-36,43).

We have found that abundant epiphytic populations of INA strains of both P. syringae and Erwinia herbicola (Löhnis) Dye occur on plant surfaces (24). We have also presented evidence that these bacteria are both necessary and sufficient to limit the supercooling ability of many frost sensitive plant species at temperatures above -5 C (17-19,24). Thus, supercooling of plant tissues will be limited by the heterogeneous ice nucleus which is active at the warmest temperature. Since the role of bacterial ice nuclei in limiting the supercooling of plant tissue has only recently been identified, the contribution bacterial ice nuclei may have made in most measurements of heterogeneous ice nucleation in plant tissue is unknown. However, bacterial ice nuclei probably are relatively unimportant in laboratory-grown plants which many investigators have shown to have little or no heterogeneous ice nucleation activity above -5 C. The nucleation activity of these plants may be due to plant components or other ubiquitous but as yet unidentified ice nucleants. Thus, the measured nucleation temperatures of these plants which are generally below -5 C represent the limit of supercooling which might be expected under conditions in which other sources of ice nuclei were eliminated. However, because bacterial ice nuclei appear to be much more efficient than other sources of ice nuclei in or on plants, they are of primary importance in limiting the supercooling ability of plants. INA bacteria do not appear to be endogenous residents on plant surfaces (19,24). Therefore, an analysis of the population dynamics of INA bacteria on leaf surfaces will be important in predicting the frost sensitivity of the host on which they reside. In addition, since ice nucleation activity does not appear to be expressed by every bacterial cell at a given time (18-20), knowledge of the nucleation activity of these bacteria on leaf surfaces will be critical to assessing the importance of epiphytic populations of INA bacteria in limiting the supercooling of plants.

We have also shown that procedures which reduce the numbers or the ice nucleation activity of INA bacteria can reduce frost injury to field-grown plants (16,17,19,20-22). Competitive non-INA bacteria have shown promise as biological control agents of frost injury (16,17). However, rational use of competitive microorganisms to alter epiphytic bacterial populations on leaves and flowers will require knowledge of the population dynamics of the target INA bacteria for proper timing of deployment. This paper will discuss these points as well as present results of control of frost injury based on alteration of epiphytic INA bacterial populations with antagonistic bacteria.

II.   MATERIALS AND METHODS

Total leaf surface bacterial populations were determined by dilution plating onto King's medium B (13) containing 100 µg/ml cycloheximide following a washing procedure previously described (23,24). Spontaneous mutants of specific non-INA bacterial antagonists resistant to 100 µg/ml of rifampicin were selected in vitro. No significant difference of growth rate in vitro and in planta or of antagonistic properties in vitro were observed between mutant and wild type strains. Inoculum of antagonistic bacteria for field applications was harvested after 3 days growth at 20 C from plates of King's medium B containing 100 µg/ml rifampicin. Inoculum was applied at a rate of approximately 5 x 10^7 cells/ml in an aqueous suspension to deciduous trees at approximately 20% bloom. Leaf surface populations of applied antagonistic bacteria were determined by dilution plating of leaf washings onto King's medium B containing 100 µg/ml cycloheximide and 100 µg/ml rifampicin.

Populations of leaf surface bacteria active in ice nucleation at temperatures above -5 C and -9 C were determined from dilution platings of leaf washings on King's medium B using a replica freezing technique previously described (24). Leaf surface INA bacteria active as ice nuclei at either -5 C or -9 C were quantitated using a droplet freezing procedure similar to that previously reported (23,42). A collection of at least 40 10-µl droplets of leaf washings or appropriate dilutions of leaf washings were placed on the surface of a paraffin-coated aluminum foil sheet floating on the surface of refrigerated constant-temperature baths maintained at -5 C or -9 C. The cumulative number of ice nuclei active at temperatures above -5 C or -9 C were calculated according to Vali (42) from the fraction of unfrozen droplets at each of these respective temperatures. When ice nuclei were not detectable in leaf washings, the leaf washings were concentrated 40-fold by centrifugation. Forty ml of a suspension of leaf washings were filtered through Miracloth and centrifuged at 30,000 g for 30 min and the pellet resuspended in 1 ml of distilled water. Forty 10-µl droplets of concentrated leaf washings were again placed on aluminum sheets at -5 C and -9 C and the fraction of droplets frozen were recorded.

The nucleation efficiency of INA bacteria on leaf surfaces or of the plant material itself was determined using a semi-automatic ice nucleus spectrometer. Sixty 100-µl droplets from each of a series of five 10-fold dilutions of leaf washings containing INA bacteria or macerates of greenhouse-grown corn tissue not colonized by INA bacteria were placed on the surface of a paraffin-coated hollow aluminum block. The block was surrounded with styrofoam for insulation and the temperature of the block was lowered at the rate of 0.2 C/min by

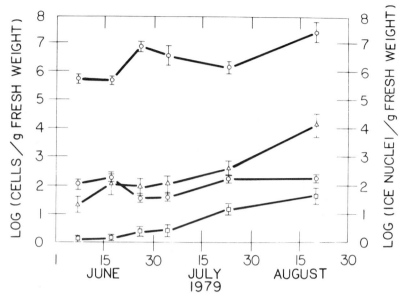

FIG. 1.   Total (O) and ice nucleation active bacteria (△) and ice nuclei active at -5 C (□) or -9 C (◇) on leaves of untreated potato grown near Tulelake, CA. The vertical bars represent the standard error of the determination of the mean of log populations.

pumping refrigerated ethanol through the block. The tempera-ture of the block was continuously recorded with a digital thermister thermometer and the temperature was displayed on a video recording of the surface of the block. The freezing tem-perature of each droplet could be determined by visual analysis of repeated observations of the video recording of freezing events. The cumulative number of ice nuclei active at a given temperature was calculated from the fraction of droplets frozen at that temperature (41,42) correcting for the dilution of bac-teria or plant material being measured.

Field plots of each plant type consisted of a randomized complete block design with 4 replications. Each replication consisted of either 200 annual plants or 4 trees. Leaf samples were taken randomly from all plants in each replication.

III.   RESULTS AND DISCUSSION

Large seasonal variations in the numbers of INA bacteria were observed on both annual and perennial plants. Initial INA bacterial populations on both potato taken shortly after emer-gence from the soil in early June (Fig. 1), or on tomato taken

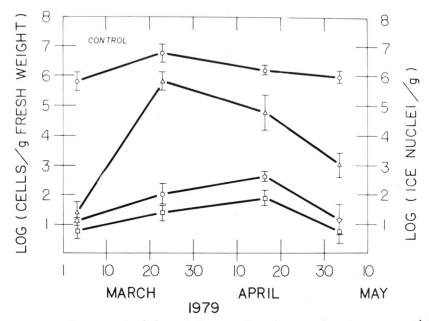

FIG. 2.   Total (0) and ice nucleation active bacteria (△) and ice nuclei active at -5 C (□) or -9 C (◇) on leaves of untreated tomato grown near Visalia, CA. The vertical bars represent the standard error of the determination of the mean of log populations.

shortly after transplanting from a greenhouse (Fig. 2), were very low, i.e., approximately 30 cells/g fresh weight. Large increases in populations of INA bacteria on potatoes and tomatoes were observed during periods of relatively cool wet weather.   In the California potato growing region, this occurred both during early June and again after August 10 (Fig. 1).   Rapid increases in populations of INA bacteria on tomato (over 1000-fold) were observed during the first 3 weeks after transplanting, followed by slow decreases in INA bacterial populations after this time, coinciding with very warm (30 C) and dry weather.   Whereas INA bacteria comprised generally less than 0.01% of the total bacteria on leaves of potato, INA bacteria comprised up to 10.0% of the total bacteria present on tomato leaves.   Strains of P. syringae, E. herbicola, and P. fluorescens were found in nearly equal numbers on the leaves of potato during all parts of the 1979 growing season.   In contrast, only INA bacteria identifed as P. syringae were found on leaves of tomato until April 15.

Numbers of ice nuclei active in ice nucleation at -5 C increased with increasing populations of INA bacteria on potato and at either -5 C or -9 C on tomato.   However, approximately

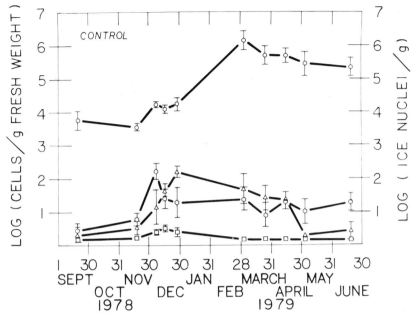

FIG. 3.   Total (O) and ice nucleation active bacteria (△) and ice nuclei active at -5 C (□) or -9 C (◇) on leaves of untreated improved Thompson navel orange grown near Exeter, CA.   The vertical bars represent the standard error of the determination of the mean of log populations.

100 ice nuclei active at -9 C per gram fresh weight of potato tissue was observed throughout the growing season.   These nuclei may be of soil origin since, unlike most California soils examined, soil in the potato plot area contained abundant nuclei active at -9 C.

Waxy-leaved subtropical species such as navel orange and avocado were found to harbor very low numbers of INA bacteria on their leaves (Fig. 3 and 4, respectively).   Under California growing conditions these plants are subject to very hot and dry weather during much of the growing season (May to October). INA bacterial populations were very low duing this period (less than 3 cells/g fresh weight of leaf tissue).   However, populations of INA bacteria increased to approximately 100 cells/g fresh weight during winter and early spring months coinciding with the period of maximum frost hazard.   Numbers of ice nuclei active at -5 C or higher were undetectable (less than 2 nuclei/g fresh weight) until after December and increased to detectable levels during the period of maximum populations of INA bacteria (Fig. 3 and 4).   Prior to December 15, P. syringae and E. herbicola comprised nearly equal proportions of the isolated INA bacteria.   After December 15, P. syringae was the

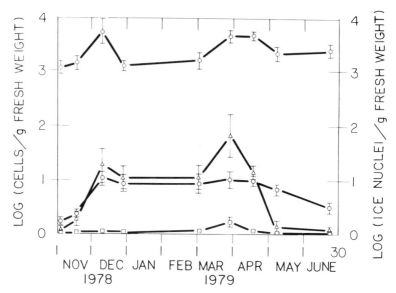

FIG. 4.   Total (0) and ice nucleation active bacteria (△)
and ice nuclei active at -5 C (□) or -9 C (◇) on leaves of
untreated avocado grown near Exeter, CA.   The vertical bars
represent the standard error of the determination of the mean
of log populations.

predominant INA bacteria on both navel orange and avocado.
Pseudomonas fluorescens was rarely found on either navel orange
or avocado.

Large seasonal variations in populations of INA bacteria
were observed on the deciduous fruit trees, Bartlett pear and
almond (Fig. 5, 6). Whereas the total number of bacteria on
these trees remained nearly constant during stages of maturation ranging from bud break to following fruit set, the number
of INA bacteria increased rapidly following bud break from 100
cells/g fresh weight to approximately 10^4 and 10^6 cells/g
fresh weight on pear and almond within a month, respectively
(Fig. 5 and 6). Populations of INA bacteria observed on almond
are the highest yet observed on any wild or agricultural
plant. Hot dry weather occurred in the almond growing area
starting in late April and was associated with the decreases in
INA bacterial populations seen on almond after mid-April (Fig.
6). Large increases in the number of ice nuclei active at -5 C
present on almond and pear foliage were also observed one month
after bud break starting in early March or early April for
almond and pear, respectively (Fig. 5 and 6). Populations of
INA bacteria (exclusively P. syringae) and ice nuclei active at
-5 C generally increased and decreased quantitatively with time
in a similar fashion on both pear and almond.

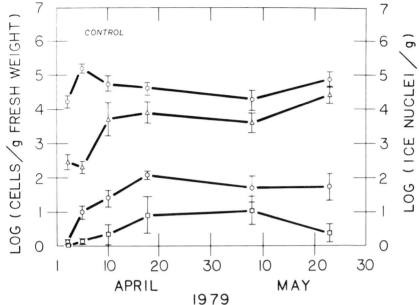

FIG. 5. Total (O) and ice nucleation active bacteria (△) and ice nuclei active at -5 C (□) or -9 C (◇) on leaves of untreated Bartlett pear grown near Lakeport, CA. The vertical bars represent the standard error of the determination of the mean of log populations.

Thus, large seasonal variations in the populations of INA bacteria were observed on all of the agricultural plants examined. Critical time periods exist when relatively low numbers of INA bacteria are present on these plants. Since frost injury at a given temperature has been shown to be related directly to the number of INA bacteria on these plants (19), procedures that reduce or eliminate increases in INA bacterial populations during periods of frost hazard could reduce the frost sensitivity (the chances of frost injury at a given temperature) of plants compared to plants which would otherwise support large epiphytic population of INA bacteria. Certain non-INA bacteria can effectively colonize plant surfaces and possibly produce antibiotic-like substances that limit the ability of INA bacteria to colonize the same surfaces. In this study, certain non-INA bacteria selected from pear leaf surfaces were evaluated for protective activity to reduce buildup of P. syringae populations by application to pear flowers and young leaves at approximately 20% bloom when minimum populations of INA bacteria were detected (Fig. 5). Populations of INA bacteria and numbers of ice nuclei active at -5 C and -9 C were reduced over 100-fold on trees treated with the antagonistic bacterium A506 (identified as Pseudomonas fluorescens

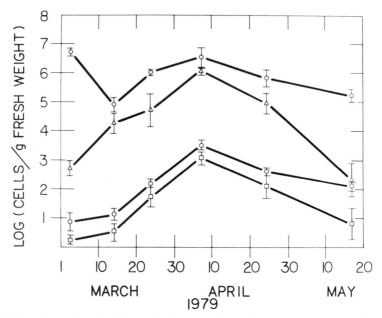

FIG. 6.   Total (O) and ice nucleation active bacteria (△) and ice nuclei active at -5 C (□) or -9 C (◇) on leaves of untreated almond grown near Turlock, CA. The vertical bars represent the standard error of the determination of the mean of log populations.

biotype A) (Fig. 7) compared to untreated trees (Fig. 5). Antagonist A506 was very effective at colonizing flower and leaf surfaces of pear.   Total populations of bacteria were significantly higher on trees treated with antagonist A506 for all times after a single application of this bacterium (Fig. 7). Antagonist A506 was the predominant bacterium found on treated trees, comprising over 70% of the total bacteria found on these trees for over 30 days following inoculation (Fig. 7).   Antagonistic bacteria were also applied to pear trees at the same time in a warmer growing area where trees were more advanced phenologically (70% bloom).   The number of INA bacteria on untreated trees in this area was much higher in early April than on the less advanced trees of the first area, and because of its much warmer and drier climate, began to decline by late April (Fig. 8).   Many antagonistic bacteria such as A501, B7, and A505 also readily colonized pear tissue which emerged subsequent to inoculation in these more mature trees (Fig. 9, 10 and 11, respectively).   Each of these antagonistic bacteria comprised the majority of the total bacteria present on treated trees for at least 30 days following inoculation (Fig. 9-11). Populations of INA bacteria and numbers of ice nuclei active at

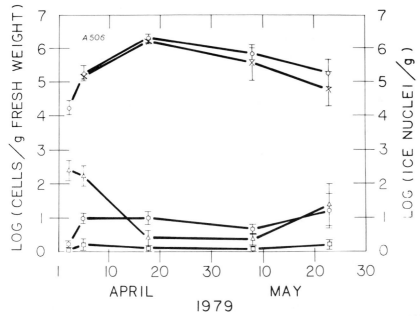

FIG. 7.   Total (O) and ice nucleation active bacteria (△) and ice nuclei active at -5 C (□) or -9 C (◇) on leaves and flowers of Bartlett pear treated with the antagonistic bacter- ium A506 (X) grown near Lakeport, CA.   The vertical bars repre- sent the standard error of the determination of the mean of log populations.

-5 C and -9 C were reduced 10 to 100-fold on trees treated with either of these antagonistic bacteria (Fig. 9-11).

In experiments where large populations of INA bacteria were found on trees at the time of inoculation, the large increase in mass of pear foliage following inoculation and the efficient colonization of this new foliage by antagonistic bacteria pre- cluded efficient colonization of this new tissue by INA bacteria, thus reducing the average population density of INA bacteria on leaf surfaces.   However, these reductions in INA bacterial populations (Fig. 9-11) were not as large as those observed on trees with initially lower populations of INA bac- teria (Fig. 8), suggesting that antagonistic bacteria will be most effective in controlling populations of INA bacteria when used as a protectant before significant leaf surface popula- tions occur.

Frost injury to immature pear fruit in a mild radiative frost of -3 C was significantly reduced on trees treated at 20% bloom with various antagonistic bacteria compared with non- treated trees (Table 1).   Antagonistic bacteria such as A510 and A506 which colonized pear leaves and flowers most effec- tively were also most effective in reducing frost injury.

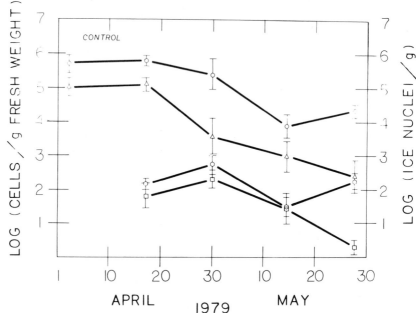

FIG. 8. Total (O) and ice nucleation active bacteria (△) and ice nuclei active at -5 C (□) or -9 C (◇) on leaves and flowers of untreated Bartlett pear grown near San Jose, CA. The vertical bars represent the standard error of the determination of the mean of log populations.

TABLE I. Frost injury to immature Bartlett pear fruit following a radiative frost of -3 C.

Treatment	Injury (fraction of fruit)
A510	0.12
A509	0.18
A507	0.27
A506	0.33
A508	0.51
Control	0.95
LSD 5%	0.12

Antagonistic bacteria were applied to pear trees at 20% bloom near Lakeport, CA, approximately 3 weeks prior to freezing temperatures. Injury represents the fraction of fruit with internal discoloration accertained by longitudinal disection of fruit harvested 1 week following freezing temperatures.

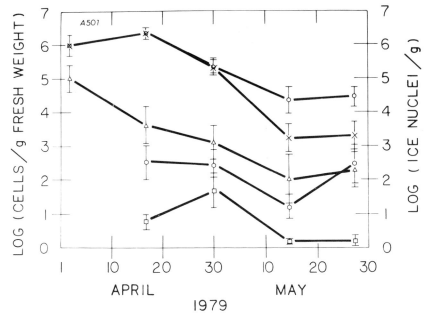

FIG. 9. Total (O) and ice nucleation active bacteria (△) and ice nuclei active at -5 C (□) or -9 C (◇) on leaves and flowers of Bartlett pear treated with the antagonistic bacterium A501 (X) grown near San Jose, CA. The vertical bars represent the standard error of the determination of the mean of log populations.

Thus, protective applications of antagonistic bacteria reduced INA bacterial populations sufficiently to achieve significant reduction in frost injury. Because numbers of ice nuclei active at -5 C or higher were also reduced by bacterial antagonists, a significant correlation was found between frost injury to pear fruit in this plot area and the logarithm of the number of ice nuclei/g fresh weight present on these trees (Fig. 12).

Knowledge of the nucleation frequency (the fraction of bacterial cells active in ice nucleation at a given temperature) of bacteria in vivo will be important in assessing the frost sensitivity of plants even if the numbers of INA bacteria present on leaves can be determined. The probability of frost injury at a given temperature appears to be related to the number of ice nuclei present (Fig. 12), not necessarily the number of bacteria present. Knowledge of the nucleation frequency of INA bacteria as a function of temperature while on leaves will be of critical importance in predicting the number of ice nuclei on plants based on measurements of bacterial populations and therefore in assessing frost control procedures attempting to enhance the supercooling ability of plants.

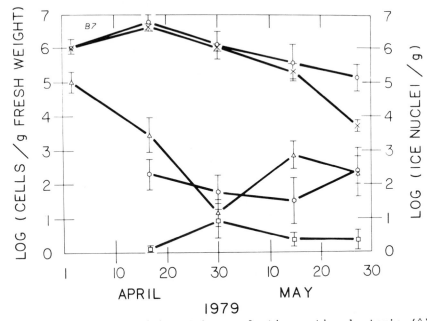

FIG. 10. Total (O) and ice nucleation active bacteria (△) and ice nuclei active at -5 C (□) or -9 C (◇) on leaves and flowers of Bartlett pear treated with bacterium B7 (X) grown near San Jose, CA. The vertical bars represent the standard error of the determination of the mean of log populations.

An estimate of the nucleation frequency of INA bacteria on several plant species under field conditions can be obtained from analysis of Figures 1-6. Because of relatively large uncertainties of the absolute value of populations of either INA bacteria or ice nuclei active at -5 C, only average nucleation frequencies can be calculated. Nucleation frequencies of approximately 100 and 300 cells/ice nucleus at -5 C were observed during the growing season on potato and tomato, respectively (Fig. 1 and 2). Because of the very low levels of ice nuclei active at -5 C on either navel orange or avocado, calculations of nucleation frequency on these hosts is difficult. The nucleation frequency of INA bacteria was approximately 100 cells/ice nucleus on both hosts for those dates where significant numbers of ice nuclei were detected (Fig. 3 and 4). The nucleation frequency on both pear and almond were lower, i.e., approximately 4000 and 650 cells/ice nucleus at -5 C on pear and almond, respectively (Fig. 5 and 6). These estimates of nucleation frequency may be overestimates because measurement of only viable INA bacteria were made. Should non-viable INA bacteria retain significant ice nucleation activity, their contribution to the total number of ice nuclei present on

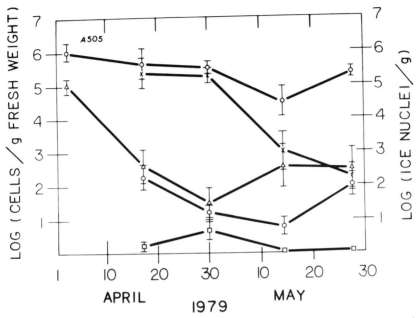

FIG. 11. Total (O) and ice nucleation active bacteria (△) and ice nuclei active at -5 C (□) or -9 C (◇) on leaves and flowers of Bartlett pear treated with bacterium A505 (X) grown near San Jose, CA. The vertical bars represent the standard error of the determination of the mean of log populations.

leaf surfaces may be important. No estimate can yet be made of the proportion of viable and nonviable INA bacteria on plants. Some evidence is available from Fig. 6 indicating that non-viable INA bacteria may efficiently retain their ice nucleation activity at -9 C for 2 or more weeks. Although viable INA bacterial populations and ice nuclei active at -5 C reached a maximum in mid-April and then declined, the numbers of ice nuclei active at -9 C had increased with increasing numbers of INA bacteria until mid-April suggesting that these ice nuclei were of bacterial origin. It is clear that nucleation frequencies of INA bacteria differ from plant to plant. This discrepancy may represent different environmental conditions found in different growing areas or different physical, chemical or nutritional environments on leaf surfaces of different hosts. INA bacterial populations under field conditions are composed of a mixture of different bacterial species and different strains of a given species. Large differences in in vitro nucleation frequency between strains of P. syringae and E. herbicola and among strains of P. syringae have been observed. Thus, genetic differences in the composition of INA bacterial

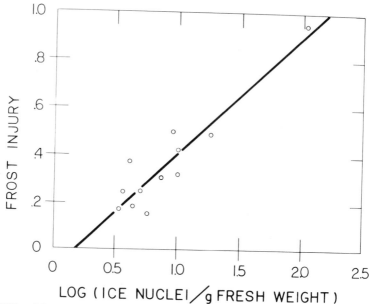

FIG. 12. Relationship between the logarithm of ice nuclei active at -5 C and frost injury at -3 C. The fraction of immature (1 cm diameter) fruit with internal discoloration from trees treated with antagonistic bacteria or bactericides prior to freezing. The line drawn represents the linear regression, y = 0.48X - .071, r = 0.903, P 0.05.

populations may also play an important role in determining nucleation frequency on leaf surfaces.

Most P. syringae strains differ significantly from those of E. herbicola in their ice nucleation frequency in vivo (Fig. 13 and 14, respectively). Although the threshold temperature for ice nucleation of P. syringae was only slightly warmer than that for E. herbicola (-1.7 C vs. -2.0 C, respectively), the nucleation frequency of P. syringae strain 31 rifl at -5 C under greenhouse conditions in vivo was approximately 100 cells/ice nucleus (Fig. 13), thus closely resembling the nucleation frequency measured under field conditions. The nucleation frequency of E. herbicola strain 26SR6-2 was only approximately 1000 cells/ice nucleus at -5 C (Fig. 14). The nucleation frequency of both P. syringae and E. herbicola increased over 1000-fold from -2 C to -5 C but then increased slowly with decreasing temperatures below -5 C. Therefore, the freezing temperature of plant tissues is expected to be highly dependent upon the number of INA bacteria present on those plants and, more importantly, the efficiency of these INA bacteria in nucleating ice. The nucleation efficiency of macerated corn tissue not colonized with INA bacteria was also found to increase rapidly with decreasing temperature (Fig. 15).

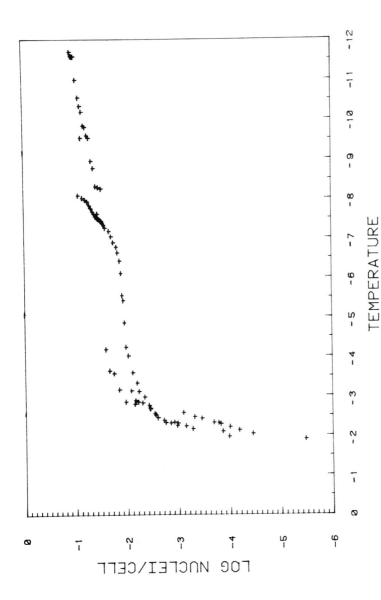

FIG. 13. Ice nucleation efficiency of Pseudomonas syringae as a function of temperature on corn leaf surfaces. The logarithm of the cumulative number of ice nuclei per P. syringae cell removed by vigorous washing of corn leaves in sterile water as a function of temperature is given.

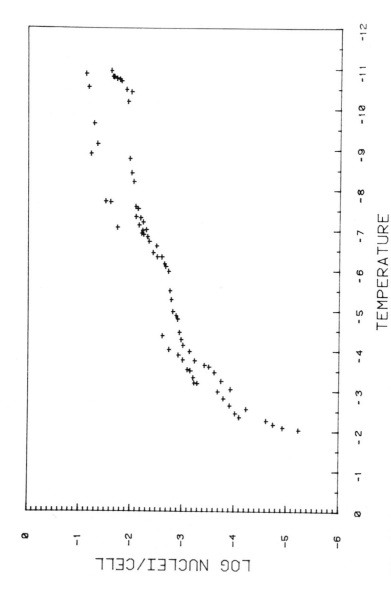

FIG. 14. Ice nucleation efficiency of Erwinia herbicola cells as a function of temperature on corn leaf surfaces. The logarithm of the cumulative number of ice nuclei per E. herbicola cell removed by vigorous washing in sterile water from corn leaves is given.

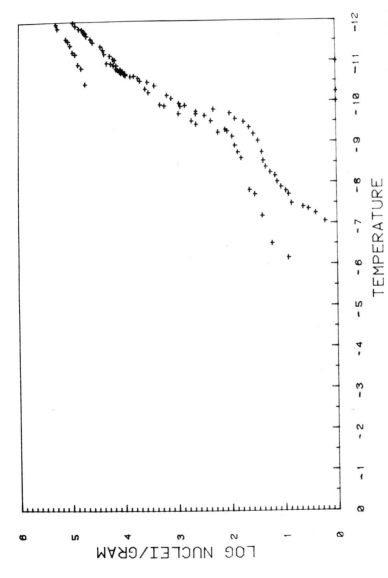

FIG. 15. Cumulative number of ice nuclei as a function of temperature in macerated greenhouse-grown corn leaves not colonized by ice nucleation active bacteria.

However, the threshold temperature of nucleation activity was approximately -6.0 C. Thus, at temperatures above -6 to -7 C, the probability of ice nucleation and thus the chances of frost injury to a given plant part will be determined by the mass of the plant part, the numbers of INA bacteria per unit plant mass, and because of its influence upon nucleation frequency, the minimum temperature of the plant part.

In summary, large seasonal variations in populations of INA bacteria occurred on all plant species examined. This finding may explain differences in frost sensitivity (limited super-cooling ability) that have been reported for certain tender plants such as deciduous fruit trees at different phenological stages following dormancy. Because of the nonubiquity of INA bacteria, it is important to assess accurately the populations and particularly the numbers of ice nuclei contributed by these bacteria in any studies of the frost sensitivity of such plants. Better assessment of populations of INA bacteria and their nucleation frequency on leaves of plants such as Solanum species (as reported here) which are capable of limited hardi-ness (i.e., to -3.5 C) will be important. In contrast to com-pletely tender plant species for which nuclei contributed by INA bacteria should be avoided, INA bacteria may play an impor-tant role in initiating ice formation at only small degrees of supercooling on plants with limited hardiness, allowing them to tolerate extracellular freezing to their hardiness potential (15). The low populations of INA bacteria (and thus ice nuclei) present on Citrus species and avocado may account for the greater degree of supercooling observed in these plants (5,7,44-48) compared with other plants for which little super-cooling has been observed (32). Success in altering the popu-lation dynamics of INA bacteria with antagonistic bacteria as reported here attest to the possibility of significant frost control by this procedure. Antagonistic bacteria may prove to be excellent agents for enhancing the supercooling ability of tender plants and thus reducing frost injury since they rapidly colonize plant tissue and do not require repeated inoculations for successful colonization. The observed differences in ice nucleation efficiency of INA bacteria in vivo point out the importance of knowledge of nucleation frequency of a variety of INA bacteria on leaf surfaces and the need for further knowledge of factors that affect nucleation frequency in vivo.

ACKNOWLEDGMENTS

   I gratefully acknowledge the valuable technical assistance
of Mr. D. Dahlbeck and Mr. D. Haefele. This work was supported
in part by the Almond Board of California, the California
Potato Research Advisory Board, the California Tomato Research
Advisory Board, Pear Zone 1, and the California Citrus Research
Advisory Board.

REFERENCES

1. Arny, D.C., Lindow, S.E., and Upper, C.D., Nature
   262:282-284 (1976).
2. Burke, M.J., Gusta, L., Quamme, H.A., Weiser, C.J., and Li,
   P.H., Annu. Rev. Plant Physiol. 27:507-528 (1976).
3. Cary, J.W., and Mayland, H.F., Agron. J. 62:715-719 (1970).
4. Chandler, W.H., Proc. Am. Soc. Hort. Sci. 64:552-572 (1958).
5. Gerber, J.F., and Hashemi, F., Proc. Am. Soc. Hort. Sci.
   86:220-225 (1965).
6. Head, G.C., J. Hort. Sci. 34:1-6 (1959).
7. Hendershott, C.H., Proc. Am. Soc. Hort. Sci. 80:247-254
   (1962).
8. Kaku, S., Bot. Mag. 77:283-289 (1964).
9. Kaku, S., Bot. Mag. 79:98-104 (1966).
10. Kaku, S., Plant Cell Physiol. 12:147-155 (1971).
11. Kaku, S., Plant Cell Physiol. 14:1035-1038 (1973).
12. Kaku, S., Cryobiology 12:154-159 (1975).
13. King, E.O., Ward, M.K., and Raney, D.E., J. Lab. Clin. Med.
   44:301-307 (1954).
14. Levitt, J., "Responses of Plants to Environmental
   Stresses". Academic Press, New York, 1972.
15. Li, P.H., Rajashekar, C., and Carter, J.V., Plant Physiology 67:62 (1981).
16. Lindow, S.E., Arny, D.C., and Upper, C.D., Proc. Am. Phytopathol. Soc. 4:169 (1977).
17. Lindow, S.E., Arny, D.C., and Upper, C.D., Proc. Am. Phytopathol. Soc. 4:169-170 (1977).
18. Lindow, S.E., Arny, D.C., Barchet, W.R., and Upper, C.D.,
   Plant Physiol. 59:4 (1977).
19. Lindow, S.E., Arny, D.C., Barchet, W.R., and Upper, C.D.,
   in "Plant Cold Hardiness and Freezing Stress" (P.H. Li and
   A. Sakai, eds.), pp. 249-263. Academic Press, New York,
   1978.
20. Lindow, S.E., Arny, D.C., Barchet, W.R., and Upper, C.D.,
   Phytopathol. News 12:138 (1978).
21. Lindow, S.E., Arny, D.C., Barchet, W.R., and Upper, C.D.,
   Phytopathol. News 12:138 (1978).

22. Lindow, S.E., Arny, D.C., Barchet, W.R., and Upper, C.D., Phytopathol. News 12:138 (1978).
23. Lindow, S.E., Arny, D.C., and Upper, C.D., Phytopathology 68:523-527 (1978).
24. Lindow, S.E., Arny, D.C., and Upper, C.D., Appl. Environ. Microbiol. 36:831-838 (1978).
25. Lucas, J.W., Plant Physiol. 29:245 (1954).
26. Maki, L.R., Galyon, E.L., Chang-Chien, M., and Caldwell, D.R., Appl. Microbiol. 28:456-460 (1974).
27. Maki, L.R., and Garvey, D.M., E.O.S. Am. Geophys. Union Trans. 56:944 (abstr.) (1975).
28. Marcellos, H., and Single, W.V., Agr. Meteorol. 16:125-129 (1976).
29. Marcellos, H., and Single, W.V., Cryobiology 16:74-77 (1979).
30. Mayland, H.F., and Cary, J.W., Adv. Agron. 22:203-234 (1970).
31. Mazur, P., Annu. Rev. Plant Physiol. 20:419-448 (1969).
32. Modlibowska, I., J. Hort. Sci. 37:249-261 (1962).
33. Olien, C.R., Annu. Rev. Plant Physiol. 18:387-408 (1967).
34. Schnell, R.C., and Vali, G., Nature 236:163-165 (1972).
35. Schnell, R.C., and Vali, G., Nature 246:212-213 (1973).
36. Schnell, R.C., and Vali, G., J. Atmos. Sci. 33:1554-1564 (1976).
37. Shaw, R.H., Plant Physiol. 19:102-104 (1954).
38. Single, W.V., Austr. J. Agric. Res. 12:767-782 (1961).
39. Single, W.V., and Olien, C.R., Austr. J. Biol. Sci. 20:1025-1028 (1967).
40. Sukumaran, N.P., and Weiser, C.J., Plant Physiol. 50:564-567 (1972).
41. Vali, G., and Stansbury, E.J., Can. J. Phys. 44:477-502 (1966).
42. Vali, G., J. Atmos. Sci. 28:402-409 (1971).
43. Vali, G., Christensen, M., Fresh, R.W., Galyon, E.L., Maki, L.R., and Schnell, R.C., J. Atmos. Sci. 33:1565-1570 (1976).
44. Yelenosky, G., and Horanic, G., Cryobiology 5:281-283 (1969).
45. Yelenosky, G., Plant Physiol. 56:540-543 (1975).
46. Yelenosky, G., in "Plant Cold Hardiness and Freezing Stress" (P.H. Li and A. Sakai, eds.), pp. 297-311. Academic Press, New York, 1978.
47. Young, R.H., Proc. Am. Soc. Hort. Sci. 88:272-279 (1966).
48. Young, R.H., Proc. Am. Soc. Hort. Sci. 94:244-252 (1969).

TYPOLOGY OF FREEZING PHENOMENA AMONG VASCULAR PLANTS
AND EVOLUTIONARY TRENDS IN FROST ACCLIMATION

Walter Larcher

Institute of Botany
University of Innsbruck
Sternwartestraße, Innsbruck, Austria

I. INTRODUCTION

The goal of stress physiology is the understanding of basic
processes involved in cell damage and resistance. Since the
essential structures of cells are fundamentally similar, a
valid theory of stress action and cell response is expected to
be a unifying concept. A valid theory, however, should also be
comprehensive, considering diversity in life forms, physiological response patterns, ecological preferences, and survival
mechanisms in various plant species. Diversity is the means by
which evolutive adaptation to the various conditions of life
proceeds. Thus, comparative studies on freezing injury and
resistance mechanisms provide information for analyzing fundamental processes as well as concrete data useful in the application of stress physiology to genetics, phylogeny, ecology,
geobotany, agriculture, horticulture, and forestry.

II. CATEGORIES OF FREEZING SUSCEPTIBILITY

A selected set of examples exhibiting variation in freezing
susceptibility and freezing tolerance is presented in Table 1.
Two basically different categories can be distinguished
according to the concept of Levitt (38,39).

A. Tissues Protected Only by Avoidance Mechanisms

Freezing injury may occur as soon as ice appears in the
plant. Tissues subject to immediate freezing injury, which in
the classical descriptive terminology were called "freezing-

417

sensitive", are protected against frost damage only by freezing
point depression and supercooling. The freezing point depression due to cell sap constituents rarely exceeds 2 to 3 K. It
can be determined by cell sap cryoscopy or by evaluation of the
freezing exotherm of previously killed tissues. The upper
threshold temperature below which extracellular or intracellular ice is formed in living plant tissues can be determined by exotherm plateau detection techniques in leaves and
bark parenchyma (5,25,28,51); in deep supercooling tissues it
can be derived from DTA measurements. Parenchymous plant tissues usually begin to freeze at temperatures between -2 and
-6°C, but specific exotherm peaks may also be found at temperatures as low as -10 to -12°C (cf. Table 1). The supercooling
capacity of plant cells depends on the presence of nucleators,
the cell water content, the degree of adaptation, and upon
histological and cytological peculiarities (cell size and
shape, hydrophobous barriers). Transient supercooling provides
protection against brief radiation frosts to an extent of 3-8 K
below the exotherm peak temperature (13,14,18,40). Flower buds
(9,54), seeds (16) and xylem parenchyma cells of certain plant
species exhibit persistent supercooling: they are able to
remain in the supercooled state during the whole winter. The
xylem parenchyma of many temperate trees supercools to the
point of homogeneous nucleation, freezing only at temperatures
of -40 to -50°C (deep supercooling) (2,3,10). It is worth
noting that specific, reproducible limits for persistent supercooling exist at temperatures well above the homogeneous nucleation point. For example, Kaku and Iwaya (19) found that xylem
parenchyma of subtropical and warm temperate trees supercooled
to -20°C, but not below that temperature. In determining the
specific temperature limit for lethal non-equilibrium freezing
(44) of living cells biomembrane integrity seems to play the
decisive role (15,41,45,49,60,67,68,69).

B.   Freezing-Tolerant Tissues

In tissues with tolerance of equilibrium freezing (44)
freezing injury appears at temperatures lower than the temperature at which ice formation begins. The extent of freezing
tolerance can be expressed by the difference between the temperature at which freezing takes place (highest exotherm peak,
persistent supercooling limit) and the temperature at which
initial injury appears (LT_i). As shown in the table, various
degrees of tolerance can be recognized, ranging from very
specific limits (e.g. leaves of Quercus ilex and Abies alba) up
to unlimited tolerance (shoots of Betula papyrifera). It is
possible to trace the transition from freezing-sensitive to
freezing-tolerant taxa within series of related plants (for
examples see 29).

Table I. Tissue freezing temperatures (T_f) and temperatures causing initial injury (T_i) in unhardened (unh) and hardened (h) plants.

Plant	Temperature below which tissue freezing occurs			Frost resistance			Freezing tolerance	
	T_f unh °C	T_f h °C	Adaptation effect K	LT_i unh °C	LT_i h °C	Adaptation effect K	LT_i h $-T_f$ h K	Ref.
Solanum tuberosum, leaves	-3	-3	0	-3	-3	0	0	6,7
Trachycarpus fortunei, leaves	-10	-11	1	-10	-11	1	0	36
Olea europaea, leaves	-5	-10	5	-5	-10	5	0	25
Quercus ilex, leaves	-5	-8	3	-5	-13	8	5	35
Abies alba, leaves	-4	-7	3	-4	-30	25	23	51
Malus sylvestris, bark	-5	-6	1	-5	-30...-60	25-55	25-55	34,42,64
Malus sylvestris, xylem	-8	-38	30	-8	-30...-35	up to 30	0	34,53,22
Betula papyrifera, xylem		-10			-196	unlimited	unlimited	10

III. CATEGORIES OF FROST HARDENING

Hardening consists in improved freezing-avoidance and/or temporal development of freezing-tolerance. Three categories are to be distinguished.

A. Tissues Without Hardening Capacity

Many extremely freezing-sensitive species such as various C_4 grasses (52) or several Solanum species (46; cf. Table 1) do not acclimate at all when exposed to low temperatures. However, very little hardening capacity may also be observed in species with considerable freezing-avoidance (e.g. leaves of the warm-temperate palm Trachycarpus fortunei; 23,36).

B. Hardening Involves Enhanced Avoidance but Does Not Induce Freezing-Tolerance

Freezing-sensitive plants can only be hardened by lowering the threshold temperature for tissue freezing and increasing the capacity of persistent supercooling. The adaptation effect is 3-5 K in leaves (e.g. Olea europaea, cf. Table 1); it is greatest in tissues, which, whilst still remaining freezing-sensitive, attain deep supercooling in the hardened state.

C. Hardening Induces Freezing-Tolerance

The hardening processes in those plants which become freezing-tolerant include both lowering of the tissue freezing temperature and development of the ability of the protoplasm to release water, permitting extracellular equilibrium freezing. Freezing-tolerance is apparently a temporary state linked with reduced growth activity in perennial herbaceous plants (11,58) and with dormancy in woody plants (50,61,62,64,66). As a rule, freezing-tolerance is lost during intensive growth. There is some evidence, however, that certain high mountain plants may retain freezing-tolerance during the growing season (Pamir: M. Tyurina, pers. comm.; Mt. Kenia: E. Beck and M. Senser, pers. comm.). In Saxifraga oppositifolia and Silene acaulis, collected at high altitudes in the Alps, Ch. Kainmüller (unpubl.) observed maintenance of tolerance only in restricted, probably inactive, regions of the shoot; actively growing parts of the cushions were freezing-sensitive.

Freezing-tolerance can also be induced or enhanced by progressive dehydration due to evaporation or to migration of water during freezing. Desiccation-induced freezing-tolerance has been observed in desiccation-tolerant thallophytes and cormophytes when in an air dry anabiotic state. Certain ferns and phanerogams are damaged by frost at -10 to -20°C when

saturated with water, but they withstand -50°C at 5% relative
water content (Fig. 1) (20). Also dry seeds (59) and pollen
possess unlimited freezing-tolerance. By freeze-dehydration
due to water translocation from supercooled tissues to nucleation centers in adjacent tissues (extra-tissue freezing) or at
surrounding surfaces (extraorgan freezing, as defined by Sakai
cf. 55), a high degree of freezing-tolerance can be attained in
buds and seeds: In freezing winter buds water migrates from
the apical meristem to the bud scales and into intercellular
spaces at the basal parenchyma (8,55,56). Segregation of water
in certain seeds from embryonal meristems to distal parts
during slow freezing has been described by Keefe and Moore (21)
and Ishikawa and Sakai (17). In woody stems, too, it is to be
expected that water is redistributed between cambium cells and
the adjacent cortical parenchyma containing large intercellular
spaces. Such a mechanism could explain the observation that in
frost-hardened woody plants the cambial zone becomes the most
freezing-tolerant tissue of the stem (33,34).

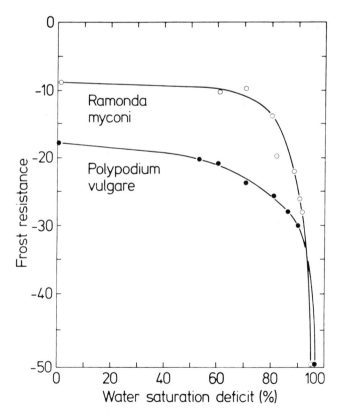

FIGURE 1. Desiccation-enhanced freezing tolerance of leaves
of drought-tolerant vascular plants. After Kappen (20).

IV.   EVOLUTION OF RESISTANCE GRADIENTS ALONG ALTITUDINAL
AND LATITUDINAL STRESS GRADIENTS

Gradual changes in the level and the mechanism of cold resistance in plants growing in different climatic zones indicate that the capability of acquiring increased cold hardiness may have developed in a stepwise manner along stress gradients (31).

Evolutionary adaptation processes, which involve biomembrane alterations, most certainly have developed in regions bordering the tropics and especially on tropical mountains. Among chilling-sensitive vascular plants acclimation series have been found in species of Passiflora occupying distinct altitudinal belts (47) and in high altitudinal races of Lycopersicon hirsutum (48). A transition from chilling-sensitive species (Solanum trifidum) to chilling-tolerant, but extremely freezing-sensitive species (e.g. Solanum commersonii) has been recognized by Palta and Li (46) and Chen and Li (6) among potato relatives with different altitudinal distribution. It may be possible to demonstrate transitional forms even in woody

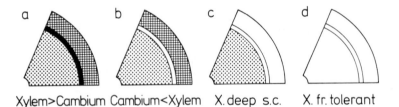

Xylem>Cambium   Cambium<Xylem   X. deep s.c.   X. fr. tolerant

FIGURE 2.   Evolutionary trends in freezing resistance of stem tissues.   Dark: Most susceptible tissue; white: most resistant tissue.

a: All stem tissues are freezing-sensitive, xylem parenchyma supercool persistently; injuries occur between -7 and -15°C. Examples: Ceratonia, Laurus, Nerium.

b: Cambium capable of developing limited freezing-tolerance when inactive; xylem and cortex remain freezing-sensitive even in winter.   Injuries occur between -15 and -20°C.   Examples: Citrus, Quercus ilex.

c: Cambium and cortex are freezing-tolerant during dormancy, xylem survives by deep-supercooling (down to -50°C). Examples: Many deciduous trees of the temperate zone, e.g. all deciduous Quercus species.

d. All stem tissues become freezing-tolerant during dormancy. Examples: Conifers and certain deciduous woody plants (e.g. Betula, Populus, Ribes) of high latitudes and altitudes.

plants of the tropics, whose seedling stages and leaves are chilling-sensitive but whose woody tissues have developed chilling resistance. Whether such species exist is still not known, since essentially all experimental research on tropical species to date has been limited to herbaceous species or to the leaves of woody plants.

Among freezing-sensitive woody plants there are several examples of altitudinal resistance gradients and ecotype differentiation (12,37,57,65). Since in freezing-sensitive plants any improvement of frost resistance is to be attributed to avoidance mechanisms, both the lowering of the tissue freezing temperature (mainly produced by accumulation of soluble carbohydrates) and the development of persistent supercooling (which probably depends on the inoculation barrier function of cell walls and biomembanes) are to be considered as effective acclimation processes. The large short-term fluctuation in temperature and the increasing frequency of low night temperatures at higher altitudes can act as a stimulus for metabolic adaptations which result in osmoregulative freezing point depression (4,30,63).

A limited capacity to become freezing-tolerant may have first evolved in evergreen woody plants of subtropic and warm temperate regions, where winter drought due to reduced water uptake from cold soils can induce a cessation of growth. Among the various organs and tissues of woody plants the first to become freezing-tolerant are inactive meristems, whereas leaves, roots, and the xylem parenchyma of shoots are the last (Fig. 2) (26,27). High freezing-tolerance can be attained only during deep dormancy. The evolution of perfect tolerance was therefore dependent on a precise synchronization of growth rhythms with climatic periodicity. In many species the timing mechanism is triggered by photoperiodic and/or thermoperiodic signals. This kind of acclimation could have evolved only in regions where unfavorable climatic conditions (low temperatures, drought) occur in definite seasons.

V. ECOLOGICAL SIGNIFICANCE OF THE DIVERSITY IN FREEZING RESISTANCE MECHANISMS

Evolution is promoted by genetic diversification and screening for fitness. Thus each step in low temperature stress can be an optimization for a given environment.

Freezing-avoidant plants appear especially adapted to habitats where slight frosts occur during periods of metabolic and developmental activity. Freezing-avoidance, which can be improved after a few days of cold weather, is a prompt and in most cases sufficiently efficient protection against episodic frosts which occur in warm temperate regions during winter, in extratropical mountains in summer, and in tropical mountains throughout the year. On the other hand, avoidance mechanisms ensure maintenance of CO_2-uptake at low temperatures (24,29,32,51) and permit dry matter production even after moderate night frost.

Where frost is a regularly recurring severe and long-lasting threat to plant life as in high latitudes and in central parts of Eurasia and North America with a continental climate, <u>dormancy linked freezing-tolerance</u> is the only effective survival mechanism. Plant families such as the palms, in which dormancy and freezing-tolerance did not evolve, are prevented from extending their natural geographical ranges to latitudes higher than 40° (30). However, since deep and stable dormancy consists in the depression not only of growth but also of metabolic activity, photosynthesis of evergreen plants in cold-temperate regions is strongly reduced during winter (33).

ACKNOWLEDGMENT

I am grateful to W. H. Hatheway for stimulating discussions and for his help with the translation of the manuscript.

REFERENCES

1. Ashton, D.H., Aust. J. Bot. 6:154-176 (1958).
2. Burke, M.J., Gusta, L.V., Quamme, H.A., Weiser, C.J., and Li, P.H., Ann. Rev. Plant Physiol. 27:507-528 (1976).
3. Burke, M.J., and Stushnoff, C., in "Stress Physiology in Crop Plants" (H. Mussell, R. Staples, eds.), p. 197-225. Wiley and Sons, New York, 1979.
4. Cappelletti, C., Ann. di Botanica 19:278-332 (1931).
5. Cernusca, A., and Vesco, A., Cryobiology 13:638-644 (1976).
6. Chen, P.M., and Li, P.H., Plant Physiol. 65:1146-1148 (1980).
7. Chen, P.M., Burke, M.J., and Li, P.H., Bot. Gaz. 137:313-317 (1976).
8. Dereuddre, J., Physiol. Vég. 16:469-489 (1978).
9. George, M.F., Burke, M.J., and Weiser, C.J., Plant Physiol. 54:29-35 (1974).
10. George, M.F., Burke, M.J., Pellett, H.M., and Johnson, A.G., HortSci. 9:519-522 (1974).
11. Gusta, L.V., and Fowler, D.B., in "Stress Physiology in Crop Plants" (H. Mussell, R. Staples, eds.), pp. 159-178. Wiley and Sons, New York, 1979.
12. Harwood, C. E., Aust. J. Bot. 28:587-599 (1980).
13. Hatakeyama, I., Biol. J. Nara Women's Univ. 10:65-69 (1960).
14. Hatakeyama, I., Mem. Coll. Sci. Univ. Kyoto. Ser. B 28:402-429 (1961).
15. Horvath, I., Vigh, L., Belea, A., and Farkas, I., Physiol. Plant. 45:57-62 (1979).
16. Ishikawa, M., and Sakai, A., Low Temp. Sci. Ser. B36:39-49 (1978).

17. Ishikawa, M., and Sakai, A., in "Plant Cold Hardiness and Freezing Stress", Vol. 2 (P.H. Li, and A. Sakai, eds.) Academic Press, New York (1982).
18. Kaku, S., Cryobiology 12:154-159 (1975).
19. Kaku, S., and Iwaya, M., in "Plant Cold Hardiness and Freezing Stress" (P.H. Li, and A. Sakai, eds.), pp. 227-239. Academic Press, New York, 1978.
20. Kappen, L., Flora 156:427-445 (1966).
21. Keefe, P.D., and Moore, K.G., Ann. Bot. 47:635-645 (1981).
22. Krasavtsev, O.A., and Khvalin, N.N., Fiziol. rast. 25:5-11 (1978).
23. Larcher, W., Planta 44:607-638 (1954).
24. Larcher, W., Planta 56:575-606 (1961).
25. Larcher, W., Protoplasma 57:569-587 (1963).
26. Larcher, W., Oecol. Plant. 5:267-286 (1970).
27. Larcher, W., Oecol. Plant. 6:1-14 (1971).
28. Larcher, W., in "Methoden der Pflanzenökologie" (K. Kreeb, ed.), pp. 51-59. VEB Fischer, Jena, 1977.
29. Larcher, W., Rhein. Westf. Akad. Wiss. Vortr. N291:49-88 (1980).
30. Larcher, W., Anz. Österr. Akad. Wiss. Math. Natw. Kl. 1980, 1-12 (1980).
31. Larcher, W., Plant Syst. Evol. 137:145-180 (1981).
32. Larcher, W., in "Components of Productivity of Mediterranean Regions, Basic and Applied Aspects" (N.S. Margaris, and H.A. Mooney, eds.). Junk, The Hague (1981).
33. Larcher, W., and Bauer, H., in "Encyclopedia of Plant Physiology", Vol. 12A, pp. 402-437. Springer, Berlin (1981).
34. Larcher, W., and Eggarter, H., Protoplasma 51:595-619 (1960).
35. Larcher, W., and Mair, B., Oecol. Plant. 4:347-376 (1969).
36. Karcher, W., and Winter, A., Principes 25:143-152 (1981).
37. Layton, C., and Parsons, R.F., Bull. Torrey Club 99:118-122 (1972).
38. Levitt, J., Frost, Drought and Heat Resistance. Protoplasmatologia VIII/6. Springer, Wien, 1958.
39. Levitt, J., Responses of Plants to Environmental Stresses. Vol. 1. Chilling, Freezing, and High Temperature Stresses. 2nd ed. Academic Press, New York, 1980.
40. Marcellos, H., and Single, W.V., Cryobiology 16:74-77 (1979).
41. Melcarek, P.K., and Brown, G.N., Cryobiology 16:69-73 (1979).
42. Mittelstädt, H., Dtsch. Akad. Landw. Wiss. Tagungsbericht 96:149-173 (1969).
43. Nath, J., and Anderson, J.O., Cryobiology 12:81-88 (1975).
44. Olien, C.R., in "Plant Cold Hardiness and Freezing Stress" (P.H. Li, and A. Sakai, eds.), pp. 37-48. Academic Press, New York, 1978.
45. Palta, J.P., and Li, P.H., in "Plant Cold Hardiness and Freezing Stress" (P.H. Li, and A. Sakai, eds.), pp. 93-115. Academic Press, New York, 1978.

46. Palta, J.P., and Li, P.H., Crop Sci. 19:665-671 (1979).
47. Patterson, B.D., Murata, T., and Graham, D., Aust. J. Plant Physiol. 3:435-442 (1976).
48. Patterson, B.D., Paull, R., and Smillie, R.M., Aust. J. Plant Physiol. 5:609-617 (1978).
49. Pike, C.S., Berry, J.A., and Raison, J.K., in "Low Temperature Stress in Crop Plants" (J.M. Lyons, D. Graham, and J.K. Raison, eds.), pp. 305-318. Academic Press, New York, 1979.
50. Pisek, A., and Schießl, R., Ber. Nat.-wiss. med. Ver. Innsbruck 47:33-52 (1947).
51. Pisek, A., Larcher, W., and Unterholzner, R., Flora B 157:239-264 (1967).
52. Rowley, J.A., Tunnicliffe, C.G., and Taylor, A.O., Aust. J. Plant Physiol. 2:447-451 (1975).
53. Quamme, H.A., Can. J. Plant Sci. 56:493-500 (1976).
54. Sakai, A., HortSci. 14:69-70 (1979).
55. Sakai, A., Plant Cell Physiol. 20:1381-1390 (1979).
56. Sakai, A., in "Plant Cold Hardiness and Freezing Stress", Vol. 2 (P.H. Li, and A. Sakai, eds.). Academic Press, New York, 1982.
57. Sakai, A., and Wardle, P., New Zealand J. Ecol. 1:51-61 (1978).
58. Sikorska, E., and Kacperska-Palacz, A., Physiol. Plant. 47:144-150 (1979).
59. Stanwood, P.C., and Bass, L.N., in "Plant Cold Hardiness and Freezing Stress" (P.H. Li, and A. Sakai, eds.), pp. 361-371. Academic Press, New York, 1978.
60. Steponkus, P.L., Dowgert, M.F., Evans, R.Y., and Gordon-Kamm, W.J., in "Plant Cold Hardiness and Freezing Stress", Vol. 2 (P.H. Li, and A. Sakai, eds.). Academic Press, New York, 1982.
61. Tumanov, I.I., Proc. Int. Hort Cgr. Bruxelles, 737-743 (1962).
62. Tumanov, I.I., Fiziologiya zakalivaniya i morozostoikosti rastenii. Izdat. Nauka, Moskva (1979).
63. Tyurina, M.M., Issledovanie morozostoikosti rastenii v usloviyakh vysokogorii Pamira. Izdat. Akad. Nauk. Tadzhik. SSR (1957).
64. Tyurina, M.M., Gogoleva, G.A., Jegurasdova, A.S., and Bulatova, T.G., Acta Hort. 81:51-60 (1978).
65. Wardle, P., and Campbell, A.D., Proc. N.Z. Ecol. Soc. 23:85-91 (1976).
66. Weiser, C.J., Science 169:1269-1278 (1970).
67. Yoshida, S., Contr. Inst. Low Temp. Sci. Sapporo, B 18:1-43 (1974).
68. Yoshida, S., in "Plant Cold Hardiness and Freezing Stress" (P.H. Li, and A. Sakai, eds.), pp. 117-135. Academic Press, New York, 1978.
69. Yoshida, S., Plant Physiol. 64:252-256 (1979).

FREEZING RESISTANCE OF TEMPERATE DECIDUOUS FOREST PLANTS
IN RELATION TO THEIR LIFE FORM AND MICROHABITAT

F. Yoshie
A. Sakai

The Institute of Low Temperature Science
Hokkaido University
Sapporo, Japan

I. INTRODUCTION

Woody plants from different climates exibit large and gradual differences in freezing resistance which correspond well
with the winter cold of their natural distribution ranges. In
species with a wide distribution range, intraspecific differences in freezing resistance are also evident. Almost all of
the provenances from colder climates were more hardy than from
warmer climates. These facts suggest that low winter temperatures rank high among the natural selection pressures that have
led to evolution of ecotypes and species adapted to cold climates. On the other hand, even in a smaller area, plants are
exposed to different low temperature stress depending on their
life form and microhabitat distribution, and may differ in
freezing resistance (1,7).
    Till (7) first determined the freezing resistance of vari-
ous organs of trees and understory plants of a temperate deciduous forest in western Europe. Although a clear relationship
between freezing resistance and frost exposure of winter buds
was demonstrated, subterranean organs showed little or no difference in winter hardiness. However, freezing resistance was
assessed with wintering plants without artificial hardening
treatment. Since the degree of freezing resistance in winter
depends on the low temperatures to which the plants were previously exposed (4), it is necessary to harden plants at low
temperatures in order to determine the full winter survival
capacity of species varying in life form and microhabitat distribution. In this paper, the freezing resistance of 48
species in a deciduous forest was assessed with and without

Table I. Freezing resistance of dormant buds, leaves, rhizomes
and roots of plants growing in Tomakomai on April 9,
1980.
* : Injured at that temperature.

| SPECIES | FREEZING RESISTANCE (°C) | | | |
	DORMANT BUDS	LEAVES	RHIZOMES	ROOTS
PHANEROPHYTES				
Quercus mongolica var. grosseserrata	-20.0			
Juglans ailanthifolia	-30.0			
Carpinus cordata	-17.5			
Cornus controversa	-25.0			
Cercidiphyllum japonicum	-25.0			
Fraxinus sp.	-25.0			
Tilia maximowicziana	-30.0			
Magnolia kobus var. borealis	-25.0			
Sorbus alnifolia	-25.0			
Acer mono var. mayrii	-15.0			
Acer palmatum var. amoenum	-25.0			
Hydrangea petiolaris	-20.0			
CHAMAEPHYTES (FOREST FLOOR)				
Daphne pseudo-mezereum var. jezoensis	-20.0	-7.5	-	-5.0*
Pachysandra terminalis	-20.0	-20.0	-7.5	-5.0*
Euonymus fortunei var. radicans	-20.0	-20.0	-	-10.0
Lycopodium obscurum	-20.0	-20.0	-	-
Lycopodium serratum var. serratum	-20.0	-15.0/-20.0	-	-10.0
Pyrola secunda	-17.5	-15.0	-5.0	-
Pyrola incarnata	-15.0	-15.0	-7.5	-
Pyrola alpina	-15.0	-12.5	-5.0	-
Pyrola renifolia	-12.5	-12.5	-5.0	-
Oxalis acetosella	-12.5	-7.5	-5.0	-5.0
Tripterospermum japonicum	-15.0	-5.0	-	-
(FOREST MARGIN)				
Lycopodium clavatum var. nipponicum	-25.0	-15.0/-25.0	-	-

Table I continued

	FREEZING RESISTANCE (°C)			
SPECIES	DORMANT BUDS	LEAVES	RHIZOMES	ROOTS
HEMICRYPTOPHYTES				
(FOREST FLOOR)				
Tiarella polyphylla	−10.0	−7.5	−10.0	−5.0
Agrimonia pilosa	−10.0	−	−7.5	−5.0*
Maianthemum dilatatum	−10.0	−	−5.0	−5.0
Chamaele decumbens	−5.0*	−	−5.0*	−5.0*
Sanicula chinensis	−5.0	−	−5.0	−5.0*
(FOREST MARGIN)				
Oenothera sp.	−12.5	−12.5	−	−7.5/−12.5
Miscanthus sinensis	−7.5	−	−7.5	−7.5
Filipendula sp.	−12.5	−	−10.0	−5.0*
Solidago virga-aurea	−12.5	−	−10.0	−10.0
Plantago asiatica	−10.0	−	−10.0	−10.0
Trifolium pratense	−5.0	−	−5.0*	−5.0*
Petasites japonicus var. giganteus	−5.0	−	−5.0*	−5.0*
Leibnitzia anandria	−7.5	−	−7.5	−5.0
Artemisia montana	−10.0	−	−7.5	−7.5
Artemisia japonica	−12.5	−	−12.5	−12.5
Anaphalis margaritacea	−7.5	−	−7.5	−7.5
Sanguisorba tenuifolia var. alba	−7.5	−	−7.5	−5.0
Lysimachia vulgalis var. davurica	−7.5	−	−7.5	−7.5
GEOPHYTES				
(FOREST FLOOR)				
Phryma leptostachya var. asiatica	−5.0*	−	−5.0*	−5.0*
Cacalia delphiniifolia	−5.0*	−	−5.0*	−5.0*
Cacalia auriculata var. kamtschatica	−5.0*	−	−5.0*	−5.0*
Sceptridium multifidum var. robustum	−5.0*	−17.5	−5.0*	−5.0*
Lilium cordatum var. glehnii	−5.0*	−	−5.0* (bulb)	−5.0*
Adoxa moschatellina	−5.0*	−	−5.0*	−5.0*

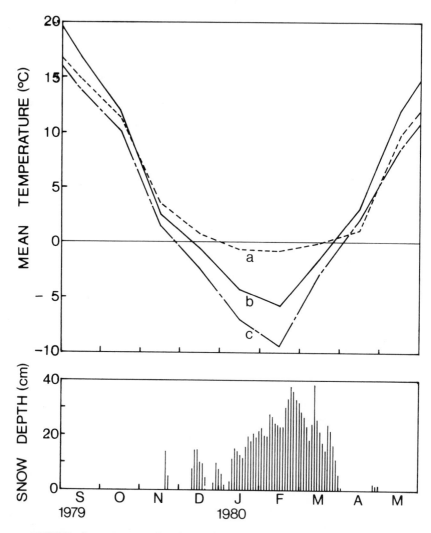

FIGURE 1.   Seasonal changes in mean monthly air temperature, soil temperatures and snow depth based on data from the
Meteorological Station of Hokkaido University Forest in
Tomakomai.   a: at soil surface in the deciduous broad-leaved
forest, b: at soil surface at the bare ground, c: at 1.2 m
height above bare ground.

artificial hardening during the period from October to December 1979 and was assessed without hardening in April 1980.

II. MATERIALS AND METHODS

The samples were collected at the Hokkaido University Forest Station at Tomakomai (N 42°40', E 141° 35'), located about 50 km southeast from Sapporo. This research station is near to the northern distribution limit of temperate deciduous forest in Japan. At this station, a temperate deciduous forest composed mainly of Quercus mongolica var. glosseserrata and Acer mono var. mayrii and plantations of Larix leptolepis and Abies sachalinensis are established. The samples list is given in Table 1, with their hardiness values in early April. Their hardiness values in October, November and December have been reported in a previous study (8).

Five uniform one-year-old twig sections, 10 cm long, of trees and five herbaceous plants, were enclosed in polyethylene bags and then progressively cooled at 5°C steps with 2 to 3 hr intervals until the selected test temperatures were reached. After remaining at the selected test temperature for 24 hrs, the samples were removed from the freezer and rewarmed in air at 0°C. Each experiment included one set without hardening and one set hardened at -3°C for 20 days. After thawing, woody twigs were placed with their bases in water and herbaceous plants were placed in polyethylene bags saturated with water vapor at room temperatures for 2 to 10 weeks. Thereafter, freezing injury was evaluated visually and microscopically. Browning, budding and rooting were used for rating injury, as follows: 0, no injury; 1, slight injury; 2, medium injury; 3, serious injury; 4, killed. Hardiness of the vegetative buds, leaves, stems, rhizomes and roots was expressed by lowest test temperature at which samples were injured slightly and/or normal budding or rooting were observed.

III. METEOROLOGICAL DATA

The annual mean monthly air and soil temperatures were cited from the Annual Report of Hokkaido University Forest Station. They were measured with resistance thermometer. The annual minimum air and soil temperatures were measured with maximum and minimum thermometer. Seasonal change of mean monthly air and soil temperatures at bare ground and soil temperature at forest floor were shown in Figure 1. The mean air temperature at bare ground in the coldest month was -8.2°C, that was almost same as the temperature at forest floor. In this area, the soil at 10 cm depth usually remained frozen as

Table II. Minimum air and soil temperatures measured by maximum and minimum thermometer at various habitats in the experimental site at Tomakomai recorded during December 1979 to March 1980.

Habitat	Minimum air temperature (°C) (at 1 cm above the litter)	Minimum soil temperature (°C)		
		0 cm	10 cm	20 cm
Deciduous broad-leaved forest	−11.9	−4.5	−1.3	−0.3
Larch forest	−12.5	−6.0	−1.5	−0.3
Fir forest	−13.0	−8.5	−4.5	−2.3
Bare ground	−24.4*	−10.9	−9.5	−6.9

* : measured by thermister thermometer.

long as four months. The annual minimum soil temperatures at the depth of 0, 10, and 20 cm at the forest floor were −4.5°C, −1.3°C and −0.3°C, respectively. At bare ground the annual minimum soil temperatures were −10.9°C, −9.5°C and −6.9°C, respectively (Table 2). Also the difference of the depth of frozen soil was observed between forest floor and bare ground. Thus, the litter layer (about 5 cm thickness) in the deciduous forest floor plays an important role to prevent the subterranean organs from the low temperatures. The organs of plants near or below the ground, especially chamaephytes are also protected by snow from the low temperatures. Snow deposited 40 cm at maximum in late February (Fig. 1).

IV. RESULTS

The winter buds of plants under the forest canopy which were collected in December and without being subjected to artificial hardening at -3°C for 20 days resisted to -25°C to -70°C in phanerophytes, -20°C to -60°C in chamaephytes, -7.5°C to -20°C in hemicryptophytes and about -3°C to -10°C in geophytes (Fig. 2-a). Little difference in hardiness of winter buds was found between hemicryptophytes at the forest floor and at the forest margin, in spite of remarkable difference in low temperature stress between both microhabitats. On the other hand, subterranean organs of hemicryptophytes growing at the forest margin were much hardier than those at the forest floor (Fig. 3-a and 3-c), suggesting the ecological significance of regeneration by subterranean organs after frost injury to leaf buds. The hardiness of rhizomes of chamaephytes at the forest

DORMANT BUD

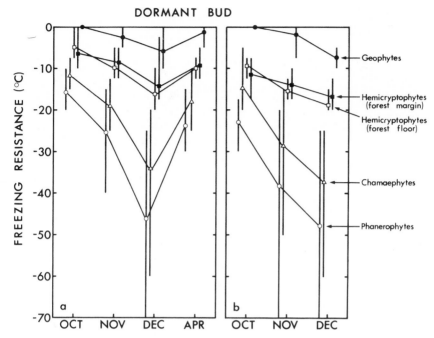

FIGURE 2. Freezing resistance of dormant buds of plants of various life forms and from different microhabitats before (a) and after (b) artificial hardening at -3°C for 20 days. Symbols show mean freezing resistance, and vertical lines show the range between maximum and minimum freezing resistance.

floor was higher than those of hemicryptophytes and geophytes at the forest floor and also higher than those of hemicryptophytes at the forest margin in December. Geophytes was the least hardy plants among life forms. These characteristic differences in hardiness among organs of plants varying in life form and microhabitat were unchanged during the experimental period and even after artificial hardening (Fig. 2-b, 3-b and 3-d).

In Pachysandra terminalis and Lycopodium serratum var. serratum, the freezing resistance of stem increased with the increase of height above the ground. On the contrary, in Oenothera sp. and Sanguisorba tenuifolia var. tenuifolia, the hardiness of root decreased with increasing distance from the soil surface.

In phanerophytes and chamaephytes, the hardiness of winter buds were invariably higher than those in hemicryptophytes and geophytes, and the winter buds became much hardier than rhizomes and roots.

In aerial organs, artificial hardening at -3°C for 20 days was more effective in enhancing hardiness in November or

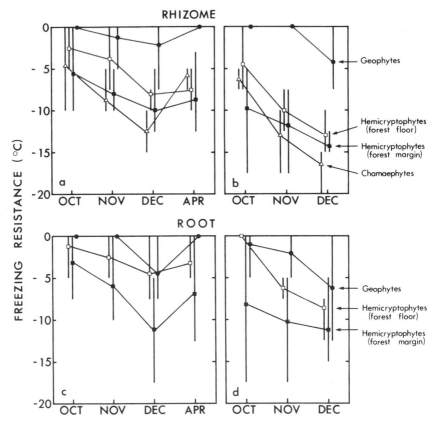

FIGURE 3. Freezing resistance of rhizomes and roots of plants of various life forms and from different microhabitats before (a and c) and after (b and d) artificial hardening at -3°C for 20 days. Symbols show mean freezing resistance, and vertical lines show the range between maximum and minimum freezing resistance.

October than in December (Fig. 4). In subterranean organs, the hardening effect was little in October, except for hemicryptophytes at the forest margin, and it became more pronounced in November or December. These observations confirm that the capacity to be hardened by subzero temperatures differs in different organs and varies with the seasons.

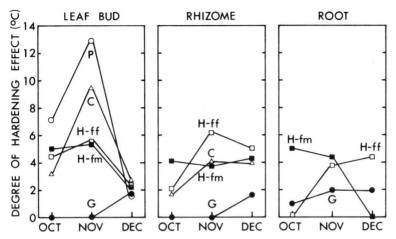

FIGURE 4. Increase in freezing resistance of leaf buds, rhizomes and roots of plants of various life forms and from different microhabitats by artificial hardening at -3°C for 20 days. G: Geophytes, H-ff: Hemicryptophytes, C: Chamaephytes under the forest canopy. H-fm: Hemicryptophytes at the forest margin. P: Phanerophytes.

V. DISCUSSION

Winter buds of phanerophytes were the hardiest and of geophytes were the least hardy. Subterranean organs of hemicryptophytes at the forest margin were much hardier than those at the forest floor. By artificial hardening in mid-winter, these characteristic differences among life forms and microhabitats were maintained. These results indicate that the difference in hardiness of plants varying in life form and microhabitat at the same deciduous forest corresponds closely to the degree of low temperature stress to which these organs or plants were exposed.

The difference in hardiness of subterranean organs among different life forms and microhabitats was much greater in Tomakomai, Hokkaido as compared with the data obtained in Gottingen by Till (7). This may be explained by a difference in the low temperature stress between the two locations. In the more severe climate of Hokkaido the difference in hardiness of plants varying in life form and microhabitat was more clearly demonstrated.

There is considerable interest in intraspecific variation in freezing resistance among climatic races and ecotypes of species with a wide distribution range. The timing of cold hardening in autumn differs significantly among races; northern races harden earlier than those from southern and coastal regions (6), which may sustain injury to branch tips in early winter. Such injury is due to improper onset of cold acclimation rather than to an insufficient inherent ability to harden. In the present study, the subterranean organs were characterized by a later onset of cold acclimation and insufficient inherent ability to harden.

Various organs and even tissues within an organ differ greatly in freezing resistance (2). Underground organs are most frost sensitive. Geophytes are overwintering underground, so that soil temperatures compared with the freezing resistance of their buds, bulbs and rhizomes is critical for the plant's survival. In the woody plants, the part of the stems just near to the ground is the most sensitive to frost (3). The freezing resistance of the perennial buds is to some extent related to the degree of low temperature stress to which they are exposed. Most of the buds which are not shielded from winter cold become as well hardened as the axes that bear them. Buds that overwinter near the ground or under the litter and snow develop moderate resistance. Cryptophytes, perennial herbs with persistent organs under the ground (geophytes) or under water (hydrophytes) are most sensitive to frost. These ecomorphs represent an evolutive adaptation to winter cold. In the case of trees, very hardy species that withstand freezing below -30°C seem not to have evolved in the Southern Hemisphere, because the mild, oceanic winters did not provide the stimulus (5).

These facts suggest that low temperature stress rank high among the natural selection pressure factors which have led to evolution of species adapted to cold climates, and that hardiness may have developed in dependence on life form and microhabitat.

REFERENCES

1. Larcher, W., "Physiological Plant Ecology," 303 pp. translated by M.A. Biederman-Thorson. Springer-Verlag, Berlin, 1980.
2. Precht, H., Christopherson, J., Hensel, H., and Larcher, W., "Temperature and Life," 779 pp. Springer-Verlag, Berlin, 1973.
3. Sakai, A., Contrib. from the Inst. of Low Temp. Sci.:1-14 (1968).
4. Sakai, A., Ecology 51:485-491 (1970).
5. Sakai, A., Paton, D.M., and Wardle, P., Ecology 62:563-570 (1981).
6. Smithberg, M.H., and Weiser, C.J., Ecology 49:495-505 (1968).
7. Till, O., Flora 143:499-542 (1956).
8. Yoshie, F., and Sakai, A., Jap. J. Ecol. 31:395-404 (1981) (In Japanese with English summary).

FREEZING RESISTANCE AND THERMAL INDICES WITH REFERENCE TO DISTRIBUTION OF THE GENUS PINUS

S. Oohata

Kyoto University Forest in Hokkaido
Department of Agriculture, Kyoto University
Kyoto, Japan

A. Sakai

The Institute of Low Temperature Science
Hokkaido University
Sapporo, Japan

I.   INTRODUCTION

In an extensive study of winter hardiness of many plants, it was found that almost all of the tree species from colder provinances were more hardy than species from warmer regions (9-14). The degree of winter hardiness corresponds well with the winter cold at the natural distribution range of many plants. These results show that low winter temperatures are important among the natural selection pressures which have led to the evolution of the frost hardy ecotypes and species. The timing of cold hardening processes in the autumn is also an important aspect of winter hardiness which differs significantly among ecotypes and climatic races of species with a large distribution area. In general, clones from high latitudes harden earlier than those from warmer climatic zones. In a comparative study with 42 pine species planted in Kyoto it was shown (8) that the timing of cold hardening differed significantly depending on provenance. The winter hardiness of pines wintering in Kyoto was closely related to the mean air temperature of the coldest month of their natural distribution area. However, the results obtained suggests that winter minimum temperature seems not to be the principal factor limiting natural distribution of hardy pines.

Kira (3) found that temperature indices calculated by summation of monthly means and expressed as warmth and cold indices show a remarkably good correlation with the distribution of forest types, and are more suitable for explaining actual distribution limits of forest zones than simple mean temperatures. In a gradient analysis of the distribution of conifers in Japan on the basis of thermic zones defined by Kira & Yoda (4) and by Kira & Yoshino (5) presented a climatic characterization of each species according to the distribution frequency in their natural habitats.

To clarify factors governing geographic distribution, climatic preferences, cold resistance and the distribution frequency of 57 representative species of the genus Pinus were investigated and compared with the thermal indices of their natural distribution area. The results obtained are discussed here in view of climatic characteristics and of evolution of cold hardiness of the genus Pinus.

II. MATERIALS AND METHODS

The winter hardiness of 9 species of subgenus Strobus and 32 species of the subgenus Pinus was assessed. Data for the winter hardiness of 16 pine species were taken from earlier reports (11,12). The samples were collected in midwinter at the Kamigamo Experimental Station, Kyoto University, and sent by air to Sapporo. About five uniform pieces of one-year-old twigs were cut from each sample tree and enclosed in polyethylene bags. All twigs were subjected to an artificial hardening regime which consisted of holding the samples at $-3°C$ for two weeks, $-5°C$ for one week, and $-10°C$ for three days to induce the maximum hardiness. The hardened twigs were cooled in $5°C$ increments at daily intervals to $-30°C$, and then to $-80°C$ in $10°C$ increments at 2 hour intervals. After equilibration for 20 hours at the selected test temperature, the twigs were removed from the freezer and thawed in air at $0°C$.

To evaluate the viability of the twigs after freezing, the thawed twigs were stored in polyethylene bags saturated with water vapor at room temperature for 30 days or 45 days. Thereafter, freezing injury was evaluated visually. Browning was used as a criterion for injury rating. Hardiness of leaves was expressed as freezing resistance LT_0, i.e., the lowest temperature at which no injury was obtained. It would have been desirable to take randomized samples of each species as an indication of variability, but this posed several difficulties. We decided, therefore, to use uniform material from one single tree. While we cannot claim that these measurements accurately represent the winter hardiness of the species, they nevertheless show good agreement with the geographical and

TABLE I. Freezing resistances in pine species of genus Pinus
and thermal conditions of their native regions.

Pinus species	Mean warmth Index (°C/ month)	Mean cold Index (°C/ month)	Mean air temp. of coldest month (°C)	Freezing resistance (°C)
Subgenus Strobus				
Subsect.				
Cembrae P. koraiensis	42.8	-78.4	-14.8	-70
P. pumila	20.0	-115.6	-18.3	-70*
P. cembra	23.7	-53.2	-7.3	-70*
Strobi P. strobus	62.2	-41.0	-6.7	-80*
P. monticola	46.6	-21.3	-1.4	-80*
P. ayacahuite	94.7	0.0	9.8	-15
P. peuce	50.6	-35.3	-5.7	-40
P. griffithii	71.0	-17.2	-2.2	-35
Cembroides P. cembroides	113.4	-1.2	8.0	-12
Gerardianae P. bungeana	81.0	-33.6	-7.2	-30
Subgenus Pinus				
Leiophyllae P. leiophylla	105.5	0.0	10.6	-8
Canarienses P. canariensis	57.4	-0.9	6.7	-22
Sylvestres P. resinosa	55.1	-51.6	-9.4	-80*
P. nigra	81.2	-10.6	1.2	-40
P. mugo	44.2	-27.4	-3.6	-70*
P. densiflora	86.4	-14.9	-0.1	-60
P. insularis	178.9	0.0	14.8	-7
P. merkusii	211.0	0.0	19.9	-10
Australes P. palstris	156.4	0.0	8.9	-18
P. taeda	146.9	-0.5	7.5	-23
P. elliottii	165.3	0.0	10.0	-22
P. caribaea	221.3	0.0	20.0	-9
Ponderosae P. ponderosa	60.6	-31.7	-4.4	-26
P. engelmannii	120.7	-0.3	8.2	-15
P. michoacana	133.7	0.0	11.8	-10
P. lawsonii	155.2	0.0	14.4	-8
Contortae P. banksiana	37.9	-100.7	-19.5	-80*
P. contorta	29.9	-51.2	-6.9	-75*
P. clausa	195.1	0.0	14.2	-14
Oocarpae P. radiata	92.3	0.0	9.5	-13
P. patula	100.7	0.0	9.8	-15
P. greggii	113.7	-0.2	8.3	-15

* Uninjured at the lowest temperature indicated.

ecological distribution. The same trend was observed in New Zealand trees (13).

In order to calculate the thermal indices for the distribution area of species, data from weather stations and distribution areas of each species reported by Mirov (6) were used. From the temperature data of selected stations, the monthly mean temperature of every month and for every 100m of elevation was calculated by application of an elevational correction basing on the lapse rate of -0.55°C/100m. Mean values of the warmth index (MWI) and the cold index (MCI) were calculated according to Kira (3) by summation of the estimated monthly mean temperatures higher or lower than 5°C as the threshold, respectively. Mean air temperatures of the coldest month (MTCM) in the natural range of all pine species were also derived from the same data material.

III. RESULTS AND DISCUSSION

The winter hardiness of artificially hardened pines is listed in Table 1 together with climatic data of their natural distribution area. As shown in Fig. 1, winter hardiness decreases curvilinearly with an increase of MWI. Most species in the warm temperate and subtropical zone are less hardy and marginally hardy to -20°C or -10°C. All pines native to the subarctic zone, where MWI ranges from 15 to 45°C/month, survive freezing to -70°C or below. The extreme degree of winter hardiness exhibited by northern conifers and some deciduous broad-leaved species, especially Salix (14) may reflect a very long history of adaptation to climates with severe winter. Such hardiness, once acquired, can persist, more or less, in a warm or temperate climate, where it is no longer appropriate. This is also the case in pine. Pines have rapidly recaptured lost land since the maximum glaciation after the Ice Age and extended to warmer climates at the present, especially at the West Coast of North America. As previously reported (8), in regions of Eurasia having the same value of MWI, winter hardiness of pine from East Asia is much higher than of those from Europe, probably as a consequence of the different continentality of the climate. From these results, it is concluded that MWI may not be intimately correlated to winter hardiness in pines.

Also the "cold index" as proposed by Kira (3), which is the corresponding negative sum for monthly means below 5°C, shows a roughly linear relation to the winter hardiness. Winter hardiness was more closely related to MTCM than MWI or MCI (Fig. 2). The climatic factor best related to winter hardiness may be the average annual minimum temperature or the extreme minimum temperature, as discussed by Flint (1). Since it is

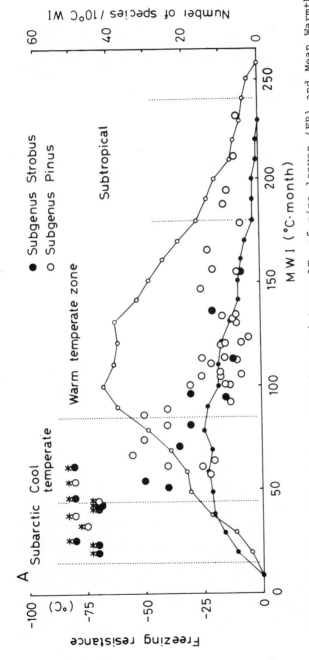

FIGURE 1. Relationship between freezing resistance LT_0 of pine leaves (FR) and Mean Warmth Index (MWI), and distribution frequency of the two subgenera of Strobus and Pinus (expressed by species number included in each 10°C-WL). * Uninjured at the lowest temperature.

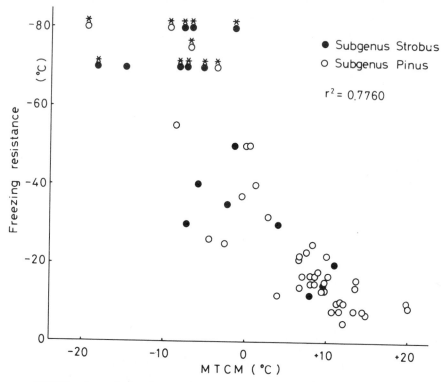

FIGURE 2.   Relation between freezing resistance of each
pine species and mean air temperature in the coldest month
(MTCM) at the natural distribution range.   * Uninjured at the
lowest temperature.

difficult to obtain these data throughout the world, the mean
air temperature in the coldest month (MTCM) was correlated with
the winter hardiness.

The distribution frequency curve (cf. Fig. 1) which
expresses the number of species included in each 10°C-WI shows
that the center of distribution of subgenus Strobus situates at
the cool temperate zone and that of the subgenus Pinus to the
warm temperate zone. The distribution of subgenus Pinus
appears symmetrical and similar to normal distribution. This
property seems to be characteristic for the distribution curve
against the WI axis.

The distribution frequency curves against the MTCM are
shown in Fig. 3. Since freezing resistance in pine species was
closely related to the MTCM at their natural ranges, Fig. 3
represents the hardiness distribution of pines belonging to two
subgenera. It becomes evident that very hardy pine species are

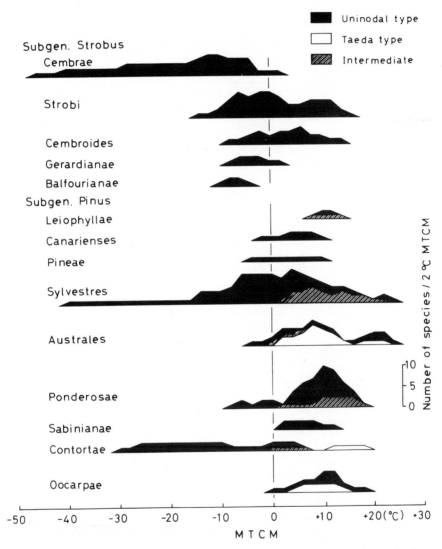

FIGURE 3. Distribution frequency of each subsection of the two subgenera <u>Strobus</u> and <u>Pinus</u> against the thermal gradient of the mean air temperature in the coldest month (MTCM). The number of species included in each 2°C-MTCM is indicated. Shoot growth pattern is distinguished into the three types: uninodal, intermediate and taeda type.

confined in their distribution to winter cold regions. The major species of the subgenus Strobus are distributed in the MTCM range from -15 to 15°C. On the other hand, most of the species of the subgenus Pinus are distributed mainly in the region ranging from 5 to 15°C MTCM, and their frequency decreases sharply in the climates with MTCM higher than 15°C.

The MTCM as shown in Fig. 3 expresses the frequency of species included in each 2°C-range of MTCM. Very hardy pine have evolved from the subsection Cembrae of the subgenus Strobus and from the subsection Sylvestres and subsection Contortae of the subgenus Pinus, and they expanded into winter cold regions, where a severe continental climate has prevented less hardy pines from reaching there. The subsection Sylvestres is the largest group of the genus Pinus which includes numerous species with wide distribution areas. The same trend can be observed in the subsection Cembrae and Contortae as shown in Fig. 3. In general, the subsection of Pinus has evolved very hardy species, indicating a great adaptability to changing environments probably due to genetic variation and plasticity. In contrast, most of the sections, which have confined at the present to a narrow discontinuous winter mild area have not evolved very hardy species. This general distributional trend of pines may suggest that the specification of widely ranging pine subsections was initiated earlier than that of large pine groups distributed only in warm temperature regions such as the subsection Strobi, Cembroides, Australes, Ponderosae and Oocarpae. To strengthen this hypothesis, further studies from various aspects are required.

Species of the genus Pinus occur in wide regions in the Northern Hemisphere from the subarctic to lower latitudes, reaching near the equator. Growth pattern of Pinus species can be distinguished into two phenological groups: One group of pines is characterized by one single period of bud flushing and shoot extension growth in spring and early summer producing shoots with one (Uninodal type e.g. P. sylvestris or P. densiflora) or more internodes (multinodal Banksiana type). The other phenorhythmo type (multinodal Taeda type, according to Oohata [7]) is characterized by successive flushes of shoot growth during the same vegetation period each producing one internode. Pines of the Taeda type respond only slightly or not to the photoperiodic stimulus and continue to grow intermittently throughout the year without dormancy, which enables certain species of this type to expand into subtropic and tropical regions. Pine species of the unimodal and of the Banksiana type native to temperate and subarctic climates develop winter dormancy, induced by short days and released by winter cold. Only these pines, which cease their shoot growth already in summer and enter into true dormancy are able to become frost hardy enough to survive the severe winter conditions at higher latitudes.

IV. SUMMARY

From an extensive study on winter hardiness of pine leaves, it appears that winter hardiness corresponds well with the degree of winter coldness at their natural distribution area. However, the observed hardiness exceeds the minimum annual temperature at their natural ranges, suggesting that winter minimal temperatures seems not to be the principal factor limiting natural distribution of every hardy species.

From a gradient analysis of the thermal environment it appears that the distribution center of Pinus is likely to be the cool temperature zone for the subgenus Strobus and the warm temperate zone for the subgenus Pinus. Very hardy species have evolved only in a few subsections of Pinus: Cembrae of subgenus Strobus and Sylvestres and Contortae of the subgenus Pinus, which include species with distribution in a wide range of different climates. Subsections which are confined to winter mild climates have not evolved very hardy species because the mild winter did not provide the stimulus for adaptation.

ACKNOWLEDGMENT

The authors wish to thank Dr. T. Kira, Osaka City University, for providing the climatic data.

REFERENCES

1. Flint, H., Ecology 53:1164-1170 (1972).
2. Kaku, S., and Iwaya, M., Oicos 33:402-411 (1979).
3. Kira, T., Kantai-Nogaku 2:143-173 (1948). (In Japanese).
4. Kira, T., and Yoda, K., in "Living Things and Environments" (J. Ashida et al. eds.), pp. 231-268. Kyoritsu--Shuppan Inc., Tokyo, 1965. (In Japanese).
5. Kira, T., and Yoshino, M., in "Nature", Contribut. in honor of Dr. K. Imanishi on the occasion of his sixtieth birthday. (M. Morishita, and T. Kira, eds.), pp. 133-161. Chuokoron-sha Inc., Tokyo, 1967. (In Japanese).
6. Mirov, N.T., in "The Genus Pinus", pp. 572. Ronald Press Co., New York, 1967.
7. Oohata, S., Hoppoh-Ringyo 31:377-382 (1978). (In Japanese).
8. Oohata, S., Hasegawa, Y., and Sakai, A., Jap. J. Ecol. 31:79-89 (1981). (In Japanese).
9. Sakai, A., New Phytol. 70:1199-1205 (1971).
10. Sakai, A., and Weiser, C., J. Ecology 54:118-126 (1973).
11. Sakai, A., Jap. J. Ecol. 25:101-111 (1975).

12. Sakai, A., and Kurahashi, A., Jap. J. Ecol. 25:192-200 (1975). (In Japanese).
13. Sakai, A., and Wardle, P., New Zeal. J. Ecol. 1:51-61 (1978).
14. Sakai, A., Low Temp. Sci. Ser. B. 36:1-19 (1978).

ADAPTATION TO COLD CLIMATE OF FERNS NATIVE TO HOKKAIDO
WITH REFERENCE TO THE ALTERNATION OF GENERATIONS

Toshiyuki Sato

The Institute of Low Temperature Science
Hokkaido University
Sapporo, Japan

I.   INTRODUCTION

Ferns, in contrast to seed plants, have two independent
phases in their life history, the conspicuous sporophytic phase
alternating with a thalluous gametophytic phase without vascular system.   In ferns the reproductive processes can be far
removed from mother sporophytes by dispersion of the spores, in
contrast to seed plants which produce the new generation on the
mother plants.   Since plant establishment depends on the climatic conditions of the sites in which the propagules are dispersed, the distribution of ferns depends upon the responses to
environmental conditions of every step of the life cycle.   In
order to maintain a normal life cycle, no serious restriction
in any part of the life history must occur.   On the Japanese
Archipelago, with increasing latitude ferns are decreasing in
number of species and changing from the evergreen to the
summer-green life-mode (7,9).   It is suggested that this may be
due to effects of cold climate on some part of the life history
of ferns.   Gametophytes were supposed to be sensitive to winter
cold rather than the sporophyte (5).   On the other hand, it has
been clearly demonstrated that the gametophytes of some Euro-
pean fern (Aspidiacea) survive freezing to -10 or -20°C; the
gametophyes became slightly less frost resistant than the
sporophytes (2,3).   Farrar (1) reported that the gametophytes
of the Appalachian fern Vittaria survived freezing to nearly
-10°C, but not sporophytes of V. lineata from Florida.   This
implies that the gametophytes of V. lineata, and presumably
other subtropical species of Vittaria, can survive in areas
where sporophytes cannot withstand winter cold.   In previous
reports, it was demonstrated that the overwintering gametophytes of some cool temperature ferns are much hardier than

their sporophytes (8,12). In this report, adaptational phenomena in different stages of the life history of ferns will be
considered with particular reference to their overwintering
capacity.

II. MATERIALS AND METHODS

Phenological observations on leaves of adult sporophytes
were made from April 1977 to March 1981 at about 350 localities
in Hokkaido. The seasonal development of the leaves was
expressed by the relative green area of the leaves as percentage of the maximal area of leaves of the last year. For the
field observations, the relative green area was judged visually
in intervals of each 10% basing on the completely extended
leaves per sporophyte.

The sporulation period was determined by the state of the
opening of sporangia using the binocular microscope in the
field and touching the fertile leaves for the detection of
sporulation. Some of the determinations of the sporulation
period was done in the same way with herbarium specimens; in
such case the retroactive change of 15 days from the collecting
data of the specimen was taken in account, because drying
specimens accelerate the sporulation date for 10 or 20 days
from the collecting date.

Growing patterns of the gametophytes were observed on
potted materials in the nursery of the Institute of Low Temperature Science, Hokkaido University, considering the sporulation
period on the species in Hokkaido. In order to determine the
temperature requirement for growing to maturity of the gametophytes, the potted material was cultivated for five months
under various temperature conditions (4, 7,10, 12, 15, 17 and
24°C) at a 12/12 h photoperiod and 300-600 Lux.

Freezing resistance of adult sporophytes was evaluated with
materials collected near Sapporo at Mt. Teine (300-500 m
a.s.l.). Freezing resistance of gametophytes was assessed with
material cultivated in the nursery. Two or three whole plants
with leaves, roots and rhizomes of sporophytes enclosed in
polyethylene bags were prefrozen at -5°C for at least 12 h.
Then they were progressively cooled at steps of 5°C with 2 h
intervals to a series of test temperatures until -40°C. After
remaining at each test temperature for at least 12 h, they were
removed from the freezer and thawed in air at 0°C. To evaluate
freezing resistance, the thawed materials were deposited in
open polyethylene bags at room temperature (about 20°C) for
either 3 to 5 months. The degree of freezing resistance is
expressed by the lowest temperature at which the tissues
remained alive. The freezing resistance of the gametophytes
was assessed using at least 10 prothalli ranging 2-5 mm in
width without sex organ. The material collected in the nursery

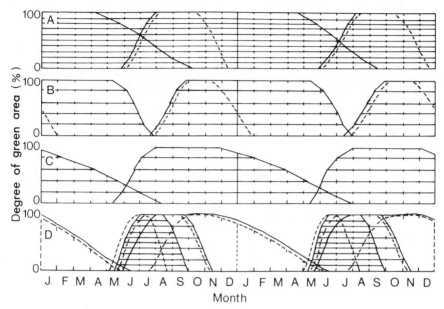

FIGURE 1.   Phenology of the life span of fern leaves in Hokkaido.   Solid lines: sterile leaves.   Broken lines: fertile leaves.   Ordinate: ratio of the actual green area to the maximal leaf extended.   Area drawn by the horizontal solid lines indicate the unfolding state of green leaves.   A; Evergreen, B; winter-green, C; semi-evergreen, D; summer-green.

was placed in 12 cm diameter Petri dishes, and frozen and thawed in the same way as the sporophytes.   To evaluate the freezing resistance of the gametophytes, the thawed material was placed on cotton in Petri dishes at room temperature for some months.   Regrowth after one month into heart-shaped thalli as large as the gametophytes have been before freezing (2-5 mm) was used as the criterion for the survival of the gametophytes.

III.   RESULTS

A.   Phenology

The life span of the leaves of adult sporophytes is summarized in Fig. 1.   Ferns can be patternized in evergreen, winter-green, semi-evergreen and summer-green.   In Hokkaido, the leaves of evergreen ferns develop during June and July and decay after the following spring.   The life span of their leaves ranges from 16 to 27 months.   One of the dimorphic evergreen ferns, <u>Plagiogyria</u> <u>matsumureana</u>, elongates the fertile

leaves half a month later than the sterile leaves and sheds
them at late October to November (Fig. 1, A). In winter-green
ferns the leaves develop during August and decay in the following June or July. The fertile leaves of Sceptridium
multifidum unfold about half a month later than sterile leaves
and decay during December (Fig. 1, B). In semi-evergreen ferns
the leaves develop during late May and June and begin to decay
from December; however, some green leaves remain until the next
spring (Fig. 1, C). In the summer-green ferns the leaves
usually develop in late May and June and they decay in
October. Among the summer-green ferns seven different types of
seasonality of leaves depending upon dimorphism can be distinguished (Fig. 1, D). Type 1: Lunatyrium pycnosorum; the fertile leaves are elongated about a month later and decay about
half a month earlier than the sterile leaves. Type 2:
Matteuccia struthiopteris; the fertile leaves elongate during
August and decay in the following spring. Type 3:
Dennstaedtia hirsuta; the sterile leaves of this species are
elongated about one or two months later than the fertile leaves
and decay in the following spring or summer. Type 4:
Cystopteris fragilis; the fertile and sterile leaves decay
during late August to September. Type 5: Osmunda japonica the
fertile leaves unfold about half a month earlier than sterile
leaves and decay soon after the sporulation in July. Type 6:
Japanobotrychium virginianum; the fertile leaves unfold and
decay about one month later and earlier than the sterile
leaves. Type 7: In the most common type including 70% of the
summer-green species, the fertile and sterile leaves develop
during late May and June and decay during October (9).

On the Japanese Archipelago, warm temperate ferns and ferns
of subtropical origin have a long sporulation period of almost
one year in Southern Honshu. In the northern distributional
ranges, the sporulation period is shortened. Most of the cool
temperate ferns in Hokkaido sporulate during late August to
early September; even in the same species the sporulation
period becomes gradually shorter at higher latitudes. About 40
ferns occuring in Hokkaido can sporulate during the converged
period late August and early September (11).

After sporulation, there are three possibilities of wintering in the early life history of ferns: 1) at the spore
stage, 2) at the gametophytic stage and 3) at the juvenile
sporophytic stage after alternation of generations. In
Hokkaido, ferns of the first group (e.g. Osmunda japonica)
sporulate during May, then the spores germinate in the summer
and gametophytes develop into juvenile sporophytes in late
August and September. The spores of many fern species are dispersed during June to September and juvenile gametophytes are
developed before winter. Juvenile sporophytes are observed in
the following summer. Thus, in many of the ferns which sporulate in late August and early September, the juvenile gametophytes of them are overwintering as simple thalli as well as

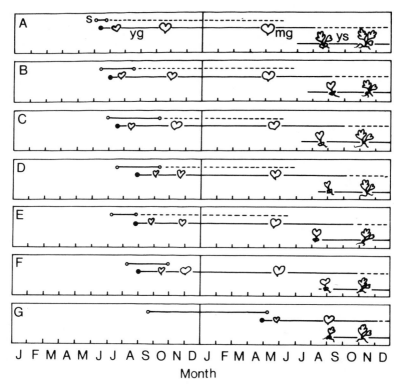

J F M A M J J A S O N D J F M A M J J A S O N D
Month

FIGURE 2. Seasonality of the early life history of seven cool temperate ferns in Hokkaido. A: Polystichum tripteron. B: P. retroso-paleaceum. C: P. braunii. D: Dryopteris crassirhizoma. E: D. sabaei. F: Asplenium incisum. G: Matteuccia struthiopteris. s: stage of sporulation. yg: stage of young gametophyte. O-O : period of spore dispersal in Hokkaido. ●-● : dispersal period of spores in this study.

the perennial sporophytes. If the spores are dispersed after October such as in Matteuccia orientalis and Dryopreris saxifraga they overwinter in the spore stage and germinate not until the following spring and early summer. In these species juvenile sporophytes appear in the following late summer and autumn (Fig. 2). About 20 ferns were needed for the maturation of the gametophytes a warmth index of 40-80 (as defined by Kira, 4).

TABLE I. Maximum freezing resistance of Pteridophyta in Hokkaido

Species Name	Freezing resistance (°C)	
	Rhizome	Leaf
1. Lunathyrium pterorachis	-5	0
2. Athyrium vidalii	-5	0
3. Stegnogramma pozoi subsp. mollissima	-5	-5
4. Plagiogyria matsumureana	-5	-10
5. Arachniodes miqueliana	-5	-15
6. Sceptridium multifidum var. robustum	-5	-40
7. Coniogramme intermedia	-7	0
8. Matteuccia orientalis	-7	0
9. Osmunda cinnamomeum var. fokiensis	-10	0
10. Osmunda japonica	-10	0
11. Phegopteris polypodioides	-10*	0
12. Arachniodes standishii	-10	-12.5
13. Dryopteris tokyoensis	-10	-12.5
14. Cyrtomium falcatum	-10	-12.5
15. Polystichum retroso-paleaceum	-10	-15
16. Polystichum tripteron	-10	-17.5
17. Struthiopteris niponica	-10	-20
18. Polystichum braunii	-10	-20
19. Equisetum palustre	-12.5	0
20. Equisetum arvense	-12.5	-3
21. Lunathyrium pycnosorum	-15	0
22. Dryopteris monticola	-15	-5
23. Dryopteris sabaei	-15	-20
24. Dryopteris crassirhizoma	-15	-22.5
25. Asplenium scolopendrium	-15	-22.5
26. Onoclea sensibilis var. interrupta	-17.5	0
27. Adiantum pedatum	-17.5	0
28. Dryopteris amurensis	-17.5	-7
29. Mecodium wrightii	-17.5	-25
30. Athyrium brevifrons	-20	0
31. Matteuccia struthiopteris	-20	-3
32. Dennstaedtia hirsuta	-20	-20
33. Asplenium incisum	-20	-25
34. Dryopteris laeta	-25	-5
35. Dryopteris austriaca	-30	-5
36. Lycopodium serratum var. serratum	-30	-30
37. Lycopodium obscurum	-30	-40
38. Dryopteris saxifraga	-30	-40
39. Athyrium yokoscense	-40*	0
40. Woodsia polystichoides	-40*	-5
41. Polystichum craspedosorum	-40*	-30
42. Equisetum hyemale	-40	-40
43. Selaginella helvetica	-40	-40
44. Lycopodium complanatum	-40	-40

TABLE I continued.

Species Name	Freezing resistance (°C)	
	Rhizome	Leaf
45. Lycopodium chinense	-40	-40*
46. Lycopodium clavatum var. nipponicum	-40	-40*
47. Lycopodium annotinum	-40	-40*
48. Polypodium virginianum	-40*	-40
49. Dryopteris fragrans var. remotiuscula	-40*	-40
50. Polypodium fauriei	-40*	-40*
51. Camptosorus sibilicus	-40*	-40*
52. Pleurosoriopsis makinoi	-40*	-40*
53. Lepisorus ussuriensis var. distans	-40*	-40*
54. Pyrrosia tricuspis	-40*	-40*
55. Selaginella shakotanensis	-40*	-40*

* Uninjured at the indicated temperature at least.

B. Freezing Resistance

The seasonal change in freezing resistance of leaves and rhizomes of sporophytes, and of the gametophytes is shown in Fig. 3. The freezing resistance of leaves increases gradually from October to December, slightly earlier than that of the rhizomes. The maximal resistance after prehardening is usually observed in December. The maximal resistance of the leaf was higher than that of rhizomes, e.g. -17.5°C in leaves and -10°C in rhizomes of Polystichum tripteron; -12.5 and -10°C of P. retroso-paleaceum; -20 and -15°C of Dryopteris crassirhizoma (Table 1). The freezing resistance of the gametophytes increased later than that of the sporophytic organs. Nevertheless, the maximal freezing resistance of gametophytes in December was much higher than that of sporophytic organs (Fig. 3). Gametophytes of Dryopteris sabaei, Polystichum tripteron and P. braunii tolerated freezing to -196°C as well as desiccation to 2.1 to 5.5% water per fresh weight in March. The freezing resistance of the gametophytes of further 10 fern species occuring in deciduous forests of Hokkaido range from -30 to -40°C (12).

The freezing resistance of the leaves, assessed on 67 fern and fern-allies species in relation to the leaf phenology is shown in Fig. 4. The average maximal freezing resistance of leaves of winter-green, evergreen, semi-evergreen and summer-green Pteridophyta is -33, -32, -19 and -2°C respectively. The leaves except summer-green ferns could acclimate from October to December and resist freezing between -10 to -40°C in

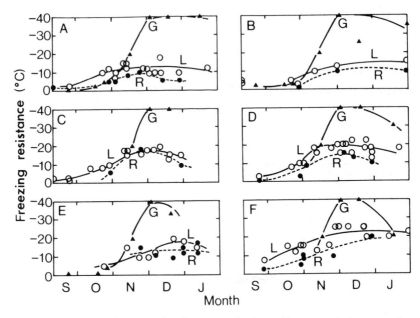

FIGURE 3.   Seasonal change of freezing resistance of six cool-temperate ferns native to Hokkaido.   A: Polystichum tripteron, B: P. retroso-paleaceum.   C: P. braunii.   D: Dryopteris sabaei, E: Asplenium insisum, L (○); freezing resistance of the leaf, R (●); freezing resistance of the rhizome, G (▲); freezing resistance of the gametophyte.

December.   Meanwhile,   as   for   the   summer-green   ferns,   their leaves   were   still   susceptible   to   temperature   lower   than   -10°C in November, though the freezing resistance of their rhizomes increased (10); their leaves showed little or no increase in acclimation; leaves of many of them remained green before by severe frost.

The freezing resistance of the rhizomes was determined for 55 species (Table 1).   In most of these species hardiness from late October to December, increased i.e. about a month later than in the leaves, and reached the maximal level in December (Fig. 3).   Among the different species maximal frost hardiness depended upon the microclimate at the habitats in which each species occurs in Hokkaido (Fig. 5).   Almost all of the ferns growing on trunks of deciduous trees and on cliffs without snow cover became resistant to temperatures below -40°C.   About 60-80% of the ferns usually occuring on the mixed-forest floor and on rocky slopes resisted -20 to -40°C.   On the other hand, about 90% of the ferns growing in a humid and humus-rich dale along the stream were damaged by frost at temperatures below -5

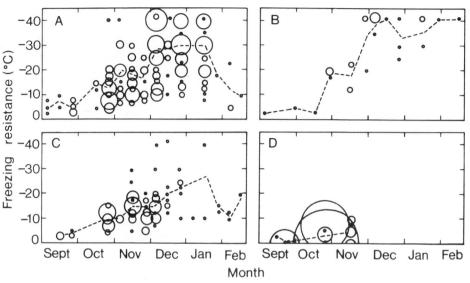

FIGURE 4.   Seasonal change in the freezing resistance of leaves of 65 ferns and fern-allies of 4 different life-modes. The diameter of the circle expresses the number of species investigated, the smallest circle showing one species; the largest circle represents seventeen species. A: evergreen. B: winter-green. C: semi-evergreen. D: summer-green.

to -17.5°C.   About 70% of the species occuring at the forest margin and on the deciduous forest floor resisted temperatures ranging -5 to -17.5°C (13).

IV. DISCUSSION

The comparison of the freezing resistance of the leaves with their phenology suggests that severe frost may be one reason for a change of the life-mode in ferns. Typical summer-green ferns, such as Osmunda, Thelypteris, Cystopteris and Gymnocarpium species shed their leaves before frost occurs. In Athyrium and Cornopteris species the decay of leaves, was accelerated by frost in late autumn. In such cases the life span of leaves is reduced by frost. Leaves of Stegnogramma pozoi and Coniogramme intermedia collected in warm temperate regions at Southern Honshu in early February resisted at most -7°C, which corresponds to the resistance level of these species attained in Hokkaido (unpublished data). Another evidence of habitat restriction by winter cold can be drawn from

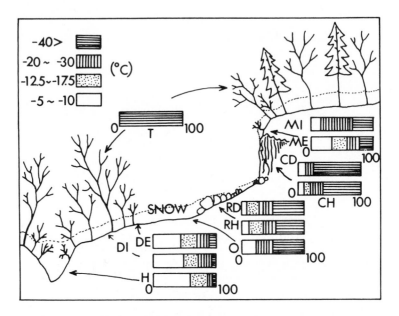

FIGURE 5.   Schematic representation of the distributional
patterns of freezing resistance of Pteridophyta growing in
Hokkaido with special reference to their habitats.   Freezing
resistance was evaluated on fifty-five species from late autumn
to winter.   Percentage was calculated as the ratio of the number of species in four ranges of freezing resistance to the
total number of species in each habitat in Hokkaido.   The habitats shown in the figure are: MI forest floor of mixed deciduous broad-leaved and evergreen coniferous forest; ME margin of
the mixed forest; CD dry cliffs; CH humid cliffs; RD dry rocky
slopes; RH humid rocky slopes; O in open places; DE deciduous
broad-leaved forest margin; DI deciduous broad-leaved forest
floor; T trunks of deciduous trees; H humid and humus rich dale
along streams.

observations on certain species which are widely distributed in
Japan such as Asplenium incisum and Mecodium wrightii.   These
species, which are resistant to -25°C in December, are usually
growing on stone walls, on cliffs and at forest margins in
Southern Honshu.   However, they are never found on cliffs or
stone walls in Hokkaido but only on rocky slopes in forests,
where they are usually covered with snow during winter.
     For summer-green ferns in Hokkaido a sporulation period at
late August and early September is characteristic.   This may be
caused by the comparably short growing season in Hokkaido.
Even in the same species, such as Coniogramme intermedia and
Stegnogramma pozoi, leaf phenology from evergreen to

summer-green behaviour may change with increasing latitudes in Japanese Archipelago (9). Also the sporulation period of these species becomes shorter at higher latitudes in Japan. Most of the ferns occuring in Hokkaido sporulate only for 2 or 3 months, as compared with sporulation period of more than half a year in warm temperate regions. Except of typical dimorphic ferns like Matteuccia and Osmunda species, ferns are forced to synchronize the seasonal growth pattern and the sporulation period with the short favourable season in Hokkaido.

After restriction of the sporulation period, the period of gametophyte growth is also shortened and the timing of the alternation of generations postponed to the following summer. The spores dispersed during June and September germinate into vegetative gametophytes before winter but, in contrast to ferns of warm temperate regions, they do not readily develop into sporophytes. Thus, in many ferns the gametophytes are over-wintering in the thalluous stage.

In order to succeed the normal life cycle, ferns in cool temperate regions need a sufficient freezing resistance at each stage of the life history. More pronounced as in the studies of Pickett (6) and Kappen (2,3), clear differences in freezing resistance of sporophytes and gametophytes could be demonstrated in the results as presented here. The phenological observation and the determination of the freezing resistance of the different stages in the life history of ferns can explain the ecological phenomena presented by Farrar (1) and Scheure (14), who reported the existance of sporophyteless species of Appalachian Vittaria and some genera of ferns. After evolving the ability of vegetative propagation of the gametophytes by gemma and gemifera, the gametophytes of cool temperate ferns growing on the forest floor may be able to survive independently from sporophytes, providing they develop sufficiently high freezing and drought resistance.

As shown in these results, the freezing resistance of the rhizomes of sporophytes was depended upon the extent of cold stress at the habitat. Usually fern species are specialized to certain microhabitats. In the sporophytic stage, only epi-phytic ferns and ferns growing on cliffs resist -40°C. Ferns occuring on the deciduous forest floor as a rule do not resist temperatures much below -20°C. However, the gametophytes of ferns of the forest floor might be able to survive in more severe habitats, where sporophytes of the same species would be killed. Even after remaining frozen at -5°C for more than five months in shade, the gametophytes could survive. The drought resistance of the gametophytes was also investigated in late winter on some fern gametophytes. In the dehydrated state with a water content 2 to 5%, the gametophytes could survive and develop into normal sporophytes after rehydration (12).

ACKNOWLEDGMENT

    I would like to express my hearty appreciation to Prof. A.
Sakai who advised and supported me to carry out this study.
Also, I am heartily grateful to Prof. W. Larcher for his great
suggestions to arrange this manuscript.

REFERENCES

1. Farrar, D.R., Amer. J. Bot. 65:1-12 (1978).
2. Kappen, L., Flora, Bd. 155:123-166 (1964).
3. Kappen, L., Flora, Bd. 156:101-115 (1965).
4. Kira, T., in "Nature from the Ecological View", pp.
   233-241. Kawade Syobo, Tokyo, 1971.
5. Nishida, M., in "Flora of Chiba Prefecture", pp. 208-222.
   Biological Society of Chiba Prefecture, Chiba, 1958.
6. Pickett, F.L., Amer. J. Bot. 1:477-498 (1914).
7. Sato, T., and Sakai, A., Low Temp. Sci, Ser. B 35:45-53
   (1977).
8. Sato, T., and Sakai, A., Can. J. Bot. 58:1144-1148 (1980).
9. Sato, T., and Sakai, A., Jap. J. Ecol. 30:369-375 (1980).
10. Sato, T., and Sakai, A., Low Temp. Sci, Ser. B 38:15-22
    (1980).
11. Sato, T., and Sakai, A., Jap. J. Ecol. 31:91-97 (1981).
12. Sato, T., and Sakai, A., Can. J. Bot. 59:604-608 (1981).
13. Sato, T., and Sakai, A., Jap. J. Ecol. 31:191-199 (1981).
14. Scheure, P.J., Science 155:1266-1268 (1966).

CRYOBIOLOGY OF ISOLATED PROTOPLASTS[1]

Peter L. Steponkus, Michael F. Dowgert, Richard Y. Evans
and William Gordon-Kamm

Department of Agronomy
Cornell University
Ithaca, New York

I. INTRODUCTION

Aspects of freezing injury and cold acclimation of higher plants have been addressed for many decades (20). The majority of the efforts have been concerned with the cold acclimation process while less attention has been directed to the nature of freezing injury. In spite of an overwhelming number of publications in the area, progress has been slow. Although numerous cellular constitutents have been analyzed with respect to a possible involvement in the cold acclimation process, there is little agreement on the significance of the measured changes (46,48). Several factors have contributed to this dilemma.

First, in studies of cold acclimation, emphasis has been placed on correlative studies of biochemical changes that occur during cold acclimation. Little attention has been given to whether the biochemical changes observed during cold acclimation are directly involved in the cold acclimation process or are associated with other developmental processes occurring at the same time (46). The alterations directly involved in the cold acclimation process include those that are ultimately responsible for the observed increase in hardiness in addition to preliminary metabolic changes which precede such direct alterations. Yet, rarely is a distinction made between the two.

[1] Department of Agronomy Series Paper No. 1409. This material is, in part, based on work supported by the National Science Foundation under Grant No. PCM-8021688 and the U.S. Dept. of Energy under Contract No. DE-AC02-81ER10917.

Direct alterations may preclude or mitigate the various stresses that occur during freezing or increase the tolerance of cryosensitive targets i.e., cellular membranes, to the specific stresses.

Second, there is a prevailing attitude that a single compound or change is responsible for the total increase in cold hardiness. Quite often, it is inferred that the greater the quantitative increase (or decrease) in a given constituent, the more likely it is associated with the cold acclimation process. Frequently, the involvement of a particular metabolite in the cold acclimation process is assessed solely on the basis of the correlation coefficient between the quantity of the metabolite and the degree of hardiness. The validity of such reasoning is quite dubious as such a correlation would exist only if the particular compound were the rate limiting step in the entire cold acclimation process.

Third, in studies of freezing injury, it is frequently inferred that various cellular stresses resulting from freezing are separate and distinct events that are mutually exclusive in a given plant. While such a delineation is helpful in organizing a treatise (18,20), it probably represents an unrealistic portrayal of the situation. An equally erroneous corollary is that freezing injury is the result of the same stress in all plants at various states of acclimation or subjected to various freeze-thaw protocols.

Fourth, while there is a general consensus that cellular membranes, especially the plasma membrane, are the site of freezing injury, in very few instances has it been established what constitutes membrane damage and how membrane structure is affected by a freeze-thaw cycle. Too frequently, a recorded biochemical change during cold acclimation is interpreted a priori with little consideration of the injurious stress or resultant lesion. As injury and death of the tissue can be effected by any of a number of membrane lesions, it is difficult to envisage how various aspects of cold acclimation can be accurately assessed or how a mechanism of injury can be proposed without insight into the nature of injury.

An alternate approach involves the elucidation and quantification of the various cellular stresses that occur during a freeze-thaw cycle and identification of the specific plasma membrane lesions that result. Then the tolerance of the membrane to the individual stresses may be quantified and differences in survival may be ascribed to changes which either mitigate or preclude the various cellular stresses or, alternatively, increase the tolerance of the membrane to these stresses. Such an approach would greatly facilitate investigations of genotypic differences in cold hardiness, biochemical aspects of cold acclimation, or the development of effective cryopreservation protocol. It is even possible that an enduring hypothesis of cryoinjury might emerge!

As part of such an approach, we prefer to view the cellular stresses that arise during a freeze-thaw cycle as a sequence of stresses -- any one of which is potentially lethal depending on the species, the degree of acclimation, and the freeze-thaw protocol. Thus, survival at progressively lower temperatures depends on the successful and sequential avoidance, mitigation, or tolerance of each stress. Hence, cold acclimation can be envisaged as a sequence of alterations that allow for each stress barrier to be overcome rather than one particular alteration which is responsible for the total increase in hardiness.

II. THE ISOLATED PROTOPLAST AS AN ARENA FOR CRYOBIOLOGY

Although qualitative aspects of cellular behavior during a freeze-thaw cycle have been known for several decades, there have been few direct quantitative studies. Although injury, manifested as the loss of semi-permeability of the plasma membrane is routinely measured by any one of several techniques, the molecular nature of injury has not been elucidated. Some efforts have been directed to post-mortum analyses of isolated membrane fractions. Our early efforts using isolated plasma membrane enriched fractions (45,51,61,63) were terminated because of uncertainties of the purity of such fractions. Instead, we elected to study the plasma membrane _in situ_ using isolated protoplasts. The first reports of such an approach appeared simultaneously in 1976 (43,62).

It is apropos to select and consider the isolated protoplast as the most appropriate arena...any sphere of struggle or exertion...in which to study freezing injury in higher plants. Isolated protoplasts behave as ideal osmometers over a wide range of osmolalities (0.30 to 3.00 osm) (52,53,64). Over this range of osmolalities, the protoplasts remain spherical facilitating quantitative microscopic studies of cell dehydration. During a freeze-thaw cycle, a protoplast contracts with the minimum volume a function of the lowest temperature attained and subsequently expands to an extent determined by the osmolality of the suspending medium. Thus, calculations of volume and surface area, water flux and permeability, and the extent of supercooling of the intracellular and extracellular solutions are possible. Such parameters are necessary for thermodynamic considerations of cellular volumetric relations during cooling and warming. In addition, various forms of injury and the time and conditions of occurrence can be readily documented.

A significant concern is whether survival of isolated protoplasts frozen _in vitro_ in a suspending medium corresponds with the survival of the intact plant. This concern is suggested by the analysis of freezing stresses by Olien (35,36,37). Survival of isolated protoplasts frozen _in vitro_ parallels the hardiness of the tissue from which the protoplasts were isolated. Following cold acclimation, the hardiness of rye seedlings (Secale cereale L. cv Puma) increases from -3°C to -26°C, as determined by freezing of iso- lated crowns; the LT_{50} of protoplasts isolated from non- acclimated leaves is -2 to -3°C while the LT_{50} for proto- plasts isolated from acclimated tissues is -25 to -28°C. Independent observations of others using protoplasts of rye (44) or an extremely hardy woody species, Robinia pseudoacacia (41), are consistent with this observation. The observations that protoplast hardiness parallels the hardiness of the parent tissue in species with similar levels of hardiness in the non- acclimated state (-3°C) but widely disparate levels of hardi- ness in the acclimated state (-25°C for rye vs. -196°C for R. pseudoacacia) strongly suggest that these observations are not the result of some fortuitous occurrence. Rather, the conclu- sion is that the stresses and resultant lesions arising during freezing of isolated protoplasts frozen in vitro are represen- tative of those occurring during freezing in situ.

III. CRYOMICROSCOPY OF ISOLATED PROTOPLASTS

Observations of biological specimens during a freeze-thaw cycle have been reported for many years with varying degrees of precision in temperature control and optical resolution (5,34,60). Although crude, the approach was rather enlightening. For instance, Molisch (34) was able to distin- guish between extracellular and intracellular ice formation as a function of cooling rate. Although sophistication of the cryomicroscopes subsequently improved (4,22,33,42,57), observa- tions were largely qualitative and primarily concerned with the location of ice formation because precise control of tempera- ture was not effected.

In 1970, Diller and Cravalho (6) detailed the construction of an electronically controlled stage which enabled them to precisely control the specimen temperature and cooling/warming rates. This system has allowed for quantitative studies of the volumetric behavior of various microorganisms (59) or mammalian cell types (7,8,17,31,38). We have constructed a similar cryo- stage for studies of higher plant cell types.

Briefly, the cryomicroscope employs an electronically pro- grammable conduction heat transfer stage whereby the tempera- ture of the protoplasts can be controlled between +56°C and -196°C at 0.05°C intervals. Cooling and warming rates can be varied between 0.1 and 7500°C/min. The light microscope is equipped with phase contrast, dia- and episcopic fluorescence and differential interference contrast optics. Although most other systems use phase contrast optics, it is our experience that the resultant image in the frozen state is unsatisfactory in comparison to that observed with differential interference

contrast optics. The microscope is interfaced with both a
35 mm photomicrographic system and a video system - either
color or high resolution black and white. A video cassette
recording system allows for the immediate and permanent
recording of the sequence of events. The 'freeze-frame' mode
allows for the opportunity to view events that occur within 33
milliseconds. A character generator records the time (to 0.01
sec) and specimen temperature (to 0.01°C) directly on the video
tape.

IV. FREEZE-THAW INDUCED LESIONS IN THE PLASMA MEMBRANE

Three distinct forms of injury to the plasma membrane of
isolated protoplasts have been documented by direct observations during a freeze-thaw cycle. Intracellular ice formation
at large fractional cell volumes is associated with the immediate physical disruption of the plasma membrane. Injury
resulting from cellular dehydratiuon is manifested as either
expansion-induced lysis during warming or the loss of osmotic
responsiveness following cooling. The incidence of any particular lesion is probabilistic and depends on the freeze-thaw
protocol (cooling rate and minimum temperature imposed), the
hardiness of the tissue from which the protoplasts were isolated (acclimated vs. non-acclimated), and the composition of
the suspending medium. For instance, the incidence of intracellular ice formation is high ($\approx$100%) in non-acclimated protoplasts cooled to -20°C at rates greater than 5°C/min, but only
moderate (<50%) in acclimated protoplasts cooled to -40°C at
16°C/min. The incidence of expansion-induced lysis is greatest
in non-acclimated protoplasts frozen to relatively warm subzero
temperatures (-3 to -5°C, the LT_{50}). In contrast, expansion-
induced lysis is seldom observed in acclimated protoplasts.
Instead the predominant form of injury is the loss of osmotic
responsiveness following cooling to temperatures below the
LT_{50} (-25 to -28°C).
   Thus, there is no single form of injury or moment of injury
that is universally applicable under all conditions. In the
following discussion the factors which contribute to the incidence of each of these three lesions, the influence of cold
acclimation, and the molecular basis of injury are discussed.

A. Intracellular Ice Formation

The incidence of intracellular ice formation is generally
considered a function of the cooling rate of the tissue. As
early as 1897, Molisch (34) observed that intracellular ice
formation occurred only upon rapid cooling. Subsequently,
Scarth and Levitt (40) suggested that during rapid cooling
water permeability of the plasma membrane limited water efflux

and that intracellular ice formation resulted from excessive supercooling of the intracellular solution. Later, Mazur (24) derived a quantitative relation between the cooling rate and the extent of supercooling in order to predict the likelihood of intracellular ice formation. From these studies one might infer that the <u>extent</u> of supercooling of the intracellular solution is the primary determinant of intracellular ice formation. Such an inference cannot be made, however, without insight into the factors which are responsible for nucleation of the supercooled solution.

There is a general consensus that intracellular ice nucleation is due to heterogeneous nucleation by the external ice mass. Although the plasma membrane is an effective barrier to the external ice at relatively warm subzero temperatures, its efficacy is diminished at lower temperatures, around -10°C (1,3,16,23,25). Thus, in the absence of foreign nuclei, cells have been supercooled to their homogeneous nucleation temperature (39), although this is not always the case (14). In spite of these considerations, the influence of the temperature dependent heterogeneous nucleation event has been largely avoided in models considering the likelihood of intracellular ice formation (26,28,29) except to note that "differences in the nucleation temperature will cause major differences in the cooling rate required to produce intracellular freezing" (30). A major reason for this omission is the lack of quantitative data on the temperature dependent nucleation event, as only recently has such a study been conducted (17). In this particular study, however, mouse ova suspended in 1 M DMSO nucleated at -40°C independently of the cooling rate over the range of 2 to 30°C/min.

In the area of cold acclimation of higher plants, the omission of any consideration of the temperature dependent nucleation event is more obvious and of longer standing. In 1938, Siminovitch and Scarth (42) reported that the incidence of intracellular ice formation was significantly reduced in acclimated tissues. This was attributed to a purported increase in water permeability of the plasma membrane following cold acclimation (21) which presumably precluded excessive supercooling of the intracellular solution. Since that time, Levitt has steadfastly championed this interpretation (18,19,20). This is in spite of other studies which have failed to confirm an increase in water permeability following cold acclimation (56,58). This is especially disconcerting since the original paper acknowledged that "On account of the complications involved in the methods so far used, the relation between water permeability and hardiness is not completely worked out. Some more satisfactory procedure will have to be evolved in order to get accurate results" (21).

Direct quantitative cryomicroscopic observations of isolated protoplasts permit the direct measurement of the incidence and temperature of intracellular ice formation and the cellular volumetric behavior from which the water permeability of the plasma membrane and the extent of supercooling of the intracellular solution can be determined as a function of cooling rate and minimum temperature imposed and the degree of acclimation. Such studies have recently beeen completed and preliminary reports have appeared previously (9,10,12). Briefly, the following observations have been made.

1. Under similar freezing protocol (cooling rate, minimum temperature imposed, and suspending media), the incidence of intracellular ice formation is consistently greater in protoplasts isolated from non-acclimated tissues (Secale cereale L. cv. Puma leaves) than in those isolated from acclimated tissues.

2. In non-acclimated protoplasts, the incidence of intracellular ice formation is strongly dependent on both the cooling rate and minimum temperature imposed. In contrast, the incidence in acclimated protoplasts is independent of the cooling rate over the range of 3 to $16°C/min$ regardless of the minimum temperature imposed over the range of -5 to -30°C.

3. In non-acclimated protoplasts, nucleation occurred at temperatures between -8 and -13°C. Over this range, the incidence of intracellular ice formation approached 100%. In acclimated protoplasts, nucleation occurred at temperatures between -8 and -29°C, with the incidence not greater than 43%.

4. There was no unique relationship betwen the extent of supercooling and the incidence of intracellular ice formation. At faster cooling rates and lower temperatures, where differences in the incidence of intracellular ice formation between acclimated and non-acclimated protoplasts were the greatest, the extent of supercooling was similar. The extent of supercooling was also similar in those protoplasts which froze intracellularly and those that did not - regardless of whether they were acclimated or non-acclimated.

5. Direct determinations of water flux during freezing indicated that there was no significant difference in the hydraulic conductance of acclimated and non-acclimated protoplasts.

Such observations raise the question as to whether intracellular ice formation is primarily determined by factors which influence the extent of supercooling of the intracellular solution or whether it is determined by the efficacy of the plasma membrane in preventing nucleation of the external ice. The answer, of course, is that both are of concern.

These observations suggest that the incidence of intracellular ice formation is primarily determined by a temperature dependent nucleation event rather than the extent of supercooling of the intracellular solution. Furthermore, the

decreased incidence of intracellular ice formation in acclimated protoplasts is the result of nucleation being deferred to
a lower temperature. This indicates that cold acclimation
increases the efficacy of the plasma membrane in preventing
nucleation by the external ice rather than increasing the water
permeability to minimize the extent of supercooling. In non-
acclimated protoplasts, the efficacy of the plasma membrane to
act as a barrier depends on the cooling rate since the incidence of intracellular ice formation is cooling rate dependent.

Previously, Mazur (25) considered several alternatives
whereby the plasma membrane may be broached by the external
ice. The most enduring alternative appears to be the notion
that ice crystals of sufficiently small radii to pass through
channels in the membrane are only stable at temperatures below
about -10°C (30). In contrast, Asahina (2) has suggested that
intracellular nucleation is the result of alterations in the
membrane. A major distinction between the two perspectives is
whether nucleation should be viewed as a consequence of the
temperature-dependent characteristics of the external ice or of
the membrane barrier.

Video recordings (50) of isolated protoplasts during intracellular ice formation have revealed several interesting
phenomena which may have bearing on this distinction. Immediately (0.03 to 2 sec) before intracellular ice formation, the
following phenomena have been observed: 1) outward flow of the
intracellular solution, 2) fluttering of the plasma membrane,
and 3) eddy-like flow patterns in the cytoplasm. Upon ice
nucleation and attendant gas bubble formation (49), gas bubbles
emerge from the protoplast in the region of the plasma membrane
where the above phenomena were observed. Upon thawing, gas
bubbles and cellular contents (cytoplasm and chloroplasts)
emerge from this same region. If considered singly, the interpretation of any one of these events is equivocal. When taken
collectively, they suggest that mechanical failure of the
plasma membrane precedes intracellular ice nucleation. This
would imply that intracellular ice formation is a consequence
of disruption of the plasma membrane rather than the reverse.
As nucleation is deferred to lower temperatures in acclimated
protoplasts, the molecular distinction may reside in an alteration in the composition of the plasma membrane which renders it
more stable at low temperatures and high solute concentrations.

B. Expansion-Induced Lysis

1. The TSAI Concept. Previously we have established that
one form of freeze-thaw injury to isolated protoplasts is the
result of surface area contraction and expansion of the plasma
membrane during freezing and thawing (64). This aspect of
freeze-thaw injury has been further elaborated in recent publications (52,53,55). Briefly, surface area contractions of the
plasma membrane incurred during cellular dehydration are

incompletely reversible and lysis occurs during subsequent expansion during warming. Thus, the expansion potential of isolated protoplasts is not characterized by a fixed maximum critical surface area. Instead, the expansion potential is characterized by an absolute Tolerable Surface Area Increment which is a constant and independent of the extent of contraction. In a population of protoplasts, the absolute magnitude of change in the mean surface area of the population which results in lysis of 50% of the population is denoted as the $TSAI_{50}$ value. This value varies among the species examined to date and is influenced by the nature of the suspending medium. Most impor- tant, however, the $TSAI_{50}$ progressively increases during cold acclimation.

The TSAI concept was originally developed from observations of populations of protoplasts contracted and expanded by direct osmotic manipulation or via freeze-thaw cycles where the protoplast dimensions were determined before and after the manipulation (64). Subsequently, we have directly verified the validity of the TSAI concept by cryomicroscopy (55). Cryomicroscopy has allowed for the direct measurement of TSAI values of individual cells and it can be demonstrated that expansion-induced lysis occurs at a surface area which is less than the initial surface area if the TSAI value is exceeded during contraction. Furthermore, cryomicroscopic studies confirm the assumption that although the $TSAI_{50}$ is an absolute constant for a population of protoplasts and is independent of the extent of contraction, the TSAI value for any particular protoplast in the population is proportional to its initial size (53). In other words, larger cells have a larger TSAI value, but it does not vary if the cell is contracted.

Recently, the incidence of expansion-induced lysis relative to other freeze-thaw induced lesions and its quantitative accountability for freeze-thaw injury have been elaborated (55). For example, in non-acclimated 'Puma' rye protoplasts the $TSAI_{50} = 1000$ μm^2 and would be exceeded if the protoplasts were frozen to -1 to -2°C (the LT_{50}) and subsequently thawed. Cryomicroscopic studies verify that expansion-induced lysis is the predominant lesion in non-acclimated protoplasts frozen to these temperatures. In acclimated 'Puma' rye protoplasts the $TSAI_{50} = 3000$ μm^2 (54). This value exceeds the extent of contraction and expansion that would be incurred if the protopalsts were to remain spherical when completely dehydrated and subsequently returned to isotonic conditions. Hence, large surface areal contractions are reversible in acclimated protoplasts. Direct cryomicroscopic studies confirm that the incidence of this lesion is low in acclimated protoplasts and does not account for injury incurred at the LT_{50}. Thus, a major stress encountered by non-acclimated protoplasts is the contraction/expansion of the plasma membrane and the resultant lesion, expansion-induced lysis, accounts for injury in non-acclimated protoplasts. Following acclimation, however, the tolerance of the plasma membrane to contractile stresses is increased and a lethal stress-strain situation is precluded. Thus, injury to acclimated protoplasts at the LT_{50} is the result of another stress-strain complex.

2. The Stress-Strain Relationship of Isolated Protoplasts. Further elaboration of the TSAI concept has been concerned with the molecular basis for the behavior of the plasma membrane during contraction and expansion. To this end, the stress-strain relationship (SSR) of isolated protoplasts has been elucidated by micropipette aspiration (11,55,67). Briefly, we have demonstrated that elastic surface area contraction and expansion of the plasma membrane of isolated protoplasts is limited to 2%. Surface area deformations which occur during osmotic manipulation or during a freeze-thaw cycle greatly exceed this value. For instance, non-acclimated protoplasts frozen to only -1 to -2°C undergo a surface area deformation of >33%. Therefore, we have proposed the existence of a reservoir into which plasma membrane material is deposited during contraction and from which material is retrieved during expansion (55). The SSR analysis has allowed for the resolution of the TSAI value in that:

$$TSAI = \frac{(\gamma_c - \gamma_r)}{k_A} + \delta A_o \ (\gamma, t)$$

where γ_c is the critical lysing tension, γ_r is the resting tension, k_A is the modulus of elasticity, and A the area of the membrane; the parameter $\delta A_o(\gamma,t)$ describes the transfer of material between the membrane and the reservoir.

The molecular scenario for the behavior of the plasma membrane undergoing surface area deformation is as follows: Upon exposure to hypertonic conditions, volumetric changes in the protoplasts due to water efflux result in relaxation of the tension (γ_r) in the plasma membrane. As γ_r is on the order of 100 $\mu N \cdot m^{-1}$, small changes in volume (<2%) which occur in milliseconds are sufficient. At zero tension, material is deleted from the plasma membrane until γ_r is reestablished and the protoplast regains sphericity. The amount of material deleted is a function of the extent of the volumetric contraction. When the protoplast is returned to isotonic conditions, rapid water influx occurs. This stretches the membrane elastically and results in a large ($mN \cdot m^{-1}$) tension. Following a surface energy law, membrane material is reincorporated into the membrane. The excursion will not be totally reversible, however, if the rate of reincorporation does not proceed as rapidly as water efflux or if all of the subduced material is not reincorporated into the membrane so that γ_c is exceeded during expansion.

3. Light and Electron Microscopic Studies. The above molecular scenario during contraction may include homogeneous (intrinsic) contraction and closer molecular packing, microscopic buckling or ruffling, or vesiculation and subduction of the membrane into the associated reservoir. Light microscopic studies of non-acclimated protoplasts labelled with Concanavalin-A conjugated with fluorescein and subjected to direct osmotic manipulation suggest that contraction results in endocytotic vesiculation (55).

Behavior of the plasma membrane of acclimated versus non-acclimated protoplasts during a freeze-thaw cycle has also been contrasted. In the absence of a specific plasma membrane marker such as Concanavalin-A, vesiculation of the plasma membrane of non-acclimated protoplasts during freezing cannot be discerned with light microscopy. In contrast, during freezing of acclimated protoplasts, extensive exocytotic vesiculation of the plasma membrane is clearly visible. Most often the extruded material appears as either tethered spheres or long filamentous strands. Although a noticeable proportion of these exclusions becomes detached from the membrane during thawing, the majority remains attached and is gradually reincorporated into the plasma membrane during expansion.

Recently, electron microscopic studies of protoplasts subjected to large surface area changes by direct osmotic manipulation have been initiated (15). In non-acclimated protoplasts subjected to either isotonic or hypertonic conditions, the plasma membrane remains smooth with no observable folding or pleating. Instead, numerous vesicles are observed in the cytoplasm of contracted protoplasts. The frequency of the endocytotic vesicles increases as the extent of contraction is increased.

C. Dehydration-Induced Loss of Osmotic Responsiveness

A second form of injury to isolated protoplasts that results from cellular dehydration is the complete loss of osmotic responsiveness following cooling. Although the protoplasts exhibit characteristic osmometric behavior during cooling, they are osmotically inactive during warming and remain contracted. Thus, the semipermeable characteristics of the plasma membrane are disrupted in the contracted state. This is the predominant form of injury observed in acclimated protoplasts cooled to the LT_{50} (-25 to -28°C). It is also the predominant form of injury observed in non-acclimated protoplasts that are cooled significantly below the LT_{50} (>-10°C).

Dehydration-induced loss of osmotic responsiveness has also been elicited by exposure to hypertonic solutions in the absence of ice or low temperatures (13). As determination of injury in severely dehydrated protoplasts is equivocal, protoplasts were pre-incubated with fluorescein diacetate and

survival was based on the retention of fluorescein in the contracted state. This eliminated the need to return the protoplasts to more dilute solutions and hence precluded any con-
founding of the incidence of injury with that resulting from
expansion-induced lysis. Preliminary reports of such studies
have previously appeared (13,65). From these studies the
following observations have been made.

1. In both non-acclimated and acclimated protoplasts,
   maintenance of semipermeability following exposure to
   hypertonic conditions is a function of both solute
   concentration and duration of exposure, and there is
   no critical concentration threshold above which damage
   occurs. A corollary is that this form of injury is
   not associated with the attainment of an absolute
   minimum critical volume.

2. In non-acclimated protoplasts, the incidence of injury
   is similar at equiosmolal concentrations of either an
   ionic (NaCl + $CaCl_2$) or non-ionic (sorbitol)
   osmoticum. As the osmometric behavior is statistically indistinguishable in the two osmotica at the
   elevated concentrations, a corollary is that the incidence of injury is similar at equal extents of cellular dehydration i.e., equivolumes and independent of
   the species of osmoticum used.

3. Cold acclimation reduces the sensitivity of protoplasts to elevated solute concentrations i.e., the
   incidence of injury in acclimated protoplasts compared
   to the incidence of injury in non-acclimated protoplasts is less at equiosmolalities.

4. Cold acclimation reduces the sensitivity of protoplasts to cellular dehydration if dehydration is
   effected by sorbitol, but not if it is effected by
   NaCl + $CaCl_2$. That is, the incidence of injury in
   acclimated protoplasts compared to non-acclimated
   protoplasts is less at equivolumes if sorbitol is the
   osmoticum, but not if NaCl + $CaCl_2$ is the osmoticum.

When considered in toto, several conclusions can be drawn
from these observations. (It should be noted that, if any one
is considered singly, the conclusions drawn may be completely
different). First, in spite of the similar incidence of injury
in non-acclimated protoplasts suspended in equiosomolal solutions of either sorbitol or NaCl + $CaCl_2$, injury is unlikely
to be due solely to the attainment of a physical minimum critical volume as suggested for other cell types by Meryman (32)
and, more recently, Williams and Shaw (66). If such were the
case, duration of exposure would have a minimal influence.
This was not observed, in that the extent of injury icreased
with time at all solute concentrations employed - long after
volumetric equilibration occurred. This conclusion, however,
is predicated on the assumption that the injury observed is not
a composite of that resulting from a cell volumetric reduction

and a second from a storage injury. Such a consideration is neither proven nor disproven at this time.

Second, the fact that acclimation reduces the sensitivity of protoplasts to elevated solute concentrations, regardless of osmoticum used, can be accounted for by osmotic adjustment. Following cold acclimation, the internal solute concentration of acclimated protoplasts is nearly doubled - as evidenced in Boyle-van't Hoff plots of osmotic behavior (47,54). Therefore, at equiosmolalities acclimated protoplasts are subjected to less cell dehydration than non-acclimated protoplasts, the difference being a direct function of the two-fold increase in internal solute concentration. Thus, the difference in the incidence of injury incurred by non-acclimated and acclimated protoplasts at equiosmolalities is attributable to the difference in the extent of cell dehydration. When $NaCl + CaCl_2$ is the osmoticum, osmotic adjustment entirely accounts for the difference in survival between acclimated and non-acclimated protoplasts. When sorbitol is the osmoticum, it only partially accounts for the difference. If one were to only contrast acclimated and non-acclimated protoplasts in $NaCl + CaCl_2$, support for a volume mediated stress could be garnered. When one considers the responses in both $NaCl + CaCl_2$ and sorbitol, the likelihood of this interpretation is again diminished because of the extent of injury in acclimated versus non-acclimated protoplasts is not equal at equivolumes in sorbitol.

Third, the influence of cold acclimation is, at least, two-fold. Whereas the incidence of injury in sorbitol differs between acclimated and non-acclimated protoplasts when contrasted at both equiosmolalities and equivolumes, cold acclimation must involve more than osmotic adjustment. This suggests that the nature of injury in acclimated protoplasts in sorbitol differs from that incurred in $NaCl + CaCl_2$. At the present time, the molecular basis of injury in either case is unknown. Although this situation makes it easier to propose a basis for the difference, such an endeavor would be unwise at this time.

V.   SUMMARY

During a freeze-thaw cycle, isolated protoplasts are subjected to a spectrum of thermal, mechanical, and chemical perturbations. Disruption of the plasma membrane is a primary consequence, but the specific form of injury depends on the freeze-thaw protocol and the hardiness of the tissue from which the protoplasts were isolated. In protoplasts isolated from non-acclimated tissues and cooled at fast cooling rates (>5 °C/min) to low temperatures (>-10°C), intracellular ice formation is the result of a temperature dependent nucleation event - possibly due to a thermally induced breakdown of the plasma membrane. Cold acclimation decreases the temperature of nucleation. When cooled at slow rates (<3°C/min) to temperatures near the LT_{50} (-2 to -5°C), injury is the result of incompletely reversible surface area contraction incurred by the plasma membrane which results in expansion-induced lysis during warming. Cold acclimation increases the tolerance of the plasma membrane to the mechanical perturbation incurred during contraction and expansion. In protoplasts isolated from acclimated tissues, the predominant form of injury at the LT_{50} (-25°C) is the complete loss of osmotic responsiveness following cooling.

It is obvious that freeze-thaw injury cannot be universally ascribed to a single stress-strain complex - even in the same species at different stages of acclimation. A corollary is that cold acclimation involves a sequence of alterations. It is unlikely that such a sequence will be elucidated by correlative studies of biochemical changes during cold acclimation. Instead, quantitative studies of cellular behavior during a freeze-thaw cycle together with the identification of specific plasma membrane lesions to establish the individual stress-strain relationships are required to correctly assess the significance biochemical changes that occur during cold acclimation. For this purpose, quantitative cryomicroscopy of isolated protoplasts is most enlightening.

REFERENCES

1. Asahina, E., The Institute of Low Temperature Science, Hokkaido University, Sapporo, Japan, 10:83-126 (1956).
2. Asahina, E., Nature 196:445-446 (1962).
3. Asahina, E., and Emura, M., Cryobiology 2:256-262 (1966).
4. Brown, M.S., and Reuter, F.W., Cryobiology 11:185-191 (1974).
5. Chambers, R., and Hale, H.P., Proc. Roy. Soc. Ser. B. 110:336-352 (1932).
6. Diller, K.R., and Cravalho, E.G., Cryobiology 7:191-199 (1970).
7. Diller, K.R., Cravalho, E.G., and Huggins, C.E., Cryobiology 9:429-440 (1972).
8. Diller, K.R., Cravalho, E.G., and Huggins, C.E., Med. Biol. Eng. 14:321-326 (1976).
9. Dowgert, M.F., and Steponkus, P.L., Plant Physiol. 63:S-76 (1979).
10. Dowgert, M.F., and Steponkus, P.L., Plant Physiol. 65:S-45 (1980).
11. Dowgert, M.F., and Steponkus, P.L., Plant Physiol. 67:S-64 (1981).
12. Dowgert, M.F., Steponkus, P.L., Levin, R.L., and Ferguson, J.R., Cryobiology 16:593 (1979).

13. Evans, R.Y., and Steponkus, P.L., Plant Physiol. 67:S-122 (1981).
14. Franks, F., and Bray, M., Cryo-Letters 1:221-226 (1980).
15. Gordon-Kamm, W.J., and Steponkus, P.L., Plant Physiol. 67:S-122 (1981).
16. Gupta, K.C., Cryobiology 12:417-426 (1975).
17. Leibo, S.P., McGrath, J.J., and Cravalho, E.G., Cryobiology 15:257-271 (1978).
18. Levitt, J., "Responses of Plants to Environmental Stress." Academic Press, New York, 1972.
19. Levitt, J., in "Plant Cold Hardiness and Freezing Stress" (P.H. Li and A. Sakai, eds.), pp. 3-15. Academic Press, New York, 1978.
20. Levitt, J., "Responses of Plants to Environmental Stresses," Vol. 1. Academic Press, New York, 1980.
21. Levitt, J. and Scarth, G.W., Can. J. Res. C. 14:267-305 (1936).
22. Luyet, B.J., and Gibbs, M.C., Biodynamica 1(25):1-18 (1937).
23. Mazur, P., Biophys. J. 1:247-264 (1961).
24. Mazur, P., J. Gen. Physiol. 47:347-369 (1963).
25. Mazur, P., Ann. NY Acad. Sci. 125:658-676 (1965).
26. Mazur, P., in "Cryobiology" (H.T. Meryman, ed.), pp. 213-315. Academic Press, New York, 1966.
27. Mazur, P., Cryobiology 2:181-192 (1966).
28. Mazur, P., Ann. Rev. Plant Physiol. 20:419-448 (1969).
29. Mazur, P., Science 168:939-949 (1970).
30. Mazur, P., Cryobiology 14:251-272 (1977).
31. McGrath, J.J., E.G., Cravalho, and Huggins, C.E., Cryobiology 12:540-550 (1975).
32. Meryman, H.T., Nature 218:333-336 (1968).
33. Modlibowska, I., and Rogers, W.S., J. Exp. Bot. 6:384-391 (1955).
34. Molisch, H., Fischer, Jena. 1-73 (1897).
35. Olien, C.R., Ann. Rev. Plant Physiol. 18:387-408 (1967).
36. Olien, C.R., USDA Tech. Bull. No. 1558 (1977).
37. Olien, C.R., in "Plant Cold Hardiness and Freezing Stress - Mechanisms and Crop Implications" (P.H. Li and A. Sakai, eds.), pp. 37-48. Academic Press, New York, 1978.
38. Rall, W.F., Reid, D.S., and Farrant, J., Nature 286:511-514 (1980).
39. Rasmussen, D.H., MaCauley, M.N., and MacKenzie, A.P., Cryobiology 12:328-339 (1975).
40. Scarth, G.W., and Levitt, J., Plant Physiol. 12:51-78 (1937).
41. Siminovitch, D., Plant Physiol. 63:722-725 (1979).
42. Siminovitch, D., and Scarth, G.W., Can. J. Res. C. 16:467-481 (1938).
43. Siminovitch, D., Singh, J., Keller, W.A., and de la Roche, I.A., Cryobiology 13:670 (1976).
44. Singh, J., Plant Sci. Letters 15:195-201 (1979).
45. Steponkus, P.L., HortScience 9:282 (1973).

46. Steponkus, P.L., Adv. in Agron. 30:51-98 (1978).
47. Steponkus, P.L., in "Genetic Engineering of Osmoregulation" (D.W. Rains and R.C. Valentine, eds.), pp. 235-255. Plenum Press, New York, 1980.
48. Steponkus, P.L., in "Encyclopedia of Plant Physiology, New Series" Vol. 12A, (O.L. Lange, P.S. NObel, C.B., Osmond and H. Ziegler, eds.), Chapter 12, pp. 371-402. Springer-Verlag, Berlin, 1981.
49. Steponkus, P.L., and Dowgert, M.F., Cryo-Letters 2:42-47 (1981).
50. Steponkus, P.L., and Dowgert, M.F., Plant Physiol. 67:S-58 (1981).
51. Steponkus, P.L., and Wiest, S.C., Cryobiology 10:532 (1973).
52. Steponkus, P.L., and Wiest, S.C., in "Plant Cold Hardiness and Freezing Stress - Mechanisms and Crop Implications" (P.H. Li and A. Sakai, eds.), pp. 75-91. Academic Press, New York, 1978.
53. Steponkus, P.L., and Wiest, S.C., in "Low Temperature Stress in Crop Plants: The Role in the Membrane" (J.M. Lyons, D.G. Graham and J.K. Raison, eds.), pp. 231-254. Academic Press, New York, 1979.
54. Steponkus, P.L., Dowgert, M.F. and Roberts, S.R., Cryobiology 16:594 (1979).
55. Steponkus, P.L., Wolfe, J., and Dowgert, M.F., in "Effects of Low Temperatures on Biological Membranes" (A. Clarke and G.J. Morris, eds.), pp. 307-322. Academic Press, New York, 1981.
56. Stout, D.G., Steponkus, P.L., and Cotts, R.M., Plant Physiol. 50:374-378 (1977).
57. Stuckey, I.H., and Curtis, O.F., Plant Physiol. 13:815-823 (1938).
58. Sukumaran, N.P., and Weiser, C.J., Plant Physiol. 50:564-567 (1972).
59. Ushiyama, M., Cravalho, E.G., Diller, K.R., and Huggins, C.E., Cryobiology 10:517-518 (1973).
60. Wiegand, K.M., Plant World 9:25-39 (1906).
61. Wiest, S.C., and Steponkus, P.L., Cryobiology 12:555 (1975).
62. Wiest, S.C., and Steponkus, P.L., Cryobiology 13:670 (1976).
63. Wiest, S.C., and Steponkus, P.L., J. Amer. Soc. Hort. Sci. 102:119-123 (1977).
64. Wiest, S.C., and Steponkus, P.L., Plant Physiol. 62:699-705 (1978).
65. Wiest, S.C., and Steponkus, P.L., Cryobiology 16:592-593 (1979).
66. Williams, R.J., and Shaw, S.K., Cryobiolgy 17:530-539 (1980).
67. Wolfe, J., and Steponkus, P.L., Biochim. Biophys. Acta 643:663-668 (1981).

THE MECHANISM OF CRYOPROTECTION OF BIOMEMBRANE SYSTEMS
BY CARBOHYDRATES

Kurt A. Santarius

Institute of Botany
University of Düsseldorf
Düsseldorf, West Germany

I. INTRODUCTION

Freezing of isolated chloroplasts and mitochondria in the
presence of potentially membrane-toxic compounds leads to
inactivation of the membrane systems. This membrane inactivation can be prevented if other solutes such as sugars and sugar
derivatives, certain amino acids and carbonic acids and various
non-physiological compounds, f.i. dimethylsulfoxide and poly-
ethylenglycol, are simultaneously present during freezing in
sufficient concentration. Besides low molecular weight solutes
macromolecules such as polypeptides and polyvinylpyrrolidone
are also effective in cryopreservation.

The mechanism by which these cryoprotectants prevent damage
to biomembrane systems during a freeze-thaw cycle has been
studied intensively during the last two to three decades. The
accumulated evidence suggests that protection may be attributed
to a colligative mechanism, i.e. cryoprotective agents prevent
unspecifically the increase in the concentration of potentially
membrane-toxic solutes during freezing up to a critical limit
which would inactivate membranes; in addition to the colligative effect specific interactions between cryoprotective compounds and biomembranes should play a role in cryopreservation
(for literature see ref. 9,10,11). However, on the other hand
it also has been reported that both low molecular weight compounds and polymers basically act colligatively, i.e. direct
biochemical stabilizing mechanisms conferred by cryoprotectants
are supposedly not involved in cryopreservation (4,5,15,17).
Therefore, it is an open question whether protection of biomembranes during freezing in the presence of various cryoprotective agents can be explained either solely by the

colligative concept or by both colligative and specific action of the solutes.

In order to get information on the mechanism of cryopreservation of biomembranes, we investigated the effect of various carbohydrates of low molecular weights such as sugars and polyhydric alcohols and of polymers such as dextrans on thylakoids in the course of a freeze-thaw cycle.

II. MATERIALS AND METHODS

Experimental data were obtained with thylakoids which were isolated from spinach leaves (Spinacia oleracea L. cv. Monatol) and used as model systems for biomembranes. Freezing took place for 3 to 4 h at about -20 to -25°C in the presence of various concentrations of carbohydrates and potentially membrane-toxic electrolytes. After thawing, the capacity for phenazine methosulfate-mediated cyclic photophosphorylation was used as an assay for membrane integrity. Freezing point depressions of the solutions of various cryoprotectants were measured with a semimicro osmometer (H. Knauer, Berlin). For determination of the sodium chloride concentration in the thylakoid suspensions a micro-chlorocounter (Labo International B.V., Marius, Krimpen, Holland) was used. - Experimental procedures used here were recently described in detail (25).

III. CRYOPROTECTION BY LOW MOLECULAR WEIGHT CARBOHYDRATES

During freezing of a thylakoid suspension water is converted to ice and the solutes become concentrated in the unfrozen part of the system which also contains the membranes. The total osmolality reached in the unfrozen solution is dependent on the freezing temperature. It is a common view that damage to isolated biomembranes in the course of a freeze-thaw cycle is due to increase in the concentration of potentially membrane-toxic solutes, f.i. NaCl, in the surroundings of the membranes during ice formation (7,8,21,28,31). If different solutes are present in the medium, the concentration reached by each compound in the unfrozen part of the system is dependent on the osmotic ratio of the components. As sugars and sugar derivatives do not affect thylakoids even at high concentrations (26), it is obvious that protection of these membranes during freezing in the presence of carbohydrates can be explained mainly by colligative action of the solutes, i.e. at a given freezing temperature the concentration of potentially membrane-toxic compounds in the unfrozen solution is lower in the presence of a cryoprotectant than in its absence. This

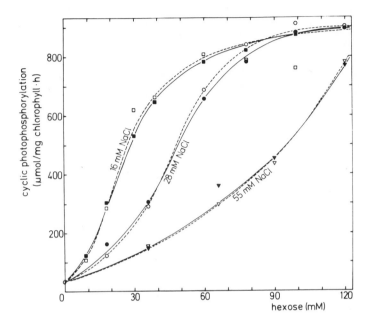

FIGURE 1. The cryoprotective effect of D(+)-glucose and D(-)-fructose on cyclic photophosphorylation of isolated thylakoid membranes in the presence of various concentrations of NaCl as a function of the hexose concentration. ----- ■/●/▲ ----- glucose; ——— □/○/△ ——— fructose. At a given NaCl concentration, differences in the cryoprotective efficiency of glucose and fructose are within limits of error.

conclusion can be drawn from the following results obtained from freezing experiments with isolated thylakoid membranes:

1) For protection the ratio between the cryoprotectant (f.i. sugar) and the potentially membrane-toxic compound (f.i. NaCl) is decisive, not the absolute concentration of these solutes in the membrane suspension before freezing, i.e. the higher the salt concentration before freezing the more sugar is necessary to reach a comparable degree of cryoprotection. This was already shown earlier (8) and is also evident from Figures 1 to 3.

2) If isolated chloroplast membranes were frozen in the presence of a constant level of potentially membrane-toxic solutes and various concentrations of different cryoprotective compounds which do not show differences in their osmotic behavior even at high concentrations, comparable degrees of membrane preservation were obtained when the cryoprotectants were present in comparable molalities in the medium. E.g., equimolal solutions of various hexoses exhibit almost identical

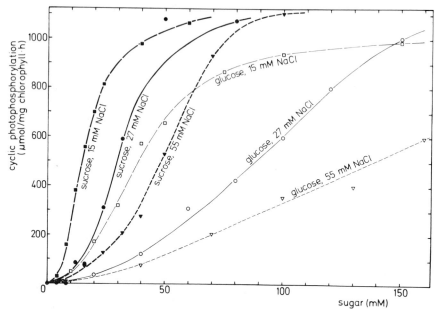

FIGURE 2. The cryoprotective effect of sucrose and D(+)-glucose on cyclic photophosphorylation of isolated thylakoid membranes in the presence of various concentrations of NaCl as a function of the sugar concentrations.

freezing point depressions. In Figure 1 it is demonstrated that glucose and fructose protect thylakoid membranes during freezing to a comparable extent.

3) If equimolal solutions of various cryoprotectants exhibit differences in freezing point depressions at higher concentrations, different amounts of these solutes are necessary for complete membrane preservation. To reach a certain degree of cryoprotection more of that cryoprotectant has to be added to the thylakoid suspension before freezing which reveals smaller deviation from ideal thermodynamic behavior (15,17). For example, with respect to increasing activity with increasing concentration sucrose exhibits much larger departure from ideal behavior than glucose (15). As can be seen in Figure 2, to maintain f.i. 50% of the capacity for cyclic photophosphorylation during freezing, on a molar basis considerably more glucose had to be present in the thylakoid suspension than sucrose (see also ref. 24,29,30,31). If the concentration of NaCl present in the medium was altered, with increasing salt level the quantity of sugar necessary for comparable degree of membrane protection had to be increased simultaneously to a much larger extent in the case of glucose than of sucrose.

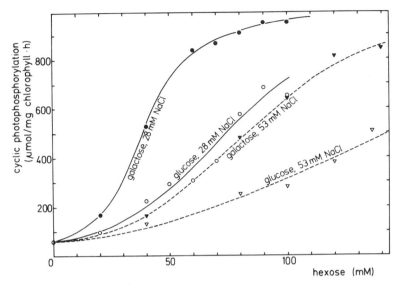

FIGURE 3. The cryoprotective effect of D(+)-glucose and D(+)-galactose on cyclic photophosphorylation of isolated thylakoid membranes in the presence of various concentrations of NaCl as a function of the hexose concentrations.

The data clearly suggest that at least in the presence of low molecular weight carbohydrates colligative action of the compounds plays a very important role in cryoprotection.

However, there are results available either already shown in the literature or obtained in our laboratory during the last months which cannot be explained solely by the colligative concept.

As mentioned already equimolal solutions of various hexoses exhibit almost identical freezing point depressions. Whereas, on a equimolal basis, glucose and fructose exert comparable degrees of membrane protection (Fig. 1), conspicuous differences in the cryoprotective efficiency of glucose and galactose were observed in the presence of various NaCl concentrations (Fig. 3). It is obvious that this result cannot be explained solely by a salt-buffering action of the sugars.

Similar data were obtained if other low molecular weight carbohydrates were used as cryoprotectants. E.g., freezing point depressions of equimolal solutions of 2-deoxy-D-galactose and 6-deoxy-L-mannose exhibit nearly comparable degrees of deviation from the thermodynamic behavior of an ideal solute. Nevertheless, at a constant level of potentially membrane-toxic compounds, fucose was considerably more effective in protection of isolated thylakoid membranes during freezing than rhamnose. Furthermore, on a molal basis, various pentoses show similar

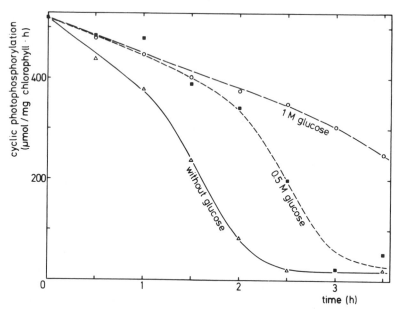

FIGURE 4. Protective effect of D(+)-glucose on cyclic photophosphorylation of isolated thylakoid membranes which were suspended at 0°C in 0.8 M NaCl in the presence and absence of sugar as a function of time.

freezing point depressions as a function of solute concentration. However, freezing experiments with isolated thylakoids in the presence of various concentrations of these carbohydrates demonstrated that f.i. D(-)-arabinose was much more effective in cryoprotection than D(+)-xylose. Also various alditols such as xylitol and meso-erythritol exhibit only small differences in deviation from ideal osmotic behavior but showed large differences in their cryoprotective efficiency. All these findings indicate that cryopreservation cannot be explained solely by the colligative concept but must involve other mechanisms such as specific interactions between solutes and membranes. - It should be mentioned that in the case of some sugars, f.i. rhamnose, ribose and xylose, solubility properties possibly could impair cryoprotective qualities: these sugars tend to crystallize at least in part during freezing when present at higher concentrations in the membrane suspension. Crystallization of solutes, however, produces extensive membrane damage (23,28).

In addition to the results obtained during freezing of isolated chloroplast membranes, also experiments performed with various cryoprotective solutes in the absence of ice formation supply data which are in contradiction to the colligative

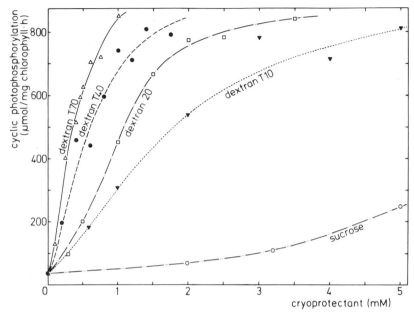

FIGURE 5. The cryoprotective effect of different dextrans and sucrose on cyclic photophosphorylation of isolated thylakoid membranes as a function of the dextran concentration. Thylakoids were frozen in the presence of 3.5 mM NaCl and various concentrations of the polymers and sucrose, respectively, as indicated on the abscissa.

concept. As mentioned already, the increase in the concentration of potentially membrane-toxic compounds during freezing is the driving force for membrane inactivation. If cryoprotectants exert only a colligative effect, one would expect that in the absence of freezing, f.i. at 0°C, membrane inactivation by exposure of thylakoids to elevated concentrations of NaCl should be independently of cryoprotectants such as carbohydrates also present in the membrane suspensions. However, it was already shown earlier (22) that inactivation of thylakoids in concentrated NaCl solutions occurred slower in the presence of sucrose than in its absence although under these conditions the sugar did not reduce the concentration of the membrane-toxic electrolyte in the surroundings of the membranes. A similar effect can be observed if instead of sucrose other carbohydrates such as glucose (Fig. 4), raffinose or glycerol were used. These findings suggest that carbohydrates are able to stabilize biomembranes by specific solute-membrane interactions against the deleterious effect of toxic compounds, i.e. protection on a noncolligative basis was observed.

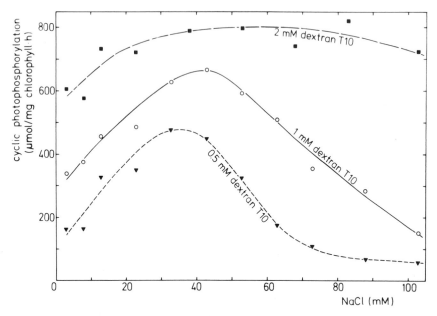

FIGURE 6. Cryoprotection of cyclic photophosphorylation of isolated thylakoid membranes by dextran T10 as a function of the NaCl concentration.

IV.   CRYOPROTECTION BY POLYMERS

In order to investigate the cryoprotective efficiency of carbohydrates with high molecular weights and the mechanism by which these polymers prevent membrane inactivation during freezing, dextrans of average molecular weights between 10,000 and 70,000 daltons have been used.

In Figure 5 it is shown that freezing of isolated thylakoids in the presence of relative low concentrations of dextrans already results in partial or complete protection of the membranes. On a molar basis, membrane stabilization increases with increasing molecular weight. This result reminds on data obtained with sugars and sugar derivatives of different molecular size (Fig. 2; cf. ref. 15,24,29,30). Therefore, the question arises on the non-ideal nature of the system at higher polymer concentrations. Freezing point depression studies on high concentrated dextran solutions are not yet done. However, detailed investigations on aqueous polyvinylpyrrolidone solutions are available. Jellinek and Fok (13) found by means of differential thermal analysis that at low concentrations the polyvinylpyrrolidone solution behaved as would be expected from the number of solute particles in solution; as a result the

freezing point depression was very small. At higher concentrations the freezing point decreased drastically indicating a
great deviation from ideal thermodynamic behavior. These
results have been confirmed with polyvinylpyrrolidone of molecular weights comparable to those of dextrans used here
(3,4,5). Thus, colligative action of polymers cannot be
excluded as a possible mechanism of protection of biomembranes
from damage during freezing.

In Figure 6 it is shown that in the presence of low concentrations of dextrans which were not sufficient for complete
membrane protection the extent of membrane preservation can be
considerably improved by the addition of either more dextran or
more potentially membrane-toxic electrolytes, f.i. NaCl. At a
given low dextran level, f.i. 0.5 or 1 mM dextran T10, with
increasing NaCl concentration first protection rose up to an
optimum and then decreased again. From Figure 6 it is also
evident that for complete membrane protection a minimum concentration of the polymer must be present in the surroundings of
the membranes during freezing. The finding that at low dextran
concentration even the addition of potentially membrane-toxic
NaCl led to an increase in membrane protection during freezing
is in contrast to results obtained with low molecular weight
carbohydrates: when sugars were used as cryoprotectants, at
low sugar concentrations an increase in protection by NaCl was
never observed, and elevated NaCl concentrations led to
increased loss of activity of cyclic photophosphorylation
during freezing (see Figures 1 to 3). Presumably the improvement of membrane preservation afforded by NaCl cannot be
explained by the colligative concept but must involve other
mechanisms.

The increase in cryoprotection caused by NaCl in the presence of low concentrations of dextrans can be also obtained if
instead of NaCl other inorganic salts, f.i. various alkali
metal chlorides, were present during freezing. Although the
extent of improvement of membrane preservation by different
electrolytes was limited by both salt crystallization and the
toxicity of the solutes, special attention should be called to
the fact that at a given low dextran concentration the increase
in membrane protection caused by different salts was only
dependent on the concentration of the electrolytes but was
almost independent from the nature of the salts. The observation that the improvement of membrane preservation by the addition of potentially membrane-toxic electrolytes was fairly
unspecific suggests that cryoprotection is rather due to a
physical mechanism but to specific interactions between the
solutes and the membrane.

As mentioned already from Figure 6 it is also evident that
at higher salt concentrations a decrease in membrane protection
was observed. The degree of membrane preservation reached in
the presence of higher NaCl concentrations is dependent on the
ratio between the cryoprotectant and the potentially

membrane-toxic compound, i.e. the higher the dextran concentration the more salt had to be present to overcome the protective effect produced by the polymer. This reminds on results obtained with sugars as cryoprotective agents (8; see also Figures 1 to 3) and might suggest that a salt-buffering action of dextran also plays a role in membrane protection.

On the other hand, inactivation of isolated chloroplast membranes during exposure to concentrated NaCl solutions at 0°C can be diminished when dextrans were also present in the suspension. This observation is in agreement with the effect of low molecular weight carbohydrates on isolated thylakoids in the absence of freezing (cf. Fig. 4) and cannot be interpreted in terms of colligative properties of the solutes because under these conditions the presence of dextran did not reduce the concentration of the membrane-toxic electrolyte in the surroundings of the thylakoids. This clearly indicates that also high molecular weight carbohydrates such as dextrans have protective qualities beyond those that can be attributed solely to colligative factors.

A part of the work done with dextrans became recently published (25).

V. CONCLUSIONS

From the data presented here it can be concluded that cryopreservation of biomembranes in the presence of carbohydrates can be explained mainly by colligative action of the solutes, i.e. the predominant mechanism of protection is due to an unspecific "dilution" of potentially membrane-toxic compounds such as electrolytes by the carbohydrates to below the critical limit at which membranes would be inactivated. This is valid particularly when low molecular weight carbohydrates such as sugars and polyhydric alcohols were used as cryoprotectants; to some extent, cryopreservation by dextrans is also due to the ability of polymers to reduce the amount of water frozen on a simple colligative basis.

However, there are data available obtained with low molecular weight carbohydrates and polymers which cannot be explained solely by the colligative concept. Therefore, additionally noncolligative mechanisms must be involved in cryopreservation. The molecular nature of noncolligative protection by sugars, polyhydric alcohols and dextrans is not yet known. Two possibilities could be taken into consideration:

1) It might be possible that specific interactions between carbohydrates and the membranes - presumably the membrane proteins (cf. ref. 33) - contribute to membrane stabilization and, thereby, protect membranes against the deleterious effect of high concentrations of electrolytes at 0°C and during freezing. E.g, it is suggested that formation of hydrogen

bonds between carbohydrates and proteins is involved in protection, which could prevent destabilization of membrane proteins during salt stress (8,12,20,32,34). Presumably, carbohydrates can influence water binding because of the affinity of their OH groups to water (27). It appears possible that they can replace bound water or some phase of water adjacent to sensitive proteins or can be bound directly to proteins (cf. ref. 6), a process which could stabilize proteins and, therefore, prevent changes in the membrane structure.

2) Under conditions of extremely high concentrations reached in the unfrozen part of the system during ice formation interactions between carbohydrates and water and possibly also potentially membrane-toxic compounds such as electrolytes may become rather more important and, thus, additionally alter the conditions in the surroundings of the membranes. E.g., it was found that at least polymers cause alterations in the water structure and hinder the crystallization of water (1,2,3,16,18, 19,35) which may contribute to their protective action. Recently, it was concluded from freezing experiments with hydroxyethyl starch that the cryoprotective action of this polymer and possibly other macromolecular additives essentially seems to depend on their ability to absorb water and to keep it thermally inert (14).

ACKNOWLE DGMENTS

    The competent technical assistance of Miss Britta Dietzel and Miss Margrit Meyer is gratefully acknowledged.

REFERENCES

1. Amrhein, E.-M., Cryobiology 12:340-352 (1975).
2. Ashwood-Smith, M.J., Warby, C., Connor, K.W., and Becker, G., Cryobiology 9:441-449 (1972).
3. Connor, W., and Ashwood-Smith, M.J., Cryobiology 10:488-496 (1973).
4. Farrant, J., Nature 222:1175-1176 (1969).
5. Farrant, J., and Woolgar, A. E., in "Ciba Found. Symp. on The Frozen Cell" (G.E.W. Wolstenholme and M. O'Connor, eds.), pp. 97-119. J. & A. Churchill, London, 1970.
6. Giles, C.H., and McKay, R.B., J. Biol. Chem. 237:3388-3392 (1962).
7. Heber, U., Plant Physiol. 42:1343-1350 (1967).
8. Heber, U., and Santarius, K.A., Plant Physiol. 39:712-719 (1964).

9. Heber, U., and Santarius, K.A., in "Temperature and Life" (H. Precht, J. Christophersen, H. Hensel and W. Larcher, eds.), pp. 232-263. Springer, Berlin/Heidelberg/New York, 1973.

10. Heber, U., and Santarius, K.A., in "Water and Plant Life" (O.L. Lange, L. Kappen and E.-D. Schulze, eds.), Ecological Studies, Vol. 19, pp. 253-267. Springer, Berlin/Heidelberg, 1976.

11. Heber, U., Volger, H., Overbeck, V., and Santarius, K.A., in "Proteins at Low Temperatures" (O. Fennema, ed.), Adv. Chem. Ser. 180:159-189 (1979).

12. Huggins, C.E., Fed. Proc. 24:S190-S195 (1965).

13. Jellinek, H.H.G., and Fok, S.Y., Kolloid-Z. & Z. Polymere 220:122-133 (1967).

14. Körber, C., and Scheiwe, M.W., Cryobiology 17:54-65 (1980).

15. Lineberger, R.D., and Steponkus, P.L., Plant Physiol. 65:298-304 (1980).

16. Luyet, B., and Rasmussen, D., Biodynamica 10:137-147 (1967).

17. Meryman, H.T., Williams, R.J., and Douglas, M.S.J., Cryobiology 14:287-302 (1977).

18. Pribor, D.B., Cryobiology 11:60-72 (1974).

19. Rapatz, G., and Luyet, B., Biodynamica 10:149-166 (1968).

20. Santarius, K.A., in "Low Temperature Biology of Foodstuffs: (J. Hawthorn and E.J. Rolfe, eds.), Rec. Adv. Food Sci., Vol. 4, pp. 135-151. Pergamon Press, Oxford, 1968.

21. Santarius, K.A., Planta 89:23-46 (1969).

22. Santarius, K.A., Plant Physiol. 48:156-162 (1971).

23. Santarius, K.A., Biochim. Biophys. Acta 291:38-50 (1973).

24. Santarius, K.A., Planta 113:105-114 (1973).

25. Santarius, K.A., Cryobiology, in press.

26. Santarius, K.A., and Ernst, R., Planta 73:91-108 (1967).

27. Santarius, K.A., and Heber, U., Planta 73:109-137 (1967).

28. Santarius, K.A., and Heber, U., Cryobiology 7:71-78 (1970).

29. Santarius, K.A., and Heber, U., in "Proc. Colloq. on the Winter Hardiness of Cereals" (S. Rajki, ed.), pp. 7-29. Agric. Res. Inst. Hung. Acad. Sci., Martonvasar, 1972.

30. Steponkus, P.L., Garber, M.P., Myers, S.P., and Lineberger, R.D., Cryobiology 14:303-321 (1977).

31. Thebud, R., and Santarius, K.A., Planta 152:242-247 (1981).

32. Ullrich, H., Angew Bot. 36:258-272 (1962).

33. Volger, H., Heber, U., and Berzborn, R.J., Biochim. Biophys. Acta 511:455-469 (1978).

34. Webb, S.J., Bound Water in Biological Integrity. Thomas Publ., Springfield, 1965.

35. Williams, R.J., and Harris, D., Cryobiology 14:670-680 (1977).

POSSIBLE INVOLVEMENT OF MEMBRANE FLUIDITY CHANGES
IN THE SURVIVAL OF PLANT CULTURED CELLS EXPOSED TO
SUBFREEZING TEMPERATURES IN THE PRESENCE OF DMSO

M. Uemura, S. Yoshida and A. Sakai

The Institute of Low Temperature Science
Hokkaido University
Sapporo, Japan

ABSTRACT

The effect of DMSO on the fluidity of mitochondria and
plasma membrane isolated from cultured cells of Jerusalem artichoke were assessed by fluorescence polarization using
1,6-diphenyl-1,3,5-hexatriene and by electron spin resonance of
5-nitroxide stearic acid in order to elucidate the cryoprotective mechanism of DMSO. Membrane samples were obtained with
purity suitable to the objective of the experiments. When
freezing was initiated in the absence of DMSO, an increase in
motion-dependent parameter in the electron spin resonance spectrum was observed in both types of membrane. However, this
phenomenon was not detected when frozen in the presence of
DMSO. In the fluorescence polarization measurements, DMSO prevented decrease in membrane fluidity at lower temperatures,
below 5°C, in both types of membrane. These results suggest
that DMSO exerts its effect on the hydrophilic regions of membrane lipids and prevents membrane fluidity from irreversible
decreasing at low temperatures at which the cells might otherwise sustain freezing injury.

I.   INTRODUCTION

It has been reported by many investigators that the occurrence of membrane lipid phase transition in chilling-sensitive
plants is the primary cause of chilling injury (14,16,17,22).
The membrane lipid phase transition will affect membrane functions such as semipermeability, selective ion-transport and

activities of membrane-bound enzymes. Although damage to the cell membrane after freezing is known to be associated with impairment of these functions and the irreversible breakdown of membrane structure, resulting in death of cells (18), little is known about molecular changes in membranes when cells are subjected to extracellular freezing. Recently, the possible role of membrane lipid phase transition in freezing injury has been proposed (3,10), but only limited direct evidence has been presented to verify this hypothesis (19).

On the other hand, the cryoprotectant, dimethylsulfoxide (DMSO) has been used for cryopreservation of plant germplasm in liquid nitrogen (24). In all species, except for some very hardy ones, the addition of one or more cryoprotective substances before freezing is essential for survival in liquid nitrogen. At present the cryoprotective mechanism of DMSO remains poorly understood.

In the present study, we tested the temperature-dependent phase transition of mitochondria and plasma membrane isolated from cultured cells derived from tubers of Jerusalem artichoke in order to elucidate the cryoprotective effects of DMSO on the cellular membranes of plant cells exposed to sub-freezing temperatures. The thermotropic transition of lipid bilayers in membranes has been examined by fluorescence polarization using 1,6-diphenyl-1,3,5-hexatriene (DPH) as a probe and by electron spin resonance (ESR) spectroscopy of 5-nitroxide stearic acid in membrane lipid domains. The effect of DMSO on the cellular membranes will be discussed from a biophysical point of view.

II.   MATERIALS AND METHODS

Cultured cells derived from tubers of Jerusalem artichoke (Helianthus tuberosus L.) were used as the experimental material. The tubers were harvested at the Hokkaido University campus in late November or early December and stored in ice until they were used. The cultured cells were induced by the method of Sugawara (21). In most cases 10 day cultures were used.

To evaluate frost hardiness, cultured cells in test tubes, in the presence or absence of 10% DMSO, were cooled at -5°C for 1 hr. To initiate freezing in supercooled cell suspensions, a small piece of ice was touched to the surface of the liquid. Frozen cells were kept at -5°C for 2 hr and then cooled to each test temperature by transferring the test tube to temperature-controlled boxes at -10, -15 and -20°C at 2 hr intervals and held at final test temperature for 12 hr at least. Frozen cells were slowly rewarmed in air at 0°C. Survival rate of frozen-rewarmed cells was evaluated by the TTC reduction test according to the method of Steponkus and Lanphear (20) with a slight modification.

The cultured cells were homogenized with a Polytron RT-20 for 60 sec at 0°C in order to isolate mitochondria and plasma membrane. The grinding medium consisted of 0.7M sorbitol, 140mM Tris-HCl (pH 7.8), 10.5mM EGTA, 20mM ß-mercaptoethanol and 3.5% (v/v) PVP. One ml of the grinding medium was used per g fresh weight of tissues. The brei was squeezed through two layers of gauze and one layer of Miracloth. The cell free extracts were successively centrifuged at 1,300 x g for 10 min to remove cell debris; pellets obtained at 14,000 x g for 15 min and at 170,000 x g for 20 min were designated as clude mitochondria fraction and clude plasma membrane-enriched fraction, respectively. Each pellet was washed twice and then resuspeneded after centrifugation in 0.7M soribitol-1mM EDTA-10mM Tris-MES (pH 7.2).

Five ml of the 14,000 x g pellet suspension were loaded onto 30 ml of a linear sucrose density gradient (15-50% w/v) and centrifuged for 2 hr at 96,000 x g. Five ml of the 170,000 x g pellet suspension were loaded onto 30 ml of Percoll (14.4% v/v)-dextran T40 (3% w/v) and centrifuged for 45 min at 94,500 x g. After centrifugation, 1.2 ml aliquots were collected and enzyme activities and protein content in each aliquot were determined as reported previously (25). After plasma membrane fractions determined as above were collected and centrifuged to remove Percoll, the pelleted material was burst in hypotonic medium consisting of 1mM EDTA-10mM Tris-MES (pH 7.2). This suspension was loaded onto a linear sucrose density gradient and then centrifuged for 13.5 hr at 96,000 x g. After collection of plasma membrane-enriched fractions, inhibition of the activity of ATPase (pH 6.5) by some inhibitors such as sodium vanadate, N,N'-dicyclohexylcarbodiimide (DCCD) and diethylstirbestrol was examined each at an adequate concentration. The mitochondria and plasma membrane-enriched fractions were diluted and centrifuged, and then they were resuspended in 0.7M sorbitol-1mM EDTA-10mM Tris-MES (pH 7.2).

Membrane suspensions (0.5 mg protein in a final volume of 2.5 ml) were incubated with 4×10^{-6}M DPH solution containing 40% ethylene glycol in the presence or absence of 10% DMSO at 25°C for 30 min. Steady-state fluorescence poralization was monitored with a microviscosimeter model MV-1a (Elscint Co. Ltd., Israel) and expressed as the fluorescence anisotropy, r, and as the anisotropy parameter $(r_0/r-1)^{-1}$; where r_0 is the maximal limiting anisotropy of DPH which was 0.362 (11). All temperature scans between about 30 to -20°C were made in descending direction.

Membrane suspensions (0.5 mg protein) were incubated for 2 hr at 25°C with 5 µl of 5-nitroxide stearic acid from a 20mM benzene stock solution containing 1mM potassium ferrycyanide to prevent reduction of the label. Labeled membranes were washed by centrifugation and the supernatant were discarded. The pelleted membranes were aspirated into glass pipets with about 0.5 mm in diameter. The pipets were sealed at the bottom and

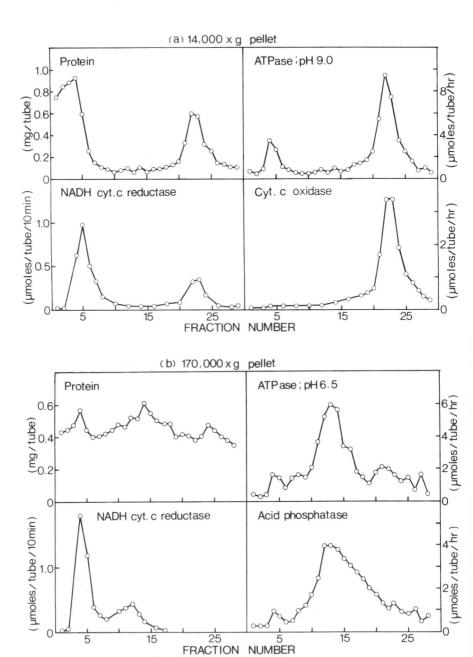

FIGURE 1. Fractionations of enzymes in the 14,000 x g pellet by sucrose density gradient centrifugation (a) and in the 170,000 x g pellet by Percoll-dextran gradient centrifugation (b).

placed into a glass sample holder with 5 mm in diameter. ESR spectra were recorded with a JEOL JES-FE3S (JEOL Co. Ltd., Japan) with a temperature-controlled unit within the microwave cavity. The field was set at 3296 gauss and the microwave power was set at 1 mW. Freezing of samples in the glass pipet was usually initiated at -5 to -6°C. The temperature was lowered approximately 3°C every 20 min. Spectra were recorded after equilibration at each temperature.

III.   RESULTS

To obtain more precise information about the cryoprotective effects of DMSO on particular cellular membranes, it is pre-requisite to use membrane samples with high purity. The distribution patterns of marker enzymes of mitochondria and plasma membrane-enriched fractions on density gradients are shown in Figures 1a and 1b, respectively.

Peaks of both ATPase (pH 9.0) and cytochrome c oxidase, which are known as marker enzymes of mitochondria, sharply banded at fraction 22 in the 14,000 x g pellet. They were coincident with protein peak.

On the other hand, peaks of ATPase (pH 6.5), which is known as a marker enzyme of plasma membrane, occurred in fractions 4, 13 and 21 in the 170,000 x g pellet. However, a major peak occurred at fraction 13. This ATPase activity was strongly inhibited by sodium vanadate, DCCD and diethylstilbestrol each at an adequate concentration (data are not shown). NADH cyto-chrome c reductase, as a marker enzyme of endoplasmic reticu-lum, was dominantly banded at near the top of the gradient and a minor peak of activity occurred at fraction 12. Thus frac-tions 12 to 14 seemed to be relatively free from contamination by endoplasmic reticulum. Although acid phosphatase, as a marker enzyme of tonoplast, was overlapping with the major peak of ATPase (pH 6.5), these could be separated with a sucrose density gradient following bursting those fractions obtained by Percoll density gradient (unpublished data). Based on these results, fractions 12 to 14 after sucrose density gradient can be presumed to be enriched in plasma membranes.

Survival of cultured cells frozen to different temperatures in the presence or absence of 10% DMSO was tested (Fig. 2). They survived freezing to -4°C in the absence of DMSO and to -8°C in the presence of DMSO. Therefore, freezing hardiness increased by 4°C with DMSO treatment.

Motion of the 5-nitroxide stearic acid spin label in both membranes in the presence or absence of DMSO compared in Figures 3a and 3b. Since the nitroxide group in 5-nitroxide stearic acid would be close to the polar head groups of phos-pholipids in membranes, the information obtained would reflect the motion of hydrophilic regions of membrane lipid bilayers.

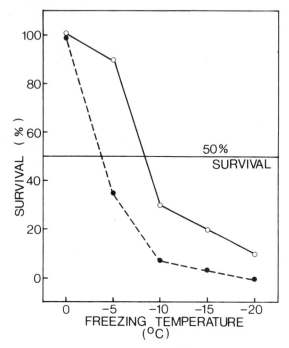

FIGURE 2. Survival of cultured cells of Jerusalem artichoke frozen to each temperature slowly in the absence (●) or
presence (○) of 10% DMSO. The temperatures of 50% survival
of cultured cells frozen with or without the addition of DMSO
were -4 °C or -8 °C, respectively.

Half the separation of the high field and low field peaks,
A⟂, a motion-dependent parameter, was plotted against temperature. When freezing of both membrane suspensions were initiated in the absence of DMSO, an abrupt increase in this parameter was observed compared to that before freezing. This
effect was caused by freezing and not by temperature effects
because no abrupt broadening of motional spectra was detected
in the supercooled state. On the other hand, this abrupt
increase in motion-dependent parameter at the time of freezing
diminished when freezing was initiated in the presence of
DMSO. Lowering the temperature after freezing produced further
motional broadening on spectra until at about -20°C, a rigid
limit spectrum was obtained with both membranes irrespectively
in the presence of DMSO. At relatively higher temperature
regime above 0°C, DMSO affects the motional behavior of the
embedded labels in membrane bilayers in such a manner as to
restrict their motions, especially in plasma membrane vesicles.
    The variation of DPH $(r_0/r-1)^{-1}$ with temperature in the
range of approximately 30 to -20°C was determined to monitor

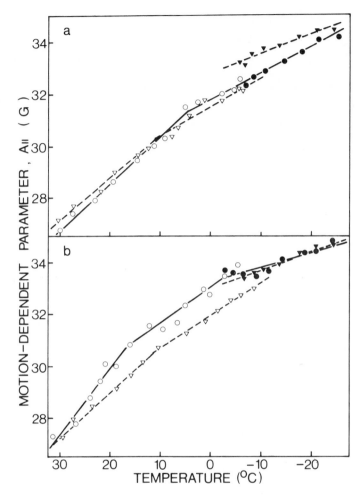

FIGURE 3.   Plots of the motion-dependent parameter, A‖,
vs. temperature in mitochondria (a) and plasma membrane (b)
spin-labeled with 5-nitroxide stearic acid.   The symbols, ▽
and ○ , show the results obtained in the absence and presence
of DMSO, respectively.   Open and solid symbols show the results
in unfrozen and frozen samples, respectively.

the thermotropic behavior of the lipid bilayers of both membranes.   Representative Arrhenius plots are shown in Figures 4a
and 4b.   In the absence of DMSO there was observed only one
break point, which may be designated as T_f (the upper temperature of membrane lipid phase transition), about 15°C in both
membranes.   However, each second break point, which may be
designated as Ts (the lower temperature of lipid phase

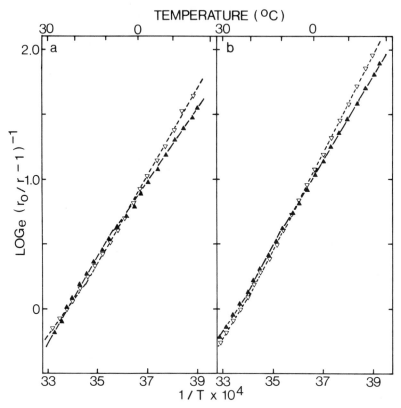

FIGURE 4. Arrhenius plots of DPH anisotropy parameter, $(r_o/r-1)^{-1}$, in mitochondria (a) and plasma membrane (b). The symbols, $\triangledown$ and $\blacktriangle$ show the results in the absence and presence of DMSO, respectively.

compared with the control membrane suspensions. The differ-ences of the fluidity parameter between membrane samples with and without addition of DMSO were coincident above 10°C, but they appeared to be hard to solidify below that temperature. This result may indicate that addition of DMSO can control the restriction of relative motional freedom in the hydrocarbon region of membranes.

IV. DISCUSSION

A number of physical methods have recently been used to demonstrate physical phase transition and fluidity of membrane lipids. The term "lipid fluidity", as applied to model transition) did not appear in both membranes. When both membranes were treated with DMSO, there was observed a marked reduction in the anisotropy parameter at lower temperatures below 10°C, bilayers and natural membranes, is used throughout this report to express the relative motional freedom of the lipid molecules or substituents thereof. We follow a heuristic usage which has become increasingly frequent in recent literature. It bears emphasis, however, that the term is broad and includes different types of motions, e.g., rotational or lateral diffusion of a molecule in an array, movements of substituent groups of a molecule, and the flow of molecules under a pressure gradient in accordance with a fluidity which is 1/viscosity of molecular array (2). In this report, "lipid fluidity" of cellular membranes is assessed by the steady-state fluorescence polarization of DPH and the ESR spectroscopy of 5-nitroxide stearic acid.

Steady-state fluorescence polarization has provided a sensitive technique for examining the lipid fluidity of a variety of artificial and natural membranes. The anistropy parameters so obtained are probe dependent and reflect the overall motional freedom of the fluorescent molecules without distinguishing the specific mechanisms affecting its motions. These specific mechanisms include alteration of the rate of rotation of the probe owing to viscous drag of the environment, anisotropic rotation, and hindered motions owing to structural factors (5-8). Furthermore, DPH appears to partition equally between crystalline gel and liquid crystalline phases of membrane lipid bilayers (9). With these limitations, the anisotropy parameter is related inversely to the relative motional freedom of the fluorophore and provides thereby an index of the fluidity of lipid molecules in the membrane bilayers.

On the other hand, spin labels have been used in a variety of different biological and model membrane systems to obtain information about molecular processes (13,15). Spin labels give information about their environment and for cases where the spin-label molecule is a minority species such that the spin label-spin label interaction is a rare event, the ESR signal is a function of the interaction between the spin labels and membrane lipid molecules. In 5-nitroxide stearic acid, it is expected that the spin label's motional behavior will be similar to the fatty acid domains of phospholipids because the spin label molecule has geometrical and chemical properties similar to that of fatty acid domains. Therefore, the analysis of the ESR spectra is direct and clear. In this experiment, the results obtained by two different techniques were compared with the cryoprotective mechanism of DMSO on mitochondria and plasma membrane exposed to sub-freezing temperatures.

It is thought in general that there are at least three locations where a cryoprotectant may act to avoid freezing injury. These are in the exterior solution, on the membrane per se, and in the cell interior. In the exterior solution, there may be a lowering of the freezing point, increasing the viscosity and, thereby, arresting the growth of ice crystals, dependent on the cryoprotective additives used. With respect to the membranes, there can occur modification of its molecular configration and/or its hydrated structure, with consequent effects on its physical stability and the functions, solute selectivity or leakiness. In the cell interior, there may be marked effects on the freezing point depression and viscosity of cytoplasm, and stability of intracellular membranes, especially if the cryoprotectant enters into cells. DMSO penetrated quickly into the cells. Therefore, DMSO may act to avoid freezing damage at any of the sites mentioned above.

It was reported previously (21) that the shapes of mitochondria and plasma membrane become irregular and rough after treatment with DMSO. This result suggests that DMSO may influence cellular membranes not only in their ultractructure but also in their physiological functions.

As to the membrane effects of DMSO, it was shown by Williams and Harris (23) that DMSO became distributed not in the lipid of the membrane but in the aqueous phase of the lipid bilayers. Barnett (1) reported that DMSO had an effect on the activity of K^+, Na^+ - ATPase, which was one of the plasma membrane-bound enzymes in animal cells and on the lipid phase transition temperature. Moreover, Lyman et al. (12) have suggested that the cryoprotective character of DMSO and other chemicals may be related to their membrane effects and that a decrease in the fluidity of cell membrane could render the cell and subcellular organs more resistant to distortion during freezing. These results suggested that DMSO exerts its effect on the polar head regions of phospholipids and/or the hydrophilic regions of membrane-bound proteins and affects the membrane fluidity. However, they were results obtained only at relatively high temperatures. These results may not explain the behavior of DMSO at low temperatures where DMSO has an important role in protecting cells from injury.

In the present study, an increased motional broadening in the ESR spectrum obtained from mitochondria and plasma membrane frozen in the absence of DMSO was observed, compared to that before freezing. This observation is in close agreement with that of Singh and Miller (19), who demonstrated in the ESR studies with rye protoplasts that the increase in molecular order near the hydrated end of the membrane phospholipids resulted not from low temperature effects but from extracellular freezing since no such changes were observed in supercooled samples. However, when both samples were frozen with DMSO, this phenomenon was not observed. Moreover, although the membrane fluidity in both membranes detected by fluorescence polarization was nearly coincident at higher temperatures above 5 to 10°C with or without DMSO, its decrease at temperatures below 5 to 10°C was supressed by treatment with DMSO. These results strongly suggest that DMSO is partitioned in hydrophilic regions in membrane phospholipids and/or membrane-bound proteins. This partitioning would cause a compensation by means of a discontinuously increasing motional broadening in the ESR spectrum when freezing is performed in the presence of DMSO and would prevent a decrease in membrane fluidity at low temperatures, under which the cells might begin to sustain specific injuries.

In the future, we must determine the cryoprotective mechanism of cryoprotectant, especially that of DMSO, gaining further insight into cellular membranes as a possible site of freezing injury of cells. More detailed studies of protoplasts which have an intact plasma membrane and of cellular membranes isolated from more hardy plant cells are under investigation from a biophysical approach.

REFERENCES

1. Barnett, R.E., Cryobiology 15:227-229 (1978).
2. Brasitus, T.A., and Schachter, D., Biochemistry 19:2703-2709 (1980).
3. Chapman, E.A., Wright, L.C., and Raison, J.K., Plant Physiol. 63:363-366 (1979).
4. Heber, U.W., and Santarius, K.A., Plant Physiol. 39:712-719 (1964).
5. Kawato, S., Kinoshita, K., and Ikegami, A., Biochemistry 16:2319-2324 (1977).
6. Kinoshita, K., Kawato, S., and Ikegami, A., Biophys. J. 20:289-305 (1977).
7. Lacowicz, J.R., Prendergast, F.G., and Hogen, D., Biochemistry 18:508-519 (1979).
8. Lacowicz, J.R., Prendergast, F.G., and Hogen, D., Biochemistry 18:520-527 (1979).
9. Lentz, B.R., Barenholz, Y., and Thompson, T.E., Biochemistry 15:4529-4537 (1976).
10. Levitt, J., Responses of Plants to Environmental Stresses," 2nd edition. Academic Press, New York, 1980.
11. Livingstone, C.J., and Schachter, D., J. Biol. Chem. 255:10902-10908 (1980).
12. Lyman, G.H., Presler, H.D., and Papahadjopoulos, D., Nature 262:360-363 (1976).
13. McConell, H.M., and McFarland, B.G., Quart. Rev. Biophys. 3:91-136 (1970).
14. McMurchie, E.J., and Raison, J.K., Biochim. Biophys. Acta 554:364-374 (1979).
15. Ohnishi, S., Adv. Biophys. 8:35-82 (1975).
16. Pike, C.S., and Berry, J.A., Plant Physiol. 66:238-241 (1980).
17. Raison, J.K., Lyons, J.M., Mehlhorn, R.J., and Keith, A.D., J. Biol. Chem. 246:4036-4040 (1971).

18. Singh, J., Protoplasma 98:329-341 (1979).
19. Singh, J., and Miller, R.W., Plant Physiol. 66:349-352 (1980).
20. Steponkus, P.L., and Lanphear, F.O., Plant Physiol. 42:423-326 (1967).
21. Sugawara, Y., Doctor Thesis, Hokkaido University, Japan (1979).
22. Uemura, M., Master Thesis, Hokkaido University, Japan (1981).
23. Wade, N.L., Breidenbach, R.W., Lyons, J.M., and Keith, A.D., Plant Physiol. 54:320-323 (1974).
24. Williams, R.J., and Harris, D., Cryobiology 14:670-680 (1977).
25. Withers, L.A., and Street, H.E., in "Plant Tissue Culture and Bio-technological Application" (W. Barz, E. Reinhard, and M.H. Zenk, eds.), p. 226-244. Springer-Verlag, Berlin, 1977.
26. Yoshida, S., Plant Physiol. 64:241-246 (1979).

THE WINTER DESICCATION DAMAGE OF TEA PLANT IN JAPAN

Hiroko Fuchinoue

Saitama Pref. Tea Experiment Station
Iruma Saitama, Japan

I.   INTRODUCTION

The winter desiccation damage, one of the four types of winter damage in tea plant, is the most serious one, because it often causes the death of leaves, twigs and stems (7). Tea plant wintering often suffers the desiccation damage in the northern economic limit areas of tea culture and also to those mountainous areas in the south where the climate is similar in Japan. The Kanto district is located in the former areas, and it is situated on the Pacific seaboard and center area in Japan. Saitama prefecture is included in the Kanto district, and main producing area of tea in this prefecture is situated in north west about 30 to 50 km from Tokyo. The meteorological features characterizing the winter in the areas are severe low temperatures, a long dry spell, very little snowfall and strong northwesterly wind. Especially, the extreme of minimum temperature shows about -10 to -15°C, and mean minimum temperature of the coldest month -4 to -6°C, in winter.

II.   MATERIALS AND METHODS

A.   Materials

For experimental materials, young tea plants of 1 to 5 years of age and adult tea plants were used. These materials were gathered in the tea plantation belonging to Saitama Pref. Tea Experiment Station, and were used in the form of cut branches or in the potted form. Tea plants in many tea plantations in the Kanto district were also observed for the purpose of this study.

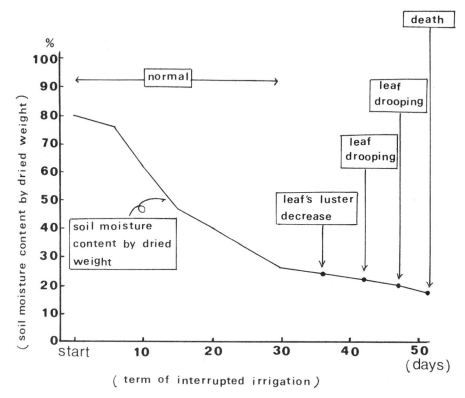

FIG. 1.   Relation between the desiccation damage and the decrease of soil moisture in tea plant. (variety: Yabukita, Age: 2 year old, potted plant).

B.   Methods

Control of soil moisture: In order to understand its influences, soil moisture was controlled by irrigation to the potted plants in the greenhouse.   In particular, irrigation was interrupted to decrease the soil moisture.   Treatment by means of low temperature: The cooling of leaves, twigs and whole crown part of potted plants was conducted in the air conditioning chamber.   In the local cooling of twigs, branches and stems, their local-surface was kept at various temperatures by a thermoelectric apparatus.   In order to control soil temperature and to freeze the soil, a thermoelectric soil temperature control apparatus was used.   This latter apparatus and the growth chamber were used in the experiment to see the interaction of air temperature, relative humidity and soil temperature.   Treatment by means of wind: A small electric fan or natural wind in the field was utilized.

The degree of damage done to the tea plants was judged on the basis of overall consideration of the moisture content, the luster, the color and the degree of drooping or withering of the leaves.

III.   EXPERIMENTAL RESULTS

A.   The Influences of Dry Spell

In the years of severe damage, the rainfall is generally very scanty from early winter to midwinter. So, the relation between the dry spell and the desiccation damage was examined first. A dry spell over one month often causes desiccation damage to tea plants. In the mountainous areas where the soil contains a lot of gravel, the soil moisture in the rhizosphere of tea plants often decreases remarkably after a long spell of dry weather, and the plants are seriously injured. But the soil moisture in the rhizosphere of the adult trees did not show water content less than available water, even after a dry spell of 2 months on the volcanic ash soil in the plain. But the rhizosphere of a very young plant is shallow, and it often suffers damage caused by the decrease in soil moisture. So, the relation between the decrease in soil moisture and the appearance of damage was examined by using young potted plants. Severe desiccation damage arose when the water content by weight of soil was below 20%. Generally, the soil moisture decreases gradually after the interruption of irrigation. After about 40 days, the soil moisture was about 20% water content by weight of soil and the leaves began to droop, to lose their luster, and then they withered. In this case, the critical water content of leaves was found to be about 70 to 80% of the normal one (Fig. 1). Thus, the desiccation damage from the decrease in soil moisture progresses very gradually in winter.

A long dry spell also brings about the decrease in relative humidity. Therefore, an experiment was conducted using cut twigs to find the relation between the relative humidity and the transpiration of tea plants. Cut twigs, with adequate water ascent maintained, were placed in a glass chamber controlled at 30 to 100% humidity for 24 hrs. The amount of water absorption at 30% humidity was about 40% higher than that at 100% humidity (5). This increase in transpiration due to the remarkable drop of relative humidity also gives rise to desiccation tendency in tea plants, if water ascent is insufficient. And this is in fact what happens. For, a long dry spell forms on the surface of soil a dry layer that measures 2 to 3 cm in thickness, and this in turn hinders the rise of the soil temperature at the rhizosphere, serving to decrease water ascent.

TABLE I. The influences of the leaf-freezing to the desiccation damage.

Low temperature treatment	Treatment of water absorption (20 C, for 20 hr)	Desiccation damage
untreated	absorption	none
	non-absorption	very little damage
-5 to -6 C	absorption	none
(3 hr)	non-absorption	little damage
-10 to -11 C	absorption	none
(3 hr)	non-absorption	desiccation-death

The experimental materials used were cut twigs in midwinter.

B.   The Influences of Low Temperature

The progress of the damage by a dry spell only is rather gradual. But in the field, the damage often appears rapidly. This damage has always been caused during midwinter or after midwinter. So, the relation between low temperature and the desiccation damage was also examined. Cut twigs were placed in a controlled chamber at -5 to -11°C for 3 hrs (Table I). After this treatment, these twigs as well as untreated ones were kept in a controlled chamber at 20°C for 20 hrs. Both treated twigs and untreated ones were divided into two groups. For one group, water absorption was allowed, and for the other, it was not. If water ascent was not enough, the twigs, which had been frozen at around -10°C suffered the most serious damage. But if water ascent was enough, they did not. The twigs frozen at -5°C showed little damage in the non-absorption condition. The untreated twigs showed the smallest damage in the non-absorption condition (5).

It is clear, then, that the remarkable extracellular freezing in the leaf-tissues causes desiccation damage after thawing, because the intercellular space is liable to lose its moisture rapidly by transpiration unless the water ascent is sufficient.

In another experiment done in midwinter, basal parts of the stems of young plants were cooled locally with a thermoelectric apparatus at -5 and -20°C for 3 hrs. It was found that the water ascent in all of the stems was blocked during the freezing. This means that the water ascent is blocked even at a not very low temperature such as -5°C. It follows that as mentioned in the introduction, the water ascent of tea plants is blocked almost every night in midwinter.

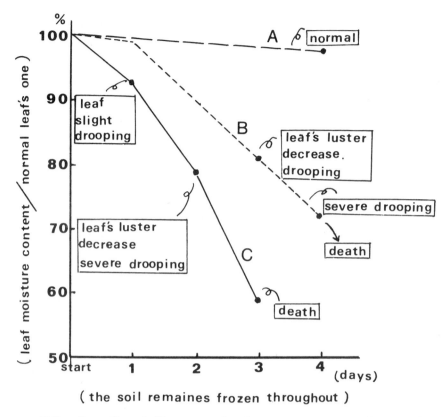

(the soil remaines frozen throughout)

FIG. 2. The influence of air temperature and relation humidity to tea leaves during soil freezing. (As experimental materials were used the potted young tea plants). A: air temperature -4°C, relative humidity 80-90%; B: air temperature 5°C, relative humidity 50%; C: air temperature 15°C, relative humidity 30 to 50%.

On the other hand, in the years when severe low temperature was recorded in December, tea plants often suffered serious desiccation damage. So, another experiment was conducted in late December. Basal parts of the stems of 4-year-old plants were cooled with a thermoelectric apparatus at -13 to -20°C for 3 hrs. The plants cooled at -13°C were dead after 110 to 120 days, by desiccation. The plants cooled at -15°C were dead after 45 to 100 days. And the ones cooled at -20°C were dead after 25 days. These plants showed damage of the same type. A similar phenomenon was observed even in the case of adult trees in the tea plantation.

Thus, the desiccation damage appeared gradually after the severe freezing of the twigs, branches and stems due to the

TABLE II. Relation between the water absorption amount and the
soil temperature.

Soil temperature	Water absorption (gr) (Individual, 24 hr)	Air temperature (°C) & relation humidity (%)
15 C	24.3 (100)	air temperatures: 20 C
8 C	19.6 (81)	humidity: 15 to 20
3 C	18.2 (75)	
15 C	17.4 (100)	air temperature: 0.3
8 C	14.0 (81)	to 18.7 C
3 C	12.5 (72)	humidity: 30 to 88
1∿2 C	8.5 (49)	

Variety: Yabukiya: 5 year old, potted plants were used.

decrease in water ascent.   The time of its appearance varied
with the degree of the freezing damage.   Especially, the
freezing of basal stems often causes fatal desiccation damage.
    In the second experiment, the soil was kept frozen, and
potted young plants were divided into 3 groups.   For each
group, the crown parts were placed in a differently controlled
chamber for 4 days (Fig. 2).   For the first group, the chamber
was controlled at air temperature of -4°C and relative humidity
of 80 to 90%, for the second group at 5°C and 50%, and for the
third group, at 15°C and 30 to 50%.   The plants in the third
group were dead after 3 days, and those in the second group
were dead after 4 days.   The critical water content of leaves
was found to be about 70% of the normal one.   However, those in
the first group were not injured, because the transpiration of
the plant was restrained by the slight freezing of leaves (6).
This observation suggests the following points.   The soil
freezing in the night only is not liable to cause damage,
because the transpiration is restrained by the freezing of
leaves in midwinter.   But if the soil remains frozen in the
day-time, the water content in leaves decreases because of the
imbalance between transpiration and water absorption, and then
the plants tend to suffer the desiccation damage.   The winter
desiccation damage of tea plants is caused not only by the
freezing of the plants themselves, but also by the frozen or
cold soil.   The damage by frozen soil generally shows rapid
progress.   In the case, the minimum soil temperature at the
depth of 10 cm is below 1 to 2°C, and the depth of frozen soil
measures about 5 to 10 cm in midwinter.   It is natural to
expect that this will hinder the water ascent of young plants.

TABLE III. The influence of wind to transpiration amount of cut twig.

Treatment of wind	Water absorption (ml) (Leaf area, 10 cm^2)	ml. of gr. leaf weight
windy-place (mean speed 5 to 6 m/sec)	0.152 (109)	4.57 (109)
windless-place	0.140 (100)	4.18 (100)

It was experimented at 20 C for 4 hr. Materials used were cut twigs.

Therefore, the next two experiments examined the relationship between their ability to absorb water and the soil temperature.

In the first experiment the soil temperature of potted young plants was kept at various levels of temperature, namely, at 1 to 15°C. The crown parts of some of these were placed in a chamber where the air temperature was 20°C, and the relative humidity 15 to 20%. The crown parts of the others were placed in a room of natural condition. After 24 hrs, the following results were obtained. The soil temperature being fixed, the plants whose crown part was placed at a higher temperature showed a higher amount of water absorption than those whose crown part was placed at a lower temperature. The air temperature being fixed, the water absorption amount increased with the rise of soil temperature. The water absorption amount at the soil temperature of 3°C was about 30% less than that at 15°C, and the amount of water absorption at the soil temperature of 1 to 2°C was about 50% less than that at 15°C. If the soil is not frozen, the water ascent decreases with the fall of soil temperature, especially below 3°C (6) (Table II).

In the second experiment, it was made clear that the rootlets of tea plants are injured at -2 to -4°C. Therefore, the young plants sometimes suffer desiccation due to the freezing damage done to their roots near the ground surface.

C.   The Influences of Strong Dry Northwesterly Wind

In the field, serious damage is often caused suddenly after the sweeping of strong wind. Especially, this damage appears remarkably in the wind-swept place. It often happens that strong wind with the mean speed of over 3 to 4 m/sec sweeps from between west and north-east for over 24 hrs in winter. Especially, northwesterly wind is frequent. Branches for which water ascent was maintained were placed in front of an electric fan, and were exposed to wind with the mean speed of 5 to

TABLE IV. The influences of age of tea plant to winter desiccation damage (February, 1974).

Age	Ratio of the damage (%)
a yearling	90.0
2 year old	63.0
3 year old	53.4
4 year old	21.9
5 year old	14.6
adult	7.6

The damage in these tea gardens of different ages was observed on 10 to 20 places severally.

6 m/sec for 4 hrs (Table III). The transpiration amount of these branches was about 10% higher than that of the branches at a wind-less place. But the moisture of the leaves exposed to the wind remained normal because water absorption was possible. Young potted plants on frozen soil were exposed to wind with the mean speed of 4 to 5 m/sec in the field, and they suffered severe damage after 17 hrs. On the other hand, potted plants whose water absorption remained normal did not show damage in the same degree of wind. If tea plants cannot maintain enough water ascent, the wind causes the desiccation damage because of the increase in transpiration. The desiccation damage by the wind appears rapidly.

D. The Influences of the Condition of Tea Plants Themselves

The degree of winter desiccation damage varies remarkably with the age of the plants. Especially, young plants of 1 to 3 years of age are more susceptible to winter desiccation (Table IV). However, in the years of severe damage, even adult trees suffer serious damage. The varieties of tea plants also showed different degrees of damage. Generally, the varieties with weaker resistance to cold suffer more serious desiccation damage.

But even among the varieties with very strong resistance to cold there were also some severely damaged. Various degrees of damage are done to different varieties depending on the meteorological factors of each year. The form of the plants (6) and the physiological characters engendered by cultural techniques also participate in determining the degree of the damage.

IV. DISCUSSION

As the meteorological factors which cause the desiccation damage to tea plants, a long dry spell, severe low temperaure and strong wind can be enumerated in the areas in question. Especially, in these areas, influences of low temperature is the most remarkable. The decrease in soil moisture causes the desiccation damage on the young tea plants, as reported by many workers (15,18,30). Sato (27) reported that the plant was killed at below 20% per wet weight of soil. Morita (16) reported that the wintering of many fruit trees caused at about 20% of maximum water holding capacity of the soil. These observations agree with the results obtained in the present study. The decrease in soil moisture is a very gradual process. Moreover it was limited only to the surface of soil, then the damage by this cause did not show up in the adult trees. In these areas, the freezing of near surface of soil often continues for a long term in winter. The depth of frozen soil is comparatively slight. In midwinter, it shows about 5 to 15 cm, because snowfalls are very rare in the areas. In early winter, the freezing of soil occurs only in the night, and it thawed completely in the day-time. However, in midwinter, the severely frozen soil which added to its thickness by the continuation of low temperature for a long term often remained frozen even in the day-time for several days. Then the water ascent from the roots in frozen soil was blocked and the water lost by transpiration was not supplied. Desiccation damage ws caused by an intensive dehydration due to the imbalance between the up take and the loss of water in leaves. The impediment of water ascent by frozen soil has also been made clear by many workers (11,12,19,22,27,30).

Remarkable drop of soil temperature also hindered water absorption of many various plants (12,14,20,32,33,34). In the tea plant, the water absorption amount decreased with the fall of soil temperature, especially, below 3°C. Yoshimure (33) reported that the water absorption amount of four various orange decreased markedly at the soil temperature below 5°C. In tea plant and orange, if the freezing of soil continues for several days, their leaves suffer serious desiccation damage. In these plants, when the soil is cooled to the temperature between 5 and 0°C, the ability of roots to absorb water decreases considerably. The ability of absorption of tea plants is better maintained than orange when the soil temperature is below 5°C. In midwinter, the soil at the depth of 10 cm often remains at a low temperature such as 0 to 5°C for a long term at the areas in question. So, the water absorption of tea plant wintering often suffers impediment. Generally, the frequency of damage due to frozen soil or the drop of soil temperature is higher than the damage due to the decrease in soil moisture. The decrease of relative humidity because of a long dry spell caused the increase of transpiration. A dry surface layer is formed by a long dry spell. This dry layer tends to keep frozen or cold soil as it is for a long time, and the tea plants suffer hindrance of water absorption. Freezing of each organ of the tea plant also caused the desiccation damage. Freezing of twigs or stems blocked water ascent at slightly low temperature such as -5°C. Sakai (25) reported that water ascent in the stems of spruce was blocked when their surface was locally kept at -1.5°C. It has been shown by many workers (3,23,31,35) that the water ascent is blocked at slightly low temperature. At the slightly low temperature of -0.5 to -3°C, the water ascent was blocked during freezing, but, the tissues did not show freezing damage after thawing. But in early winter, a very low temperature caused freezing damage on some tissues of branches or stems. And because of this, desiccation damage appeared gradually, according to the degree of the damage. Especially, the freezing damage of basal stem often caused very serious desiccation damage. If water ascent is not enough, the freezing of leaves at about -10°C also caused the damage, because the extracellular freezing in the leaf-tissues causes the desiccation damage after thawing, since the moisture of intercellular space is liable to be lost rapidly by transpiration. This phenomenon shows up even without the freezing damage after thawing. The sweeping of strong wind also caused the desiccation damage if water ascent was hindered. The increase of transpiration by wind has been made clear by many workers (9,10,13,14,17.26,28,29). But on the other hand, some workers reported that wind brought about the decrease in transpiration (8,9,10,29). The limit of wind speed tending to increase the transpiration of tea plants has not made clear. However, it is conjectured from the reports on other plants (9,14,16,17,29) that the transpiration will increase while the maximum speed is below 15 to 20 m/sec. It has been found in some of the experiments reported here that the critical moisture of leaves was about 70% of normal leaves. Sakai (30) reported that in the case of ezo spruce the critical water content of leaves was about 25 to 30% per wet weight. This figure corresponds to the critical limit of about 70% of normal leaves. So in this respect, the tea plant and ezo spruce are almost identical.

V.　SUMMARY

Tea plant wintering often suffers desiccation damage in the Kanto district which is located in the northern economic limit areas in Japan. Saitama pref. where this experiment was enforced, includes the Kanto district, and main producing area of tea in this pref. is situated in north west about 30 to 50 km from Tokyo. As the meteorological features which cause the desiccation damage to tea plant, a long dry spell, severe low temperature and strong wind can be enumerated in the areas. Especially, in these areas, influences of low temperature is the most remarkable. The water ascent of tea plant decreases with the decreasing soil temperature, especially below 3°C. In the plants where the soil remained frozen, leaves were killed when subjected at 5 to 15°C for 3 to 4 days, while little or no damage was observed when remained at -4°C. The leaves frozen at -10°C or below are liable to lose water after thawing in dry winter. The water ascent of tea plant was blocked with slight low temperature which the stem was frozen locally, such as -5°C, in midwinter. In late December, the basal part of the stem of young plant frozen around -15°C for 3 hrs, which were partially sustained damage, became more susceptible to winter desiccation than normal one. The rootlets were also damaged by freezing at -2 to -4°C which accelerates winter desiccation. A dry spell over one month often causes desiccation damage to young tea plants. Severe damage arose when the water content by weight of soil was below 20%. If the water ascent is not enough, tea plant also sustains desiccation damage when swept by strong wind over 3 to 4 m/sec for over 8 to 10 hrs. Young potted plants on frozen soil were exposed to wind with the mean speed of 4 to 5 m/sec in the field, and they suffered serious damage after 17 hrs. The critical water content of leaves in tea plant was found to be about 70% of the normal one. The degree of winter desiccation damage differs with variety, age and cultural technique. Especially, the young plants of 1 to 3 years are more susceptible to winter desiccation.

REFERENCES

1. Bialoglowski, J., Proc. Amer. Hort. Sci. 34:96-102 (1937).
2. Cameron, S.H., Proc. Amer. Hort. Sci. 38:75-79 (1941).
3. Eguchi, K., Y. Maeda and A. Sakai, Proc. Symposium Jap. Forest. Soc. 65:192-193 (1967).
4. Fuchinoue, H., Jap. J. Crop Sci. 37(2):309 (1968).
5. Fuchinoue, H., Jap. J. Crop Sci. 39(Extra issue 1):103-104 (1970).
6. Fuchinoue, H., and I. Yagi, Jap. J. Crop Sci. 41(Extra issue 1):129-130 (1972).
7. Fuchinoue, H., Jap. J. Crop Sci. 43(Extra issue 1):37-38 (1974).
8. Hasebe, T., and O. Takechi, J. Agr. Meteorology 28(2):21-26 (1972).
9. Iizuka, I., Memoirs of the Ehime Univ., Sect. II, Ser. B, 4(1):173-178 (1960).
10. Iizuka, I., Memoirs of the Ehime Univ., Sect. II, Ser. B, 4(1):179-183 (1960).

11. Konda, K., and K. Muto, Res. Bull. College Exp. Forests Fac. Agr. Hokkaido Univ. 20:393-403 (1959).

12. Konda, K., and K. Muto, Proc. Symposium Jap. Forest. Soc. 73:153-156 (1962).

13. Konakahara, M., T. Taniguchi, K. Iguchi, and S. Kubo, Bull. Shizuoka Pref. Citrus Exp. Station 5:19-30 (1965).

14. Kubo, S., Bull. National Inst. Agr. Sci. (Jap) Ser. A, 13:1-80 (1966).

15. Matsumoto, G., and S. Katsuya, J. Sci. Soil Manure, Jap. 14(4):238-242 (1940).

16. Morita, Y., J. Jap. Hort. Sci. 18:155-165 (1949).

17. Martin, E.V., and F.E. Clements, Plant Physiol. 10:613-635 (1935).

18. Negishi, K., and T. Sato, J. Jap. Forest. Soc. 36(3):66-71 (1954).

19. Okanoue, M., and O. Sasaki, J. Jap. Forest Soc. 42(9):339-342 (1960).

20. Parker, J., J. For. 48:278-279 (1950).

21. Sakai, A., Low Temp. Sci. Ser. B 18:1-14 (1960).

22. Sakai, A., T. Takatoi, and T. Watanabe, J. Jap. Forest Soc. 45(12):412-420 (1963).

23. Sakai, A., and M. Saito, J. Jap. Forest Soc. 49(5):198-204 (1967).

24. Sakai, A., Low Temp. Sci. Ser. B 15:15-35 (1968).

25. Sakai, A., Ecology 51(4):657-664 (1970).

26. Sano, H., S. Tanaka, K. Kakinuma, J. Agr. Meteorology 16(2):80-82 (1960).

27. Sato, T., and J. Namura, J. Jap. Forest Soc. 35(3):1-3 (1953).

28. Sato, T., Res. Bull. College Exp. Forests Fac. Agr. Tokyo Univ. 51:27-35 (1955).

29. Tsuboi, Y., and S. Kubo, J. Agr. Meteorology 17(2):67-70 (1961).

30. Ueda, S., Jap. J. Crop Sci. 8:3-29 (1936).

31. Wilner, J., Canadian J. Plant Sci. 41:309-315 (1961).

32. Yamazaki, T., Res. Report Kochi Univ. 1(32):1-5 (1952).

33. Yoshimura, F., Res. Report Kochi Univ., 10 Sci II (2):9-16 (1961).

34. Yoshimura, F., Res. Report Kochi Univ., 10 Sci II (6):33-37 (1961).

35. Zimerman, M.H., Plant Physiol. 39:568-572 (1964).

TRANSPORT OF XYLEM WATER IN A THERMAL GRADIENT:
AN ADDITIONAL FACTOR INVOLVED IN THE DEVELOPMENT
OF WINTER DESICCATION INJURY

Steven C. Wiest

Department of Horticulture
Kansas State University
Manhattan, Kansas

I.   INTRODUCTION

Winter desiccation injury is an especially serious problem
to nurserymen overwintering containerized plants under protective polyethylene structures in cold climates. This is particularly true because containers are most commonly held above
ground with little or no protective mulch -- hence, temperatures of the root media can undergo relatively extreme fluctuations. The purpose of enclosing these plants within polyethylene structures is primarily to maximize absorption of
incoming infrared radiation by the internal mass (plants, container media, soil, etc.) during the day and minimize convective heat loss at all times. Radiative heat loss is reduced
only slightly by polyethylene as it is almost transparent to
long-wave infrared radiation (15). Information to be presented
herein will demonstrate that, while this technique of overwintering containerized plants may be an effective means of
avoiding freezing injury to the root system, it unfortunately
tends to provide an environment quite conducive to winter
desiccation injury.
The most prevalent concept of the mechanism by which plants
are injured by excessive exposure to conditions favoring water
loss during the winter (i.e., winter desiccation as used in
this paper) is that water in the rhizosphere freezes, thus
becoming unavailable for uptake by roots due to its immobil-
ity. However, an analysis by Sakai (22) indicates that the
phenomenon of desiccation injury is more complex than this.
For example, wind, or the lack thereof, is a crucial determinant of the severity of winter desiccation injury. Laboratory
freezing of a segment of the xylem of cryptomeria appeared to

PLANT COLD HARDINESS AND FREEZING STRESS
511

severely inhibit detectable water ascent, although plants growing in areas protected from wind experienced little or no desiccation damage even though the soil and xylem water was presumably frozen for lengthy periods of time (21,22). Zimmermann (30) demonstrated that freezing a segment of red oak xylem to -2 to -3°C inhibited, but did not totally preclude, xylem water movement as detected by the thermal profile along the stem. More recently, Pellett et al. (19) suggested that the lack of ^2H_2O movement through frozen soil constituted evidence that water is unable to move to or through plants when the soil liquid is frozen. However, ^2H has a stronger hydrogen bond energy than ^1H (1) and, since the principle cohesive bond in ice I is the hydrogen bond, it is conceivable that the "label" used by Pellett et al. (19) produced erroneously low rates of water movement due to the above mentioned isotope effect. In addition, ^2H may have become a principal species of the "bound water" fraction in the soil and/or plant due to its relatively large hydrogen bond energy.

That water can, in fact, redistribute under frozen conditions has been conclusively demonstrated by a number of workers. For example, Miller (17) demonstrated that ice can behave as a semipermeable membrane. The mechanism by which liquid water becomes redistributed across such an "ice sandwich" (a layer of ice separating two liquid regions) is likely quite different from the mechanism(s) by which water traverses cellular membranes. However, the net result -- redistribution of water across the semipermeable layer -- is the same. A major driving force for water redistribution under conditions favoring frost heaving is due to the existence of a temperature gradient in the soil. As will be demonstrated, the same phenomenon must be considered to completely account for the inavailability of water to transpiring plants under conditions favoring winter desiccation.

II. DERIVATION OF THE INFLUENCE OF TEMPERATURE ON ψ OF WATER UNDER TENSION

Most models of plant water relations are concerned with isothermal systems (see for instance 6, 11, 20). Plants exposed to conditions conducive to winter desiccation are not, however, in an isothermal state. Rather, air temperatures inside typical nursery polyethylene houses are frequently greater than 10°C higher than container temperatures even in midwinter (29). Since plants are poikilothermic, the xylem of containerized plants inside these structures must be under a nearly identical temperature gradient. The consequence of this temperature gradient on the movement of water through the xylem can only be evaluated after first considering the influence of temperature on the water potential of a solution under tension.

Experimental evidence corroborates the assertion that the vapor pressure over water under tension responds essentially the same to a change in temperature as does that over bulk water under atmospheric pressure (7). Therefore, the response of ψ to temperature can be derived as follows.

The purely mathematical identity:

$$\frac{dV}{V} = \frac{1}{V} \left(\frac{\delta V}{\delta T}\right)_P dT + \frac{1}{V} \left(\frac{\delta V}{\delta P}\right)_T dP \tag{1}$$

integrates to:

$$P_f = \beta^{-1} \int \alpha dT - \beta^{-1} \ln(V_f/V_i) + P_i \tag{2}$$

where subscripts i and f indicate initial and final states of the system (i.e., prior to and after changing the temperature of the system), respectively. By definition (14):

$$P = P^{sat} \exp(V \psi /RT). \tag{3}$$

One can empirically derive the temperature dependence of P^{sat} using published values (26) and the Clausius-Clapeyron equation (16) to be:

$$P^{sat} = \exp(14.68 - 5404/T). \tag{4}$$

Substituting equations 3 and 4 into equation 2, and imposing the boundary condition that $\psi = 0$ for pure bulk water at all temperatures, the following temperature dependence of ψ can be derived:

$$\psi f = \frac{RT_f}{V_f} \ln(1 + \exp(5404(\frac{1}{T_f} - \frac{1}{T_i}))(\exp(V\psi_i/RT_i) - 1)). \tag{5}$$

Equation 5 indicates that as temperature decreases, ψ will also decrease (i.e., become more negative).

III. LOCATION OF THE PHASES OF H_2O IN A PARTIALLY FROZEN CONTAINER MEDIA

A major tenet of the prevalent concept of winter desiccation injury is that liquid water is unavailable for uptake by roots. While this may be likely under some field conditions in both very cold and moderately cold climates (22), the assumption has not been rigorously proven for the typical environment of containerized plants. The question of whether this assumption is probable has only recently been addressed (28). Results of that inquiry, while by no means unequivocal, indicate that the liquid water in a partially frozen soil solution

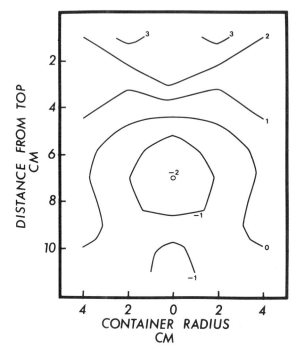

FIG. 1. Diagramatic representation of the difference in regional weight percent moisture content of media in 3.8 liter containers frozen to -18°C and thawed three times. Values associated with the contour lines represent the difference in moisture content of (frozen - unfrozen) media. Differences in percentage moisture were plotted by linear interpolation between the midpoints of adjacent sample locations and assumes the distribution of H_2O to be radially symmetrical. Reprinted from ref. 28 with permission of the publisher.

probably resides in the lower portion of the container near the sides. Evidence for this assertion is two-fold.

First, exposure of a container to a succession of freeze-thaw cycles results in the net movement of H_2O away from the lower center of the container media and towards the sides and upper portion of the container (Fig. 1). As described previously (28), it is likely that the observed redistribution of H_2O was due to the existence of temperature gradients existing within the container media during freezing and thawing. As described above, this temperature gradient, and its influence on ψ (equation 5), could provide the driving force for mass flow by any phase of the H_2O. This observation is by no means unprecedented. For example, frost-heaving is probably mechanistically quite similar to the above-observed

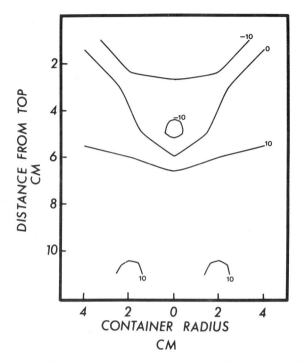

FIG. 2. Diagramatic represention of the difference in fluorescein content of frozen (3 excursions to -18°C) and unfrozen container media as a function of the location within the container, assuming the distribution to be radially symmetrical. Values associated with the contour lines represent the difference in nmols·g dry wt^{-1} of (frozen - unfrozen) media. Placement of the contour lines was obtained as described in the legend to Fig. 1. Reprinted from ref. 28 with permission of the publisher.

phenomenon. Cary and Taylor (4) have modelled unsaturated soil moisture movement in response to a temperature gradient via irreversible thermodynamics and the reciprocity rule of Onsanger. Principles thus derived have recently been confirmed under field conditions (3). Hoekstra (12) has also demonstrated moisture movement in response to a temperature gradient in a partially frozen soil via γ-ray attenuation.

Second, the location of the condensed phase of water (liquid vs. ice) was determined by the principle of solute exclusion by ice (28). In other words, solutes are not incorporated to any significant extent into growing ice crystals (9). Therefore, solutes will be forced out of areas where ice forms first and most extensively. Figure 2 demonstrates that a succession of freeze-thaw cycles results in the movement of a

relatively inert solute, sodium fluorescein, toward the bottom
and sides of the container media. It is therefore reasonable
to suggest that ice formed first and/or most extensively in the
upper and central-interior regions of the container media and
solutes such as fluorescein, being excluded from the growing
ice crystals, diffused (movement was not largely via mass flow,
see 28) away from the growing ice front(s).

These results indicate that when liquid is present in a
partially frozen container, it is likely to be localized at the
lower, outer edge of the media -- the vicinity in which many
water-absorbing roots are located. Because liquid water might
not, in fact, be completely unavailable to roots in a partially
frozen environment, a more comprehensive analysis of the
effects of microclimates conducive to winter desiccation injury
on the physiochemical behavior of water in the plant was
conducted.

IV.   INFLUENCE OF A TEMPERATURE GRADIENT
ON PLANT WATER RELATIONS

Kramer (13) enumerated a number of possible causes of
decreased water uptake by roots exposed to low temperatures.
He included such phenomena as 1) decreased rate of root elongation, 2) decreased cellular water permeability and 3) increased
viscosity of water. All of these phenomena, as well as at
least the partial phase change of H_2O from liquid to solid
may account for the purportedly high root resistance under conditions conducive to winter desiccation injury.

In order to determine whether the influence of low temperature on water uptake by roots was able to account for the total
resistance to water flow, an artificial environment conducive
to winter desiccation was established. Well-watered containerized 3 year old plants of Ilex crenata Thunb. 'Convexa' x I.
crenata 'Stokes' were placed in a chest-type freezer (held at a
preselected temperature) such that the tops of the 3.8 liter
containers were level with the top of the chest. Container
surfaces were covered with insulation to prevent convective
cooling of the shoots. Ice was not visually detectable in
media of any of the containers during the 21 hour study. Since
a quantification of the amount of ice formed in a partially
frozen container was unable to be performed using readily
available techniques, the experimental setup was designed to
encourage desiccation while avoiding uncertainties in available
water content and solute content of the media solution. Shoots
were exposed to a 20°C, ca. 50% relative humidity environment
with lighting provided by overhead fluorescent room lights.
This environment provided a potometrically-determined potential
transpiration intensity of 1.6 $\pm$ 0.7 ul hr^{-1} cm^{-2}.
Figure 3 demonstrates that this experimental environment did

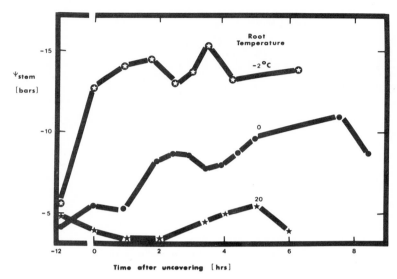

FIG. 3. Artificial induction of winter desiccation in <u>Ilex</u>
<u>crenata</u>. Shoots of containerized plants were exposed to 20°C,
ca. 50% relative humidity while container media was maintained
at the indicated temperature. Shoots were covered with greenhouse black cloth for 12 hr to limit transpiration while the
plant parts were attaining thermal equilibrium with their ambient environment. The ordinate represents the time dependent
change in stem tip ψ determined with a pressure bomb.

effectively induce decreased stem ψ and, thus, probably approximated field conditions conducive to winter desiccation
injury. Plants held isothermally at 20°C maintained a quasi-
steady-state stem ψ, oscillating about -4 bars (-0.4 MPa).
However, when plants were subjected to a temperature gradient
with roots 0°C and stems 20°C, the steady-state stem ψ
decreased, approaching -10 bars (-1 MPa) within 6 hours. When
plant roots were exposed to -2°C, a steady-state stem ψ of -14
bars (-1.4 MPa) was rapidly attained (Fig. 3). To determine
whether the low temperature-induced increase in root resistance
could account for the observed decrease in stem ψ, the temperature dependence of root water permeability was determined. The
driving force for water movement through decapitated rooted
cuttings of <u>I. crenata</u> 'Convexa' x <u>I. crenata</u> 'Stokes' was provided by an adjustable vacuum of -0.47 to -0.81 bars of suc-
tion. Roots were immersed in degassed water or solutions of
0.1 to 0.5 M NaCl maintained at various temperatures and the
rate of movement of solution through the cutting was determined. Water flux measurements were fitted to equation 6 of
Fiscus (8):

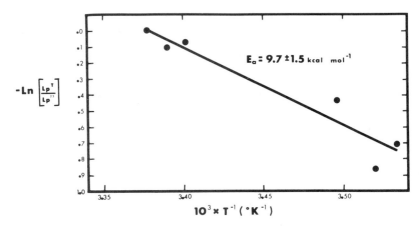

FIG. 4.   Temperature dependence of apparent root hydraulic conductivity.   Arrhenius plot of root L_p determined by equation 6.   Cuttings of I. crenata were decapitated, connected via tygon tubing and a 0.1 ml Mohr pipet to an adjustable vacuum pump, and roots were immersed in various degassed solutions of 0, .1, .2 or .5 M NaCl at the indicated temperature. L_p was determined from Jv since ΔP and π^o were known.   Each data point represents the mean L_p from at least 4 plants exposed to at least 3 of the above mentioned solutions under at least 2 vacuum settings. L_p at 23°C was 1.54 x 10^{-11} cm s^{-1} Pa^{-1}, which is similar to L_p = 1.8 x 10^{-11} cm s^{-1} Pa^{-1} determined for sunflower roots (2).

$$J_v = L_p (\Delta P - \sigma^2 \pi^o) \tag{6}$$

for each temperature tested.   The parameters L_p and σ were thus determined.   The reflection coefficient σ was apparently temperature independent, and determined to be ca. 0.12 between 10° and 24°C.   Hydraulic conductivity (L_p), however, was strongly temperature dependent, with an apparent activation energy of 9.7 ± 1.5 kcal mol^{-1} (Fig. 4).   This is comparable to the 9 kcal mol^{-1} activation energy observed for root water conductance in sunflower (13).

V.   COMPARISON OF OBSERVED STEM ψ WITH THAT EXPECTED
FROM INCREASED ROOT RESISTANCE

The classical Ohm's law analogue of steady-state plant water status (25);

$$\frac{\psi_s - \psi_1}{r_p} = \frac{\psi_1 - \psi_a}{r_1}$$

rearranges to:

$$\frac{r_p^{20}}{r_p^0} = \frac{r_1^{20}}{r_1^0} \frac{(\psi_s^{20} - \psi_1^{20})}{(\psi_s^0 - \psi_1^0)} \tag{7}$$

assuming that $\psi_a \ll \psi_1$. More precisely, the term $(\psi_1 - \psi_a)$ should be replaced with $(e_1 - e_a)$ and the dimension of r_1 appropriately modified since a phase transition from liquid to vapor is involved (18). Nevertheless, equation 7 is correct if $e_1 = 1$ since $e_a^{20} = e_a^0$ is defined in the present scenario. The ratio of leaf resistance of plants whose roots were held at 20°C, determined with a diffusion porometer (Model LI-65, Licor Inc., Lincoln, Nebraska), was 0.74 $\pm$ 0.09. Taking the approximate steady-state values of $\psi_1^{20} = -4$ bars and $\psi_1^0 = -10$ bars (Fig. 3) and assuming $\psi_s^{20} = \psi_s^{20} = -2$ bars since the medium was well-watered, equation 7 reveals that r_p^{20}/r_p^0 must have been 0.18 $\pm$ 0.02. If the major contributor of the resistance to water flow from soil to leaf (r_p) was root resistance at root temperatures of 20° and 0°C, then r_p^{20}/r_p^0 should be estimable by L_p^0/L_p^{20}. However, an activation energy of 9.7 $\pm$ 1.5 kcal mol^{-1} translates into a ratio of L_p^0/L_p^{20} of 0.24 to 0.34 (with a mean of 0.28), a range which is significantly higher than the 0.18 $\pm$ 0.02 calculated from equation 7. This disparity could mathematically occur either if plant resistance at 20°C was lower than root resistance at 20°C or if plant resistance at a root temperature of 0°C was higher than root resistance at 0°C. The latter possibility is the only realistically possible one. The most likely explanation of this disparity resides in the possibility that additional significant resistance(s) develops in the plant subjected to a temperature gradient with stems 20°C and roots at 0°C.

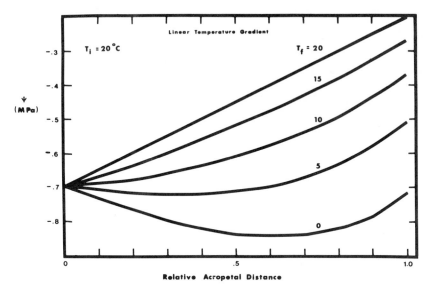

FIG. 5. Theoretical influence of a temperature gradient on
an initially linear xylem ψ gradient. If an initially linear
xylem segment ψ gradient isothermally exposed to 20°C were suddenly subjected to a linear temperature gradient ranging from
20°C at the acropetal end to the temperature indicated at the
basipetal end, the xylem ψ gradient will spontaneously become
that shown on the figure according to equation 5.

VI. IDENTIFICATION OF AN ADDITIONAL RESISTANCE TO WATER FLOW WHEN A PLANT IS SUBJECTED TO A TEMPERATURE GRADIENT

Increased plant resistance to water flow (excluding root
resistance per se) at 0°C soil temperatures could develop in a
number of ways. For example, the development of a vapor phase
between the root and the soil solution would cause a resistance
to water flow which increases with decreasing temperature.
This phenomenon would occur because, as will be shown in the
following analysis, vaporization of water is energetically less
favorable at 0°C than at 20°C. A slight hindrance to vaporization occurs because the latent heat of vaporization increases
as temperature decreases, at least below the critical temperature of 374°C for H_2O (27). A more significant effect of
temperature on evaporation rate resides in the influence of
temperature on the vapor pressure deficit of the soil air, the
major driving force for evaporation (18). For example, if the
soil air adjacent to the root has a water vapor pressure of
4 mm Hg, the major driving force for evaporation from a saturated soil solution at 20°C would be, from the Dalton equation

(10), 17.5 - 4 = 13.5 mm Hg. Under identical conditions, but
at 0°C, the major driving force for evaporation would be 4.6
- 4 = 0.6 mm Hg, or only about 1/20 the value of 20°C. Hence,
soil resistance to water flow immediately adjacent to the root
under these conditions would be ca. 20 times higher at 0°C than
at 20°C. This could result in a significant low temperature-
induced reduction in water availability to the roots. Nevertheless, it is unlikely that the development of a vapor phase
between the root surface and the soil solution occurred in the
present study since the container media was well-watered
immediately prior to the study.

Since the experimental plants were subjected to a temperature gradient, the possibility that the influence of temperature on ψ developed in section II could effect plant water
relations was investigated. The theoretical result of imposing
a temperature gradient, with temperature decreasing as acropetal distance increases is demonstrated in Fig. 5. Assuming
that an isothermal plant at 20°C has a steady-state xylem ψ
gradient increasing from -0.7 MPa at the top of a stem segment
to -0.3 MPa at the base of a segment, the instantaneous imposition of a linear temperature gradient, with temperature
decreasing from 20°C at the top to the indicated temperature
(T_f) at the base of a segment, can have a notable effect on
the xylem ψ gradient (Fig. 5). The slope of the lines in
Figure 5 are indicative of the driving force for water movement
up the xylem. A positive slope indicates a positive driving
force for water movement up the xylem. As the slope approaches
zero, the driving force lessens proportionately. A negative
slope in Fig. 5 indicates that a driving force for water movement in the reverse direction, i.e., down the xylem, exists.
Under conditions described in Fig. 5, if the basipetal portion
of the stem segment is exposed to 10°C while the acropetal portion is maintained at 20°C, with intermediate portions of the
segment forming a temperature gradient linear with distance,
there exists an overall positive driving force for water movement up the xylem, although the driving force approaches zero
at the acropetal end of the segment. If the basipetal end of
the segment is exposed to 0°C, again with a linear temperature
gradient increasing to 20°C at the acropetal end, the resultant
ψ gradient indicates that water will tend to flow up the xylem
along the bottom 1/3 of the segment, while in the upper 2/3 of
the segment the driving force favors water movement down the
xylem. If a leaf is positioned at the acropetal end of the
segment under these conditions, there will be two opposing
forces on the water in the leaf. One driving force favors
transpiration if ambient vapor pressure is less than the vapor
pressure of water in the leaf and if stomatal resistance is not
infinite. Hence, water will tend to leave the leaf via sto-
mates. In addition, the temperature-induced ψ gradient in the
xylem will create a driving force for water movement down the
xylem towards the center of the segment; i.e., water will flow

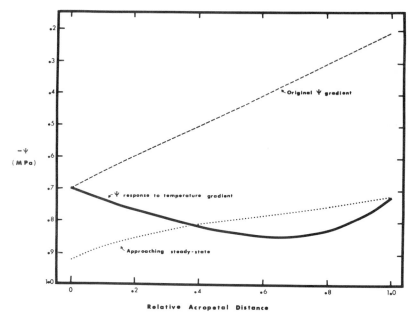

FIG. 6. Simulated redistribution of xylem water in response to a temperature gradient. An initially 20°C isothermal stem segment with a linear xylem ψ gradient ranging from $-.7$ to $-.2$ MPa was hypothetically subjected to a linear temperature gradient ranging from 20°C at the acropetal end to 0°C at the basipetal end. The redistribution of xylem water in response to the temperature-induced ψ gradient was simulated by assuming Poiseuille flow and constraining ψ at the basal end of the segment to remain constant and the rate at which water was acropetally removed from the segment to remain constant. The ψ distribution shown in the curve labelled "approaching steady-state" possesses a negative second derivative. Therefore, the true steady-state ψ at the acropetal end of the segment will be slightly lower than that shown.

in a direction reversed from normal, isothermal flow. Under these conditions there will exist two forces favoring water flow from the leaf, with no source for influx to the leaf. Hence, the xylem ψ gradient will only be able to attain a steady-state if leaf ψ decreases. The above analysis demonstrates that leaf ψ will decrease even in the absence of transpiration as a result of the xylem ψ gradient alteration induced by the imposed temperature gradient. This phenomenon may at least partially explain the lack of effectiveness of plastic antitranspirant coatings for avoiding winter desiccation injury in some species (24).

VII. SIMULATION OF XYLEM WATER REDISTRIBUTION INDUCED
BY THE IMPOSITION OF A TEMPERATURE GRADIENT

The redistribution of xylem ψ, a spontaneous result of the imposed temperature gradient, was simulated in order to ascertain the sole influence of the temperature gradient on the final steady-state ψ at the acropetal end of the stem segment described in section VI. Major assumptions and boundary conditions imposed on this theoretical system were a) that xylem water flow in $\underline{I.}$ $\underline{crenata}$ can be approximated by Poiseuille's law of viscous flow, which is most appropriate for diffuse-porous hardwoods (23) such as $\underline{Ilex}$ (5); b) that sufficient water influx was available to the basal end of the segment to maintain its ψ constant; c) that water redistribution occurs exclusively according to the difference equation:

$$\Delta\psi_i = \frac{A}{\eta_T} \ (2\psi_i - \psi_{i+1} - \psi_{i-1})$$

where A is a constant dependent on the geometry of the system and the time interval and $\Delta\psi_i$ denotes the change in ψ at location i in the chosen time interval; and d) that water is lost from the acropetal end of the segment (in an acropetal direction) at a rate such that the original ψ gradient would be maintained were a temperature gradient not imposed.

Fig. 6 presents the simulation of water redistribution in a stem segment which initially possessed a linear ψ gradient from -0.7 MPa at the top to -0.2 MPa at the base. When a temperature gradient from 20°C at the top to 0°C at the base was imposed on the system, the resultant steady-state ψ gradient ranged from ca. -0.92 MPa at the top to -0.71 MPa at the base of the segment.

If a similar temperature gradient is imposed upon an initially isothermal (20°C) stem segment with a linear ψ gradient ranging from -0.3 MPa at the acropetal end to -0.1 MPa at the basipetal end, the final steady-state ψ at the acropetal end will be ca. -0.55 MPa (under conditions defined in Fig. 7). Hence it is obvious that the imposition of a temperature gradient on a previously isothermal stem segment can have a nontrivial influence on the ψ to which acropetal tissues are exposed.

The two examples above demonstrate that the imposition of a 20° to 0°C temperature gradient, with all other factors held constant, will lower steady-state ψ values in acropetal tissues by ca. 0.22 MPa. By taking this phenomenon into account in equation 7 (section V), L_p^0/L_p^{20} is calculated to be 0.74 ($\frac{2}{5.8}$) = 0.26, quite close to the observed mean L_p^0/L_p^{20} of 0.28 (Fig. 4). It therefore appears that both the effects of a) low temperature on root L_p and b) a temperature

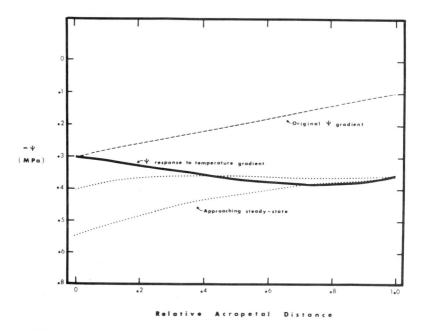

FIG. 7. Simulated redistribution of xylem water in response to a temperature gradient. Conditions were identical to those described in Fig. 6 except that the initial xylem ψ gradient ranged from $-.3$ to $-.1$ MPa. Note the qualitative similarities to Fig. 6.

gradient on the xylem ψ gradient are influential in effecting reduced steady-state leaf ψ under conditions favoring winter desiccation.

Many earlier studies have investigated the influence of reduced temperature on xylem water movement by cooling a localized portion of the stem (see 22, 30, and references cited therein). Few or no visible effects of such localized cooling (above the freezing point of the xylem water) have been observed on xylem water movement (22,30). The influence of temperature on xylem ψ and mathematical analysis of its consequences to water movement through the xylem developed in the present paper were applied to this discontinuous system to determine how well present theory describes known data. Figure 8 (solid line) diagrams the physiochemical effect of localized cooling of a portion of a stem segment on xylem ψ. The spontaneous redistribution of xylem water will result in the pre-steady-state ψ gradient diagrammed by the broken line (Fig. 8). This represents the most extreme transient deviation of acropetal ψ from the initial linear ψ gradient. The stem segment ψ gradient depicted in Figure 8 (broken line)

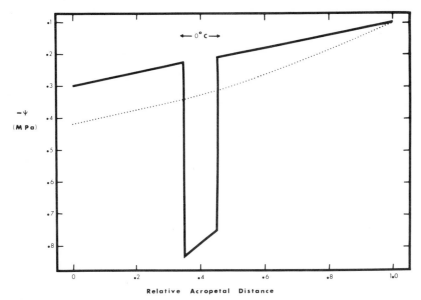

FIG. 8. Simulated redistribution of xylem water in response to localized cooling (0°C) of a portion of s stem segment initially 20°C. The imposition of localized cooling caused a transient decrease in xylem ψ throughout the segment. The broken line indicates the xylem ψ gradient when ψ at the acropetal end was at its lowest. The positive second derivative indicates that the steady-state xylem ψ gradient will occur when ψ at the acropetal end of the segment is less negative than that indicated on the broken line.

approached a steady-state by decreasing ψ throughout the segment. The final steady-state ψ at the acropetal end of the segment approached ca. -0.35 MPa in this simulation. Therefore, the theory developed in the present paper reasonably accounts for the previous observations (22,30) that local cooling of a stem segment perturbs steady-state xylem water relations only slightly. The steady-state ψ gradient in a stem exposed to a continuous temperature gradient is perturbed much more drastically (Fig. 6, 7) than is that of a locally cooled stem segment (Fig. 8).

VIII. SUMMARY

A theory to account for the high apparent plant resistance to water flow under meteorological conditions favoring winter desiccation injury has been developed. Whole plant resistance to water flow when intact I. crenata roots were exposed to 0°C while shoots were 20°C appeared to be higher than that expected from the decreased root L_p at low temperatures. However, xylem water flow under these conditions cannot be considered to be an isothermal process. When the additional effect of temperature on the ψ gradient in the xylem is considered, and the resultant reestablishment of a steady-state ψ gradient mathematically simulated, the observed apparent plant resistance can be accounted for.

Practical consequences of the theory developed here are several. The inhibition of acropetal xylem water flow by a temperature gradient may explain the purported ineffectiveness of antitranspirants for avoiding winter desiccation damage of several species. Theory predicts that antitranspirants will not avert lowered steady-state leaf ψ, although they may mitigate the extent to which leaf ψ decreases. Rather, avoidance of the establishment of a significant temperature gradient should be a primary goal for the prevention of winter desiccation injury.

IX. SYMBOLS USED

e	relative humidity x 0.01	σ	reflection coefficient
J_v	volume flux of water	ψ	water potential
L_p	hydraulic conductivity	Subscripts:	
P	pressure	a	air
p^{sat}	saturation vapor pressure	l	leaf
R	universal gas constant	p	plant
r	resistance to flow	s	soil
T	temperature (°K)	Superscripts:	
V	partial molal volume	0	root temperature of 0°C
α	$(\frac{\delta V}{\delta T})_P$	20	root temperature of 20°C
β	$(\frac{\delta V}{\delta P})_T$		
$\pi^°$	external osmolality		
η_T	viscosity at temperature T		

REFERENCES

1. Armand, P.A., and Hess, J.L., Carnegie Inst. Wash. Yearbook 78:168-171 (1979).
2. Black, C.R., J. Exp. Bot. 30:947-953 (1979).
3. Cary, J.W., Papendick, R.I., and Campbell, G.S., Soil Sci. Soc. Amer. J. 43:3-8 (1979).
4. Cary, J.W., and Taylor, S.A., Soil Sci. Soc. Amer. Proc. 26:417-420 (1962).

5. Core, H.A., Cote, W.A., and Day, A.C., "Wood Structure and Identification," 2nd ed. Syracuse Univ. Press, Syracuse, 1979.
6. Cowan, I.R., J. App. Ecol. 2:221-239 (1965).
7. Dorsey, N.E., "Properties of Ordinary Water-Substance in All Its Phases: Water-Vapor, Water, and All the Ices." Reinhold Publ. Co., New York, 1940.
8. Fiscus, E.L., Plant Physiol. 59:1013-1020 (1977).
9. Fletcher, N.H., "The Chemical Physics of Ice." Cambridge Univ. Press, Cambridge, 1970.
10. Hillel, D., "Soil and Water, Physical Principles and Processes." Academic Press, New York, 1971.
11. Gardner, W.R., Annu. Rev. Plant Physiol. 16:323-342 (1965).
12. Hoekstra, P., Water Resour. Res. 2:241-250 (1966).
13. Kramer, P.J., Plant Physiol. 15:63-79 (1940).
14. Kramer, P.J., Knipling, E.B., and Miller, L.N., Science 153:889-890 (1966).
15. Lee, R., Oecol. Planta. 1:301-326 (1966).
16. Lewis, G.N., Randall, M., Pitzer, K.S., and Brewer, L., "Thermodynamics," 2nd ed. McGraw-Hill Book Co., New York, 1961.
17. Miller, R.D., Science 169:584-585 (1970).
18. Monteith, J.L., and Campbell, G.S., J. Therm. Biol. 5:7-9 (1980).
19. Pellett, H.M., Carter, J.V., Hummel, R.L., and Parsons, L.R., J. Amer. Soc. Hort. Sci. 103:792-794 (1978).
20. Philip, J.R., Annu. Rev. Plant Physiol. 17:245-268 (1966).
21. Sakai, A., Ecology 51:657-664 (1970).
22. Sakai, A., Cont. Inst. Low Temp. Sci B15:15-35 (1968).
23. Siau, J.F., "Flow in Wood." Syracuse Univ. Press, Syracuse, 1971.
24. Smith, E.M. (ed), Proc. Woody Ornamentals Winter Storage Symp. Ohio Coop. Ext. Ser. and Ohio State Univ., Columbus, 1977.
25. van den Honert, T.H., Disc. Faraday Soc. 3:146-153 (1948).
26. Weast, R.G., (ed), "Handbook of Chemistry and Physics," 59th ed. CRC Press, West Palm Beach, 1978.
27. Weidner, R.T., and Sells, R.L., "Elementary Classical Physics," Vol. 1. Allyn and Bacon, Inc., Boston, 1965.
28. Wiest, S.C., J. Amer. Soc. Hort. Sci. 105:620-624 (1980).
29. Wiest, S.C., Good, G.L., and Steponkus, P.L., J. Amer. Soc. Hort. Sci. 101:687-692 (1976).
30. Zimmermann, M.H., Plant Physiol. 39:568-572 (1964).

Part IV
Strategies for Improving Freezing Survival

SUPERCOOLING AND PRUNUS FLOWER BUD HARDINESS

E. L. Proebsting and P. K. Andrews

Irrigated Agriculture Research and Extension Center
Washington State University
Prosser, Washington

I. INTRODUCTION

The ultimate objective of plant cold hardiness research is
to reduce the freezing injury of economic crops. One benefit
derived from this type of seminar is that the fundamental concepts are brought to the attention of production-oriented
researchers. Since the 1977 International Plant Cold Hardiness
Seminar held at the University of Minnesota, there has been
important progress in adapting these concepts to the low temperature protection of horticultural crops.   Significant
advances have occurred in the protection of deciduous,
temperate-zone tree fruits through our increased understanding
of the role supercooling plays in plant cold hardiness.
Successful transfer of basic knowledge to field conditions
often demands a higher level of accuracy and precision than was
needed in the original research.   Critical temperatures for
operating low temperature protection systems provide a clear
example of why this is so. Decisions are made that affect not
only individual farmers' livelihood but also those of the whole
community who depend upon the dollars generated by fruit production to produce jobs and sales. In this context 1°C can be
a large error in the estimation of critical temperature.

II. DEEP SUPERCOOLING IN PRUNUS FLOWER BUDS

In the central part of Washington State, in the United
States, the principal limiting factor in the productivity of
many deciduous tree fruits, especially of the genus Prunus, is
freezing injury of the reproductive parts during winter and

TABLE I. Yield of low temperature exotherms and T_{50} from flower buds of commercially cultivated Prunus cultivars.

Cultivar	Species	Date	# buds	# exo	T_{50}
'Redhaven' peach	(P. persica)	1/22/81	20	19	-19
'Bing' cherry	(P. avium)	1/21/81	20	49	-21
'Goldrich' apricot	(P. armeniaca)	12/30/80	22	21	-17
'Early Italian' prune	(P. domestica)	12/30/80	21	23	-22
'Montmorency' cherry	(P. cerasus)	1/27/81	20	33	-25
'North Star' cherry	(P. cerasus)	12/21/80	10	28	-25
	(P. tomentosa)	11/28/80	--	31	-24
'Sioux' plum	(P. besseyi)	1/20/81	20	20	-20
'Nonpareil' almond	(P. amygdalus)	1/28/81	20	3	-19
'Methley' plum	(P. salicina)	2/3/81	20	31	-9
'Ivanovka' plum	(P. salicina)	11/19/80	20	19	-14
Sel. 900	(P. mahaleb)	1/30/81	20	17	-21

spring. Freezing injury to the flower primordia of five over-wintering species of Prunus has been associated with low temperature exotherms following deep supercooling (13,14). Burke and Stushnoff (2) have shown that the flower buds of many native North American and cultivated Prunus species exhibit deep supercooling.

A preliminary report presented at the 1977 US-Japan Cold Hardiness Symposium in Minnesota suggested that deep supercooling is related to flower bud morphology, since only taxa with racemose inflorescences did not produce low temperature exotherms (15). Intra-specific anomalies were also observed in which, for example, the flower buds of P. cerasus cv. 'Meteor' produced low temperature exotherms while those of P. cerasus cv. 'Northstar' did not (2).

A survey at Prosser, Washington, USA, of commonly cultivated Prunus taxa revealed low temperature exotherms in all samples (Table 1). The low yield of exotherms from almond may have been associated with the loss of low temperature exotherms that appears to occur in the very early stages of visible bud swelling.

Burke and Stushnoff (2) reported that most of the native species were capable of rapid cold acclimation while frozen but that most cultivated taxa (e.g. 'Mt. Royal' and 'Deep Purple' plum, and 'Meteor' cherry) were not capable of rapid acclimation. Although it may not be considered rapid, such response to the frozen state has long been recognized as an important element in the wintertime critical temperatures of peach and

TABLE II. Effect of temperature preceding test on T_{50} of P. avium as determined by DTA.

Date	Hours exposed	Rate of temp fall	24-Hour storage temperature (°C)			
			-4°	+1°	+10°	+20°
			T_{50} (°C)			
Jan. 13	32	1°/hr	-26	-19	-20	-20
Jan. 14	32	1°/hr	-24	-19	-20	-20
Jan. 15	32	6°/hr	-23	-13	-15	-15
Jan. 19	64	7°/hr	-22	-14	-13	-14

TABLE III. Effect of 24 hours exposure to subfreezing or suprafreezing temperatures on T_{50} of several commercial cultivated Prunus species (P. avium as standard).

Date	Cultivar	Species	24 hour storage temperature (°C)	
			-4	+1
			T_{50} (°C)	
12/23/80	'Bing' cherry	(P. avium)	-22	-17
	'Sioux' plum	(P. besseyi)	-27	-21
1/22/81	'Bing' cherry	(P. avium)	-24	-21
	'Redhaven' peach	(P. persica)	-20	-19
	'Goldrich' apricot	(P. armeniaca)	-17	-14
	'Early Italian' prune	(P. domestica)	-25	-21
1/27/81	'Bing' cherry	(P. avium)	-24	-20
	'Montmorency' cherry	(P. cerasus)	-30	-25
1/29/81	'Bing' cherry	(P. avium)	-23	-17
	'Redhaven' peach	(P. persica)	-20	-18
1/31/81	'Early Italian' prune	(P. domestica)	-25	-22
	'Goldrich' apricot	(P. armeniaca)	-22	-20
2/5/81	'Bing' cherry	(P. avium)	-22	-20
	'Redhaven' peach	(P. persica)	-18	-18

sweet cherry (6,7,11). In work at Prosser low temperatures above freezing had little effect on hardiness (Table 2). The ability to acclimate while frozen appeared to be common to all the imporant commercial Prunus cultivars (Table 3). The rate at which buds acclimate while frozen is an important hardiness characteristic of a species (11). The response to the frozen state changes with development of the bud (7), constituting another important characteristic that has not yet been described adequately.

III.   CRITICAL TEMPERATURE DETERMINATION FOR ACCLIMATED PRUNUS FLOWER BUDS

Increasing numbers of peach and sweet cherry growers in Washington State depend on orchard heaters and/or wind machines to protect the flower buds during those winter nights when temperatures reach -20 to -25°C. Successful protection against freezing injury depends upon an accurate estimate of critical temperatures. In the past, critical temperatures for flower buds have been determined by exposing the buds to several temperatures under standard conditions in a freezing chamber, followed by examination of the buds to determine survival percentages (6,10). These critical temperatures have been made available to fruit growers in the Yakima Valley of Washington, either by tests conducted in a commercial laboratory or in the University laboratory at the Irrigated Agriculture Research and Extension Center at Prosser. These data are then broadcast with the National Weather Service fruit frost forecast. The principal disadvantage of this method is the 18-hour, minimum, delay between sampling and critical temperature determination. Accumulated evidence suggests that changes in hardiness occur very rapidly in response to air temperatures above or below 0° to -2°C (7,8). Therefore, it is often necessary to predict current critical temperatures from previously determined values, plus estimated hardiness changes calculated from temperatures during the intervening period.

The step from laboratory data to predicting resistance of the buds in the field is not simple. The conditions of the test must approximate those of the orchard, especially with respect to the rate of temperature decrease and to supercooling of the sample at the first exotherm. The temperature measured in the test must be transferable to the temperature measured in the orchard. Accuracy of the field forecast must be quite high. We expect to be accurate within 1°C during dormancy and in the bloom-postbloom periods.

The assumption that injury to the florets of peach and sweet cherry is associated with a specific low temperature exotherm has made possible rapid critical temperature determination. This technique has greatly reduced the time between sampling and hardiness evaluation, thus providing more current critical temperatures to fruit growers.

Before DTA could be used to determine Prunus flower bud hardiness for use in the orchard, it was necessary to increase the sample size to a number which would represent the field population under consideration. Quamme et al. (14) had shown that the sample size could be increased by connecting several thermocouples in series. Proebsting and Sakai (12) had found that distinguishable exotherms could be obtained from peach flower buds by placing up to 25 buds in close proximity to a single thermocouple junction. It is also possible to replace the dry reference bud, often used in DTA, by a live sample, thus producing exotherms in both directions from the baseline. Since only an event peak is necessary for critical temperature determinations and because simultaneous opposing exotherms are infrequent this modification of DTA maximizes instrument utilization.

While numerous low temperature exotherms were distinguishable with DTA from overwintering peach florets, exotherms from sweet cherry florets were often too small to distinguish from background interference. This was presumably due to the small mass of the individual sweet cherry florets. Thermal contact between the buds and the sensor was improved by using a Peltier plate, whose two surfaces were 9.0 cm^2, and thermal conducting paste. This has increased the sensitivity of DTA for determining T_{50} by 30 to 60 times over copper-constantan thermocouples alone.

There is no standard reference value for flower bud hardiness. Exposing buds to successively lower temperatures, at 1°C per hour rate of fall, then examining them for browning (10) is the standard we use. Numerous opportunities to compare lab data with field freezes have shown good correspondence.

The DTA method yielded T_{50} values that were 3°C higher than those from the standard method. Because the standard method has represented field conditions adequately it was important that the differences between the methods be quantified. The thermocouples in the package of buds in DTA read 1.5°C higher than did a U.S. National Weather Service pattern minimum recording thermometer placed nearby in the same chamber. Furthermore, in DTA the buds were exposed to a uniform temperature decrease because they were placed in a thermos bottle. In contrast the conventional test exposed buds to cycling temperatures as the freezer turned on and off. The minimum temperature from the cycling, the value used to represent exposure to cold, was about 1.5°C lower than the average temperature. The sum of these deviations was about 3°C higher for the DTA determination of T_{50} compared with the standard determination of T_{50}.

IV. DEACCLIMATING PRUNUS FLOWER BUDS

Acclimated peach and cherry flower primordia withstand temperatures as low as -20 to -34°C without injury. During this period the temperature-survival curve is relatively steep (9). As visible swelling of the flower buds begins, some primordia

TABLE IV. Number of low temperature exotherms measured from a 20-bud sample of 'Redhaven' peach flower buds, temperature lowered 1°C per hour.

Date	No. of observations	Avg. # of exotherms /20 buds	Date	No. of observations	Avg. # of exotherms /20 buds
9/19	1	1	12/1-12/31	14	18.7
10/6	1	1	1/1-1/31	10	18.5
10/9	1	15	2/1-2/14	8	19.1
10/13	1	6	2/15	1	11
10/17	1	18	2/20	1	16
10/18	1	20	2/23	1	2
10/20-10/31	5	19.4	2/24-3/10	5	0.2
11/1-11/30	6	19.2			

are injured between -5 and -9°C, while the remainder still withstand -20°C without injury. This results in a much flatter temperature-survival curve (9).

During dormancy a properly performed DTA should yield 90 to 100% of the low temperature exotherms from a sample of 20 peach buds. Prior to early October the buds did not deep supercool (Table 4). Within one or two weeks this property developed. After October 17, 90% of the buds or more produced low temperature exotherms. As abruptly as they developed, the low temperature exotherms disappeared in late February. Bud mortality at these later stages was not directly associated with the temperature at which ice formed, though an increasing number of buds were killed near the temperature of the first exotherm.

Bud mortality at the first exotherm was associated with rates of temperature fall in excess of 5°C per hour to temperatures in the -5 to -10°C range throughout dormancy.

It was initially assumed that after the flowers had lost their capacity to deep supercool they were completely tender and that freezing injury would occur at the moment of ice formation. However, it was observed that peach flowers showed significantly less injury when cooled to -7 to -9°C, if ice nucleation was induced above -4°C (Table 5). One explanation for this phenomenon is that when extracellular nucleation occurs at a high temperature the ice crystals are too large to penetrate the plasma membrane, with a concomitant loss of intracellular water. This allows the cells to dehydrate sufficiently to reduce supercooling, approach equilibrium with the extracellular ice, and avoid intracellular freezing.

Shortly before first bloom, peach flowers completely lose their resistance to freezing injury and become fully tender.

TABLE V. Injury to peach flower buds cooled to near $-8°C$ before and after petal exposure as related to the mean nucleation temperature.

	Mean nucleation temperature (°C)		Percent injured	
Treatment	Before petal exposure	After petal exposure	Before petal exposure	After petal exposure
Dry				
Excised	-6.3 d[z]	-4.6 c	100 a[z]	99 a
Attached	-5.8 d	-3.6 ab	94 a	100 a
Wet				
Excised	-3.1 ab[z]	-3.8 b	35 b[z]	100 a
Attached	-3.3 ab	-2.9 a	29 b	98 a

[z]   Means followed by the same letter do not differ at the 5% level of significance.

TABLE VI. Mean nucleation temperature of developing peach flower buds as influenced by attached stem tissue.

	Nucleation temperature	
Stage of development	Buds attached	Buds excised
First swelling	-6.0a[z]	-6.0a
Calyx green	-5.7a	-6.3a
Calyx red	-5.8a	-6.5a
First pink	-5.2a	-4.6a
First bloom	-4.8a	-5.2a
Full bloom	-2.4b	-4.3a
Post bloom	-1.9a	-5.0a

[z]   Pairs followed by same letter do not differ at 5% level of significance.

From this point, through full bloom and later, the flowers and fruitlet are injured whenever they freeze. In DTA studies, the beginning of complete frost tenderness corresponded to a stage of development marked by exposure of the petals through the calyx. The only means of resistance to freezing injury after petal exposure is avoidance of ice formation. Excised buds retained the ability to supercool at full bloom and later (Table 6). Other than current commercial methods of frost protection (e.g. orchard heaters, wind machines, etc.), this can only be accomplished by enhanced supercooling.

V. SUPERCOOLING AS A FROST PROTECTION METHOD

Lindow et al. (5) have shown that two species of epiphytic bacteria, Pseudomonas syringae and Erwinia herbicola, may act as efficient ice nucleating agents at temperatures between -2 and -5°C. Ice nucleation active (INA) strains of these bacteria have been shown to incite frost damage to corn when compared to controls without INA bacteria present (1). Injury to the leaves was proportional to the logarithm of the INA bacterial populations (5). A survey has shown that INA strains of these bacteria are present on many plant species from widely separated areas of the United States.

The elimination or reduction of INA bacterial populations from orchards offers a potential method for encouraging freezing avoidance. If even a small degree of supercooling could be permitted during the advanced stages of flower development, substantial freezing injury might be avoided. Although this method would not be expected to be a complete solution to the spring frost problem it would permit supercooling to temperatures lower than minima normally encountered during the spring. Methods involved in elimination or reduction of INA bacterial populations are compatible with current frost protection technique. At present, no reports have been published that show more than minor increases in T_{50} of deciduous tree fruit crops by any treatment to reduce INA bacterial populations.

We are concerned with two factors inherent in the development of peach flowers and, perhaps, all woody perennials, that may limit the usefulness of present techniques of bacteria control for preventing frost injury. The first is that before the buds become completely tender less injury would be expected following freezing of the buds if INA bacteria were allowed to nucleate ice crystals at high temperature. The other is that supercooling is less likely to occur on new tender, growing tissues if they are attached to the older woody stem tissues than if they are excised from the plant.

The possibility should be considered that other INA particles or sites may be present on the surface or within the tissue which may be active in the same temperature range as INA bacteria. Unidentified leaf- and ocean-derived nuclei have been found to be INA at temperatures as warm as -4°C and -3°C, respectively (17). Ice nucleation has been observed to originate in the stele of conifer needles (16). Kaku and Salt (3) suggest that ice nucleation takes place at sites associated

TABLE VII. Mean ice nucleation temperature, range and standard deviation of ice nucleation temperatures of four peach fruit attached to a single 2.5-cm stem segment compared with four fruit attached to separate 2.5-cm segments.

	Replicate					
	1	2	3	4	Mean	Range
Four fruit attached to single segment						
Mean	-2.2°	-6.0°	-5.0°	-4.2°	-4.3	3.8°
Range	0°	0.5°	0°	0.8°	0.3	
Single fruit attached to four segments						
Mean	-3.3°	-3.8°	-3.8°	-3.3°	-3.6°	0.5°
Range	3.4°	1.6°	3.4°	2.8°	2.8°	

with the cell walls of the vascular elements. We have found that ice crystallization, initiated in the vegetative tissue of Prunus, is readily propagated through the stem and into the attached flowers.

Flowers and small fruit supercooled readily when detached from the vegetative tissues but not when attached (Table 7). After nucleation had been initiated crystallization apparently progressed through the vascular system to all buds on a stem. Evidence for this is that the stem segment, not the fruit, was the unit that controlled nucleation. In Table 7 the range over which ice in the fruit nucleated was close to 0°C if the fruit was attached to the same segment, whereas in the same number of fruit but with each fruit attached to a separate stem segment the range was near 3°C. Conversely, the group of single segments with four fruit attached behaved similarly to a single replicate of segments with single fruit attached, with a similar mean and range.

Unidentified non-bacterial INA particles may be a limiting factor in the effectiveness of INA bacterial control of freezing injury. Also, it may be necessary to reduce INA bacteria populations on the surface of woody perennials to much lower numbers than on annual seedlings, because the mature vascular system may permit ice propagation throughout the shoot system from a single ice locus.

VI. SUMMARY

The presence of low temperature exotherms and their association with freezing injury in <u>Prunus</u> flower buds has permitted rapid critical temperature determinations. The use of the Peltier plate as a DTA sensor has improved thermal contact and sensitivity to the exotherms of samples large enough to make a good estimate of T_{50}. During deacclimation, critical temperature determination by DTA is complicated by significant changes in the freezing properties of the flower buds and by the large variation within the sample. After the flowers become completely frost tender, the enhancement of supercooling by eliminating INA particles or sites offers a potentially energy efficient method of frost protection for deciduous tree fruits. INA particles or sites, active at high temperature and associated with woody tissues but not identified, may be a limiting factor in controlling frost by controlling bacteria.

REFERENCES

1. Arny, D.C., Lindow, S.E., and Upper, C.D., Nature 262:282-284 (1976).
2. Burke, M.J., and Stushnoff, C., in "Stress Physiology in Crop Plants" (H. Mussell and R.C. Staples, eds.), pp. 197-225. John Wiley & Sons, New York, 1979.
3. Kaku, S., and Salt, R.W., Can. J. Bot. 46:1211-1213 (1968).
4. Lindow, S.E., Arny, D.C., and Upper, C.D., Appl. Environ. Microbiol. 36:831-836 (1978).
5. Lindow, S.E., Arny, D.C., Upper, C.D., and Barchet, W.R., in "Plant Cold Hardiness and Freezing Stress" (P.H. Li and A. Sakai, eds.), pp. 149-153. Academic Press, New York, 1978.
6. Proebsting, E.L., Jr., Proc. Amer. Soc. Hort. Sci. 74:144-153 (1959).
7. Proebsting, E.L., Jr., Proc. Amer. Soc. Hort. Sci. 83:259-269 (1963).
8. Proebsting, E.L., Jr., HortScience 5:422-424 (1970).
9. Proebsting, E.L., Jr., and Mills, H.H., Proc. Amer. Soc. Hort. Sci. 78:104-110 (1971).
10. Proebsting, E.L., Jr., and Mills, H.H., Wash. Agr. Exp. Sta. Circ. 548 (1971).
11. Proebsting, E.L., and Mills, H.H., Amer. Soc. Hort. Sci. 97:802-806 (1972).
12. Proebsting, E.L., and Sakai, A., HortScience 14:597-598 (1979).
13. Quamme, H.A., J. Amer. Soc. Hort. Sci. 99:315-1518 (1974).
14. Quamme, H.A., Layne, R.E.C., Jackson, H.O., and Spearman, G.A., HortScience 10:521-523 (1975).

15. Rajashekar, C., and Burke, M.J., in "Plant Cold Hardiness and Freezing Stress" (P.H. Li and A. Sakai, eds.), pp. 213-225. Academic Press, New York, 1978.
16. Salt, R.W., and Kaku, S., Can. J. Bot. 45:1335-1346 (1967).
17. Schnell, R.C., and Vali, G., J. Atmos. Sci. 33:1554-1564 (1976).

MODIFICATION OF FROST RESISTANCE OF FRUIT PLANTS BY APPLIED GROWTH REGULATORS[1]

T. Holubowicz, J. Pieniazek and M.A. Khamis[2]

Department of Pomology
Academy of Agriculture
Poznan, Poland

ABSTRACT

The experiments were conducted at the experimental fields of the Pomology Department of Poznan Agricultural University from 1971 to 1981 on the effect of foliar applications of growth regulators[3]: ABA, Alar, CCC, DEPEG, Embark 2-S, Ethrel, GA and NAA to peach and apricot trees, and Antonovka apple seedlings on their frost resistance. The growth regulators were also incorporated in the growth medium on which the apple shoot callus and Antonovka explants were cultured and exposed to low temperatures.

---

1   This work was partly supported by Grant PL-ARS-59 FG-Po-336/JB-1/USDA Agric. Res. Program Public Low 480 in 1975-1981.
2   Present address: College of Moistour of Agric. University of Zagazig, Egypt.
3   ABA, Abscisic acid; Alar, succinic acid - 2,2-dimethylhydrazide; CCC, 2-chloroethyl-trimethylamonium chloride; Embark 2-S, trade name (3M); Ethrel, 2-chloroethyl-phosphonic acid; GA, gibberellic acid = Gibrescol (Commercial mixure, POLFA, Poland - 94% GA_3, 4.6% GA_1, and GA_2); DEPEG, Dodecyl ether in polyethylene glycol; NAA, napthaleneacetic acid.

I.   INTRODUCTION

The trials on modification of frost hardiness in fruit plants were begun in 1970 (3,4) when Weiser's hypothesis on fall acclimation to low temperatures was proposed whereas a great role was ascribed to the synthesis of frost hardiness promoter(s) via changes in light spectrum and temperature, and the resulting activation of specific enzymes (25). There were many investigations originated since then based on the possibility of modifying the natural balance of growth regulators at the end of the active growth stage of trees and perennials by exogenous application of synthetic growth regulators. It was thus hoped to affect the innate range of resistance to low temperatures (1). Also there were many reports (14,15,23) indicating the positive or adverse effects of growth regulator applications. For instance, spraying some annual, perennial plants including trees with growth retardants such as CCC might result in an increase of frost hardiness whereas applying stimulators such as GA lowered the cold hardiness (12,13,26) or made the plants more hardy (1,20).

The results obtained by Kuiper (16) in 1967 with the application of antitranspirants to strawberries, and the observed increase in their cold tolerance should be also mentioned.

In our investigations an attempt was made to examine in more detail the relationship of some indicators such as content of soluble proteins, amino acids, etc. to the process of acquiring the frost tolerance and its loss. During the seventies there were many papers published confirming or denying the existance of such relationship.

II.   EXPERIMENTAL DATA AND COMMENTS

The first experiment was made with apple explants cultured in vitro (21). In this investigation it was shown that explants treated with ABA were more resistant to freezing than those treated with GA. Nevertheless, the frost tolerance of the explants treated with GA was higher than that of control (Fig. 1). The applied growth regulators to the medium produced changes in the content of water soluble protein, free amino acids and amides.

On the basis of results obtained with apple explants cultured under the conditions adverse to the inducement of natural acclimation to frost (24°C with continuous light) it can be said that both ABA and GA were able to increase the frost resistance of explants cultured in vitro (Fig. 1).

As it can be seen the preliminary results were very encouraging. In the next series of experiments (5,22) the experiment was repeated, and another one which was set up with Antonovka

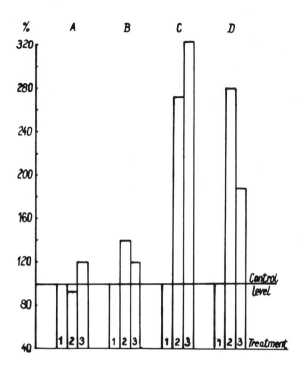

FIGURE 1. The effect of growth regulators (% of control) on: A - weight of explants, B - resistance to frost, C - content of water soluble protein, D - content of free amino acids. Treatments: 1 - control, 2 - ABA 12 mg/l, 3 - GA 15 mg/l (21).

seedlings grown in the greenhouse were sprayed with the solution of growth regulators. Antonovka seedlings at the age of 4 and 7 months, and those 2 years old were sprayed with 20 mg/l of ABA and 500 mg/l of GA. The seedlings sprayed just with distilled water were taken as controls. After application the seedlings from all the treatments were left in the heated greenhouse for 40-60 days and then divided into 3 groups: 1) left in the heated greenhouse, 2) hardened in cold chamber for 5 days at 0°C, and 3) hardened in cold chamber for 1 day at 0°C; then, after gradual lowering of the temperature to -10°C in 24 hr, kept at -10°C for 3 days, and then gradually returned to 0°C in one day. The roots of the seedlings exposed to -10°C were protected against the cold temperature, and they were maintained at 0°C.

The experiment with explants partly confirmed the results obtained in the earlier investigations (21). It showed (Table 1) that it was possible, even in conditions adverse to

TABLE I. Development of cold resistance and the water soluble protein content in apple explants during 20 days of culture on media containing growth regulators (22).

| | Number of days from the beginning of culture to analysis | | | | | |
| | 2 | | 12 | | 20 | |
Treatments (in ppm)	T_{50} [a]	WSP[b]	T_{50}	WSP	T_{50}	WSP
Control	-4.4	105	-5.6	285	-10.0	267
ABA 12	-4.8	110	-5.6	521	-12.0	860
GA 15	-4.9	96	-3.7	366	-7.5	529

[a] T_{50} = Killing point in °C.
[b] WSP = Water soluble protein μg/g fresh weight.

TABLE II. The effect of growth regulators and the temperature treatment on cold resistance (T_{50} in °C) and water soluble protein content (μg/g fresh weight) in Antonovka seedlings (5,22).

| Temperature treatments | Growth regulators treatments (ppm) | Age of seedlings | | | | | |
| | | 2 years | | 7 months | | 4 months | |
		T_{50} [a]	WSP	T_{50}	WSP	T_{50}	WSP
Heated	Control	-32.2	181	-8.4	193	-7.2	397
greenhouse:	ABA 20	-13.3	215	-7.6	220	-6.6	210
20°C	GA 500	-14.6	264	-7.2	228	-8.6	389
Cold room:	Control					-19.6	242
0°C 5 days	ABA 20					-18.0	300
(roots 0°C)	GA 500					-24.6	342
Cold room							
0°C - 1 day	Control	-16.8	392	-9.2	660	-8.0	989
-10°C - 3 days	ABA 20	-31.8	314	-7.0	612	-8.0	872
0°C - 1 day	GA 500	-11.2	604	-5.8	631	-14.0	1059

[a] T_{50} = killing point in °C.

cold acclimation, to increase the frost hardiness in 18 days of culture. It seems therefore that for explants the aging of the tissue was the most important factor for the development of the cold resistance. ABA probably accelerated this process by inhibiting growth and stimulating the aging of the explants. The increase in water soluble protein was also highest in the explants from ABA medium.

The explants cultured on medium containing GA also doubled their initial cold resistance. Contrary to our previous findings (21) they were less resistant than the control at the end of the 20 day culture. However the growth of GA treated explants was stimulated, whereas that of controls was negligible and that of ABA completely arrested.

The results obtained with Antonovka seedlings in the greenhouse sprayed with ABA and GA had shown (Table II) that the most hardy plants to frost were the control. The highest level of hardiness was attained by 2-year-old seedlings, followed by those of 7 month old and the lowest by 4-month old.

Spraying with growth regulators resulted, in most cases, in lowering of frost hardiness of Antonovka seedlings. There was, however, an exception in the 4 month old Antonovka seedlings which after spraying with GA had developed a higher level of resistance to low temperature regardless of the external conditions at which they were maintained. Sprays of ABA had not affected the level of frost hardiness. Table II shows the changes in the content of soluble proteins. As it can be seen, 4 month old seedlings had the highest content of water soluble proteins. The content of proteins was not related to the treatment with growth regulators but to the age of the plants. However, the seedlings sprayed with GA showed, in three out of seven cases, a higher content of water soluble proteins in comparison with the control and those sprayed with ABA. In conclusion, it can be said, that no relationship between the content of water soluble proteins and the level of frost resistance could be found. The greatest effect on the increase of frost hardiness was found in 4 month old seedlings when they were kept for five days at the temperature of 0°C. This hardening resulted in over twofold increase in frost hardiness of the seedlings in comparison to the seedlings left in greenhouse. Nevertheless, this increase in hardiness was not related to the change in the content of water soluble proteins. Our subsequent investigations have also confirmed the dominant effect of hardening on the seedlings at 0°C (6). However Antonovka seedlings are not the best materials to modify their highly genetically coded frost hardiness. In our further research we were using less frost hardy apple cultivars such as Belle de Boscoop, and peach seedlings, which will be discussed later.

Since it was hard to see any effect of ABA sprays on Antonovka seedlings, a study was initiated on the application of ABA to the medium on which stem callus was cultured (11).

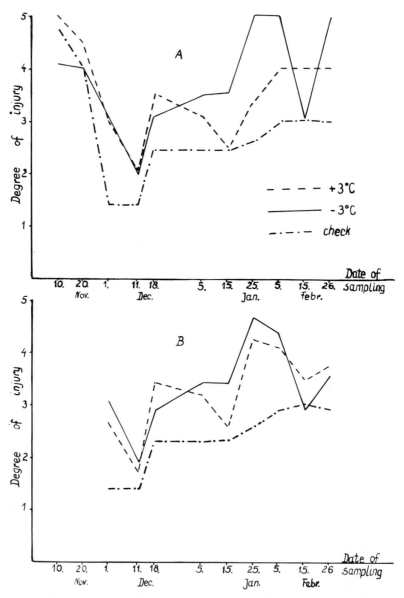

FIGURE 2. Frost tolerance of one-year-old Belle de Boscoop apple shoots collected on October 22 (A) and November 19 (B) stored at +3° ---, -3°C —— and compared with control shoots -·-· taken from the orchard on each date of freezing test (10).

The detailed microscopic analysis has shown that the addition of ABA to the medium inhibited the enlargement of the cells, and the rate of their mitotic divisions. They were also smaller than the cells of the control callus after the same period of growth. It was also found that the smaller cells were more resistant to low temperature than large and fast growing cells. Freezing tests made with callus composed of ABA treated cells and control callus (grown on ABA free medium) pointed to an only temporary higher level of frost tolerance which has disappeared after 19 days. What's more the cultures on ABA supplemented medium became less frost hardy than the controls.

It was also established that the best method for evaluation of the survival of the callus cells after freezing was plasmolysis and deplasmolysis method (11). For the assessment of the frost hardiness of shoots referred to the survival test worked out by Hołubowicz and Bojar (7,8) and Holubowicz (2).

In 1975 to 1980 another series of experiments pertaining to the modification of frost hardiness level by foliar application of growth regulators to fruit trees was undertaken.

The attempts were made to determine the rate of acclimation rhythm, and changes taking place in resistance during the autumn. The current-year shoots of Belle de Boscoop apple cultivar were frozen in the laboratory. This cultivar is considered to be very frost sensitive in Poland (7,8). It was found that the changes in the level of frost tolerance observed from the end of October were, to a large degree, independent of the environmental conditions i.e. proceeding at their own, coded pattern (Fig. 2). It was concluded from these experiments that in field conditions the application of growth regulators should be done earlier than the endogenous, coded acclimation rhythm becomes fixed i.e. in July and August. In the next set of experiments (17,18,19) the following chemical preparations were applied: ABA, Alar, DEPEG, CCC, Embark, Ethrel, GA and NAA. Peach trees, most sensitive species to frost in the West of Poland were used. One experiment was made with apricots. The treatments of the experiments are presented in Table III. The evaluation of the effect of the sprays was made in several ways: 1) the growth rate of current year shoots and the date of cessation of elongation were recorded. These data were essential for the determination of the uptake of the growth regulators through the foliar sprays, and the response of the trees. 2) the date, intensity and the succession of the leaf fall from trees sprayed with different chemicals and at different concentrations was observed. The eventual relationship between the time of leaf fall and the resistance to low temperature was investigated. 3) the changes in the content of lignin and their effect on the level of frost tolerance in the autumn were studied. 4) current year shoots of peach trees were artificially frozen in the laboratory and their level of frost hardiness was estimated by survival test (2).

TABLE III. The characteristic of design of some experiments made with foliar growth regulator applications and their influence on frost resistance of one-year-old fruit tree shoots (17,18,19).

	Number of experiments and time of duration			
	Exp. 1, 1976-1977	Exp. 2, 1976-1977	Exp. 3, 1976	Exp. 4, 1977-1978
Treatments with growth regulators (ppm)	1. Check 2. CCC_{1000} 3. CCC_{2000}	1. Check 2. CCC_{1000} 3. $Alar_{1000}$ 4. ABA_{40} 5. $Ethrel_{1000}$	1. Check 2. $Alar_{1000}$ 3. CCC_{2000} 4. $Ethrel_{1000}$ 5. ABA_{40} 6. GA_{200}	1. Check 2. CCC_{1000} 3. CCC_{2000} 4. $Alar_{1000}$ 5. $Alar_{1500}$ 6. ABA_{40}
Date of application	July 6, 13, 20, 27 August 3	July 13, 20, 27 August 3, 10	August 6, 13, 20	June 30 July 7, 14
Observation and analyses	intensity of growth: July 4,13,20,27,Aug. 3,10,17, Sept. 1,7	The same as in Exp. 1 +	intensity of leaf fall: Oct. 5,15,25 Nov. 5,14	intensity of growth: July 7,14,21,28

Date of observation or measurement	leaf fall: Oct. 6,16 26, Nov. 5,15 freezing tests: Oct. 18, Nov. 2,11 Dec. 1,15 lignin content: Oct. 18, Nov. 2,16 Dec. 1 fruit set (next year) Yield: September	dry matter content: Nov. 2, 16, Dec. 1 phospholipid con- tent: Oct. 10, Nov. 15, March 15 (next year)	freezing tests: Oct. 18, Nov. 2,15, Dec. 1,15 lignin content: Oct. 18, Nov. 2,16, Dec. 1	leaf fall: Sept. 27, Oct. 19 Nov. 2 freezing tests: Sept. Oct. 19, Nov. 11, 29 lignin content: Sept. 29, Oct. 19 Nov. 11, 30 phospholipid con- tent: Sept. 27, Oct. 19, Nov. 9

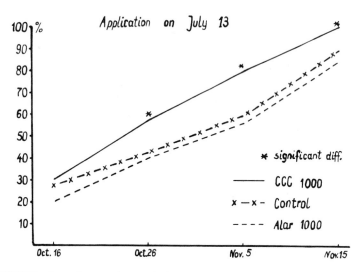

FIGURE 3.   Intensity of leaf fall from one-year-old shoots of 2-year-old peach trees sprayed with CCC (1000 ppm) on July 13, expressed as percent of leaves counted on October 6 = 100%. (17).

The results could be summarized as follows: 1) Rakoniewicka peach cultivar shoots elongated until the end of September.   Under field conditions the first inhibition of growth of control trees was noticed in the middle of July and depended on the age of a tree.   Spraying the trees with the growth regulators from June 30 to August 10 at 7 day intervals resulted in very negligible effect on the shoots growth and leaf fall.   Two-year-old trees sprayed on July 27 with Alar (1000 ppm) inhibited their growth for one week but this inhibition of growth had no effect on leaf fall.   Also Alar (1500 ppm) sprayed on August 10 although it inhibited elongation of shoots for much longer time it did not affect the rate of leaf fall.   Three-year-old trees sprayed with CCC (2000 ppm) on July 20 had growth inhibited for one week but this did not affect the leaf fall.   Five-year-old trees responded similarly to Alar (1500 ppm) treatment made on July 17.   On the other hand, the 2-year-old trees sprayed with CCC (1000 ppm) on July 13, and 3-year-old trees on July 6 did not inhibit the growth of shoots but had accelerated the leaf fall from one year old shoots (Fig. 3).   2) Application of growth regulators in 1976-1977 (CCC 1000, 2000, Alar 1000, 1500; ABA 40 ppm) from June 30 to August 10 at 7 day intervals had a significant effect on the lignification process of one-year-old peach shoots.   It may be said generally that the lignification of sprayed shoots was stimulated and they had a higher content of

TABLE IV. Freezing injuries of one-year-old shoots of 3-year-old peach trees sprayed with CCC in summer 1976 and frozen artificially at -20°C from October 1976 to January 1977 (scale 1-5; 1 - uninjured, 5 - killed).

Date of freezing tests	Treatments in ppm		
	CCC_{1000}	CCC_{2000}	Check
October 18	4.7 a[2]	4.8 a	5.0 a
November 2	3.2 cd	3.5 bc	2.4 e
November 11	1.9 f	2.0 f	1.9 f
December 1	3.5 bc	3.8 b	3.9 b
December 15[1]	3.2 cd	3.1 cd	4.0 b
January 5	2.6 e	2.6 e	2.6 e
January 20	2.9 de	3.1 cd	3.1 cd

[1] The most effective date of application, according to the received interactions was between July 20 and August 3.
[2] Means followed by the same letters do not differ at 5% level of significance.

TABLE V. Treatments and date of application of mixture of Ethrel, NAA and Alar.

No. of treatment	Date of application		
	Ethrel October 10, 1978	NAA October 21, 1978	Alar June 26, 1979
1 (Check)	0	0	0
2	0	0	2000
3	1000	0	0
4	0	100	0
5	1000	0	2000
6	0	100	2000
7	1000	100	2000
8	1000	100	0

lignin. The effect of sprays depended not only on the date they were given but also the age of the trees. The older the trees were the earlier the sprays should be applied to obtain a similar effect. The growth regulators and their concentration could be graded according to their effectiveness: CCC 2000, 1000; Alar 1000, 1500; and ABA 40 ppm. 3) Alar and CCC improved the acclimation process in the fall of one-year-old peach shoots but this effect was of short duration (Table IV). There was no beneficial effect of ABA on the hardening process in the fall. Ethrel sprays had resulted in severe dieback injuries of the tops and the middle part of the shoots.

It was found after 3 years of experiments that CCC and Alar produced inconsistent results and the modification of the level of acquired frost tolerance depended more on the climatic conditions (factors) in the fall than on the growth regulator used.

The applied growth regulators were absorbed by leaves. This was shown not only in these experiments but also in earlier findings (3,4). It was also shown that ABA was taken up by leaves by its effectiveness on fruit set, fruit drop and yield.

The most promising results were obtained after CCC application but it was impossible to relate the time of spraying to the increase and/or decrease in the frost hardiness developed. The concentrations of 500-3000 ppm were used at 7 days intervals from the end of June to the end of September so that the possibilities of varying the time of application or the concentration were completely exhausted.

There were also investigations set up from 1978 to 1980 on the effect of foliar sprays with Embark 2-S at the concentration of 1.8% on the frost tolerance of peach trees. The preparation was recommended by 3M Company for trials as a defoliant. The obtained results have shown, applied from the middle of September to mid-October, adverse effects on the growth of shoots in the consecutive year. It did not bring the expected increase in frost hardiness. The trials with Embark 2-S were thus abandoned.

The most interesting and encouraging results were obtained in the following two experiments: 1) when the sprays of Ethrel, NAA and Alar were given in a mixture or applied one after another after short or longer intervals. It was the repetition of a slightly modified pattern of the experiment carried out at USDA field station in Wenatchee, Wash. USA. 2) DEPEG application. The chemical was obtained from du Pont de Nemours.

The results obtained from the above two experiments are given below: 1) The experiments with Ethrel, NAA, and Alar are shown in Table V during 1978 to 1979. Five-year-old peach trees (cv. Rakoniewicka) planted at the distance of 2.5 x 3.5 m at Przybroda Experimental Station were sprayed in the fall of 1978 and summer of 1979 according to the following treatments (3 replications) as shown in Table V. Five liters of solution were used per tree spraying the leaves throughly from top to bottom. The samples were collected on November 6, 21, December 14, 1978, January 6, and April 4, 1979. However only the first three samples were collected for artificial freezing tests due to a sudden drop of temperature to -20.5°C on December 31, 1978.

The samples collected on January 6 and April 4, 1979 were evaluated in the laboratory with the method of survival test (10) without exposing them to any further freezing. The recovery from December 1978 frost was evaluated on June 22 and September 19, 1979. After evaluation Alar was applied in the second part of June when the recovering trees produced new growth. The next evaluation of frost hardiness of the regenerated trees was made on February 11, 1980 (Table VI). The lowest temperature of the 1979/1980 winter was recorded on January 29, 1980 and it amounted to -20.5°C like in the previous winter but it took place a month later than in 1978.

It was found that NAA and NAA + Alar sprays were beneficial to the development of frost hardiness and to the subsequent regeneration of the injured trees during the vegetative season. On the other hand Ethrel sprays, like in the earlier investigations, affected adversely on the development of frost hardiness and the subsequent recovery from injury. 2) The experiment with DEPEG was set up with 5-year-old peach trees of Siewka Rakoniewicka cv. in 1978. There were 5 following treatments: (1) Control, (2) sprayed with 1% solution of DEPEG on September 15, 1978, (3) sprayed twice; September 15 and 29, (4) sprayed once on September 29, and (5) sprayed twice on September 29, and October 3, 1978. The samples for freezing tests were collected on September 29, October 12, November 2, and 21, December 12, 1978. Suddenly on December 31 the temperature dropped to -20.5°C. The shoots for the next freezing tests were collected and the injury rated with survival tests without subsequent freezing in the laboratory after exposure to the field temperature of 1979/1980 winter. The observations on the recovery of the injured trees were made on April, June 22 and August 9, 1979. On February 4, 1980, samples of shoots were again collected for injury ratings without subjecting the shoots to the laboratory freezing.

Application of DEPEG sprays exerted a positive effect on the cold resistance level of the treated current year shoots. This was especially evident when the treatment was given at the end of September and beginning of October (Table VII). However, the effect of DEPEG spraying was of short duration and it varied in different years (Table VII). On the other hand the observations on the recovery of the injured trees after 1979/1980 winter had shown a beneficial effect of sprays given on October 13, 1978 which was seen in the rate of regrowth of tree crown in the spring of 1979 (Table VIII).

It was a big surprise for us to see that trees sprayed at the beginning of October 1978 looked the best in June and

TABLE VI. The intensity of regeneration of 5-year-old peach trees sprayed with Ethrel, NAA and Alar following injuries (-20.5°C) on December 31, 1978 (percent of new shoots on damaged crown).

Treatments (ppm)	Date of observation and evaluation of injuries					
	Intensity of recovery in percent of new shoots on damaged crowns			Mean yield kg/tree in 1979	Number of dead trees October, 1979	Survival after freezing test - Feb. 11, 1980[2]
	June 22 1979	September 19 1979	Mean for 2 dates			
1. Check - unsprayed	75.1 ab[3]	50.5 abc	63.2 ab	1.6	0	2.2
2. Alar 2000, June 26, 1979	75.1 ab	50.0 abc	63.0 ab	5.9	0	2.3
3. Ethrel 1000 October 10, 1978	31.8 bc	9.5 c	19.4 c	0	1	2.1
4. NAA 100 October 21, 1978	73.0 ab	74.7 ab	73.8 a	1.0	0	2.0
5. Ethrel 1000, October 10, 1978 + Alar 1000 June 1979	28.4 bc	8.8 c	17.5 c	0.1	2.0	2.7
6. NAA 100 October 21, 1978 + Alar 2000 June 26, 1979	86.8 a	57.3 abc	73.4 a	1.8	0	1.6

| 7. Ethrel 1000 October 10, 1978 + NAA 100 October 21, 1978 + Alar 2000 June 26, 1979 | 58.6 abc | 26.5 bc | 42.1 abc | 0.3 | 1 | 1.5 |
| 8. Ethrel 1000 October 10, 1978 + NAA October 21, 1978 | 31.2 bc | 30.0 bc | 30.6 bc | 0 | 0[1] | 2.0 |

1 all the trees showed severe damaged branches;

2 all flower buds injured

3 Means followed by the same letters do not differ at 5% level of significance.

TABLE VII. The effect of DEPEG sprays on the frost resistance
of one-year-old peach shoots in 1977 and 1978.

Date of sampling	Rank of survival	Date of application	Degree of injury at			
			-10	-15	-20	-25
September 29	1[b]	Sept. 15	2.0[a]	3.7	5.0	5.0
1977	2	Check	2.3	5.0	5.0	5.0
October 13	1	Sept. 15	1.5	4.5	5.0	5.0
1977	2	Sept. 15 + 29	1.5	4.5	5.0	5.0
	3	Check	1.5	4.8	5.0	5.0
November 2	1	Check	1.0	2.0	3.5	5.0
November 21	1	Sept. 29 + Oct. 13	1.0	1.8	2.5	4.5
1977	2	October 13	1.0	2.0	2.5	4.0
	3	Sept. 15 + 29	1.0	2.0	2.5	4.5
	4	Sept. 29	1.2	2.0	3.0	4.5
	5	Sept. 15	1.0	1.8	3.2	4.5
	6	Check	1.0	2.0	3.5	4.8
December 12	1	Oct. 13	1.0	1.2	2.2	4.0
1977	2	Sept. 29	1.0	1.2	2.2	4.2
	3	Sept. 29 + Oct. 13	1.0	1.8	2.2	4.2
	4	Check	1.0	1.8	2.2	4.2
December 14	1	Sept. 29 + Oct. 13	1.0	1.2	3.0	5.0
1978	2	Sept. 29	1.2	1.2	3.0	5.0
	3	Check	1.2	1.2	3.2c	5.0

[a]   Scale: 1-5; 1 - uninjured, 5 - killed
[b]   Comparison between control trees and less injured sprayed with DEPEG.
[c]   Line indicates T_{50}

August 1979 and gave the largest average crop (Table VIII).
Current year shoots from the trees sprayed in mid-September to
the beginning of October 1978 were still more resistant in
February 1980 than the control shoots.

The obtained results indicate that DEPEG at 1% solution
induced a short lived increase in cold resistance and that this
effect can be detected even two years later.

TABLE VIII. Regeneration in 1979 of the crowns of 5-year-old peach trees exposed to -20.5°C natural frost on December 31, 1978 (% of recovered crowns means of 4 replications).

Dates of treatments in 1978	Mean value from observations on June 22 and August 9, 1979	Mean yield in 1979	Degree of injury after artificial freezing test on February 11, 1980
1. Check - unsprayed	62.8 ab[1]	1.5	2.0
2. Sept. 15	69.4 ab	0.0	2.2
3. Sept. 15 + Sept. 29	45.0 b	0.6	1.5
4. Sept. 29	60.7 ab	0.3	2.1
5. Spte. 29 + Oct. 13	62.6 ab	0.5	1.6
6. Oct. 13	73.4 a	3.8	2.0

[1] Means followed by the same letters do not differ at 5% level of significance.

III.   CONCLUSIONS

The applied growth regulators such as ABA, Alar, CCC, DEPEG, Ethrel, Embark 2-S, GA, NAA exerted either a positive or negative effect on the level of cold hardiness of the shoots of apple, peach and apricot trees.

The addition of growth regulators to the culture medium on which the explants were grown or spraying the trees with them produced various results. There were instances of an increased or decreased level of frost hardiness unrelated to the applied growth regulator, the concentration used or the date of application. No positive effect of ABA sprays on the increase of the level of hardiness was obtained in our conditions. Ethrel sprays evoked the serious injury of the shoots, and as in case of Embark 2-S which produced adverse effects not only on the frost hardiness but growth of trees, the sprays had to be discontinued.

Alar and CCC sprays at various concentrations produced in most cases positive effects on the acceleration of acclimation process of trees but these effects were of short duration, and not consistent from year to year. The more reliable results were obtained with CCC than when Alar was applied. The most effective period for CCC application at 1000 and 2000 ppm was from July 20 to August 3. Good results on the level of frost hardiness were also obtained when sprays were applied Ethrel and NAA (100 ppm) in October, and Alar (2000 ppm) in June of the following year. Encouraging results were also obtained with 1% solution of DEPEG applied to peach trees at the end of September and beginning of October. The effect of this chemical was long lasting, and it exerted beneficial effect on the regeneration of trees during the next two growing seasons.

The applications of GA produced in most cases a drastic decrease of the hardiness level.

Observations, measurements and analyses failed to show a consistent interrelationship between the resistance to low temperatures of current-year peach shoots and the content of lignin. The acclimation process of shoots was not related to the time of leaf fall. The greatest influence on the development of hardiness was the environmental conditions, mainly the weather.

So far, the application of the above mentioned growth regulators failed to bring the expected or consistent results on the modification of the level of frost hardiness. It was shown, however, that the applied growth regulators to the leaves were absorbed and various growth responses produced. It is hoped that in the future the availability of new chemicals, similar or better than DEPEG, will justify the continuation of the research to improve cold hardiness in fruit trees.

REFERENCES

1. Alden, J., and Hermann, P.K., Bot. Rev. 37:37 (1971).
2. Holubowicz, T., Acta Horticulturae 81:119 (1978).
3. Holubowicz, T., and Boe, A.A., J. Am. Soc. Hort. Sci. 94:661 (1969).
4. Holubowicz, T., and Boe, A.A., J. Am. Soc. Hort. Sci. 95:85 (1970).
5. Holubowicz, T., Pieniazek, J., Pacholak, E., Kasprzyk, M., PTPN, Kom. Nauk Roln. i Nauk Lesn. 37:99 (1974).
6. Holubowicz, T., Pieniazek, J., PTPN, Kom. Nauk Roln. i Nauk Lesn. 41:99 (1976).
7. Holubowicz, T., and Bojar, K., Fruit Sci. Report 3:2 (1976).
8. Holubowicz, T., and Bojar, K., Fruit Sci. Report 3:8 (1976).
9. Holubowicz, T., and Bojar, K., Fruit Sci. Report 4:9 (1977).
10. Holubowicz, T., and Bojar, K., Fruit Sci. Report 4:19 (1977).
11. Holubowicz, T., Radajewska, B., Kasprzyk, M., Wisniewska, J., and Pieniazek, J., PTPN, Kom. Nauk Roln. i Nauk Lesn. 49:107 (1980).
12. Irving, R.M., Plant Physiol. 44:801 (1969).
13. Irving, R.M., J. Am. Soc. Hort. Sci. 94:419 (1969).
14. Kacperska-Palacz, A., Blaziak, M., and Wcislinska, B., Bot. Gaz. 130:213 (1969).
15. Kentzer, T., Roczn. Nauk Roln. Ser. A 93:511 (1967).

16. Kuiper, P.U.C., Mededeel. Landbouwhoogesh. Wageningen 67:1 (1967).
17. Khamis, M.El.-W.A., and Holubowicz, T., Fruit Sci. Report 6:93 (1979).
18. Khamis, M.El.-W.A., Holubowicz, T., and Skowronski Z., Fruit Sci. Report 6:101 (1979).
19. Khamis, M.El.-W.A., Holubowicz, T., and Bojar, K., Fruit Sci. Report 6:107 (1979).
20. Modlibowska, I., Cryobiology 5:175 (1968).
21. Pieniazek, J., and Holubowicz, T., Biul. de L'Acad. Pol. Sci. ser. biol. Cl.V. 19:749 (1971).
22. Pieniazek, J., and Holubowicz, T., Acta Horticulturae 34:247 (1973).
23. Toman, F.R., J. Agr. Food Chem. 16:771 (1968).
24. Toman, FR., and Mitchell, H.L., Phytochemistry 7:365 (1968).
25. Weiser, C.J., Science 160:1269 (1970).
26. Yong, R., J. Am. Soc. Hort. Sci. 96:708.

SEASONAL VARIATIONS IN PHYSIOLOGICAL FACTORS
IMPLICATED IN COLD HARDINESS OF CITRUS TREES[1]

G. Yelenosky and C. L. Guy[2]

Agricultural Research, Science and Education Administration
United States Department of Agriculture
Orlando, Florida

Abstract. Seasonal trends in physiological factors associated with leaves of citrus in the field were studied in relation to cold hardiness. The factors were total sugars, proline, amino acids (minus proline), water-soluble proteins,
starch, ATP and water content; xylem-pressure potential; and
osmotic potential of expressed sap. Sugars, proline, and water
content, and osmotic potential of expressed sap, all of which
have been correlated with cold hardiness of citrus trees in
controlled-temperature environments, were correlated with mean
lethal temperatures and with average air temperatures in the
field. But data variability largely precluded any immediate
use of these physiological factors as indexes of citrus cold
hardiness in practical citriculture.

I. INTRODUCTION

Tree-environment relationships are becoming increasingly
important in citriculture as costs in production rapidly
increase. One major area of concern is in freeze protection.

---

[1] Mention of a trademark or proprietary product does not constitute a guarantee or warranty of the product by the U.S.
Department of Agriculture, and does not imply its approval
to the exclusion of other products that may also be
suitable.
[2] Present address: Laboratory of Plant Hardiness, Department
of Horticultural Science and Landscape Architecture, University of Minnesota, St. Paul, MN 55108.

Estimated heating costs to protect citrus groves during the 1976-77 winter in Florida alone exceeded $5 million per freeze night (2). Such economic pressures intensify our need for guidelines to use basic observations in solving problems. Plant-environment relationships are basic to applied agriculture, and field studies often complement studies under more controlled conditions (18,27,35). Citrus trees, as cold-tender, subtropical evergreens, have considerable potential to survive freezes under both natural and artificial conditions (39). Multiple factors are implicated in citrus freeze survival. These factors range from inherent varietal differences in citrus cold hardening (49,50) to differences in comparative anatomy (26). Physiologically, increases in total sugars (primarily sucrose) and proline in combination with decreased water potentials contributed to the survival of citrus trees exposed to -6.7°C (40). Photosynthesis is implicated in effective cold hardening (37,38,48), and water stress can substitute within limitations for low-temperature cold hardening (43). In citriculture, physiological data are needed so that it would be possible to develop and identify different degrees of cold hardiness in trees (52), to formulate and develop cryoprotectants (39), and to encourage and promote energy-efficient practices in freeze protection (23). Toward this end, we have measured several physiological factors in citrus trees under natural field conditions during 22 continuous months, and now report our findings.

II.   MATERIALS AND METHODS

A.   Citrus Trees

Trees, arbitrarily selected for study, were 3-year-old 'Valencia' sweet orange (Citrus sinensis (L) Osbeck) on 4-year-old sour orange (C. aurantium L.) rootstock in a 98-tree research planting in central Florida. Trees were developed from buds from a single sweet orange tree within 1 mile of the planting. The budwood was healthy, as indicated by virus indexing and visual observations for major disorders in fruit, leaves, and wood of the source tree. Buds were implanted into 1-year-old rootstocks.

Trees received no special attention beyond regular grove practices in fertilization, insect and disease control, and watering with overhead sprinklers to prevent visible water stress due to inadequate rainfall. A hygrothermograph, accurate to 0.5°C, recorded air temperatures in a standard weather shelter 4-1/2 ft above ground level. Biochemical and physical determinations were made on leaves from 5 single-tree replicates randomly situated in the research planting.

B. Sampling

Leaf samples consisted of 60 leaves per tree, 15 fully mature leaves of approximately equivalent age from each of the 4 cardinal sides, 0.6 to 1.8 m above ground level. Leaves were sampled within the hour before sunrise every 3 weeks or less from September 14, 1977, to July 18, 1979. Samples were transported in ice-cooled, plastic bags to laboratory facilities 30 minutes away.

C. Physiological Factors

Field-sampled leaves were sponge-cleaned, blotted dry, and partitioned for different analyses. Ten leaves were used to sequentially determine water content, carbohydrates, and amino acids. Ten leaves were also used to determine water-soluble proteins and ATP. Three leaves were used for xylem-pressure potential and 20 leaves for expressed-sap and osmotic potential. The remaining 17 leaves were used in leaf-freezing tests for cold hardiness determinations. Biochemical determinations were duplicated.

1. Water Content. Ovendry procedures (29) were used to dry leaf tissues overnight, after which water content was calculated.
2. Carbohydrates. Carbohydrates in ovendried leaves ground in a Wiley mill were extracted with 76% (v/v) ethanol in a Soxhlet extractor. Impurities were removed with ion-exchange resins (36), and sugars were determined by the anthrone and Somogyi-Nelson methods outlined by Hodge and Hofreiter (15). Sucrose content was calculated as total sugar content minus reducing sugars.
Starch in the ethanol-insoluble residue was solubilized in boiling water for 15 minutes and then extracted by the method of MacRae (21), in the presence of amyloglucosidase (Sigma Chem. Co.). The resultant glucose was purified with ion-exchange resins (36) and quantitated with glucose oxidase reagent (Sigma Chem. Co.).
3. Amino Acids. Free amino acids were part of the initial extraction from finely ground, ovendried leaves in 76% (v/v) ethanol for 6 hours in a Soxhlet extractor. Ethanol was evaporated and the residue taken up in glass-distilled water. The water fraction was passed through Dowex 50-8X, 100-200 mesh ion-exchange columns, and the amino acids were eluted with 4N NH_4OH. Total amino acids and proline were determined by ninhydrin procedures (6,28,33).
4. Proteins. Water-soluble proteins were extracted from finely ground, freeze-dried leaves with mortar and pestle. Minimum volume of 0.5 ml of 0.1 M phosphate buffer (pH 7.2) containing 0.25 M sucrose, 1mM DTT, and 0.5 g of PVP was used per gram of dry tissue. Soluble proteins were separated from

cell debris by centrifugation at 12,000 x G for 30 minutes. Additional purification included dialysis overnight and TCA precipitation prior to determining proteins by the Lowry method (20).

5. ATP. ATP was extracted from finely ground, freeze-dried leaves in boiling water (1). Leaf samples were not exposed to light prior to freeze-drying. ATP was determined according to St. John (32), with an internal standard in a Packard model 3320[1] scintillation counter. Counter settings were those of Bewley and Gwozoz (3) and Malek and Bewley (22).

6. Xylem-Pressure and Osmotic Potential. Xylem-pressure potential of leaves was determined by a pressure-bomb technique (4). Osmotic potential was determined in a freezing-point osmometer on sap from leaves pressed with 2 metric tons after the leaves had been cleaned, surface-dried, and thoroughly frozen (19).

7. Cold Hardiness. Cleaned but surface-wet leaves were placed side by side in thin, polyethylene wraps suspended in temperature-controlled freeze facilities (38). The wetness of surfaces prevented supercooling, and leaf temperatures inside polyethylene wraps were within $\pm$ 0.3°C of ambient air. Freeze tests started at -2.2°C, and thereafter the temperature was abruptly decreased 1.1°C every 2 hours. Frozen leaves were removed at 2-hour intervals after the start of the test until 4 samples were collected (i.e., -2.2°, -3.3°, -4.4°, and -5.5°C samples). One-half of the leaves removed were placed under mist in a greenhouse and, after 36 hours, visually rated for freeze injury as indicated by permanent water soaking and mean lethal temperature (LT_{50}) (47). The remaining leaves were immersed overnight in glass-distilled water and rated for freeze injury by a conductivity method (17).

III.   RESULTS AND DISCUSSION

The average range and respective low and high temperatures between tissue sampling dates are profiled according to season in this study (Fig. 1). The summer months were characterized by average high temperatures in the low 30's°C. General cooling trends developed in late fall and progressed into the winter season when temperatures averaged 6°C and colder. Regardless of season, leaf temperatures probably were the warmest on the south side of the trees during relatively clear days (7). Coldest temperatures were in January and February, when -4°C caused slight freeze damage to the trees. Injury was confined to less than 5% of the leaves and probably did not significantly affect the seasonal variation of physiological factors in the citrus leaves. Minimum temperatures of -4°C for 4 hours or less usually do not cause much freeze damage in

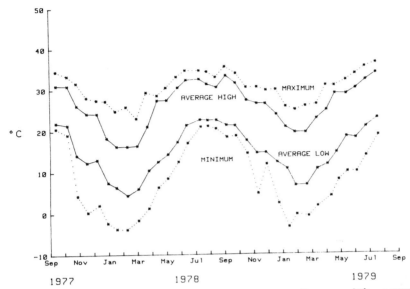

FIG. 1. Average high and low temperatures with corresponding maximums and minimums near Leesburg, Florida.

citrus groves, especially if trees are in a quiescent state of growth (8,49).

The temperature profiles in this study indirectly reflect seasonal trends in cold hardening of citrus trees since cold hardiness in citrus trees is largely temperature dependent (37,46,51). During the winter, differences of 12°C to 16°C between average positive and minimum negative temperatures indicate relatively high-risk periods for freeze injury (Fig. 1). The risk is high because, generally, relatively warm conditions are abruptly interrupted with temperature decreases to critical freeze levels (41). Studies in partially controlled environments suggest that cold-hardening in citrus trees does not develop much at temperatures above 15.6°C (49). Thus, December and January in this study were favorable months for cold-hardening. Otherwise, the -4°C might have caused more damage than was found. The positive effects of temperature on decreasing freeze injury to citrus trees in the field are vividly demonstrated in the yearly survival of plantings in the state of Georgia, latitude 31° 15'N, (41) as well as in commercial citrus-producing areas of the United States (9,45).

Physiological factors that showed marked increases in citrus trees during cooler temperature regimes in partially controlled-environment studies also increased during cooler temperatues in the field. Both sugar and proline concentrations correlated well (r>0.8) with average temperatures between sampling dates in this study (Table I). Other factors

Table I. Correlation coefficients (r)[a] of physiological factors in citrus leaves with field air temperatures[b] and leaf LT_{50} [c] from September 1977 to August 1979, near Leesburg, Florida.

| | r | |
Factor	Temperature	LT_{50}
Total sugars	-0.894*	-0.733
Free proline	-0.847	-0.754
Water content	0.866	0.800
Osmotic potential	0.861*	0.825
Starch	0.273	0.149
Free amino acids (-proline)	-0.037	-0.188
Water-soluble proteins	-0.348	-0.171
Xylem-pressure potential (pre-dawn)	0.326	0.040
ATP	-0.065	-0.202
Temperature	--	0.866

[a]   r values with asterisk indicate exponential correlations; others indicate linear correlations.
[b]   Average 3-week temperatures between sampling dates.
[c]   Mean lethal temperature determined from leaf freeze tests.

significantly associated with field temperatures were leaf water content (r >0.8) and osmotic potential of expressed sap from the leaves (r>0.8). Both of these factors, as well as sugar and proline concentrations, are implicated in cold hardiness in citrus trees (40,42,51). The seasonal changes of the 4 factors are profiled in Fig. 2, and reflect significant correlations with average field temperatures (Table I).

The seasonal fluctuations in total sugars somewhat resembled those fluctuations reported to be associated with changing air temperatures in partially controlled environments (44). In this field study, total sugars averaged between 40 and 50 mg/g of ovendry leaf weight during the summer and as much as 125 mg during the winter. Total sugars remained below 100 mg during the winter of 1978-79, which was about 2°C to 3°C warmer than the 1977-78 winter. The rate of sugar development was maximum and sustained from late November or early December to February, which are critical freeze months in U.S. citrus-producing areas. The rate of increase averaged about 1 mg/g dry wt/day during the 1977-78 season and 0.7 mg/g/day during the warmer 1978-79 season. These rates probably do not reflect the maximum capacity of the leaves to accumulate total sugars, since faster rates have been shown under artificial conditions

(44). Peak accumulations were followed by equally rapid rates of sugar depletion in this field study. New growth flushes noted during warming conditions in February indicated the transition of the trees from a quiescent to an active stage of growth. Active growth probably contributed much to the rapid decrease in sugars between February and March. The total sugar fraction contained primarily sucrose with lesser amounts of glucose and fructose as determined by paper chromatography. Less visible was raffinose detected in one winter sample, along with compounds tentatively identified as maltose and stachyose. Sucrose is the primary sugar whose concentration correlates with cold hardiness of citrus trees (40).

Similarly, proline concentration is more closely correlated with citrus cold hardiness than is the concentration of any other amino acid (40,42). Proline concentrations showed definite seasonal trends in this field study, but with considerably more variability than sugar concentrations (Fig. 2). Proline concentrations more than doubled from summer (<5 mg/g of ovendry-leaf weight) to February (10 to 12 mg). These maximum concentrations are about 70% of the maximum that has been induced under controlled temperatures (42). Proline accumulation is enhanced by high concentrations of sugars (31).

Free proline accumulation in citrus leaves is also favored by decreases in water content (42). Seasonal trends in tissue water content were found in this field study, although visible water stress was prevented by overhead irrigation (Fig. 2). This observation of seasonal trends in water content of citrus leaves under apparent non-water-stress field condition further indicates the need for continued investigation in temperature-water relations of citrus trees, not only in stomatal aspects, but also in hydraulic resistance in xylem elements. Although soil temperatures were not taken in this study, limited data indicate that temperature affects the flow of water through roots of citrus trees (24).

Decreases in water content with concomitant increases in sugars probably contributed most to the winter increase of osmotic concentration in expressed sap from the citrus leaves (Fig. 2). This hypothesis is supported by the good correlation (r>0.9) between the observed and calculated osmotic pressures (Table II), the latter based on the water and sugar contents of the leaves. Data indicate almost a 1-$\underline{M}$ concentration (22.2 atmospheres and a calculated 1.8°C freezing point depression) during the winter of 1977-78. This concentration is less than that (29.3 atmospheres) found in February in studies showing both air and soil temperatures to be inversely related to osmotic concentration of expressed sap from citrus leaves (12,13). In some of the colder areas of the world where citrus is produced, this inverse relationship is not readily apparent (34). Our field study did show the inverse relationship between osmotic concentration and field temperatures (seasonal profiles and significant correlation).

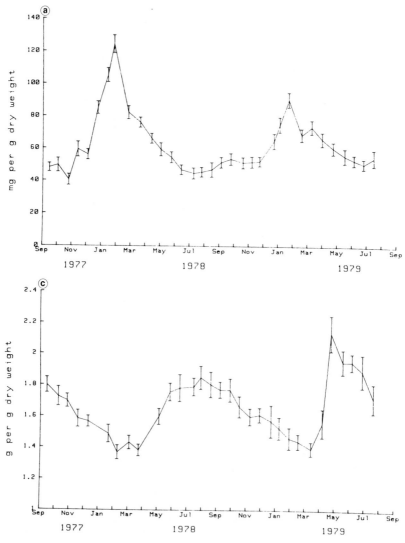

FIG. 2.   Profiles of seasonal trends in (a) total sugars, (b) free proline, (c) water content, and (d) osmotic potential of expressed sap (O.P.) and xylem-pressure potential (ψ_ω) in 'Valencia' orange leaves near Leesburg, Florida.

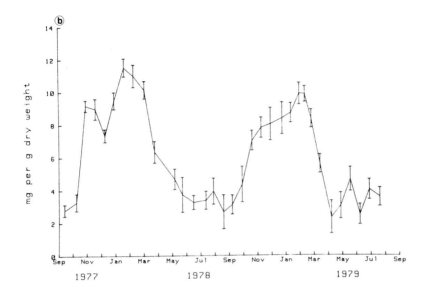

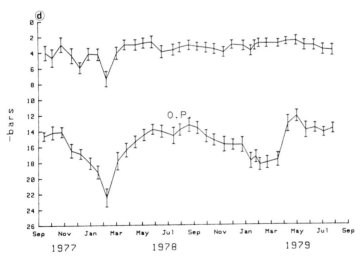

Field temperatures correlated rather poorly (r<0.4) with xylem-pressure potentials, starch content, amino acids (minus proline), water-soluble proteins, and ATP concentrations in the citrus leaves (Table I). Predawn xylem-pressure potential in leaves did not satisfactorily index citrus cold hardiness in this study. But xylem-pressure potential is valuable in indicating the extent of forced dehydration and freeze injury of young citrus trees (40) and in indicating the chances for the survival of black locust (Robinia pseudoacacia L.) seedlings

Table II. Calculated and observed values of osmotic pressure of expressed sap from citrus leaves during different months near Leesburg, Florida.

Date	Osmotic pressure (atm)		
(1977-78)	Observed[a]	Calculated[b]	Diff.
September 14	14.5	13.7	-0.8
October 5	13.8	14.3	0.5
October 26	13.9	14.2	0.3
November 16	16.1	15.9	-0.2
December 7	16.3	15.8	-0.5
December 18	13.6	13.7	0.1
January 18	18.9	19.5	0.6
February 8	22.2	22.6	0.4
March 1	17.0	18.9	1.9
March 22	16.1	19.2	3.1
May 3	14.2	16.0	1.8
May 24	13.6	14.1	0.5
June 14	13.7	13.6	-0.1

Linear correlation, a = 2.225, b = 0.828, r = 0.926.
[a]   Osmometry of expressed sap.
[b]   Based on water and sugar contents in leaves.

with freeze-damaged stems (5). The poor correlation between starch content in the leaves and seasons or field temperatures in this study was partly attributed to the multiple growth flushes, which are characteristic of citrus trees in Florida, and the carbohydrate changes caused by the growth, development, and harvesting of the fruit (14,16). Somewhat better association of starch content with seasons was found in lemon leaves (10). Free amino acids, minus proline, were not significantly correlated with seasons or field temperatures, but proline itself continued to show potential as an indicator of stress in field-growing citrus trees. Water-soluble proteins in citrus leaves also are considered poor physiological indicators of temperature trends in the field (11,48). But water-soluble proteins did show seasonal variations in cortical tissues of locust (Robinia) trees (25). Very little is known about ATP concentrations in citrus leaves in the field. In this study, ATP concentrations ranged from 80 nm/g of freeze-dried leaf weight in April to 260 nm in January. But because of the large variability in ATP measurements, we could not relate them meaningfully with seasons or field temperatures. It may require relatively uniform and prolonged cooling temperatures to detect ATP relationships in the field. Such temperature conditions were indicated by ATP buildup that occurred during cold acclimation of winter rape (Brassica napus L.) grown in temperature-controlled rooms and that was largely due to light and dark processes (30). In our study, conductivity or leakage determinations on frozen leaves were also highly variable and correlated poorly (r<0.4) with air temperatures in the field.

The LT_{50} in controlled freezes correlated well (r=0.866) with the average temperatures in the field (Table I). Consequently, similar correlations were found between physiological factors and field temperatures as between physiological factors and LT_{50} (Table I). In controlled-temperature studies, LT_{50} values were used to index citrus cold hardiness (47). However, such measurements apparently would be difficult to extrapolate to practical citriculture.

Regardless of significant correlations found in this field study, the variability encountered demonstrates the need for extensive work to develop a system for using physiological factors as practical field indexes of citrus cold hardiness. A 1°C difference is often critical during damaging freezes in Florida and can have a large economic impact in grove management decisions. Of all the physiological factors in this study, the concentration of expressed sap from leaves seemed to have the greatest potential for practical use. The determination is relatively simple, quick, and inexpensive, and the values correlate well with LT_{50} values.

REFERENCES

1. Amir, J., Kahn, V., and Unterman, M., Phytochemistry 16:1495 (1977).
2. Bartholic, J., HortScience 12:291 (1977).
3. Bewley, J.D., and Gwozoz, E.A., Plant Physiol. 55:1110 (1975).
4. Boyer, J., Plant Physiol. 42:133 (1967).
5. Brown, G.N., Bixby, J.A., Melcarek, P.K., Hinckley, T.M., and Rogers, R., Cryobiology 14:94 (1977).
6. Cocking, E.C., and Yemm, E.W., Biochem. J. 58:12 (1957).
7. Cooper, W.C., Proc. Fla. State Hort. Soc. 79:66 (1966).
8. Cooper, W.C., and Peynado, A., Proc. Amer. Soc. Hort. Sci. 74:333 (1959).
9. Cooper, W.C., Rasmussen, G.K., Peynado, A., Hilgeman, R.H., Cahoon, G.A., and Opitz, K., Proc. Fla. State Hort. Soc. 76:97 (1963).
10. Dugger, W.M., Jr., and Palmer, R.L., Proc. First Int. Citrus Symp. 1:339 (1969).
11. Ghazaleh, M.Z.S., and Hendershott, C.H., J. Amer. Soc. Hort. Sci. 90:93 (1967).
12. Haas, A.R.C., and Halma, F.F., Hilgardia 5:407 (1931).

13. Halma, F.F., and Haas, A.R.C., Bot. Gaz. 86:102 (1928).
14. Hilgeman, R.H., Dunlap, J.A., and Sharples, G.C., J. Amer. Soc. Hort. Sci. 90:110 (1967).
15. Hodge, J.E., and Hofreiter, B.T., in "Methods in Carbohydrate Chemistry" 4 (R.L. Whistler, ed.), p. 380. Academic Press, New York, 1964.
16. Jones, W.W., Embleton, T.W., Steinacker, M.L., and Cree, C.B., Proc. Amer. Soc. Hort. Sci. 84:152 (1964).
17. Ketchie, D.O., Proc. First Int. Citrus Symp. 2:559 (1969).
18. Levitt, J., in "Responses of Plants to Environmental Stresses," 2nd edition, Vol. I. (J. Levitt, ed.), p. 166. Academic Press, New York, 1980.
19. Lewis, F.J., and Tuttle, G.M., Amer. Bot. 34:405 (1920).
20. Lowry, O.H., Rosenbrough, N.J., Farr, A.L., and Randall, R.J., J. Biol. Chem. 193:265 (1951).
21. MacRae, J.C., Planta 96:101 (1971).
22. Malek, L., and Bewley, J.D., Plant Physiol. 61:334 (1978).
23. Proebsting, E.L., Jr., HortScience 10:463 (1975).
24. Ramos, C., and Kaufmann, M.R., Physiol. Plant. 45:311 (1979).
25. Sakai, A., and Yoshida, S., Cryobiology 5:160 (1968).
26. Salazar, C.G., J. Agric. Univ. PR 50:316 (1966).
27. Svec, L.V., and Hodges, H.F., Can. J. Plant Sci. 52:165 (1972).
28. Singh, T.N., Paleg, L.G., and Aspinall, D., Aust. J. Biol. Sci. 26:45 (1973).
29. Smith, D., Research Report No. 41. College of Agric. and Life Sciences, Univ. Wisconsin, Madison, 1969.
30. Sobczyk, E.A., and Kacperska-Palacz, A., Plant Physiol. 62:875 (1978).
31. Stewart, C.R., Plant Physiol. 61:775 (1978).
32. St. John, J.B., Anal. Biochem. 37:409 (1970).
33. Troll, W., and Lindsley, J., J. Biol. Chem. 26:45 (1955).
34. Vasil'yev, I.M., in "Wintering of Plants" (J. Levitt, ed. of English translation), p. 139. Amer. Inst. Biol. Sci., 1961.
35. Weiser, C.J., Quamme, H.A., Proebsting, E.L., Jr., Burke, M.J., and Yelenosky, G., in "Modification of the Aerial Environment of Crops" (B.J. Barfield and J.F. Gerber, eds.), p. 55. ASAE Monograph, 1979.
36. William, K.T., Bevenue, A., and Washauer, B., Assoc. Off. Agric. Chemists 33:986 (1950).
37. Yelenosky, G., HortScience 6:234 (1971).
38. Yelenosky, G., Plant Physiol. 56:540 (1975).
39. Yelenosky, G., Proc. Int. Soc. Citriculture 1:199 (1977).
40. Yelenosky, G., J. Amer. Soc. Hort. Sci. 103:449 (1978a).
41. Yelenosky, G., in "Plant Cold Hardiness and Freezing Stress" (P.H. Li and A. Sakai, eds.), p. 297-311. Academic Press, New York, 1978b.
42. Yelenosky, G., Plant Physiol. 64:425 (1979a).
43. Yelenosky, G., J. Amer. Soc. Hort. Sci. 104:270 (1979b).

44. Yelenosky, G., and Guy, C.L., Bot. Gaz. 138:13 (1977).
45. Yelenosky, G., and Young, R., Proc. Fla. State Hort. Soc. 90:49 (1977).
46. Young, R., Proc. Amer. Soc. Hort. Sci. 78:174 (1961).
47. Young, R., Proc. Amer. Soc. Hort. Sci. 88:272 (1966).
48. Young, R., J. Amer. Soc. Hort. Sci. 94:252 (1969).
49. Young, R., HortScience 5:411 (1970).
50. Young, R., and Hearn, C.J., HortScience 7:14 (1972).
51. Young, R., and Peynado, A., Proc. Amer. Soc. Hort. Sci. 86:244 (1965).
52. Young, R., Yelenosky, G., and Cooper, W.C., Proc. First Int. Citrus Congr. 3:145 (1977).

FREEZE INJURY AND PROTECTION OF CITRUS IN JAPAN

I. Ikeda

Akitsu Branch, Fruit Tree Research Station
Ministry of Agriculture, Forestry and Fisheries
Akitsu, Hiroshima, Japan

Japan is one of the major citrus producing countries in the world. The temperature in winter season in Japan is lower than that of other countries for growing citrus, and satsuma mandarin, which is a cold hardy variety, ranks first among leading varieties. Recently, overproduction of satsuma mandarin stimulated the renewal of top working to later maturing varieties which include less cold hardy citrus. Consequently, cold hardiness of citrus becomes one of the more important problems in the citrus industry of Japan. Severe freezes in 1963, 1977, and 1981 severely damaged the fruit and killed many trees. Tree and fruit losses from freeze injury in 1981 were estimated at more than 350 million dollars. Severe losses from freeze have stimulated research on cold hardiness and protection against freeze damage. Recently, breeding and selection of cold hardy citrus and improvement of agricultural techniques against unusual climatic conditions were started at Akitsu Branch, Fruit Tree Research Station and elsewhere (3,4,5,6,10, 11). This report describes recent efforts to develop more cold hardy citrus and protection against freeze damage of citrus in Japan.

I. DIFFERENCE IN COLD HARDINESS OF CITRUS VARIETIES

Several cold waves hit Western Japan from late December 1976 until early March 1977 and in February 1981, and the minimum temperatures as low as -9.1°C and -10.4°C were registered respectively at Akitsu Branch, Fruit Tree Research Station. Under field conditions, a wide range in cold hardiness was observed among citrus varieties. Thus, following these freezes, the damage and the recovery of the trees were researched. The author rated the cold hardiness of 200 citrus varieties (3). Citrus trees were 2 to 9 year old budded trees

TABLE I. Freeze injury and recovery rating of citrus varieties on trifoliate rootstock after the 1977 freeze.

Cold hardiness	Variety	April			August		
		Leaf killed (%)	Defoliation (%)	Wood killed (%)	Recovery rating	Dead tree (%)	Yield of next year
Most hardy	Trifoliate	0[a]	5.0[b]	0[c]	0[d]	0	H[e]
	Yuzu	0	1.0	0	0	0	H
	Troyer	0	5.0	0	0	0	L
Hardy	Miyagawa	3.0	3.0	1.0	0.2	0	H
	Kiyomi	3.3	3.3	2.0	0	0	H
	Clementine	4.0	3.8	2.0	1.0	0	H
Somewhat hardy	Sanbôkan	4.5	5.0	2.5	0.9	0	H
	Kishûmikan	2.3	2.2	1.0	1.0	0	H
	Minneola sf.	3.7	3.0	2.0	1.0	0	H
Intermediate	Hassaku	5.0	5.0	4.0	2.6	0	L
	Amanatsu	5.0	5.0	5.0	3.7	0	L
	Lee	5.0	4.0	3.0	2.8	0	L
Somewhat less hardy	Valencia	5.0	5.0	5.0	4.1	0	L
	Suzuki navel	5.0	5.0	5.0	3.6	0	L
	Natsumikan	5.0	3.6	2.4	3.6	20	L

Less	Iyo	5.0	5.0	5.0	6.1	33	N
	Dancy	5.0	3.2	4.3	6.3	50	N
	Ponkan	4.5	5.0	4.5	5.9	50	N
Least hardy	Banpeiyu	5.0	5.0	5.0	6.0	63	N
	Lisbon	5.0	5.0	5.0	7.3	82	N
	Cleopatra	5.0	2.0	4.0	6.8	75	N

a 0 = no damage   5 = 100% leaf killed
b 0 = no leaf fall   5 = 90 to 100% leaf drop
c 0 = no damage   5 = secondary scaffold branch killed
d 0 = no damage   8 = tree dead
e H = heavy crop   L = light crop   N = no crop
f S = seedling

TABLE II.   Rank of cold hardiness in citrus varieties.

Cold hardiness	Variety
Most hardy	Trifoliate, Hiryu, Troyer citrange, Yuzu, Sudachi, Kabos
Hardy	Wase satsuma, satsuma mandarin, Kiyomi, Okitsu No. 20., Clementine, Keraji, Citrangequat, Henkamikan, kumquat, Ichang lemon, Umatilla sdlg., Minneola sdlg.
Somewhat hardy	Sanbokan, Kobayashimikan, Hyûganatsu, Funadoko, Sekwasha, Kishûmikan, Kôji, Tachibana, Banôkan, Kawachibankan, Minneola, Pixie, Kara, Valencia sdlg., Neovalencia sdlg., Fukuhara orange sdlg., Suzuki navel sdlg.
Intermediate	Hassaku, Yamamikan, Daidai, Chinotto, Pope Summer, Pera Natal, Red Siletta, Page, Osceola, Fairchild, Willowleaf sdlg.
Somewhat less hardy	Amanatsu, Natsumikan, Naruto, Tankan, Parson Brown, Fukuhara orange, Hamlin, Valencia, Trovita, Tarocco, Baianinha, Tange navel, Suzuki navel, Meyer lemon, Anseikan, Marsh, Redblush, Fremont, Fortune, Seminole, Nova, Lee.
Less hardy	Iyo, Kinukawa, Juffa, F.N. Washington navel, Shamouti, Calamondin, Tanikawa buntan, New Zealand grapefruit, Crabo, Ponkan, Dancy, Murcott, Encore, Wekiwa, Beauty, Pearl, Robinson.
Least hardy	Banpeiyu, Mato buntan, Hirado buntan, Lisbon lemon, citron, Mexican lime, Cleopatra.

sdlg. = seedling

on trifoliate rootstock located from 105 to 155 meters above sea level. Freeze injury and recovery of trees were evaluated between March and August in 1977 and between March and June in 1981. Cold hardiness was evaluated on the amount of defoliation, leaf and wood killed, tree recovery and percent of dead trees. Fruit yields were surveyed the year immediately before and after a freeze.

The results of the survey for varieties are shown in Tables I and II. Citrus varieties were summarized into the following 5 groups according to their cold hardiness:

    (1)   The most cold hardy group: Trifoliate, Yuzu and their hybrids.

(2) Hardy group: Wase satsuma, satsuma mandarin and their hybrids Kiyomi and Okitsu No. 20, small fruit mandarin such as Tachibana, Clementine and Kishûmikan, kumquat and its hybrid citrangequat, nucellar seedlings of orange and American cross hybrids.

(3) Intermediate group: Hassaku, sour orange, some varieties of orange and navel and introduced breeding varieties.

(4) Less hardy group: Natsumikan, Tankan, grapefruit, Iyo, Ponkan, almost all of orange and navel and introduced cross hybrids, and some varieties of shaddock.

(5) The least hardy group: Shaddock, lemon, lime and citron.

In this study, cold hardiness of these varieties was similar to those of varieties previously reported (9,11,12,13). Nucellar seedlings of orange and cross hybrids showed more cold hardiness than their parents. Hybrids of satsuma, Kiyomi and Okitsu No. 20, were as hardy as the parent satsuma.

II. HEREDITY AND BREEDING OF COLD HARDY CITRUS

Reports related to the breeding of cold hardiness are few in Japan. Citrus breeding was begun about 40 years ago, and new varieties with cold hardiness were selected in the process of citrus breeding. In the 1977 freeze, 1,511 hybrid seedlings from 23 combinations segregated for cold hardiness when they were exposed to natural freezing condition (3). Some of the progenies were more cold hardy than their parents, and the progenies of Hassaku x Clementine, and Tanikawa buntan x Mukakukishu were more cold hardy than those of other combinations (Table III and Fig. 1). This indicated the inheritance of cold hardiness in Clementine and Mukakukishu. From the progenies of Hassaku x Clementine, two promising seedlings were selected for cold hardiness and good fruit quality. In another test, hybrid seedlings from the combinations of Imamura unshu (satsuma mandarin) x Fukuhara orange, Page and Nova exhibited obvious segregation for cold hardiness when they were exposed to natural freeze condition in 1981. Some seedlings were more cold hardy than the female parent (Fig. 2). Other female parents which segregated a high percentage of cold hardy seedlings were Umatilla, Kiyomi and Okitsu No. 20. These varieties are of a cold hardy group and produce only zygotic seedlings. Kiyomi is a F_1 hybrid of Miyagawa wase satsuma x Trovita orange and Okitsu No. 20 is a F_1 hybrid of Ueda unshu (satsuma mandarin) x Hassaku. These two hybrids were as hardy as satsuma mandarin in the 1977 and 1981 freezes (3).

TABLE III. Freeze injury segregation of 10-year-old hybrid seedlings exposed to -9.1°C in the 1977 freeze.

			No. of plants	Percentage of seedling					
				0[a]	1	2	3	4	5
Parents of cross						Recovery rating			
Iyo	x	Valencia	42			7.1	11.9	31.0	50.0
		Person Brown	103		1.9	5.8	17.5	22.3	52.4
		Fukuhara	25		4.0	4.0	16.0	36.0	40.0
		Sampson	32			12.5	12.5	43.8	31.5
Hassaku	x	Valencia	31		6.5	19.4	19.4	29.1	25.8
		Trovita	159	0.6	13.2	18.9	20.1	27.7	19.5
		Fukuhara	38			7.9	29.0	39.5	23.7
		Iyo	50		12.0	22.0	26.0	12.0	28.0
		Clementine	207	2.4	22.2	26.1	20.8	17.9	10.6
Kinukawa	x	Trovita	68		1.5	13.2	16.2	17.7	51.5
		Person Brown	69		4.4	10.1	23.2	18.8	43.5
		Hirakishu	14		7.1	7.1	7.1		78.6
Tanikawa	x	Valencia	3				66.7		33.3
buntan		Trovita	179		0.6	8.4	21.2	32.4	37.4
		Person Brown	142			5.6	28.2	33.8	32.4
		Iyo	113			13.3	19.5	18.6	48.7
		Mukakukishu	19	5.3	26.3	21.1	26.3	10.5	10.5
Hirakishu	x	Trovita	47		4.3	12.8	6.4	19.2	57.5
Naruto	x	Sampson	27		11.1	18.5	3.7	18.5	48.2
		Trovita	60			3.3	3.3	18.3	75.0
Okitsu #20	x	Trovita	16				18.8	31.3	50.0
Clementine	x	Iyo	15			13.3	20.0	20.1	46.7
		Sampson	52		1.9	17.3	11.5	21.1	48.1

[a] 0 = no damage   5 = tree dead

Data suggest that cold hardiness is a heritable factor and hybrid seedlings with cold hardiness can be obtained from crosses with cold hardy satsuma mandarin. Therefore, satsuma mandarin is one of the suitable parents for breeding cold hardy citrus hybrids.

In another study, Yoshida (10) found that hybrid seedlings of lemon or pummelo (cold susceptible) crossed by Yuzu (cold hardy) exhibited obvious segregation for cold hardiness when they were exposed to optimum artificial freezing conditions after natural hardening. Most seedlings had cold hardiness existing between those of their parents, but some seedlings were as hardy as the cold hardy parent and the inheritance of cold hardiness was demonstrated. According to our results and

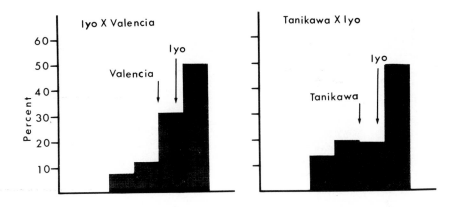

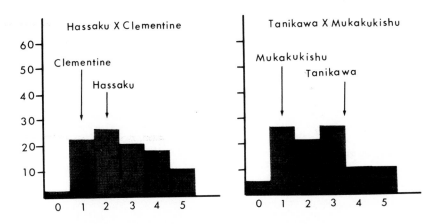

FIG. 1. Freeze injury segregation of 10 year-old hybrid seedlings exposed to -9.1°C in 1977. Plants rated 0 were the most cold hardy. Arrows indicate cold hardiness of the parents.

those of other researchers, production of cold hardy varieties through breeding and selection may be one of the most useful methods of reducing citrus losses during freezes.

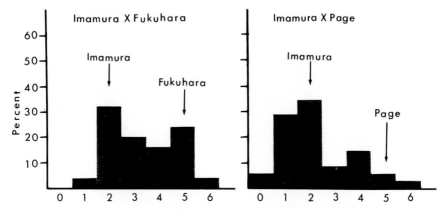

FIG. 2.   Freeze injury segregation of 4 year-old hybrid seedlings exposed to -10.4°C in 1981.   Plants rated 0 were the most cold hardy.   Arrows indicate cold hardiness of the parents.

III.   THE INFLUENCE OF VARIOUS ROOTSTOCKS ON THE COLD HARDINESS OF THE SCION

In 1968, a citrus rootstock trial was started to find superior rootstock for navel orange at Akitsu Branch (1). The cold hardiness of the rootstock seedlings was evaluated when they were exposed to natural freezes with minimum temperature of -5.7°C and -6.2°C in 1969 and 1970 respectively. Two to three year old rootstock seedlings of trifoliate, trifoliate hybrid No. 576 (Natsumikan x trifoliate), Troyer citrange (navel orange x trifoliate) and Yuzu were the most cold hardy, whereas, those of Rough lemon were the least cold hardy (Table IV).

Three scion navel orange varieties were budded on 15 rootstocks in 1969, and 32 top-root combinations were planted in the field. In 1977, the effect of the rootstock on the cold hardiness of navel orange trees was investigated following the severe freeze with minimum temperature of -9.1°C (Table V). Seven year old scion trees grown on trifoliate, trifoliate hybrid No. 576 and Tachibana rootstock were distinctly cold hardy and showed better recovery. On the contrary, the scion trees on Rough lemon and Konejime were very vigorous but less cold hardy and showed poor recovery. The navel trees on Natsumikan and Kinkoji were very weak and most severely injured. As a result, the influence of the cold hardiness of

TABLE IV. The order of cold hardiness of several rootstock seedlings and budded tree on them.

Item	Hardy	Cold hardiness			Sensitive
Seedling	Trifoliate " No. 576 Troyer Yuzu	No. 807 >	Tachibana Natsumikan Kinkoji >	Cleopatra Konejime Trovita Valencia >	Rough lemon
Budded Tree	Trifoliate " No. 576 Tachibana	No. 807 >	Cleopatra Valencia >	Troyer Trovita Yuzu Rough lemon >	Konejime Natsumikan Kinkoji

TABLE V. Freeze injury to navel orange scion on several rootstocks at Akitsu Branch.

| | Freeze injury to indicated scions | | | | | |
| | Hamachi navel | | Suzuki navel | | Tange navel | |
Rootstock	Recovery rating	Dead trees (%)	Recovery rating	Dead trees (%)	Recovery rating	Dead trees (%)
Trifoliate	3.5[a]	17	2.3	0	3.2	17
" No. 807	1.5	0	3.7	0	2.0	0
No. 576	2.2	0	2.3	0	5.0	33
Tachibana	2.0	17	2.7	0	–	–
Valencia	3.8	33	–	–	3.0	17
Cleopatra	4.7	50	3.8	33	2.7	17
Trover	4.5	50	4.7	50	5.2	33
Trovita	4.5	50	–	–	–	–
Yuzu	4.7	50	4.8	17	–	–
Rough lemon	5.2	67	4.0	23	–	–
Konejime	5.2	67	5.0	50	4.7	50
Natsumikan	5.0	50	6.8	83	–	–
Kinkoji	6.5	53	7.0	100	7.0	100

a 0 = no damage   7 = tree dead

trifoliate and Rough lemon rootstocks on the cold hardiness of
the scion was obviously demonstrated in this trial.

In this rootstock trial, we must not neglect virus disease. Because, Tristeza virus will affect the rootstock and
weaken the vigor and cold hardiness of the scion tree when the
budded scion is infected with the virus. For example, the
seedling of Yuzu was the most cold hardy but the scion tree on
it was very weak and less cold hardy because of the infection
of Tristeza virus (2).

IV. FREEZE INJURY AND PROTECTION

Cold injury of citrus trees is divided into cold wind damage and freeze injury. Cold wind damage is a phenomenon of
defoliation caused by the monsoon in winter. When the wind
velocity exceeds the critical wind speed of 7 m/s, the defoliation will increase rapidly (6).

Minimum temperature below the critical temperature kills
leaves and stems. The cold hardiness of the fruit is less
hardy than leaves or stems. When the temperature is lower than
-3.5 to -4.5°C, fruit granulation due to pulp freezing occurs
(8). When air temperature in an instrument screen reached
-4.6°C, bitterness increased in Hassaku fruit on the tree and
the average naringin content was 29.9% per mg of pulp juice
squeezed with a hand squeezer (7).

In 1976, when minimum temperature was -5.7°C in late
December, bitterness increased and granulation was observed in
almost all of the fruit on the tree. Some leaf damage is
expected when temperatures drop below -6°C for several hours.
Below -8°C, twig dieback will appear, and below -10°C, almost
all leaves will be killed (5,11). Seasonal variations of cold
hardiness have been investigated and maximum cold hardiness
occurred in January or February (5,6).

The principal countermeasure to avoid freeze injury is to
first select a warm location for planting citrus trees. Two
meteorological elements are especially important in producing
high quality citrus fruit with minimum risk of freeze injury.
One is the annual mean temperature and the other is the minumum
temperature in winter season. Ministry of Agriculture,
Forestry and Fisheries have made a standard of favorable climatic conditions for planting citrus fruit trees based on different studies of freeze injury (Table VI).

In many protection methods against the freeze injury,
covering methods are common in Japan (Fig. 3). Direct covering
with rice straw mats resulted in a warming of about 1.5 to
2.0°C when the leaf temperatures of uncovered trees were -5 to
-6°C (6). Warming effect of indirect covering with straw mats
was about 3.2°C when the leaf temperature of uncovered trees

TABLE VI. Recommended climatic condition for planting citrus tree.

Species or variety	Annual mean temperature	Minimum temperature in winter	
Satsuma mandarin (Citrus unshiu Marc.)	above 15°C	above -7°C	
Hassaku (C. hassaku Hort. ex Tanaka)	15.5°C	-5°C	not below -4°C before harvest
Iyo (C. iyo Hort. ex Tanaka	"	"	
Navel orange (C. sinensis Osbeck)	"	-4°C	not below -3°C before harvest
Amanatsumikan (C. natsudaidai Hayata)	16°C	"	
Fukuhara orange (C. sinensis Osbeck)	"	"	
Pummelo (C. grandis Osbeck)	"	"	not below -3°C before harvest
Hyûganatsu (C. tamurana Hort. ex Tanaka)	16.5°C	"	
Tangelos (C. paradisi x tangerina)	"	-3°C	
Ponkan (C. reticulata Blanco)	17°C	"	
Tankan (C. tankan Hayata)	18°C	"	

was below -7°C (Table VII). Rice straw mats are considered very effective in preventing freeze injury (5).

Vinyl-house growing of citrus plants has been used on an estimated 437 ha in 1979, and resulted in a fruit yield of 17,548 tons. This is a specific covering method developed in Japan, and is expected to increase fruit quality by 1) avoidance of freeze injury, 2) improvement of appearance, 3) hastening fruit maturation period, and 4) promoting fruit growth. Wase satsuma, navel orange, introduced hybrids such as Murcott, Encore, Minneola, Seminole, etc. are all successfully grown in vinyl-house coverings of citrus in Japan.

FIG. 3. Protection of freeze injury by direct straw mat covering in Shizuoka Prefecture (Konakahara).

TABLE VII. Warming effect of indirect covering methods of the leaf temperature of citrus trees (Konakahara).

Material	Warming effect			
	0°C to -3°C[a]	-3°C to -5°C	-5°C to -7°C	Below -7°C
Cheesecloth	1.1	1.1	1.3	1.7
Korucloth	2.1	2.1	2.3	2.8
Tetrontricot	2.2	2.3	2.5	3.0
Rice straw mat	2.0	2.4	2.6	3.2

[a] Minimum leaf temperature of uncovered trees.

V.   SUMMARY

Differences in cold hardiness among 200 citrus were determined under field conditions after the 1977 and 1981 freezes. Varieties were divided into 5 groups according to their cold hardiness and satsuma mandarin was the most cold hardy scion type. F_1 hybrid seedlings from some combinations exhibited segregation for cold hardiness under the natural freeze conditions. Clementine, Mukakukishu and satsuma mandarin were suitable parents for breeding cold hardy hybrids. Trifoliate and Tachibana rootstocks indicated more cold hardiness than other rootstocks when used for navel orange trees. Rice straw mats covering and vinyl-housing are two of the more promising cold protection methods available to citrus growers in Japan.

REFERENCES

1. Ikeda, I., Nakatani, M., and Kobayashi, S., Bull. Fruit Tree Res. Stn. Japan, E-2, 39-57 (1978).
2. Ikeda, I., Nakatani, M., and Kobayashi, S., Bull. Fruit Tree Res. Stn. Japan, E-3, 25-47 (1980a).
3. Ikeda, I., Kobayashi, S., and Nakatani, M., Bull. Fruit Tree Res. Stn. Japan, E-3, 49-65 (1980b).
4. Ikeda, I., Nakatani, M., and Kobayashi, S., Annual Rep. Fruit Tree Res. Stn. Akitsu, 14-15 (1981).
5. Improvement of Agricultural Techniques against Unusual Climatic Conditions. A data of Agr. For. and Fish. Rec. Council 294-395 (1981).

6. Konakahara, M., Special Bull. Shizuoka Pre. Cit. Exp. Sta. 3:1-164 (1975).
7. Maotani, T., and Hase, Y., Bull. Fruit Tree Res. Stn. Japan, E-3, 67-74 (1980).
8. Nakagawa, Y., Honjo, H., and Konakahara, M., J. Agr. Met. 31(4):195-198 (1976).
9. Yelenosky, G., and Young, R., Citrus Industry 58(1):5-14 (1978).
10. Yoshida, T., Bull. Fruit Tree Res. Stn. Japan, B-8, 1-11 (1981).
11. Yoshimura, F., Memoirs Fac. Agr. Kochi Univ. 18:79-134 (1967).
12. Young, R., and Olson, E.O., Proc. Amer. Soc. Hort. Sci. 83:333-336 (1963).
13. Young, R., and Hearn, C.J., Hort. Sci. 7:14-18 (1972).

INHERITANCE OF COLD HARDINESS OF TEA PLANTS IN CROSSES BETWEEN VAR. SINENSIS AND VAR. ASSAMICA

T. Toyao

Laboratory of Genetics and Physiology
National Research Institute of Tea
Kanaya, Shizuoka, Japan

The project for cross breeding of tea plants (Camellia sinensis (L.) O. Kuntze) was begun in 1929 in Makurazaki, Kagoshima, Japan. The principal objectives of the project were to introduce the quality and flavour of black tea and the vigor of growth in var. assamica into Japanese tea clones.

The breeding materials collected from the tropical, subtropical, and temperate regions in Asia were used as basic materials for the production of a number of inter-varietal hybrids. Because there are less hardy plants among those belonging to var. assamica, the breeding of tea clones requires preliminary information about the variation and the inheritance pattern of cold hardiness.

A very wide variation in cold hardiness in tea plants was reported (8,9). It is believed from some experiments that winter injury to tea plants principally depends upon the leaf hardiness in midwinter, while in a warmer region of Japan the winter injury is found to be partially affected by bark splitting injury of stem which is closely related to the stem hardiness in late autumn (10).

In the present study, an attempt to clarify the inheritance pattern and to estimate the magnitude of heritability values of cold hardiness in a leaf and stem, was carried out in progenies using various hybrid combinations.

TABLE I.   The crosses and the offspring.

	Year of cross	Cross matings	Offspring per cross	Total No. of offspring
Exp. I[a]	1960, 1961	37	8-33	832
Exp. II[a]	1964	40	11-14	474
Exp. III[b]	1969	20	13	260

[a]   The offspring were chosen at random from each cross in the field for individual selection work.

[b]   The field planting consisted of a randomized block design with 3 replications by the cuttings of offspring and their parents. In each offspring, 13 cross seedlings were chosen at random.

I.   MATERIALS AND METHODS

A.   Materials

Parental materials used in this study consist of three groups of tea clones which belong to var. sinensis, var. assamica, and selections from hybrids between both varieties. Plants in the first group which were selected in Japan and China generally show a high tolerance to cold stress, and those in the second group which have been introduced from India are susceptible to cold, and those in the third group have an intermediate hardiness.

Controlled pollinations among these parental clones were made, and the hybrid seedlings and their parents were planted at the Makurazaki Branch. Data were collected from approximately five- to six-year-old adult trees. The number of crosses and offspring used in this study are listed in Table 1.

In both Exp. I and II, parents from various sources were used for cross experiments, so that they showed a wide range of hardiness. In Exp. III, 20 Japanese and Assam hybrid clones were crossed with a common parent Yabukita, which is one of the most superior tea clones in Japan.

B.   Cold Hardiness

For Exp. I and II, one-year-old twigs with mature leaves were collected from each tree in midwinter in 1968 (Exp. I) and 1971 (Exp. II). The twigs were placed into a freezer at 0°C for one hour, and then treated at -7°C   -11°C for two hours. Cold injury to the leaves was scored two or three days after freezing by comparison with the check clones which had been described in a previous report (8). The hardiness score ranged from 1 to 10; the healthy and uninjured leaves were scored as 1 while completely damaged ones were scored as 10.

For Exp. III, the twigs were treated at -9°C in December, 1979 and at -11°C in January, 1980. Cold injury of the leaves was scored from 1 to 10 in a similar manner as in Exp. I and II. For stem hardiness, freezing injury was graded from 1 (uninjured) to 5 (dead) by estimating the dead area of the inner surface of the stem cambial tissue.

C. Other Traits

Leaf characters, earliness of sprouting time, fermentation ability of fresh shoots and tannin contents were measured to estimate the relationship between the cold hardiness and their traits.

D. Heritability and Genetic Correlation (1)

Regression and correlation coefficients between the parental and progeny's hardiness scores were calculated from the three experimental data.

Heritability estimates (h^2) were obtained by the midparent-offspring regression coefficients (b) directly, or by doubling the regression coefficients (b') based on single parent values:

$$\text{Offspring and midparent} \quad\quad b = h^2$$
$$\text{Offspring and single parent} \quad\quad b' = 1/2\, h_2$$

Genetic correlation coefficients between the cold hardiness (character X) and another trait Y were calculated from the following equation:

$$r_{xy} = cov_{xy}/\sqrt{cov_{xx} \times cov_{yy}}$$

where cov_{xy} is the "cross-covariance" between X and Y, between parent and offspring, and cov_{xx} or cov_{yy} is the offspring-parent covariance of X or Y character.

II. RESULTS

A. Varietal Difference of the Cold Hardiness among the Parental Clones

Clonal difference of cold hardiness in the parents used in this study was found to be widely variable except for the leaf hardiness in Exp. III (Table 2). The scores of the hardiest parent in Exp. I and II were found to be 1.2 and 1.9, while the tenderest one as 9.5 or 10.0, respectively. On the other hand,

TABLE II. Variation in cold hardiness in parent clones.

Hardiness		Score of hardiness										
	Exp. No.	1	2	3	4	5	6	7	8	9	10	Total
Leaf[a]	I	3	3	2	8	5	2	4	3	4		34
	II	1	2	4	7	6	5	5	1	2	1	34
	III	1	1	8	4	5	2					21
Stem[b]	III	2	3	7	7	3						21

[a]   Hardiness score of mature leaves; 1 (uninjured) - 10 (dead).
[b]   Hardiness score of stem; 1 (uninjured) - 5 (dead).

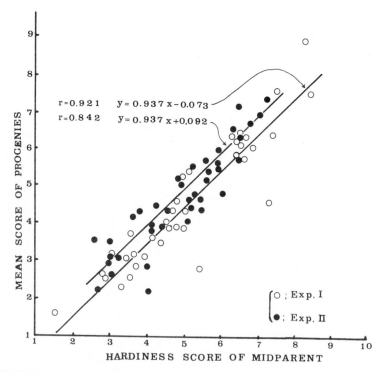

FIGURE 1. Regression of progeny on midparent in cold hardiness score of leaves (Exp. I and Exp. II). Hardiness score: 1 (uninjured)— 10 (dead).

the stem hardiness of parents ranged also widely from the hardiest score 1 to the tenderest score 5.

The clonal difference in cold hardiness of leaves was variable as compared with the variation in the collected materials at the Makurazaki Branch. In previous observations (9), it was shown that the "sinensis" plants among the collections introduced from nearly all of the tea growing regions in Asia were scored from 1 to 3, while the plants belonging to var. assamica were scored from 7 to 10.

Parental variation of leaf hardiness in Exp. III is found to be in a narrower range, because the parents consisted of Assam hybrids (var. sinensis x var. assamica) and some Japanese clones.

B. Heritability Estimated by Parent-Offspring Regression of Leaf Hardiness

As shown in Fig. 1, mean value of progeny corresponded closely to their midparental value; therefore, the correlation coefficients as well as the regression coefficients were very high. Regression coefficients of the mean score of hybrid progeny on their midparental one were estimated at 0.937 in both Exp. I and II. The values of the above regression coefficient are equivalent to the estimates of the heritability which express the reliability of selection on the basis of the phenotypic value.

Doubled values of the regression of progeny on single parents crossed with a common parent also correspond to ther heritability values. Estimates from a doubled value of the regression coefficient of the progeny on single partents were 0.893, 0.982 and 0.610 according to the analyses of variance - covariance in Exp. I, II and III, respecitvely (Table 3). In Exp. III, the heritability value was somewhat lower than those in Exp. I and II. It may have been caused by the hardiness range of the parents being narrower than those of Exp. I and II.

As a result of the relationship study between the parent and offspring as mentioned above, the cold hardiness of the mature leaf in midwinter is concluded to be highly heritable. In that case, it is considered that when the parental variation ranged from extremely hardy to extremely tender, the average heritability value seemed to exceed 0.9. Therefore the degree of hardiness in the progeny could be efficiently predictable from that of their parents.

Furthermore, the results thus far obtained showed no significant differences between the reciprocal cross combinations of the parents in the hardiness scores of the progeny.

C. Heritability of Cold Hardiness of the Stem

The stem hardiness which affects the bark splitting injury of young trees which occurs in late autumn was also found to be

TABLE III. Heritability of cold hardiness of leaves estimated by regression of progeny on single parents.
- Variance-covariance analysis[d] -

Source		Exp. I				Exp. II				Exp. III		
	df	X^a	XY^b	Y^c	df	X	XY	Y	df	X	XY	Y
Total	38	4.81	1.32	2.01	48	3.74	1.61	1.61				
Between common parent	8	3.27	-2.45	5.11	11	4.73	1.36	1.86				
Within common parent	30	5.22	2.33	1.19	37	3.44	1.69	1.17	19	1.69	0.51	0.35
Regression coefficient	$b = 0.446 \pm 0.027$				$b = 0.491 \pm 0.040$				$b = 0.305 \pm 0.012$			
Heritability	$h^2 = 0.893 \pm 0.055$				$h^2 = 0.982 \pm 0.080$				$h^2 = 0.610 \pm 0.024$			

a   Variance of parental values.
b   Covariance of parent and progeny.
c   Variance of progeny.
d   Variance-covariance analysis were derived from a part of crosses in Exp. I and II. The data were used repeatedly regardless of parental sex.

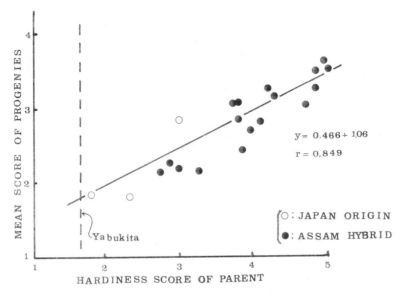

FIGURE 2. Regression of progeny on single parent crossed with a common parent, Yabukita, in stem hardiness (Exp. III). Hardiness score: 1 (uninjured) — 5 (dead).

higher in the heritability value than 0.9. The estimate was obtained from doubled value of the regression coefficient of 20 progenies upon their parents crossed with a common parent, i.e. clone Yabukita, in Exp. III (Fig. 2).

As mentioned above, it was ascertained that the tea plant winter hardiness measured mainly by the leaf and stem hardiness, was very heritable as shown by the estimates for the two characters being exceeding 0.9.

D. Inheritance Pattern of the Cold Hardiness Through the Inter-Tree Variation

Individual trees in each cross showed a certain variation in both the leaf and stem hardiness. Fig. 3 and 4 give the distributions of frequency of progenies in some crosses in Exp. I and III. The variation within a cross is as small as 1.0 in terms of the standard deviation in both the leaf and stem hardiness. The standard deviations are little variable regardless of the magnitude of the parents' hardiness score and regardless of the difference in the score of both parents. It appeared that the frequency of the leaf hardiness score in a cross showed a monomial and continuous distribution with the mode of the distribution being close to the midparental value in almost all crosses.

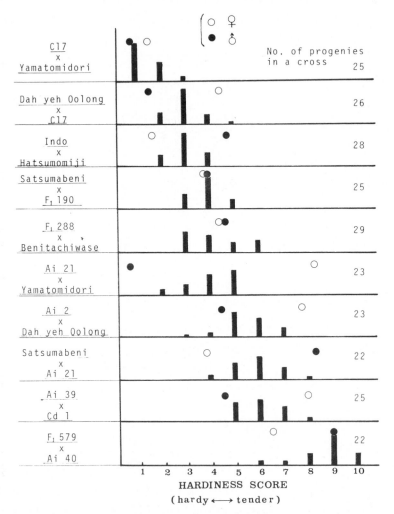

FIGURE 3.   Distribution of the frequency of the progenies for cold hardiness of mature leaves within a cross (Exp. I).

According to the above observations, it has been corroborated that the leaf hardiness in the tea plants is regulated by a considerable number of genes, i.e. polygenes, instead of a few major genes.

The variation of stem hardiness in each cross also showed a monomial distribution and the mode was close to the midparental value.   The pattern of progeny distribution for stem hardiness, however, differed from that of leaf hardiness.   When a hardier parent and a tenderer one for stem hardiness were mated, the distribution of the progenies showed a wide variation between

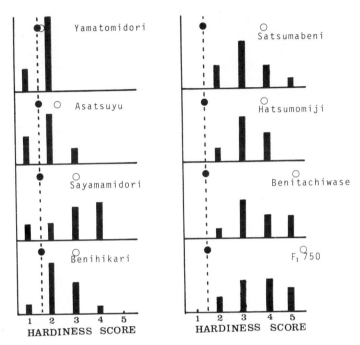

FIGURE 4.   Distribution of the frequency of the progenies
for cold hardiness of stems within a cross (Exp. III). O : a
single parent. ● : a common parent, Yabukita.

the score 2 to 5 (Fig. 4). For example, we have four clones
selected from the same progeny of a clone Yabukita. It is
interesting to find that both of them are very hardy to stem
freezing injuries, while the remaining two, which are sisters
of the former two, were very tender. The reason why the
hardiest clone and the tenderest one were selected from the
same progenies may have been due to a discontinuous distribution of extreme types. Therefore, it may be suggested that the
hardiness of the stem is controlled by a small number of genes.
    In leaf hardiness, on the contrary, extremely hardy and
extremely tender trees were not found in the same progenies
even the crossing was made between the tenderer clone Ai 21 and
the hardiest clone Yamatomidori (see Fig. 3).
    In summing up the inheritance pattern of cold hardiness, it
is suggested that leaf hardiness is most likely controlled by
polygenes, while stem hardiness seems to be regulated by a
relatively small number of genes.

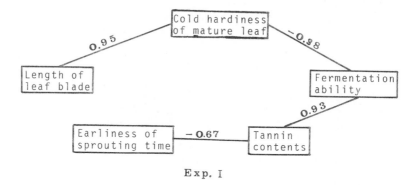

Exp. I

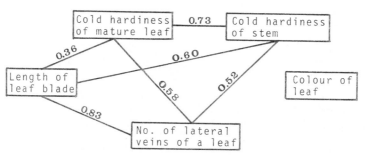

Exp. III

FIGURE 5.   Genetic correlation coefficients[a] derived from parent-offspring regressions between cold hardiness and other traits (Exp. I and III). [a] Correlation coefficients showed only the values exceeding 0.3.

TABLE IV. Cold hardiness scores of leaves and critical temperatures for the check clones (8).

Clones	Origin	Hardiness score	Critical temperature
Yabukita	Japan	1	−9°C
Chin-shin Oolong	China	3	−7°/−9°C
Hatsumomiji	Assam hybrid	5	−5°/−7°C
F_1 522	Assam hybrid	7	−3°/−5°C
Ai 50	Assam hybrid	9	−2°C

E. Genetic Correlation between Cold Hardiness and Other Traits

Genetic correlation coefficients between the cold hardiness and length of leaf blade were estimated as 0.95 (Exp. I), 0.36 and 0.60 (leaf and stem hardiness, Exp. III) on the basis of parent-offspring regression (Fig. 5). As a matter of fact, the trees with larger leaves were found to be less hardy than the trees with smaller ones. No other characters were found to be genetically correlated with the cold hardiness.

On the other hand, genetic correlation between the leaf and stem hardiness was calculated to be 0.73. But, it should be remembered that the correlation was not absolutely high. An example may be mentioned; clone Okumusashi, a very hardy Japanese one with regard to leaf injury in midwinter, is not resistant to bark splitting injury which is affected by the stem hardiness.

III. DISCUSSION

The great contrast in the cold hardiness between Camellia sinensis var. sinensis and var. assamica is interpreted to be due to the difference in the critical temperature for leaf freezing injury in winter. Our previous paper (8) described the critical temperature for several check clones as shown in Table 4.

Tests for local adaptability of tea clones had been carried out at many experimental locations in Japan. It was found in those tests that our hardiness score by an artificial freezing experiment could show the northern boundaries of the cultivated tea clones (9).

This paper reported that the great difference in the leaf hardiness score among tea varieties is controlled by polygenes. Consequently, it is suggested that many of the hardier genes of alleles related to the cold hardiness of tea plants had been accumulated in the hardy plants, i.e. ones belonging to var. sinensis, through a very long history of adaptation to winter cold climate.

A species distributed over a wide range of area usually shows a great variation in cold hardiness which approximately corresponds to the minimum temperature or mean temperature in the coldest month in their native habitats (2,6,7). With regard to the pattern of inheritance for cold hardiness, a few investigations have been reported, i.e. in apple trees (3), and in conifers (4). In those species cold hardiness had been revealed to be controlled by polygenes, and the present experimental data derived from inter-varietal hybrids have suggested that the leaf hardiness of tea plants is regulated by polygenes, while the stem hardiness by some major genes. Also, the effects of the genes which reflects leaf and stem hardiness were confirmed to have a heritable performance with an additive effect because of the high heritability values.

A close relationship was observed between the resistance to the bark splitting injury and the stem hardiness determined by an artificial freezing test. The mechanism for the occurrence of the bark splitting was ascertained (10). That is, when a stem is not yet hardy enough to withstand the autumn frost, a large ice crystal was formed in the bark tissue of the basal stem by the water supply from roots, results in the bark splitting lengthwise and injury to cambium. The kind of ice segregation in the bark tissue occurs at a low temperature of about -3 to -5°C continuing for a few hours in early winter.

Furthermore, resistance to bark splitting injury is affected by the age of the trees, the application rate of fertilizer, and, especially, the clonal difference of stem hardiness in late autumn (10).

Sakai (5) investigated the occurrence of frost damage in the basal stem of young trees of several species and pointed out that the basal stem was unusually sensitive to frost damage. Watkins and Spangelo (11) found that the additive variance for the stem damage of apples was very large, occupying approximately 100% of the total genetic variance, and they concluded that the potential parents could be successfully screened phenotypically.

In Japan, the tea plants often suffer winter injury. It has been found from our studies that the cold injury in tea plants, particularly in the warmer regions of Japan, involves two different phenomena, that is the leaf injury and the other the bark splitting injury. Thus, the tea breeders are requested to improve tea clones in these two respects. The investigation reported in this paper may give a bright prospect to this problem.

IV. SUMMARY

Crossing experiments between two varieties were made, and the hardiness of the parent and their progenies was investigated by an artificial freezing test and compared with the check clones. The following results were obtained with respect to their inheritance of cold hardiness:

(1) By the regression of the hybrid progeny on their parents, it was estimated that the cold hardiness of the mature leaf in midwinter was highly heritable. The heritability values exceeded 0.9 on the average when the variation ranged from extremely hardy to extremely susceptible to cold stress. Consequently the degree of hardiness in the progeny would be efficiently predictable from those of their parents.

(2) Stem hardiness which affects the bark splitting injury in late autumn was also found to be higher than 0.9 in the

heritability value estimated from the doubled value of the regression coefficient of the progeny upon the parents crossed with a common parent.

(3) Individual trees in each cross showed a certain variation in both leaf and stem hardiness. The inter-tree variations in leaf hardiness were interpreted as a typical polygenic inheritance, while those in stem hardiness seemed to be controlled by a relatively small number of genes.

(4) Genetic correlation between leaf and stem hardiness was calculated to be 0.73 on the basis of parent-offspring regression. It was also found that leaf size was genetically correlated with both of the leaf and stem cold hardinesses.

ACKNOWLEDGMENTS

The author is grateful to Dr. Kan-Ichi Sakai, Professor Emeritus of Kagoshima University, for many helpful discussions and suggestions, and to Professor Akira Sakai, Institute of Low Temperature Science, Hokkaido University for his valuable advice.

REFERENCES

1. Falconer, D.S., in "Introduction to Quantitative Genetics" (Oliver and Boyd, eds.), pp. 164-185. Edinburgh and London, 1967.
2. Flint, H.L., Ecology 53:1163-1170 (1972).
3. Lantz, H.L., and Picket, B.S., Proc. Amer. Hort. Sci. 40:237-290 (1940).
4. Rudolf, T.D., and Nienstaet, H.P., J. Fore. 60:138-139 (1962).
5. Sakai, A., Contr. Inst. Low Temp. Sci. B, 15:1-14 (1968).
6. Sakai, A., Paton, D.M., and Wardle, P., Ecology (In Print).
7. Smithberg, M.H., and Weiser, C.J., Ecology 49:495-505 (1968).
8. Toyao, T., Kayumi, S., and Katsuo, K., Study of Tea 35:23-39 (1967).
9. Toyao, T., and Kayumi, S., Study of Tea 45:11017 (1973).
10. Toyao, T., Kayumi, S., Katsuo, K., and Matsushita, S., Bull. Nat. Res. Inst. of Tea 9:42-57 (1974).
11. Watkins, R., and Spangelo, L.P.S., Theo. Appl. Gen. 40:195-203 (1970).

ENERGY FORESTRY AND FROST HARDINESS

Lars Christersson

Energy Forestry Project
Swedish University of Agricultural Sciences
Uppsala, Sweden

I.   INTRODUCTION

Sweden is one of the countries which are today completely dependent upon imported energy. Although there is a certain amount of water power it has no oil, no coal and no gas. Up to 70% of the total energy required is supplied by imported oil. Sweden is a vast country with a lot of sunshine during spring, summer and autumn, when the energy of the solar radiation falling on every square metre is equivalent to 100 liters of oil.

Some of this solar radiation is utilized by the highly developed agricultural and silvicultural industries, which are big business in Sweden and occupy all suitable land areas. But there are also areas which agriculture and silviculture cannot utilize, for example wet lands, bogs and mires, peat lands, areas subjected to flooding, abandoned farming land, very dry sandy soils etc. The reason why these lands are not used in ordinary agri- and silviculture is that they are too wet, too dry, too acid or too frosty. At today's oil prices, if suitable energy crops could be found, wet lands could be economically drained, dry land irrigated and acid land limed, whereas frost remains a major problem. It is not only the frost during the winter but also, particularly, the frost during the vegetation period, which creates great problems for biomass production in these areas. An example from a sphagnum mire in the southern part of Sweden is given in Figure 1. The summer frost is so-called radiation frost and its occurrence and frequency is the same all over Sweden (10).

In searching for suitable crops for these areas the most promising results have been obtained with species and clones of the genus <u>Salix</u> and <u>Populus</u> (14). These species are planted by cuttings and are intended to be harvested every second - fifth

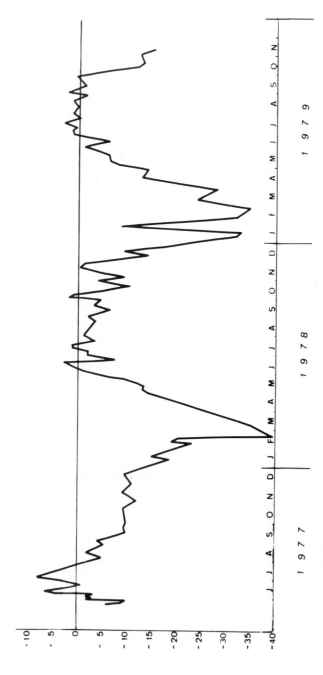

FIG. 1. Temperature variation 50 cm above the ground for a mire in southern Sweden. The curve shows the minimum temperature every week for two and a half years.

year. The utilization of fast-growing species of deciduous trees for energy production has been given many names, such as energy forestry, short rotation forestry or agriforestry, and is defined as systematic, rational cultivation of fast-growing clones and species of deciduous shrubs and trees, carried out, without creating serious environmental problems, on lands unsuitable for ordinary agriculture or silviculture.

In Sweden there are thousands of hectares of such land. Use of this land for energy forestry would represent a considerable contribution to the national economy and security of Sweden. The amount of energy from an energy forestry crop from 1.5 million hectares would be equivalent to one third of the total oil imports. Calculations show that more than that area of suitable land is available. However, whether such production would be successful depends, among other things, whether it is possible to find a solution to the problem of damage by summer frost.

II. FIELD OBSERVATIONS

Because the energy forestry project was directed to land unsuitable for agriculture or silviculture the problem of summer frosts was recognized at the outset (13). For economic reasons little attention has so far been given to solving or avoiding this problem. However, some field observations have been made and these showed great differences in frost hardiness between various clones of Salix and Populus.

The frost damage is always at the top of the shoots, where the apical meristems are killed. The extent of the injury by summer frost can vary from 1 cm to 1 m from the top, very seldom is the whole plant killed. Field observations have so far shown that clones differ in their susceptibility to damage.

It has also been observed that after frost injury during summer the axilary buds on the shoots of some Salix clones broke almost immediately, whereas the buds of other clones did not break at all during the current vegetation period.

Unexpected variation in the frosty hardiness during the vegetation period has also been observed. During spring 1981 some clones of S. viminalis started sprouting about the first of May. Ten days later the night temperature dropped to -10°C. The flowers and the leaves of the opening buds were killed but not the apical meristems of the growing branches, nor the cambium of the branches and stems. After a week, when the dead flowers and leaves had been shed, there was no sign of damage to the shoots and they grew normally. During summers of previous years, the same clones, in the same places, were damaged by summer temperatures of -2 - -3°C to an extent of 20 - 40 cm of every leading shoot.

During winter, clones of Salix native in southern Sweden were killed in central or northern Sweden by the lowest temperatures. The damage sometimes was unexpectedly distributed. On the same bush some shoots were completely killed whereas others were not damaged at all. No correlation with age, direction, distance from neighbors etc. could be found.

The origin of the various clones was very important for survival by spring and autumn frost. If a clone was moved too far to the north, bud break was too early in the spring, and bud set was too late in the autumn to avoid frost damage. Clones native in the north could not utilize the full length of the vegetation periods if they were moved too far to the south.

III. LABORATORY TESTS

The work of testing different clones and species in the laboratory for frost hardiness has just started and only very preliminary results have been obtained.

The method was as follows: Seedlings of different clones of Salix species were grown from minicuttings in water culture in a phytotron and fertilized using a special technique (4,7), which allowed predetermination of different growth rates. The seedlings were grown in 24 hours light at a temperature of 20°C. The starting point of the different experiments was chosen in such a way that all seedlings were 15 cm tall at the same time when the freezing test began.

The method used for determination of the frost hardiness of the plant material has been described in detail by Christersson (1,2). The shoots of the plants were placed individually in test tubes, which were immersed in a bath of alcohol. The seedlings were upside down during the test.

The roots were wrapped in damp blotting paper to prevent desiccation and were kept at a temperature of 20°C throughout the test. The temperature of the bath was lowered at a cooling rate of 4°C per hour to the predetermined temperature. This temperature was maintained for 2 hours and then increased by 4°C per hour to 3°C. Small pieces of ice were put in each test tube for inoculation (8).

After the test the seedlings were put back in water culture and transferred for one day to a chamber with a photoperiod of 8 hours light and a temperature of 3°C. Then the seedlings were transferred to darkness at 20°C for a further 12 hours before being placed in a glasshouse. Survival was estimated after 5 weeks. Living seedlings (+) were considered to be those in which the apical bud continued to grow and dead seedlings (-) those whose apical bud did not grow.

From the results of the first test (Table I) it is apparent that all seedlings, irrespective of growth rate, were killed at about -2°C to -3°C. Christersson and Sandstedt (3) showed that

TABLE I. Frost hardiness determination of different species and clones of Salix during active growth. The growth rate is expressed as per cent increase in fresh weight per day. For details, see text.

Species	Clone No.	Growth rate	Lowest temperature (°C)					
			Control	-1	-2	-3	-4-	-6
Salix viminalis	683	5%	+		+-	--	--	-
					++	-	--	-
		10%	++		--	--	--	-
			+		++	--	-	-
		optimum	++	+	++	--	--	-
			++		+-	--	--	-
Salix smithiana	666	5%	+	+	++	--	---	--
					++	--	--	-
					--			
		10%	+	+	+	+-	---	-
					++	--	--	-
					++			
		optimum	+	+	++	--	---	-
			+		++	--	--	-
Salix purpurea	077	5%	+	+	++	--	---	-
					---	--	--	-
		10%	+		++	+-	--	-
			+		---	--	--	-
		optimum	+	+	+-	--	--	-
			+		++	--	+-	-
Salix viminalis	082	5%	+	+	++	--	+-	-
					+-	--	--	-
		10%	++	+	++	--	---	-
					++	--	--	-
		optimum	++	+	++	--	--	-
					++	--	--	-
Salix aquatica	056		++		+++	+++	---	---
					+++	+++		

ice crystal formation in pine needles in nature always took place at about -2°C. The results of the present investigation suggests the same, indicating that ice crystal formation takes place in seedlings at about -2°C to -3°C and, as distinct from pine (1), ice formation always kills young growing tissues of these species, as it does in many other species (15). Field observations support this theory.

IV. PLANS FOR FUTURE WORK

The plant material to be tested for different kinds of frost hardiness comprises collected clones and species, native or exotic, of the genera Salix, Populus, Betula and Alnus, together with clones of Salix and Populus produced by artificial pollination. The work will proceed along the following lines:

A. Determination of the Ice Crystal Formation Temperature

The ice crystal formation temperature and the capacity to supercool to different degrees will be determined in different parts of the seedlings during different stages of growth. The tests will be performed partly in artificial experiments with or without inoculation of ice crystals by different methods (11) and partly under natural conditions. It may be possible to determine whether supercooling of different organs is a property which could be of value for minimizing damage by summer frosts and which could be utilized in a breeding program. Very preliminary results suggest a difference between Salix and Betula in supercooling temperature of about 2 to 3°C. Even as small a difference as one degree may be of importance.

B. Recording the Effect of Ice Crystal Formation

Until now, all results indicate that there is frost damage in growing plant tissues of fast-growing clones and species as soon as ice crystal formation takes place. A systematic investigation of fast-growing and slower-growing plants of the particular genus will show whether there is any difference between them. The possibility of finding fast-growing clones and species which tolerate ice cyrstal formation in the growing tissues during the vegetation period is thought to be very small.

C. Determination of the Delay in Bud Break after Frost Injury

Field observations reveal that surviving axilary buds of some clones break much more rapidly than others after frost damage. If, as suspected, growing tissues are killed by ice crystal formation, use of clones with rapidly breaking buds might be a way of avoiding too much inhibition of production by a single summer night frost.

Investigation of A, B, and C will take place simultaneously, using the method previously described, with the addition of equipment for TA and DTA registration. Small thermocouples (0.1 mm in diameter will be used) (5). The best method of application of these small thermocouples in or on the tissues and buds will be decided by pilot experiments.

D. Description of the Morphology and Anatomy of Different Species and Clones

As freezing avoidance is assumed to be the main characteristic enabling survival of summer frosts, the morphology and anatomy of the leaves and stems are factors which may be of great importance. This has been shown to be valid for potatoes, in which a close correlation between frost hardiness and highly developed palisade tissue was demonstrated (9). Development of hairs and trichomes on leaves and stems might also be of importance for freezing avoidance. The lignification of the secondary cell wall should be investigated in detail in order to determine the importance of the properties of the cell wall for freezing avoidance.

The anatomy and morphology of clones and species are genetically determined and positive correlations between any factors and frost hardiness should be utilized in pollination and selection program.

E. Determination of Winter Hardiness

Winter hardiness of all plant material will be determined. Both laboratory tests and field observations will be made. Special attention will be paid to the existence of a second exotherm (6).

The old question about the possibility of improving frost hardiness development by fertilizer treatment will be considered. At the same time, the significance of the growth rate before the hardening processes on development of frost hardiness will be determined.

F. Description of the pH Fenology of the Plant Material

The conditions for bud break and bud set will be recorded for all clones and species used. Existing pilot experiments set up all over the south and central of Sweden will be utilized. The plant material in these field experiments consists of 25 fast growing clones of <u>Salix</u> in four replicates. Bud break and bud set will be recorded. Complementary experiments will be performed in the phytotron. The beginning and end of cambium activity are of special importance for frost hardiness in the spring and autumn (12). The results of this investigation will be the basis for deciding where the different clones might be grown in Sweden.

G.   Practical Methods to Avoid Summer Frost Injuries

During radiation frosts the temperature is always lowest at the ground surface and in areas which are not covered by a canopy. For that reason it would be possible to adopt ordinary silvicultural methods to change the microclimate of the seedlings by using a ploughing technique, where the air with the lowest temperature is collected at the bottom of the furrow, and the cuttings are planted at the top of the furrow-slice. It is thus possible to avoid exposing the growing part of the seedlings to temperatures below zero. However furrow-slice can be very dry during drought periods at the beginning of the summer, when the root system is not completely developed. This problem can be overcome by using tall cuttings.

The establishment of energy forest plantations in areas with summer frosts is the major problem. After successful establishment it should be possible to avoid summer frosts by special harvesting techniques, e.g. by harvesting every second row and using the remaining rows as a protective canopy to inhibit radiation. This technique could probably be improved by using special spacing.

REFERENCES

1. Christersson, L., Physiol. Plant. 25:273-278 (1971).
2. Christersson, L., Physiol. Plant. 44:288-294 (1978).
3. Christersson, L., and Sandstedt, R., Can. J. For. Res. 8:480-482 (1978).
4. Eriksson, T., Physiol. Plant. 51:423-429 (1981).
5. Fricks, H., and Christersson, L., Physiol. Plant. (in press) (1981).
6. George, M.F., Burke, H.M., Palett, and Jonsson, A.G., Hort-Science 9(6) (1974).
7. Ingestad, T., and Lund, A-M. Phsyiol. Plant. 45:137-148 (1979).
8. Levitt, J., Responses of Plants to Environmental Stresses. Academic Press, New York, 1972.
9. Li, P.H., and Palta, J.P., in "Plant Cold Hardiness and Freezing Stress" (P.H. Li and A. Sakai, eds.). Academic Press, New York, 1978.
10. Perttu, K., in Technical Report No. 8. Energy Forestry Project. Swedish University of Agricultural Science, Uppsala, 1979.
11. Rajashekar, C., Li, P.H., and Carter, J.V., in "Plant Cold Hardiness and Freezing Stress" Vol. 2, (P.H. Li and A. Sakai, eds.). Academic Press, New York, 1982.
12. Sennerby-Forsse, L., Technical Report No. 24. Energy Forestry Project. Swedish University of Agricultural Science, Uppsala, Sweden, 1981.

13. Sirén, G. Tidskrift 118:305-310. Stockhom, Sweden, 1929.
14. Sirén, G., and Sivertsson, E., Research Notes No. 83.
    Royal Coll. of Forest., Stockholm, Sweden, 1976.
15. Weiser, C.J., Science 169:1269-1278 (1979).

BREEDING WILD AND PRIMITIVE POTATO SPECIES TO OBTAIN FROST-RESISTANT CULTIVATED VARIETIES

Nelson Estrada R.

Tuberous Crops Program
Colombian Institute of Agriculture (ICA)
Bogota, Colombia

I.   INTRODUCTION

Wild and primitive frost hardy potato species possess definite features that differentiate them from nonhardy species.
They have anatomical characters and physiological processes
that protect them from frost damage.

Among the morphological characters are: a) two or more
layers of palisade parenchyma cells instead of one palisade
layer commonly found in nonhardy species, b) smaller cell size,
c) thicker cell walls, d) greater stomatal indexes, and e)
rossette habits.

Among the physiological attributes, major differences exist
between these two groups of potatoes, for example: cell lipids
content, starch-sugar transformation capacities, tolerance to
certain amount of liquid water frozen, pigmentation of stems
and leaves (3,4,6).

These characteristics have permitted these wild species and
their descendants, produced by natural crosses with the cultivated ones, and here called primitive cultivated species, to
survive and evolve in the cold stress conditions frequently
present at over 3,000 meters of altitude in the South American
highlands.

This paper presents aspects considered important and
observed as genetically transmissible when breeding for frost
resistance. They include: a) effect on frost resistance of
two or more palisade layers of parenchyma cells, b) induced
resistance by leaf and stem pigments, c) frost damage recovery,
and d) yield of selected clones under very severe frost stress.
In addition, a comparison, statistically measured, between
field and laboratory testing using the same clones and the same
plants grown in the same environmental conditions, is discussed.

Table I. Records of crosses and selections made for frost resistance at the International Potato Center.

	1974	1975	1976	1977	1978	1979	total
Species used	11	16	9	12	8	7	16
Seeds	5,050	10,400	31,500	27,300	33,950	33,400	141,600
Seedlings	1,515	4,300	17,860	15,000	25,300	26,770	90,745
Resistant seedlings	60	100	410	320	3,400	4,100	8,390
Resistant seedling percentage	4.0	2.3	2.3	2.1	13.4	14.9	9.2

TABLE II. Localities and altitudes where potato clones were tested.

Locality	Altitude (m)
Usibamba, Peru	3,600
Belen Exp. Station, Bolivia	3,900
Toralapa, Exp. Station, Bolivia	3,700
Tibaitata Exp. Station, Colombia	2,600
San Jorge, Exp. Station, Colombia	3,200
Toluca, Exp. Station, Mexico	2,600

II.   MATERIALS AND METHODS

The list of species, 17 wild and 7 cultivated used in crosses, was previously described (4). During 12 years, about 120,000 seedlings were obtained by crosses and back crosses. They were initially tested in growth chambers at temperatures of -5°C for 2 hours, at the age of 2 months, using facilities at the International Potato Center, Lima, Peru and at Tibaitata Agricultural Experiment Station in Bogota, Colombia. Table I shows the data obtained while working at CIP.

The surviving clones, from 2 to 15%, were planted in the following years in the field at altitudes varying from 2,600 to 3,900 meters in adequate time for exposing them to the usually expected frosts.

The localities used for the clonal testing are given in Table II.

The most promising clones were maintained and re-tested in the field during several years at various localities to confirm the initial results, to make more precise observations of their characters, and to use them in a continuous (several cycles) breeding and selection program in order to generate newer clones with improved resistance and other desirable characters.

A scale from 1 to 9 was used to evaluate the degree of clonal resistance in the field or in the growth chamber tests (2), only three grades were used for young seedlings tests: resistant, tolerant or susceptible.

A.   Laboratory vs. Field Testing

On November 10, 1980, a set of 380 clones were planted in the field, at Tibaitata Experiment Station, Bogota, Colombia. These included 360 clones selected for cold resistance and good yield from previous selection cycles and 20 non-selected, frost

TABLE III. Low tempeatures registered at Tibaitata Experiment Station, Bogota, Colombia during the months of December 1980, January and February 1981.

Date	Temp. °C	Duration (hr & min)	
12/10.80	-1.6	1	0
12/11/80	-1.0	1	15
1/4/81	-2.7	1	30
1/5/81	-3.0	1	45
1/10/81	-3.2	0	50
1/15/81	-2.8	2	00
1/16/81	-2.3	0	50
1/20/81	-7.0	1	30
1/28/81	-1.5	1	15
2/2/81	-1.4	0	45
2/5/81	-0.5	0	40
2/8/81	-1.5	1	00
2/11/81	-2.8	1	10

TABLE IV. Rain at Tibaitata Experiment Station, Bogota, Colombia during the growth season and test of potato clones, in 1980-81.

Month	Rain (mm)
November/80	53.6
December/80	56.6
January/81	3.3
February/81	23.2
March/81	41.8
Total	178.5

susceptible clones used as biological checks during the expected frosts.

The weather conditions, including lack of rain and presence of frosts, were very severe and can be observed in Tables III and IV. No irrigation was applied. The natural rain which occurred was about 1/3 amount of what the potato crop should have normally. It is believed that some significant positive correlation may exist between frost and drought resistance, since many clones were able to grow well under such drought stress conditions.

The heaviest frost occurred on January 20, seventy days after planting, when the plants were at their maximum growth, had not yet started any foliar maturity process, and were only starting the tuberization process.

Readings on degrees of damage were taken one day after the heavy frost of January 20, 1981. The next day healthy samples of the surviving basal leaves of 55 clones exposed at the same field, were taken to be tested in growth chambers. They received comparable temperature shocks as in the field, which were of -6°C during 3 hours daily for 3 consecutive days. Moist paper covered the leaves to favor ice nucleation and to prevent supercooling of the leaf tissue. Two replicates were made. All the samples after the treatment indicated that they were under frost stress (Table V).

The X^2 test for statistical significance between both methods of testing, field and growth chamber, was made.

It should be indicated that the majority of the tested bred material did not have such a high degree of frost hardiness to be able to survive the kind of tests here reported in the field and laboratory. The field conditions were too stressful, and out of control. For this reason to validate the artificial tests they had to be of comparative stress. Nevertheless, a considerable proportion of the clones were only partially damaged, and the results permitted a fair evaluation of the most resistant clones which were saved for future cycles of breeding.

B.   Palisade Layers in Leaf Parenchyma Cells

Palta and Li (5) and Li and Palta (3) reported that two layers of palisade parenchyma cells was a common occurrence in wild hardy species. This is also confirmed by Tiwari et al. (7) who found an association between increased palisade cell length and frost tolerance, specially in the cultivated species. The xeromorphic character of xerophytes according to Esau (1) is associated among other features with a palisade tissue more strongly developed, than the spongy parenchyma. The association of a strong palisade layer and frost resistance appears logical since frost damage is caused physiologically by a freeze-induced dehydration process.

A total of 146 clones mostly originated from wild species crosses, also including S. tuberosum and S. andigena varieties, primitive frost resistance species, and advanced selected hybrid clones with variable degree of resistance, were observed at the microscope. The purpose was to check any association between the character of double or triple palisade layers of parenchyma cells and frost resistance. At the same time the relation of this anatomical character and the genetic background of the clones was studied. Microscopic pictures were taken to illustrate this feature.

TABLE V. Frost damage readings in 55 clones tested at the field and growth chamber, in January 1981, as described in materials and methods.

Clones	Parentage*	Readings	
		Field −7°C, 1-1/2 hr.	Growth chamber −6°C, 3 hr.
Baraka	tuberosum, cultivar	9	9
Nicola	" "	9	9
Atzimba	" "	9	9
BMJ 69-1	" "	9	8
Ccosi imilla	andigena	7	6
702870	tbr x adg	9	7
79-903-21	(tbr x adg) x (tbr x adg) x ajh	7	6
904-2	(adg x phu) x (ajh x stn) x cha	6	5
904-3	" "	7	5
907-5	(ajh x stn) x (stn x phu)	7	5
907-7	" "	9	7
908-6	" "	7	6
913-1	(phu x phu) x tbr	8	8
915-1	tbr x (tbr x adg)	8	6
915-2	"	9	7
918-15	"	7	8
918-18	"	9	9
921-13	(adg x phu) x tbr	7	9
922-3	(tbr x adg) x tbr	9	9
922-6	"	8	9
922-11	"	6	9
923-5	"	7	9
924-7	(tbr x adg) x tbr	8	9
924-12	"	9	9
926-3	tbr x cha	7	7
928-1	stn x phu	8	6
928-5	"	6	8
79-928-21	"	8	7
931-8	adg x tbr	6	8
932-4	"	7	6
932-11	"	6	8
933-10	"	6	8
934-1	(adg x tbr) x adg	6	6
935-1	"	5	6
Rev x Sipena	(tbr x adg) x adg	7	8
Yari	ajh	3	4
37403210	ajh x stn	4	6
374080-1	tbr x adg	7	8
375057-45	(ajh x stn) x phu	6	8
375089-31	tbr x (tbr x phu) x clb	6	7

Table V. continued.

375597-15	(tbr x adg) x adg	8	8
375517-6	(tbr x adg) x (clb x adg)	6	7
375528-2	tbr x clb	8	9
376944-3	(tbr x adg) x (ajh x stn)	7	9
379048-1	(tbr x adg) x (tbr x adg) x ajh	7	8
379072-1	(clb x adg) x tbr	6	8
379094-1	(clb x adg) self	5	7
379101-1	(adg x clb) self	4	6
379102-2	(clb x adg) self	6	6
379105-1	(tbr x adg) (ajh x stn)	6	8
379107-1	(ajh x stn) self	5	8
H-5	(tbr x adg) x (acl x phu)	6	7
C-215	(tbr x adg) x cha	7	7

* See Table VI.

Table VI. Tuber bearing Solanum species given in the text and their abbreviations.

Species	Abbreviation	Chromosomes, 2n
1. S. acaule	acl	48
2. S. ajanhuiri	ajh	24
3. S. andigena	adg	48
4. S. boliviense	blv	24
5. S. bulbocastanum	blb	24
6. S. chacoense	chc	24
7. S. chaucha	cha	36
8. S. commersonii	cmr	24
9. S. curtilobum	clb	60
10. S. juzepczukii	juz	36
11. S. phureja	phu	24
12. S. stenotomum	stn	24
13. S. stoloniferum	sto	48
14. S. tuberosum	tbr	48

C. Leaf and Stem Pigmentation

Previous author's observations indicated that there was an association between stem and leaf pigments and frost hardiness among certain genetic materials; however, not necessarily all the species or clones with resistance were pigmented, i.e. S. acaule is generally non-pigmented.

FIGURE 1.   Frost damage effect after a heavy frost (-7°C)
at Tibaitata Exp. Station.
    a) S. tuberosum cv. Baraka, grade 9.
    b) S. tuberosum cv. Atzimba, grade 8.
    c) (S. tuberosum x S. andigena) x S. andigena, grade 7.
    d) S. tuberosum x S. andigena, grade 6.

FIGURE 2.   Frost damage effect after a heavy frost (-7°C) at Tibaitata Exp. Station.

a) S. ajanhuiri x S. stenotomum, grade 5.
b) S. curtilobum x S. andigena, self, grade 4.
c) S. andigena x S. curtilobum, self, grade 3.
d) S. ajanhuiri cv. Yari (S. megistracrolobum x S. stenotomum), grade 2.

To test this hypothesis, 9 families of diverse genetic background, originated from crosses between different diploid clones were planted at the International Potato Center, Lima, Peru, in August 1979. A total of 1,101 plants were obtained which showed clear differences in stem and leaf pigmentation. At the age of 2 months they were exposed to frosts in the growth chambers at -5°C for two hours. The number of seedlings pigmented before the test and the surviving ones after the test were recorded to obtain the percentage of each surviving group in each family.

The x^2 test was conducted to observe if there was any statistically significant difference between both characters for frost survival.

III.   RESULTS AND DISCUSSION

A.   Field and Laboratory Testing

Data on readings obtained from a sample of 55 selected clones tested both in the field and the growth chamber are presented in Table V, and Figures 1 and 2. The x^2 test for significance of paired samples between both methods of frost testing indicated no statistical difference between field and growth chamber tests. An experimental t value of 1.65 was obtained. The t table values were: 2.67 (1%), 2.00 (5%), and 1.70 (10%). These results provide good confidence for laboratory tests using detached leaves, and supplying adequate moisture conditions to induce ice nucleation. These tests are very useful to check the resistance of a limited number of clones, when there is no security or possibility of having adequate field frosts for testing. It is also advantageous because the clones are not lost in the test, which can be repeated several times, if desired.

B.   Palisade Layers and Resistance

There was a high consistency between double or triple palisade layers of parenchyma cells in the leaves and high frost tolerance. Likewise, when the palisade layer was simple but its cells were quite long, filling up 50% or more of the whole leaf thickness, there was good frost tolerance. This is well observed in the andigena variety Ccosi imilla (Fig. 6).

From 146 clones observed at the microscope, 67 showed double or triple palisade layers and 79 had only one layer. However, 43 clones from the 79, had longer cells than usual in their palisade tissue as indicated above, and just 36 clones had palisade layers which occupied 40% or less of the leaf thickness.

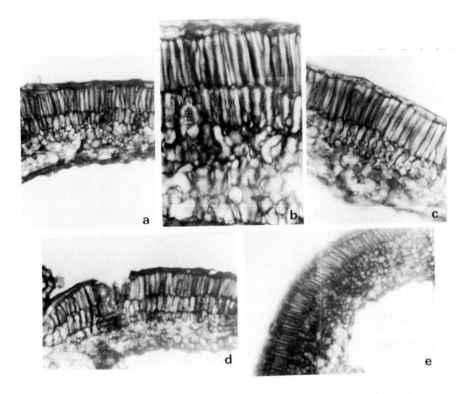

FIGURE 3. Leaf cross sections from clones showing diverse developments of palisade cell layers.
  a) S. acaule with 3 palisade layers.
  b) S. acaule (Och 1208) x (S. commersonii x S. stenotomum).
  c) S. chacoense WRF 88.10.
  d) ICA 79-136-2, S. ajanhuiri x S. stenotomum with 3 palisade layers.
  e) S. tuberosum x S. andigena cv. Monserrate, with a simple and short cell palisade layer.
  All of them except cv. Monserrate exhibit a high degree of frost resistance.

It was also found that all the clones with double or triple palisade layers had in their background wild frost hardy species such as S. acaule, S. megistracolobum, S. boliviense etc. or primitive cultivated frost tolerant species such as S. juzepczukii, S. ajanhuiri or S. curtilobum, which had likewise wild species in their origin. Figures 3, 4, 5 and 6 illustrate these aspects with microscopic views of parenchyma cells of several clones observed.

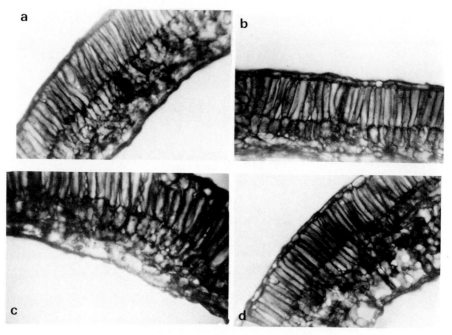

FIGURE 4. Leaf cross sections from clones possessing high frost resistance and showing 2 palisade alyers of parenchyma cells.

a) S. tuberosum x S. ajanhuiri.

b) (S. stoloniferum x S. andigena) x S. andigena x (S. acaule x S. phureja).

c) (S. tuberosum x S. andigena) x S. ajanhuiri.

d) (ajh x stn) x [(acl x blb) x phu] x tbr x tbr = CIP 374032-12 x (WAC ABPT-30 x Gineke).

C. Leaf and Stem Pigmentation

Table VII presents the data on the families and seedlings tested, their parentage and the records of pigmented seedlings in the surviving population. After the analysis of the x^2 test for paired samples, an experimental t value of 2.90 was obtained. The t values of the table were: 3.35 (1%), 2.75 (2.5%) and 2.30 (5%). This indicated that there was a significant difference at the levels of 95% and 97.5% between pigmented and non-pigmented seedlings. Figure 7a illustrates this effect showing that the only 4 surviving plants in the large family tested at -5°C were pigmented in contrast with the ones non-pigmented and killed.

These results suggest that, other factors being equal, pigmented plants are more protected against frost damage than the

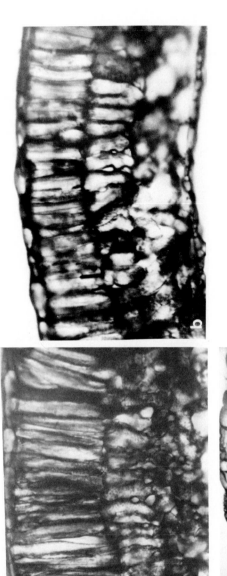

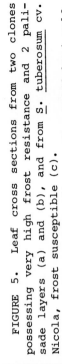

FIGURE 5. Leaf cross sections from two clones possessing very high frost resistance and 2 palisade layers (a) and (b), and from S. tuberosum cv. Nicola, frost susceptible (c).

a) CIP 379101-1, (S. andigena x S. curtilobum) self.
b) CIP 374032-10, (S. ajanhuiri x S. stenotomum).
c) S. tuberosum cv. Nicola, whose palisade tissue appears less developed and less organized.

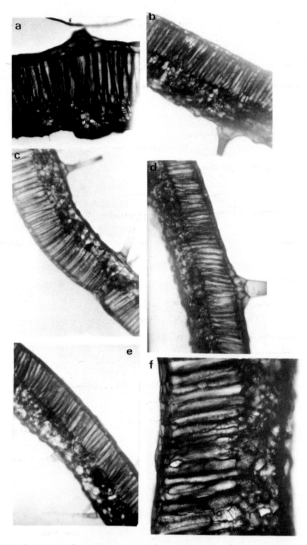

FIGURE 6.   Leaf cross sections from clones possessing high
frost tolerance and showing only one palisade layer of parenchyma cells.   However, they fill up from 50% to 70% of the leaf
thickness tissue.
a) <u>S. andigena</u> cv. Ccosi imilla.
b) CIP 379045-1 (<u>S. tuberosum</u> x <u>S. andigena</u>) x <u>S. curtilobum</u>.
c) CIP 378573-1 (<u>S. andigena</u> x <u>S. andigena</u>) x <u>S. andigena</u>.
d) CIP 378559-1 (<u>S. tuberosum</u> x <u>S. andigena</u>) x <u>S. stenotomum</u> x
   (<u>S. ajanhuiri</u> x <u>S. stenotomum</u>).
e) CIP 379106-1 (<u>S. tuberosum</u> x <u>S. andigena</u>) x (<u>S. ajanhuiri</u> x
   <u>S. stenotomum</u>), self.
f) CIP 379094-1 (<u>S. curtilobum</u> x <u>S. andigena</u>), self.

TABLE VII. Diploid families tested in the growth chamber at -5°C for 2 hours, and their survival.

Family N°	Parentage	Total seedlings	Surviving seedlings	survival %
378630	ajh x stn x (ajh x stn) x stn			
	pigmented	101	22	21.8
	green	33	1	3.0
378631	(ajh x stn) x (ajh x stn) stn			
	pigmented	109	58	53.2
	green	29	3	10.3
378632	(ajh x stn) x stn x (ajh x stn)			
	pigmented	52	19	36.5
	green	22	4	18.2
378633	(ajh x stn) x stn x (ajh x stn)			
	pigmented	42	13	31.0
	green	35	4	11.4
378634	(ajh x stn) x (mga x stn)			
	pigmented	47	2	4.3
	green	14	0	0.0
378635	(stn x stn) x (ajh x stn)			
	pigmented	37	11	29.7
	green	117	3	2.6
378636	(phu x stn) x (ajh x stn)			
	pigmented	98	4	4.1
	green	43	0	0.0
378637	(stn x stn) x (ajh x stn)			
	pigmented	40	5	12.5
	green	99	0	0.0
378638	(ajh x stn) phu x phu			
	pigmented	105	9	8.6
	green	77	2	2.6

non-pigmented ones. These characters appear to have been influencing natural selection in the Andes highlands of South America where it is frequent to find strong pigments in many native varieites of S. andigena, S. phureja and S. stenotomum, probably caused by natural selection as a result of frequent exposition to frost effects.

D. Importance and Capability for Frost Damage Recovery

It is not infrequently found, that some clones or varieties, not highly resistant but tolerant to frost, have however much capacity to recover, or re-grow and can develop well after the frost effect, being able at the end of producing an

acceptable crop. This faculty for recovery seems to be controlled genetically and is of great practical value for the farmer. In field conditions, many times only one or two succesive frosts occur and a very satisfactory weather for the crop development follows. Varieties or clones with outstanding capacity to recover would be very adequate in these circumstances. The author found among the clones tested several of which showed markedly this character, i.e. 374080-1, 375517-6, Revolucion x Sipena (Fig. 7b).

FIGURE 7.

← a-1) The only four surviving plants of this family after -5°C frost, were pigmented.

a-2) Three different families with the same male parent, showing different reaction at -3°C.

b) Yield of five plants of a hybrid with good frost recovery capability after -7°C field frost.

c) Selected frost tolerant clones growing at CIP-Huancayo-Peru Station at 3.200 mts.

d) Yield of ten plants which had moderate frosts (-4°C) at Belen-Bolivia-Station. Left, selected hybrid 374080-5; right, S. andigena cv. Imilla, regional check.

e) A commercial field at Tibaitata Station of the frost tolerant improved cv. Nevada, after -5°C frost, having a yield of 14 ton/ha.

TABLE VIII. Yield from 5 plants belonging to clones tested in the field at severe frost conditions (-7°C) described in Table III.

Clone	Parentage	Yield, (Kgs)
Achat	tbr (control)	0.200
Atzimba	tbr (control)	0.140
India-832	tbr (control)	0.400
Ccosi imilla	adg	0.800
Revol x Sipena	(tbr x adg) x adg	1.300
374080-1	tbr x adg	1.400
375057-9	(ajh x stn) x phu	0.800
375517-6	(tbr x adg) x (clb x adg)	1.650
375597-15	tbr x adg	0.600
377688-8	tbr x adg	1.500
379041-1	(phu x phu) x adg	1.300
379072	(clb x adg) x tbr	0.800
79-934-2	adg x tbr	1.250

E. Comparison of Resistant vs. Susceptible Clones in Yield

Table VIII shows the yield of several resistant as compared with S. tuberosum susceptible clones after being exposed to the severe frosts of January 20, 1981. The best clones are now at an advanced stage permitting them to be tested with farmers for important characters like yield, frost resistance and quality (Figs. 7c and 7d).

IV.   SUMMARY AND CONCLUSIONS

Four steps appear fundamentally important to obtain cultivated potato clones which may possess acceptable yields, good
tuber quality, and frost resistance: a) To select species or
clones with high frost resistance, b) To cross these genetic
sources of resistance with the best available cultivated types,
c) To transfer by testing and selection the genetic characters
controlling frost resistance into advanced cultivated clones,
and d) To employ massal methods of testing and screening which
permit selection among large populations of the most resistant
types.

Points of a), b) and c) are not very difficult to fulfill.
Point d) is crucial and extremely important, in order to
develop significant results.   In addition, it needs to have
complementary methods.   The logical sequence to follow in order
to make tests practical is to: a) Test thousands of seedlings
in growth chambers at low temperatures.   This method at least
eliminates the most susceptible material, b) Retest detached
leaves of adult selected clones grown in the growth chambers,
again at low temperatures, and c) Conduct field tests at various localities to balance the effect of the microclimate, environmental and unpredictable weather conditions of frost
occurence.

Table I supports the previous statements since progress was
achieved in grouping genes for resistance in populations after
several cycles of testing and selection.   In this table one can
see that the percentage of resistant population was very low
during the first 4 cycles of testing and selection (2% to 4%).
On the other hand, the resistant population was much higher
(13% and 14%) in the 5th and 6th cycles of selection.

Several morphological characters such as strong palisade
tissue of parenchyma cells, and leaf and stem pigmentation,
appeared quite associated with frost hardiness and could be
transmitted by crossing and selection.   However, many endogenous factors involved in frost hardiness have been described,
including the kind and content in lipids, types of proteins and
enzymes, starch-sugar relationship, growth regulators involvement, protoplast ability for water loss, plasma membrane structure and its changing capacity.   Likewise, exogenous factors
which protect the plant, i.e. stomatal density, leaf thickness,
rossette habits in herbaceous plants which capture soil heat,
and possible inhibiting substances for epiphytic ice nucleation
bacteria, with no doubt contribute to a degree of cold
hardiness.

This complexity is reflected in the apparent high quantity
of genes controlling frost resistance, and explains why so many
generations of crosses and use of recurrent selection are
required to achieve frost resistance combined to many other

horticultural and quality characters in determined individuals or clones.

Finally, it was also observed that when frost resistance is incorporated into advanced clones, it was reflected in higher tuber yields than in susceptible clones. See Table VII and Figures 7c and 7d. These improved frost resistant clones were obtained after several backcrosses to high quality clones permitting the elimination of undesirable characters of wild and primitive species, such as long stolons, poor tuber quality, tuber bitterness, late maturity, and low yields.

ACKNOWLEDGMENT

A full recognition is given to the International Potato Center in Peru, and to ICA, IBTA, and INIA, respectively institutes of agriculture of Colombia, Bolivia and Mexico for the facilities and support provided during this research. The encouragement and scientific advice received from Doctors C.J. Weiser, P.H. Li and their associates at the University of Minnesota were invaluable. Thanks are expressed to Ing. Ramon Pineda from ICA, Colombia for his cooperation to process and observe leaf tissues.

REFERENCES

1. Esau, K. Anatomy of Seed Plants. p. 277-281. John Wiley & Sons, New York (1961).
2. Estrada, R.N., in "Plant Cold Hardiness and Freezing Stress" (P.H. Li and A. Sakai, eds.), p. 333-341. Academic Press, New York, 1978.
3. Li, P.H., and Palta, J.P., in "Plant Cold Hardiness and Freezing Stress" (P.H. Li and A. Sakai, eds.), p. 49-71. Academic Press, New York, 1978.
4. Mendoza, H.A. and Estrada, R.N., in "Stress Physiology in Crop Plants" (H. Mussel and R.C. Staples, eds.), p. 227-262. John Wiley & Sons, New York, 1979.
5. Palta, J.P., and Li, P.H., Crop Sci. 19:665-671 (1979).
6. Pomeroy, M.K., and Siminovitch, D., Can. J. Bot. 49:787 (1971).
7. Tiwari, S.P., Sukumaran, N.P., and Upadhya, M.D., J.I.P.A. 5(3):175-178 (1978).

RECENT ADVANCE OF CRYOPRESERVATION OF APICAL MERISTEMS

A. Sakai and M. Uemura

The Institute of Low Temperature Science
Hokkaido University
Sapporo, Japan

Since the constituent cells of the shoot apex are less differentiated and have more uniform ploidy than those of mature
tissues, plants regenerated by shoot apex culture show the
recovery of true-to-type progenies. Furthermore, the shoot
apex has a greater ability to regenerate the whole plant than
do plant cultured cells.

During the past four years, successful development of whole
plants from excised shoot apices after freezing in liquid
nitrogen has been reported for some plant species including
carnation (11,12,13), strawberry (7,9), pea (6,13), tomato (3)
and potato (4).

This study was designed further to examine some factors
contributing to the survival of shoot apices frozen in liquid
nitrogen, and also to list up the results on successful
development of whole plants from excised shoot apices after
freezing in liquid nitrogen, to find a routine method applicable to shoot apices.

I. MATERIALS AND METHODS

Sample apices were treated with 10% DMSO solution at 0°C
for 1 hr. In some cases, excised sample apices were precultured on the culture medium with 5% DMSO for 2 days at 26°C
before freezing. Sample apices suspending in 0.25 ml freezing
solution were immersed in a -5°C ethanol bath for 15 min and
then transferred to a -10°C bath. To induce freezing in the
sample suspension, each tube was cooled with dry ice for a few
seconds. Frozen suspensions then were successively cooled down
in ethanol baths to about test temperatures by 5 or 10°C increments at 15 min intervals. We refer to this freezing method as

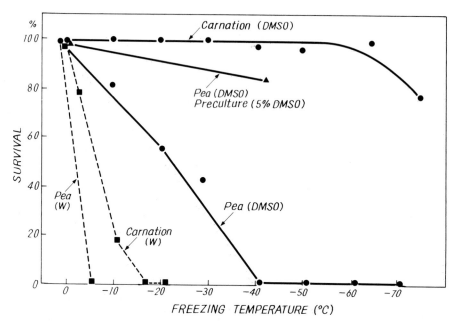

FIGURE 1. Survival of the carnation, pea and strawberry runner shoot apices frozen at different temperatures in the presence of 10% DMSO and rewarmed rapidly. W: Suspended with water only.

"step-by-step" method. Frozen suspensions were immersed in liquid nitrogen following prefreezing at selected test temperatures for 10 min. Bunches of 3 shoot apices were transferred to sterilized plastic 2 ml-ampoules (1.3x4.2 cm) along with exactly 0.5 ml of 10% DMSO solution. Carefully controlled freezing was achieved in a CRCF-1 set of a Biological Freezing System (Union Carbide Co. Ltd.). The samples were continuously frozen at cooling rates varying from 0.5 to 50°C/min until the selected test temperatures. The suspensions frozen in liquid nitrogen were all rewarmed rapidly in water at 37°C (rewarming rate: 550°C/min). These frozen-rewarmed sample apices were washed several times in each culturing solution over a period of 1 hr. The apices were then cultured at 26°C under continuous light (3,500 lux). The temperature of the sample was determined with 0.1 mm copper-constantan thermocouples and recorded with an oscilloscope. The cooling rates of the frozen suspension immersed in liquid nitrogen and the rewarmed rate thereafter were the time required for the temperature to fall from -10 to -40°C and to rise from -40 to -10°C, respectively.

Apices which sustained freezing injury lost their original green and whitened. Apices which retained their green color and which showed signs of development, excluding callus formation, 30 days after thawing were termed surviving apices.

II. RESULTS AND DISCUSSION

Most of the shoot apices frozen slowly without cryoprotectants then rewarmed slowly in air at room temperature were marginally hardy to -5°C. However, a marked difference in survival of shoot apices cooled slowly in the presence of DMSO was observed among species. Carnation shoot spices survived slow freezing to -70°C in the presence of DMSO. On the other hand, in the shoot apices of pea, only a few apices remained alive below -30°C (Fig. 1). Thus, the most important thing for success of freeze preservation of apical meristems is preconditioning before freezing. Preculture with DMSO for 2 days at 26°C proposed by Kartha et al. (6) is a prominent method of improving sensitivity to freezing of apical meristems. In the precultured shoot apices of pea, strawberry, asparagus, survival rate after slow freezing to -40°C was greatly increased (Fig. 1), which enabled many shoot apices to survive freezing in liquid nitrogen following prefreezing to -40°C (Table 1 and Fig. 1).

Preconditioning, cryoprotectants, cooling and rewarming conditions which enabled shoot apices to survive freezing to the temperature of liquid nitrogen are listed in Table 1. In the shoot apices of carnation, potato and tomato, relatively rapid cooling gave better survival. In carnation shoot apices, Seibert et al. (11) showed 80% survival after freezing to the temperature of liquid nitrogen. In their experiments, carnation shoot apices which floated on 0.5 ml of freezing solution were rapidly cooled by pouring liquid nitrogen, while dipping the vial directly into an open Dewar flask filled with liquid nitrogen. The mean cooling rate was estimated 400°C/min in this method and the maximal cooling rate was 1,100°C/min in the temperature range between -10 to -70°C (10,11). In the potato shoot apices, survival was observed only when the most rapid rate (1,000°C/min) was employed by direct immersion of shoot tips on the point of a hydermic needle into liquid nitrogen (3). In tomato shoot apices, however, intermediate cooling rate from 20 to 50°C/min gave better survival than the rapid cooling of 800°C/min (2).

To clarify the effect of cooling rates on survival of shoot apices, carnation and hardened pea shoot apices were cooled to -40°C at different rates in the presence of 10% DMSO and were then immersed in liquid nitrogen. These apices were all rewarmed rapidly. The survival rates gradually decreased to

TABLE I. Successful development of shoot meristems frozen to the temperature of liquid nitrogen.

Species or cultivar	Pretreatment	Cryoprotectant	Cooling rate, Prefreezing temp. (PR)	Thawing rate	Ref.
Carnation (Stem) (Dianthus caryophyllus)	Hardening (4°C, 4-10 days)	DMSO 5-10%	50°C/min	Rapid	11
"	none	DMSO 5-15%	0.5°C/min 1. PR:-40°C 2. Continuous to -70°C	"	12
Strawberry (Runner-tip) (Fragara x ananassa) (Mass propagation of plantlets)	none	DMSO 10-16% + 3% Sucrose	Ultrarapid(5×10^4°C/min) PF:-15 to -20°C	Rapid	9
	Preculture 5% DMSO, 2 days	DMSO 5%	0.5-1.0°C/min PF:-40°C	"	7
Pea (Seedlings) (Pisum sativum)	Preculture 5% DMSO, 2 days	DMSO 5%	0.5-1.0°C/min PF:-40°C	Rapid	6
"	1. Preculture 5% DMSO, 2 days 2. Hardening at 0°C, 10 days	DMSO 5%	"	"	13
Wild potato (Solanum goniocalyx)	none	DMSO 5%	Ultrarapid	"	3
Tomato (Seedlings) (Lycopersicon esculentum)	none	DMSO 15%	22-55°C/min	Rapid	2
Asparagus (Stem) (Asparagus officinalis)	Preculture 4% DMSO, 2 days	DMSO 8-16%	0.5°C/min PF:-40°C	Rapid	Kumu et al. (unpubl.)

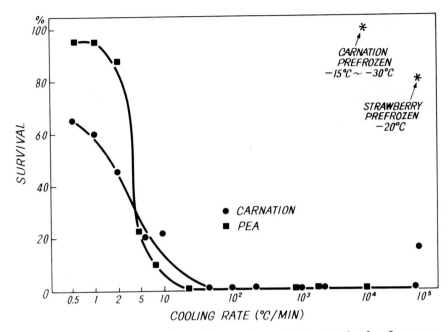

FIGURE 2. Effect of cooling rate on the survival of carnation and hardened pea shoot apices in the presence of 10% DMSO. *Ultrarapidly cooled shoot apices after prefreezing.

zero as the cooling rate increased from 0.5 to 50°C/min. Even at higher cooling rates of 50 to 5×10^4°C/min, no inverse in survival was observed as it has been for very hardy thin twig cortical tissues, when rewarmed rapidly (8). Strawberry runner shoot apices treated with DMSO solution were placed on a coverglass and the surrounding solution was removed. They were cooled ultrarapidly by direct immersion in liquid nitrogen (cooling rate: 10^5°C/min) after prefreezing below -20°C. Most of the shoot apices developed normal plantlets (9). In shoot apices of carnation no survival was observed even when cooled ultrarapidly without prefreezing. However, when apices prefrozen below -15 to -30°C were cooled ultrarapidly at 10^4°C/min by direct immersion into liquid nitrogen, all remained alive on subsequent rapid rewarming and developed normal young plants in light on the Murashige-Skoog's culture medium as did in shoot apices of strawberry runner (Fig. 2). Thus, this ultrarapid cooling and rewarming method combined with prefreezing seems to be useful for the cryopreservation of shoot apices from various plants.

In the carnation shoot apices, which survive slow freezing to -70°C or below in the presence of DMSO, their meristematic dome cells still remained alive after freezing to the temperature (12). In pea shoot meristems, however, after 5 days culture following freezing in liquid nitrogen, Haskins and Kartha (5) observed that most of the surviving cells were located in the primordial leaf tissues and lateral area of the dome. Thus, it is evident that in this materials the undifferentiated cells in the area of the meristematic "dome" are more sensitive to the freeze preservation procedures than are cells elsewhere in the isolated meristem. These surviving cells or groups renewed growth and differentiated and regenerated into whole plantlets (5). The same result was observed in the potato shoot apices by Grout and Henshaw (4).

We have observed several shoots developing from a single isolated and cultured carnation shoot apices following freezing and rewarming. These facts indicate that the surviving cells or groups of cells have the capacity for active cell division and structural organization into shoot meristems. It also indicates that the original meristematic dome does not have to be alive in its entirety or even in part for regeneration from frozen-rewarmed meristems into whole plants to take place. Frozen-rewarmed shoot apices can be grown without callus-mediated process of organogenesis. Thus, it is minimizing

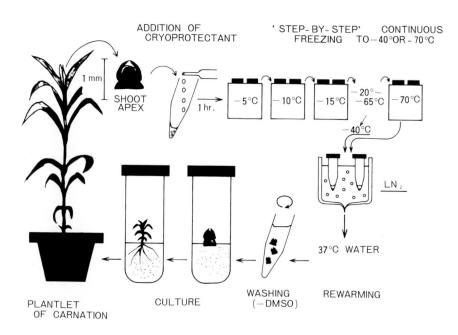

FIGURE 3. Our routine method of freeze preservation for plant apical meristems.

risks of gentic change, which means the system is likely to be suitable for prolonged storage of valuable plant germplasm. Isolated shoot apices are the most suitable explants for cultures which have greater genetic stability of plant germplasm as they have greater genetic stability than either callus or suspension cultures (1). This stability is believed to be a consequence of the organized growth of shoot apices in culture, and the success of any cryopreservation technique will depend upon the resumption of the organized growth after rewarming (4). The production of unorganized callus tissues by rewarmed shoot apices cultures would increase the possibility of genetic change during the culture period and plantlets regenerated from such a callus might not be genotypically identical to the original explant.

We need to develop a simple and reliable routine method of freeze preservation for plant germplasm. Further studies are required to more explain differences in susceptibility to freezing between different species and to clarify the action mechanism of DMSO as a cryoprotectant. Figure 3 shows our routine method of freeze preservation for plant apical meristems.

REFERENCES

1. D'Amato, F., in "Crop genetic resources for today and tomorrow", (O.H. Frankel and J.G. Hawkes, eds). pp. 333-348. Cambridge University Press, London, 1975.
2. Grout, B.W.W., and Henshaw, G.G., Ann. Bot. 42:1227-1229 (1978).
3. Grout, B.W.W., and Henshaw, G.G., Ann. Bot. 46:243-248 (1980).
4. Grout, B.W.W., Wescott, R.J., and Henshaw, G.G., Cryobiology 15:478-483 (1978).
5. Haskins, R.H., and Kartha, K.K., Can. J. Bot. 58:833-840 (1980).
6. Kartha, K.K., Leung, N.L., and Gamborg, O.L., Plant Sci. Lett. 15:7-15 (1979).
7. Kartha, K.K., Leung, N.L., and Pahl, K., Amer. Soc. Hort. Sci. 105:481-484 (1980).
8. Sakai, A., and Yoshida, S., Plant Physiol. 42:1695-1701 (1967).
9. Sakai, A., Yamakawa, M., Sakata, D., Harada, T., and Yakuwa, T., Low Temp. Sci. Ser. B 36:31-38 (1978).
10. Seibert, M., Science 191:1178-1179 (1976).
11. Seibert, M., and Wetherbee, P.J., Plant Physiol. 59:1043-1046 (1977).
12. Uemura, M., and Sakai, A., Plant & Cell Physiol. 21:85-94 (1980).
13. Uemura, M.. Master Thesis, Hokkaido University (1981).

RESPONSES OF SEVERAL LINES OF RICE AND DATE PALM CALLUS
TO FREEZING AT -196°C

B. J. Finkle and J. M. Ulrich
Western Regional Research Center
Agricultural Research Service
U.S. Department of Agriculture
Berkeley, California

B. Tisserat
Fruit and Vegetable Chemistry Laboratory
Agricultural Research Service
U.S. Department of Agriculture
Pasadena, California

The science of freezing plant tissues in the living state
has reached a stage where, in recent years, many but by no
means a majority of our cultivated species have been reported
to survive freezing at very cold temperatures (14). Freezing,
using low temperatures such as -196°C, makes possible the long
term stable storage of plant tissues, organs, and embryos, with
subsequent regeneration into whole plants. Thus, cryogenically
treated plant materials could be stored in banks dedicated to
germplasm preservation and other long-term tissue storage purposes. Yet, two questions with respect to the freezing treatments arise concerning the expectations from such banks: 1)
How much variability is there among different plant species,
and among clones within the same species, in their survival
response? 2) Do any genetic (biochemical or morphogenic)
changes take place in the tissue as a result of the freezing-
thawing procedures? We will address these questions by
describing 1) a degree of variability even among close lines of
a single species, rice, in their growth responses to several
cryogenic treatments and 2) a constancy of isozyme patterns in
extracts of leaves that were regenerated from thawed callus of
date palm after exposure to -196°C, indicating that protein
synthesizing pathways were not altered even by low temperature
freezing. Our observing the survival, after deep-frozen treatment, of the sub-tropical arid-grown date palm emphasizes the
somewhat anomalous question that was evident even in the early

ISBN 0-12-447602-3

CRYOPROTECTIVE ADDITIONS:

PEG-GLUCOSE-DMSO (P-G-D, 10-8-10% W/V)

    ADDED TO CELLS

FREEZING:

3°/MIN. TO -4°C

SEEDING AT -4°C

1°/MIN. TO -30°C

THEN INTO L.N. (-196°C)

THAWING:

SWIRL IN +40°C BATH

WASH WITH 3% SUCROSE MEDIUM AT R.T.

DETERMINING SURVIVAL:

TRANSFER TO GROWTH MEDIUM,

OR, ADD TRIPHENYLTETRAZOLIUM CHLORIDE (TTC)

TO DETERMINE VIABILITY INDEX

    (530 nm ABSORBANCE)

FIG. 1. Procedure for cryogenic treatments and subsequent evaluation of survival.

work of Maximow with tropical <u>Tradescantia</u> (4): can it be assumed that a tropical or warm climate plant tissue is less resistant to freezing by a cryogenic methodology than tissue from a cold climate species?

This paper will describe the freezing of eight closely-related callus lines of rice (<u>Oryza</u> <u>sativa</u> L.); some differences in response to cryogenic manipulations among these genetically close, including large differences in growth response to the temperatures of addition and post-thaw washing out of cryoprotectant; the freezing and regeneration into plants of callus cultures of two lines of date palm (<u>Phoenix</u> <u>dactylifera</u> L.); and the use of analytical data to test the normalcy of date palm plants regenerated after freezing.

I. GENERAL METHODS

Methodology for callus culture freezing and thawing treatments and subsequent survival evaluation is illustrated in Figure 1. Portions (1-2 grams) of callus grown on agar medium were treated with a cryoprotective mixture of polyethylene glycol (Carbowax 6000, Fisher Scientific Co., Fairlawn, N.J., U.S.A.), glucose, and dimethylsulfoxide (DMSO) (PGD, 10%-8%-10% w/v), first adding diluted mixture, followed by full-strength PGD and added over a period of about 30 minutes (2,12). The samples were treated in conical glass freezing tubes capped with heavy aluminum foil, and frozen in a controlled freezer (Cryo-Med, Mt. Clemens, MI, U.S.A.). The freezer was programmed, following the initiation of crystallization at -4°C, to lower the temperature at the rate of 1° per minute to -30°C, after which the tubes were plunged into liquid nitrogen for 4 minutes, or longer for storage experiments. Thawing was performed in a 40°C bath with rapid swirling of the tubes just to the point of ice disappearance, then followed by washing of the callus pieces with a simplified nutrient medium at either ice temperature or at room temperature as will be described below. After the washing step, small, standardized spoonfuls of callus (about 25 mg fresh weight in the case of rice) were planted on medium in petri plates and observed for growth at 28°C.

II. RICE CALLUS

A. Cryogenic Treatments and the Effects of Washing Temperature

The cryogenic potential of eight lines of rice supplied from the laboratory of G. W. Schaeffer was tested. All of the rice lines were derived from the same original callus line BL-2 initiated from anther tissue. Mutants of standard line BL-2

TABLE I. Growth response of rice lines to freezing treatments.

Line	BL-2	A-4	A-5	A-7	A-13	A-15
Growth (weeks)	10	4	8	9	10	8
Treatment			Fresh wt (mg)			
Control (0°)	664	571	938	804	1249	1121
	100	100	100	%̲ 100	100	100
+ PGD* (0°)**	84.8	108.8	%̲ 98.7	44.9	87.4	74.0
" (-23°)	45.2	92.5			46.8	23.3
" (-30°)	35.1	59.0			15.1	8.6
"(-196°)	15.2	60.1	39.1	25.7	15.6	7.1

* PGD = Polyethylene glycol (10% w/v) - Glucose (8%) - DMSO (10%)

** PDG was added at 0°, washed out at 22°.

were selected for their ability to grow on a medium containing S-aminoethyl-L-cysteine (8). We describe several large differences in growth in response to the treatments to which these closely related cultures were subjected, including their varied responses to 1) contact with cryoprotectants, 2) freezing-exposure temperature, and 3) the temperature at which the cultures were washed after thawing.

1. Growth Response After Exposure to Cryoprotectants. Table I illustrates several growth responses of rice lines to treatments with PGD. The results of three different factors in the cryogenic procedure will be described. In order to simplify the presentation, the table displays the effects of cryogenic treatments on the subsequent growth of only six (of the eight) lines of rice callus that we used.

There is a wide difference in response of lines to the addition of PGD (without freezing), ranging from callus growth that was not significantly different from the respective non-PGD controls (represented by Lines A-4 and A-5, and A-11 not shown) to growth that showed appreciable growth inhibition from PGD treatment (A-7 and A-15). Overall, most lines were characterized, however, as fairly resistant to toxic effects of PGD, similar to that reported for callus of alfalfa (13). In contrast, sugarcane suspension cultures are more sensitive to

PGD toxicity, as was indicated by the death of many cells after exposure to cryoprotective compounds such as glucose or DMSO in a 10% concentration range, even without freezing (2). We have reported a beneficial cryoprotective effect (decreased toxic effect) from using mixtures of cryoprotective compounds during freezing (2,12), hence our use of the PGD combination as the cryoprotective mixture of current experiments.

2. Response to Freezing and Thawing. Lines of rice differ in response to freezing also. Line A-15 of rice callus resembles callus cultures of less freeze-tolerant species of plants that survive poorly when frozen at -30° or -196°C. Other culture lines also, namely BL-2 and A-13, A-11 not shown, displayed little growth after immersion in liquid nitrogen. In contrast, still other lines (namely A-4, A-5, and A-7, and A-1 not shown) demonstrated extensive survival of cells at -196°C; e.g., Line A-4 cells exposed to -196°C showed 60% growth compared to unfrozen controls.

3. Effects of Washing Temperature. We have made an interesting and, we believe, significant finding that the temperature at which the PGD is washed out, after thawing, can markedly affect the survival of callus. Most often cryoprotective compounds are both added and removed at ice temperature, for reasons of both tradition and biochemical prejudice about the generally increased stability of biological systems at cold temperatures. Figure 2 shows the effects of both the temperature at which PGD is added to rice callus Line A-5 and the temperature at which the PGD is washed out.

a. Unfrozen A-5. A comparison of several types of unfrozen controls showed little or no difference in their growth. The untreated, unfrozen control callus pieces are at the center of the dish and the pieces that were treated with cold PGD solution, and also not frozen, are at the eleven o'clock and one o'clock positions in the figure; at the eleven o'clock position, the treatment was 0°C addition of PGD, 0°C washing out; at the one o'clock position, 0°C addition but 22°C washing out. It may be concluded about the non-frozen treatments of rice Line A-5 that there was little effect from adding PGD (compared to not adding PGD) and little or no effect of the temperature used in washing out PGD (0°/0° vs. 0°/22° treatments).

b. Frozen A-5. In contrast with the rapid and uniform growth of the unfrozen tissue, when samples were frozen to -196°C, recovery depended upon the washing-out temperature. With Line A-5, adding PGD at ice temperature followed by washing out at room temperature was the only combination that permitted growth of this callus line after freezing at -196°C. That addition/washing temperature regime (0°/22°) made the

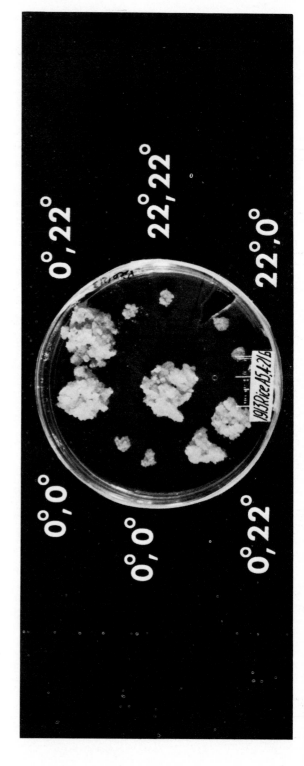

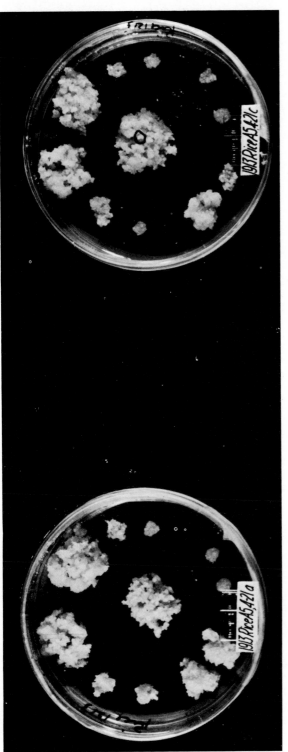

FIG. 2. Photographic record of Rice A-5 calli taken after 7 weeks of growth on agar medium. Temperatures of addition and removal, respectively, of PGD are labeled, at six positions around the top petri dish only. The sample at the center of each petri dish is an untreated, unfrozen control. PGD in the treated but unfrozen calli (top positions on dish) was added at 0°C and removed by washing at either 0°C (11 o'clock position) or 22°C (1 o'clock position). The remaining four sets of samples, in duplicate, in each dish were frozen to -196°C as described.

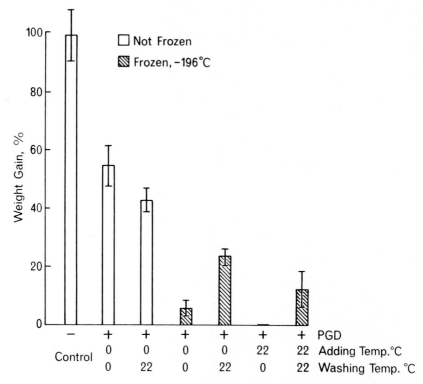

FIG. 3. Histogram of weight gain of unfrozen and frozen (-196°C) Rice A-7 calli after 9 weeks of culture on agar, normalized to weight gain by untreated control, versus temperatures of addition and removal of PGD. Vertical bars represent standard deviations based on three replicates.

difference between callus survival and non-survival of the -196°C freezing treatment.

c. Unfrozen A-7. Various effects of washing temperature on the callus growth of another rice Line (A-7) are presented in the histogram (Figure 3). The growth of Line A-7 was diminished more than was that of Line A-5 by exposure to PGD (not frozen).

d. Frozen A-7. When A-7 with added PGD was frozen, there was a further decrease in the survival rate. Again, the highest survival rate occurred from the 0°/22° addition/wash temperature sequence. There were lesser degrees of growth after 22°/22° and, still less after 0°/0° treatments. The most harmful sequence was a 22° addition of PGD, followed by a 0°

washing out. Table I shows that other rice lines, even given the beneficial 0°/22° addition/wash treatments, still demonstrated poor survival in liquid nitrogen. We have seen similar results of washing temperature on sugarcane callus and, with Persidsky (6), have reported beneficial effects on human granulocyte blood cells treated with dimethylsulfoxide when they were washed at room temperature, as compared with an ice cold washing treatment.

B. Discussion of Rice Callus

Our data from eight closely related lines of rice callus have indicated considerable difference in growth response to the treatments of these cultures. Specifically, the lines displayed large growth differences among themselves in response to exposure to the added cryoprotective mixture PGD, to freezing temperature, and to the temperatures at which PGD was added and washed out. Carefully controlling some of these factors, e.g., by using a 0°/22° addition/wash regime, made it possible for cultures to survive freezing in liquid nitrogen where, otherwise, they were killed by the freezing event. Of course, age, culture condition, and other selective characteristics of the rice culture are also contributing factors in the degree of cell culture survival (5,7,8). We may observe that the original culture Line BL-2 showed intermediate survival and growth behavior in its response to the various treatments, a finding to which we can not, of itself, ascribe any particular interpretation.

III. DATE PALM

A. Cryogenic Treatments, Regeneration, and Plantlet Analysis

The date palm callus cultures have quite different properties than the lines of rice. For one thing, the type and consistency of callus is different. Date palm callus, derived from lateral bud meristem and cultured on nutrient medium containing 100 mg/l 2,4-D and 0.3% charcoal (10), consists of a mass of minute, preformed embryoids (9). We have succeeded in freezing this callus tissue to the temperature of liquid nitrogen, storing for three months or more, and regenerating the entire plant. This capability will make available an entirely new vegetative method for clonal storage and propagation of data palm trees, instead of the slow field propagation method classically employed.

We will present a simpler picture of the freezing response here than with the rice cultures, although here, too, we observed a difference in survival response between the two lines of date palm studied.

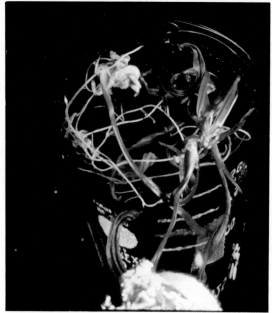

FIG. 4. Responses of thawed date palm callus subjected to cryogenic treatments. A) Four months growth of unfrozen date palm callus (0°C) and calli thawed after programmed freezing to -15°, -23°, -30° and -196°C. B) Plantlets produced from callus stored at -196°C for 3 months.

1. Treatment and Results. In experiments with a laboratory line of date palm callus culture (Line 1) we observed a callus line that is very resistant to all cryogenic treatments (10). Its normally slow rate of growth was not disturbed by treatment with the PGD cryoprotective mixture. There was also no visually discernible effect on the growth following any administered freezing temperature, even including immersion in liquid nitrogen following slow freezing to -30°C (Figure 4A). At four months of post-thawing growth, in light, shown in the figure, small plantlets were regenerated about equally in each of the tubes. After storage above liquid nitrogen for 3 months, normal-appearing plantlets developed, as illustrated in Figure 4B, and were established in soil as free living plants.

With another line of date palm, Line 2, variety Medjool, we again observed survival at -196°C but to a lesser extent. Table II shows the effects of treatments on both the growth in weight and the number of surviving embryos. The embryos were counted after an initial screening and sizing procedure. The table shows the data at 3, 6, and 9 weeks of growth of treated samples of callus grown on agar plates, and then the effects of subculturing 10 mg pieces of the 9 week-old callus onto fresh agar medium. Three sets of conditions were tested on Medjool calli: -PGD (0°), +PGD (0°) and +PGD (-196°), each treatment in triplicate. The results indicate that a) PGD addition without freezing, as with Line 1, had no adverse effect on growth or on the number of embryos regenerated from the original culture, but that b) a substantial number were killed by freezing at -196°C. Those that survived, however, grew in a rapid, healthy manner. This was demonstrated by the subcultured pieces; both the gain in weight and the number of embryos developed were comparable to controls. It appears that those cells, or embryoids, which survived the -196°C treatments were morphogenically unimpaired and relatively quickly gave rise to normal plantlets.

2. Isozyme Analyses of Regenerated Plants. Another aspect of normality was tested in the plantlets that developed. An analysis of the types and amounts of isozyme proteins found in leaf cell extracts has revealed differences in the protein-synthesizing mechanism among cultivars of date palm (11). Presumably variation in the protein-synthesizing mechanism within a cultivar can also be detected if gene or chromosome damage has occurred. To test for changes in the protein-synthesizing pathways after freezing Line 2 callus, isozyme patterns of extracts of triplicate leaf samples from plantlets regenerated after treatments were examined for enzyme polymorphism, using gel electrophoresis in 0.04 M sodium borate buffer pH 9 (11). Table III shows the electrophoretic separation of five enzymes. The pattern is characteristically different for each of the varieties tested, but no differences appear among the extracts of the leaves of var. Medjool (Line 2), whether the

TABLE II. Effects of cryogenic treatments on the growth (fresh weight) and asexual embryogenesis of date palm callus tissue.

Time in culture (wks)	Minus Cryoprotectant		Plus Cryoprotectant		Liquid Nitrogen	
	Original culture	Subculture	Original culture	Subculture	Original culture	Subculture
	g fresh weight/culture					
3	0.069 + 0.004	0.307 + 0.019	0.095 + 0.004	0.174 + 0.063	0.048 + 0.003	0.092 + 0.008
6	0.197 + 0.015	0.572 + 0.093	0.207 + 0.053	0.799 + 0.114	0.054 + 0.009	0.475 + 0.010
9	0.606 + 0.068	1.219 + 0.197	0.582 + 0.123	1.188 + 0.231	0.090 + 0.028	1.307 + 0.218
	embryos/culture					
3	18.0 + 3.2	117.0 + 39.0	14.7 + 4.1	186.0 + 22.2	12.3 + 3.3	84.3 + 41.1
6	133.0 + 18.0	439.3 + 71.9	148.3 + 40.8	523.7 + 103.7	41.7 + 8.8	338.7 + 24.2
9	451.3 + 63.0	545.6 + 75.7	528.3 + 174.9	543.0 + 54.1	87.3 + 27.1	691.0 + 50.5

TABLE III. Relative positions (R_p values) of isozymes from leaves of _Phoenix_ _dactylifera_ L. cultivars or specimens.

Enzyme	Deglet Noor[a]	Khad-rawy[a]	Dayri L7,BC$_3$	Medjool Field[a]	-C[b]	+C[c]	LN[d]
			50				
			53				
Alcohol dehydrogenase	56	56	56	56	56	56	56
Esterase	77	77	77	77	77	77	77
	86	86	86				
Peroxidase	50	50	50	50	50	50	50
Phosphoglucose isomerase	7	7		7	7	7	7
	12	12		12	12	12	12
	20	20	20	20	20	20	20
	67	67	67	67	67	67	67
Phosphoglucose mutase	70	70	70	70	70	70	70
	73		73	73	73	73	73

a   Mature leaf samples from verified cultivars grown in the field.
b   Specimen leaf sample from untreated Medjool cultures; C means cryoprotectant.
c   Specimen leaf sample from cryoprotectant treated Medjool cultures.
d   Specimen leaf sample from cryoprotectant treated Medjool cultures, frozen to -196°C.

originating callus samples had been frozen in liquid nitrogen or not frozen.

B. Discussion of Date Palm

With date palm callus we see again, as with rice callus, a variability of response to the freezing treatment between the palm lines tested. Line 1 callus was highly resistant to freezing, even at -196°C, whereas the Medjool line was much less so. In both lines, those cells that survived grew well and produced normal appearing plants. When surviving Medjool callus was regenerated into plantlets after freezing at -196° and extracts were made of the young leaves, isozyme patterns of

TABLE IV. Plant species that have survived freezing.

Frozen to -196°C	Frozen to other temperatures	
Alfalfa (3 lines)		
Apple	Elm	-30°C
Asparagus	Grape	-15°C
Carrot*	Grapefruit	-15°C
Palm (2 lines)	Phaseolus	-15°C
Rice (8 lines)	Tomato	-30°C
Soybean		
Strawberry*		
Sugarcane		
Wheat		

* Carrot suspension culture; strawberry meristem; all others callus cultures.

the extracts appeared identical with those of leaves of the original cultured line and, also, of field grown trees of this variety. This observation suggests an identity of genetic makeup of the surviving highly stressed frozen cells and the unfrozen source cells of the experiments. It should be noted that date palm callus is highly heterogeneous in nature, composed of a variety of cell types and structures (9). It would be interesting to relate the callus composition and degree of uniformity within the callus to the freezing potential of the callus cell population.

An interesting physiological note can be drawn from these findings. One might expect that callus from a warm-climate plant might be more sensitive to freezing by cryogenic methods than callus from a temperate zone plant. Yet we find no basis for this assumption from our research results. We and others have found that cultures of tropical and semi-tropical species, including date palm, sugarcane (12), and cassava (1) will withstand freezing to -196°C by cryogenic methods that are fatal to some temperate species, as illustrated in Table IV from our own experiments. With one of the lines of date palm (Line 1), cell survival appeared to be close to 100% (10). These observations on survival might indicate (see findings by Breidenbach, and others) that the primary aspects of the trauma of cellular freezing (probably structural) are likely to be at a different level of cell functioning than the metabolic and/or structural "chilling injury" effects from nonfreezing low temperatures (3), and that these different traumatic temperature events do not necessarily affect the responses of different target species in parallel. Furthermore, the degree of injury from several treatments may not run in parallel even between lines of the same target species, as was seen in the presented examples of date palm lines, and rice.

The capability of using small bits of growing tissue for storage over indefinitely long periods of time, without genetic deterioration, makes available a new, compact method for species, cultivar and clone preservation where the tissue can, at will, be manipulated to undergo rapid proliferation. The described sequence of techniques makes possible the preservation and proliferation of thousands of date palm plants from each of many small pieces of callus tissue taken from each lateral bud meristem.

IV. GENERAL CONCLUSIONS

By the general methods described we have been able to freeze several species to -196°C with survival of tissue and with the regeneration of whole plants in some cases. Success in regeneration was described for date palm (Fig. 4), and another example is illustrated in Figure 5 that shows an alfalfa plant regenerated from callus frozen at -196°C (13). The date palm and alfalfa callus tissues have gone a long way toward maturity following the freezing treatment at -196°C. In Table IV are listed results from our laboratory describing, in the first column, species of callus tissue that have been found to survive freezing in liquid nitrogen and, in the second column, species that have survived freezing but only at some higher frozen temperature as noted. Examples of the latter freeze-sensitive category are lines of grape, grapefruit, and phaseolus bean that have not demonstrated growth after freezing at temperatures below -15°C. Still other species, among them taro · and coast redwood, have not survived freezing, in our hands, at any temperature. We see an apparent marked difference in freezing capability among species. But we have also described here that rice and date palm lines, even close mutant lines of rice, showed large differences in their cryogenic capabilities. We are dealing here with some very freeze-sensitive factors which may sometimes be resolved by simple types of procedural manipulation, such as a warmer temperature wash (22° instead of 0°) after thawing the frozen tissue. In Table IV some of the rice lines described and, also, asparagus callus were recently shifted from the right-hand column of the table to the left-hand column, with -196°C freezing capability, by making relatively small changes in procedural details. Many of the freezing experiments on the tested species indicated in Table IV were performed using an ice-cold wash, before we had learned of the improved effect from washing out the cryoprotectants at room temperature. Hence we do not yet know if their recovery capability could be changed even by this one factor of

Figure 5

post-thaw washing temperature. In any case, we should not allow ourselves to assume, in cryogenic explorations, that the very important tropical crop tissues cannot be cryogenically frozen: it might be better to try them out. A great deal of work needs to be done here toward understanding the impact of physiological and treatment factors, sometimes even small details, that may be important toward the recovery of frozen cells.

Dr. M. E. Zavala, in our laboratory, has initiated ultra-structural studies concerning the effects of cryoprotective compounds themselves, when administered at different temperatures to suspension cultures of sugarcane. She is studying these effects both with cells that are not taken through the freezing process and those that are frozen. From an ultra-structural approach to the procedural factors affecting cell damage we hope to learn what can be changed in the freezing-thawing process that will overcome the sensitivity of plant cells to such procedures.

Referring again to Table IV, it appears that, to date, without conscious selection on our part, about one-half of the species of callus we have tested for freezing capability have demonstrated recovery from freezing in liquid nitrogen. We can visualize that perhaps an even much larger proportion of plant tissues will be made capable of surviving through manipulations that will be developed. Perhaps we can learn enough about a tissue's growth stage, population pattern, and chemical and osmotic sensitivities to know, in advance, what modifications of procedure (such as custom-fitting cryoprotective mixtures to the particular tissue) might assure viability of a given tissue in the face of the freezing-thawing trauma. Or we could learn to do specific examinations or assays on the tissue that would indicate how best to proceed. Understanding and dealing with the freezing thawing phenomena of plants at this level would be an approach that could make the idea of widespread, near-universal plant germplasm and tissue banks a reality. One can also visualize that a dynamic comprehension of the structural and metabolic factors affecting freezing recovery could help us to understand, on a new level, how to prevent the economically devastating freezing losses of whole plants or their fruits in the field.

FIG. 5. Alfalfa plant, in flower, regenerated from callus frozen at -196°C. (Sixteen months growth).

ACKNOWLEDGMENTS

We wish to thank G.W. Schaeffer and Frank T. Sharpe, Jr., for supplying so generously the lines of rice we studied, and to Bruce E. Mackey for his generous help with statistics. We express appreciation to Maxim D. Persidsky for his suggesting the possibility of improving tissue recovery from cryoprotective agents through using a higher-temperature washing procedure.

REFERENCES

1. Bajaj, Y.P.S., Crop Improv. 4:198-204 (1977).
2. Finkle, B.J., and Ulrich, J.M., Plant Physiol. 63:598-604 (1979).
3. Lyons, J.M., Raison, J.K., and Steponkus, P.L., in "Low Temperature Stress in Crop Plants" (J.M. Lyons, D. Graham, and J.K. Raison, eds.), pp. 1-24. Academic Press, New York, 1979.
4. Maximow, N.A., Ber. Deut. Botan. Gesel. 30:504-516 (1912).
5. Ono, K., "Sympos. on Methods for Crop Breeding". Tropical Agric. Res. Ser. No. 22, Tropical Agric. Res. Cntr., Yatabe (1979).
6. Persidsky, M., Ulrich, J.M., and Finkle, B.J., Abst., Soc. of Cryobiologists, Vancouver (1980).
7. Sala, F., Cella, R., and Rollo, F., Physiol. Plant. 45:170-176 (1979).
8. Schaeffer, G.W., and Sharpe, F.T., Jr., In Vitro 17:345-352 (1981).
9. Tisserat, B., and DeMason, D.A., Ann. Bot. 46:465-472 (1980).
10. Tisserat, B., Ulrich, J.M., Finkle, B.J., HortScience 16:47-48 (1981).
11. Torres, A.M., and Tisserat, B., Amer. J. Bot. 67:162-167 (1980).
12. Ulrich, J.M., Finkle, B.J., Moore, P.H., and Ginoza, H., Cryobiology 16:550-556 (1979).
13. Urich, J.M., Finkle, B.J., Rains, D.W., and Stavarek, S., Abst. Int. Botan. Congress, Sydney (1981).
14. Withers, L.A., "Tissue Culture Storage for Genetic Conservation". Tech. Rpt., Int. Board Plant Genetic Resources, Food and Agric. Org., United Nations, Rome (1980).

Part V
Other Temperature-Related Stresses

MOLECULAR STRATEGIES OF ADAPTATION TO HIGH TEMPERATURES:
BIOCHEMICAL BASIS OF UNUSUAL HEAT STABILITIES OF CELL
COMPONENTS FROM AN EXTREME THERMOPHILE

Tairo Oshima

Mitsubishi-Kasei Institute of Life Sciences
Tokyo , Japan

I.   INTRODUCTION

The author and his colleagues have studied the molecular
basis of thermophily using an extreme thermophile, Thermus
thermophilus, isolated from a hot spring (1). We isolated
enzymes, proteins, nucleic acids, lipids, and cell organellae
such as membranes and ribosomes from the thermophile. They are
more resistant to heat than the corresponding ones from mesophilic organisms. Summing up our studies (13-15) on the structural basis for the unusual stabilities so far carried out, the
author wishes to propose two general principles; (i) the structural changes which are important for the improved stability
are often subtle, and (ii) a variety of tactics are used to
make cell components heat stable and in many cases a molecule
is stabilized by the combination of two or more different
tactics.

Though the tactics used to make thermophile cell constituents thermostable are diverse and different from molecule to
molecule, organelle to organelle, the author has proposed that
they can be classified into three groups (14,15), that is,
intrinsic, protector, and biochemical modification mechanisms,
as summarized in Table I.

II.   ENZYMES

Many enzymes and proteins of the thermophile seemed to be
stabilized by increased intramolecular interactions caused by
replacing amino acid residues. One of the typical examples is
phosphoglycerate kinase from T. thermophilus (7,17). The

TABLE I. Some examples of molecular mechanisms of adaptation to
       high temperature.

	Constitutive	Temperature inducible
Intrinsic	Enzyme; stabilized by amino acid replacements Nucleic acid; stabilized by replacing A-U pairs with G-C pairs	not known
Protector	Enzyme; stabilized in the presence of Ca^{++} ion	Protein biosynthesis; stabilized by poly-amines
Biochemical modification	not known	tRNA; stabilized by thiolation of a thymidine residue

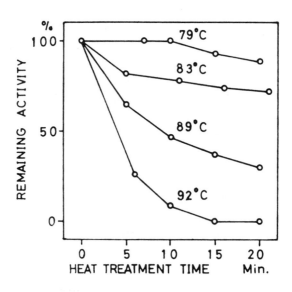

FIG. 1.   Heat stability of phosphoglycerate kinase from T.
thermophilus. Taken from ref. (7).

TABLE II. Comparison of enzymatic properties of phosphoglycerate kinases from T. thermophilus and yeast.

	Source	
	T. thermophilus	Yeast
Molecular weight	43,000	47,000
Subunit	1	1
pI	5.0	7.2
Opt. pH	broad	broad
Km (GAP)	1.8 mM	1.3 mM
Km (ATP)	0.28 mM	0.48 mM
α-helix content	29%[1]	24%[2]
cystein[3]	1	1
Heat stability[4]	90°(10)	62°(15)

[1] Estimated from CD spectrum
[2] Estimated from X'-ray crystallographic study
[3] Residue per mole of enzyme
[4] 50% loss of activity, time in ( ).

enzyme is a simple protein without any prothetic group, and is
highly stable to heat in the absence of the substrate, metal
ions, or other proteins as shown in Fig. 1 (7). Except heat
stability, the enzymatic properties are similar to the mesophilic counterparts in many respects including molecular size,
kinetic parameters, amino acid composition and secondary structure contents as shown in Table II. Similar observations were
made for other enzymes from thermophiles (1,3,17). The facts
suggest that the important changes for the unusual heat stability do not affect the molecular architecture of the active site
of the thermophilic enzyme.

   If an enzyme of a thermophile is stabilized only by the
intrinsic mechanism, a mesophilic organism may produce the heat
stable enzyme when a gene which codes for the enzyme is cloned
into the mesophile and the gene is expressed by the organism.
To demonstrate this possibility, a gene for 3-isopropylmalate
dehydrogenase from T. thermophile was ligated into a E. coli
plasmid called pBR322 and was then transformed into E. coli
(26). The cells of E. coli which harbor the thermophile gene
produced the heat stable isopropylmalate dehydrogenase as shown
in Fig. 2. As shown in the figure, the heat stability of the
enzyme produced by the cloned E. coli was similar to that of
the enzyme extracted from the cells of T. thermophilus. The
observation suggests that the thermophile enzyme is stabilized
by the intrinsic mechanism.

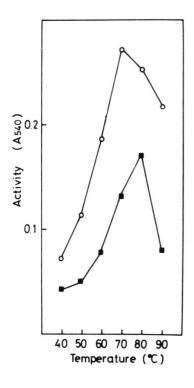

FIG. 2. Heat stability of 3-isopropylmalate dehydrogenase produced by T. thermophilus (—■—) and E. coli cloned with the thermophile gene (—○—). Reproduced from ref. (26).

Some enzymes from thermophiles are stable only in the presence of specific ligand(s). For instance, an α-amylase from a thermophile is stable only when Ca^{++} and Na^{+} ions are present (4,24). This is a typical example of the protector mechanism. Since sufficient amounts of Ca^{++} and Na^{+} ions are present in the normal culture conditions, a mesophilic recipient into which DNA coded for the α-amylase had been transformed, produced the heat stable α-amylase (25).

III.   POLYAMINES

Polyamines are required for in vitro protein synthesis catalyzed by a T. thermophilus extract (8,9). The requirement was essential and specific. More detailed study revealed that polyamine is necessary to form an active 30S ribosomal complex (10). This is another example of the protector mechanism.

To study roles of polyamines in protein synthesis in vitro, polyamine composition was analyzed. The extreme thermophile,

T. thermophilus contained two new tetraamines, thermine (18) ($NH_2(CH_2)_3NH(CH_2)_3NH(CH_2)_3NH_2$), and thermospermine (19) ($NH_2(CH_2)_3NH(CH_2)_3NH(CH_2)_4$ NH_2) as major polyamine components, and three triamines (including two new amines), norspermidine ($NH_2(CH_2)_3$ $NH(CH_2)_3NH_2$), spermidine ($NH_2(CH_2)_3NH(CH_2)_4$ NH_2) and sym-homospermidine ($NH_2(CH_2)_4NH(CH_2)_4$ NH_2) as minor components (20). These novel polyamines are also found in other extreme thermophiles such as T. aquaticus or Sulfolobus acidocaldarius (except sym-homospermidine), however, not found in moderate thermophiles (Bacillus stearothermophilus) or mesophilic bacteria.

Polyamine composition of the thermophile depends on environmental factors, especially growth temperature (21). When the cells were grown at lower temperature such as 55° to 65°C, the cells contained more triamines. At higher temperature such as 80°C or higher, the cells contained more tetraamines, especially thermine. In addition, the cells contained another new polyamine, caldopentamine $NH_2(CH_2)_3$ $HN(CH_2)_3$ $NH(CH_2)_3$ $NH(CH_2)_3NH_2$ (22). It can be speculated that the cells adapt to the environmental temperature by changing the polyamine composition.

IV.   THERMOPHILE tRNA

The third type of mechanism, biochemical modification, can be seen in tRNA from T. thermophilus. As described below, a part of the unusual stability of the tRNA is given by the modification of a nucleotide base.

tRNA from the thermophile has a higher melting temperature than those from mesophilic organisms. Two structural features of the heat stable tRNA were found to be related to the heat resistance. (i) High G-C pair content in the base paired region, a physico-chemical study suggested 90% G-C pair conent for $tRNA_{mix}$ of the thermophile (28). (ii) The presence of a modified base, 5-methyl-2-thiouridine (or 2-thioribothymidine) instead of unmodified ribothymidine in T loop (29). The implication of the thiolated base in the unusual stability was indicated by biochemical (30), physico-chemical (2,31) and organic chemical (32) studies.

tRNA specific for formyl methionine from the thermophile ($tRNA_f^{Met}$) was isolated, sequenced and compared with the chemical structure of the corresponding tRNA from Escherichia coli (23). Only one base replacement was found; that is, U-51 in E. coli tRNA was replaced by C in the thermophile tRNA giving one more G-C pair in the T-arm of the molecule. The G-C pair content in the base paired region of the thermophile tRNA specific for formylmethionine is 90% as suggested by a physico-chemical study (28).

Except the base replacement, three modified bases were found in the thermophile tRNA$_{II}^{Met}$, 2-thioribothymidine in T loop, 1-methyladenosine in T loop and 2'-0-methylguanosine in D loop. The latter two modifications might be neutral in terms of the unusual stability since the same or similar modifications are often found in eukaryote (mesophilic) tRNAs. Recently an enzyme which catalyzes the 2'-0-methylation of the specific guanosine residue in D loop was isolated from the thermophile (5). The melting temperature of E. coli tRNA methylated by the enzyme was identical to that of the unmodified tRNA. This experimeht directly demonstrated that the methylation at the specific guanosine residue had no effect on the thermal stability.

The nucleotide sequence of T. thermophilus tRNA$_{II}^{Met}$ is identical to that of E. coli tRNA$_{f}^{Met}$ except the above mentioned four residues (one replacement and three modifications). Thus the thermophile tRNA$_{II}^{Met}$ is so far the only molecule of thermophiles for which unusual stability is fully explained by changes in chemical structure. The molecule is more stabilized by an extra intramolecular hydrogen bond given by replacing a uridyl residue at position 51 with a cytydyl residue and increased stacking interaction brought by thiolating a ribothymidine residue. These changes are quite subtle compared to the sum of the intramolecular hydrogen bonds (roughly 60), or the number of atoms composed of the molecule, one (oxygen) of which is replaced by a sulfur in the thermophile tRNA. It might be of interest to point out that these changes are concentrated in a part quite far from anti-codon loop and amino acid accepting terminal; both sites are of direct importance for the function of tRNA, as shown in Fig. 3.

V.   PSYCHROPHILIC tRNA

Euphausia, a shrimp in Antarctic Sea, lives only in a cold environment. The author and his colleagues (34) found that tRNA from Euphausia has a lower melting temperature than that from mesophilic eukaryotes. To explain the unusual thermal stability of the tRNA, its chemical structure was studied.

We sequenced some tRNA molecules isolated from Euphausia. In contrast to our expectation that the molecule contains low G-C pair content and/or a modified base, we found unusual base pairs such as A-C or G-U as shown in Fig. 4. Since except for these unusual base pairs, no significant difference was found between the psychrophilic and mesophilic eukaryote tRNAs, these pairs seemed to be the only structural basis for the unstability. In a sense, tactics for low temperature adaptation are qualitatively different from those for high temperature adaptation.

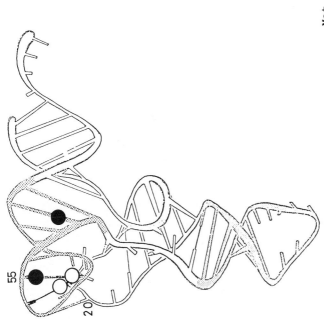

FIG. 3. Chemical structure of tRNA$_f^{Met}$ from T. thermophilus. The locations of different nucleotide residues from the corresponding tRNA of E. coli are indicated with circles. Black circles indicate the sites of important changes for the unusual stability to heat. Reproduced from ref. (23).

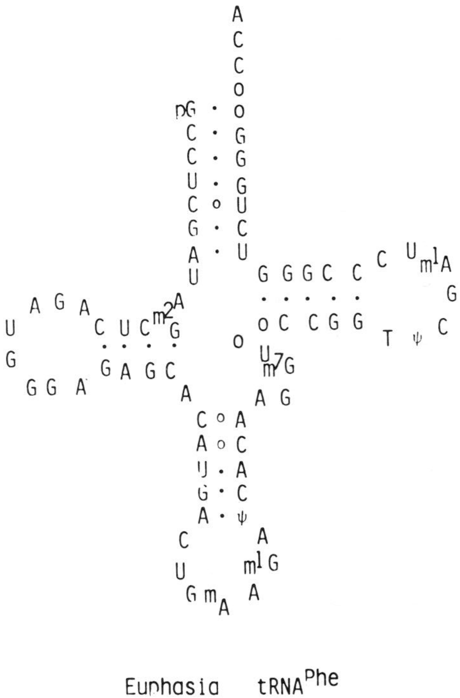

Euphasia tRNA^{Phe}

FIG. 4. Nucleotide base sequence of tRNA^{Phe} from Euphausia. Taken from ref. (34).

TABLE III. Temperature dependence of microviscosity of T. thermophilus membrane (27).

Growth temperature	Higher inflexion point of Arrhenius plots of the microviscosity
50°C	55°C
55	60
60	70
75	75
80	80

VI.   MEMBRANES AND LIPIDS

Cell membrane from T. thermophilus is another example of molecular acclimation to environmental temperature. Chemical structure of membrane lipids was studied by M. Oshima and collaborators (11,12). The major lipid component is a novel glycolipid, galactofranosylgalacto-pyranosyl-N-methylhexadecanoylglucosaminylglucosyldiglyceride. Fatty acid residues attached to the glycerol group are mostly C_{17} and C_{15} iso and anteiso acids.

Thermal transition temperature of the membrane or lipid suspension was studied by using a differential scanning calorimeter and a fluorescence polarization technique (27). A thermogram recorded by a scanning calorimetry showed a broad endothermic peak centered around the growth temperature. Arrhenius plots of the microviscosity of the membrane estimated from the fluorescence polarization measurements and fluorescence life time measurements with a single photon counting apparatus, are generally characterized by two inflexion points. The lower point could be the transition temperature from the solid phase to intermediate, and the higher one from intermediate to liquid crystalline phase.

We found that the upper inflexion point is a function of the growth temperature of cells from which membrane was prepared. The transition temperature from intermediate to liquid crystal is close to, but slightly higher than the growth temperature as shown in Table III (27). Phase transition temperature of membranes from a moderate thermophile, Bacillus stearothermophilus was also reported to be dependent on the growth temperature (6). These findings suggest the homeoviscous regulation of membrane is an essential feature of adaptation to high temperatures.

It seemed that the homeoviscous regulation is achieved by changing fatty acid composition of membrane lipids. The molar ratio of C_{17}/C_{15} acids correlated with the growth temperature (12). The ratio increased when the growth temperature was raised, presumably giving rise to a membrane with higher transition temperatures. In other words, the membrane is acclimated to the environmental temperature by changing its fatty acid composition so that it can be in appropriate physical state at the temperature.

REFERENCES

1. Arai, K., Ota, Y., Arai, N., Nakamura, S., Henneke, C., Oshima, T., and Kaziro, Y., Eur. J. Biochem. 92:509-519 (1980).
2. Davanloo, P., Sprinzl, M., Watanabe, K., Albani, M., and Kersten, H., Nucl. Acids Res. 6:1571-1581 (1979).
3. Fujita, S.C., Oshima, T., and Imahori, K., Eur. J. Biochem. 64:57-58 (1976).
4. Hasegawa, A., Miwa, N., Oshima, T., and Imahori, K., J. Biochem. 79:35-42 (1976).
5. Kumagai, I., Watanabe, K., and Oshima, T., Proc. Natl. Acad. Sci. USA 77:1922-1926 (1980).
6. McElhaney, R.N., and Souza, K.A., Biochim. Biophys. Acta 443:348-359 (1976).
7. Nojima, H., Oshima, T., and Noda, H., J. Biochem. 85:1509-1517 (1979).
8. Ohno-Iwashita, Y., Oshima, T., and Imahori, K., Arch. Biochem. Biophys. 171:490-499 (1975).
9. Ohno-Iwashita, Y., Oshima, T., and Imahori, K., J. Biochem. 79:1245-1252 (1976).
10. Ohno-Iwashita, Y., Oshima, T., and Imahori, K., in "Enzymes and Proteins from Thermophilic Bacteria" (H. Zuber, ed.), pp. 333-345. Birkhäuser-Verlag, Basel, 1976.
11. Oshima, M., and Yamakawa, T., Biochemistry 13:1140-1146 (1974).
12. Oshima, M., and Miyagawa, A., Lipids 9:476-480 (1974).
13. Oshima, T., and Imahori, K., Intern. J. Syst. Bacteriol. 24:102-112 (1974).
14. Oshima, T., Sakaki, Y., Wakayama, N., Watanabe, K., Ohashi, Z., and Nishimura, S., in "Enzymes and Proteins from Thermophilic Microorganisms" (H. Zuber, ed.), pp. 317-331. Birkhäuser-Verlag, Basel, 1976.
15. Oshima, T., "Thermophilic Bacteria", University of Tokyo Press, Tokyo (1978). (in Japanese).
16. Oshima, T., in "Strategies of Microbial Life in Extreme Environments" (M. Shilo, ed.) pp. 455-469. Dahlem Konferenzen, Berlin, 1979.

17. Oshima, T., in "Enzyme Engineering" Vol. 4 (G.B. Broun, G. Menecke, and L.B. Wingard, Jr.) pp. 41-46. Plenum Publishing Co., New York.

18. Oshima, T., Biochem. Biophys. Res. Commun. 63:1093-1098 (1975).

19. Oshima, T., J. Biol. Chem. 254:8720-8722 (1979).

20. Oshima, T., and Baba, M., Biochem. Biophys. Res. Commun. 103:156-160 (1981).

21. Oshima, T., in "Biochemistry of Thermophily" (S.M. Friedman, ed.), pp. 211-220. Academic Press, New York, 1978.

22. Oshima, T., Submitted (1982).

23. Oshima, T., in "Molecular Evolution, Protein Polymorphism and the Neutral Theory" (M. Kimura, ed.), pp. 349-356. Japan Scientific Societies Press, Tokyo.

24. Pfueller, S.L., and Elliott, W.H., J. Biol. Chem. 244:48-54 (1969).

25. Sinomiya, S., Yamane, K., and Oshima, T., Biochem. Biophys. Res. Commun. 96:175-179 (1980).

26. Tanaka, T., Kawano, N., and Oshima, T., J. Biochem. 89:677-682 (1981).

27. Wakayama, N., and Oshima, T., J. Biochem. 83:1687-1692 (1978).

28. Watanabe, K., Seno, T., Nishimura, S., Oshima, T., and Imahori, K., Polymer J. 4:539-552 (1973).

29. Watanabe, K., Oshima, T., Saneyoshi, M., and Nishimura, S., FEBS Letters 43:59-63 (1974).

30. Watanabe, K., Shinma, M., Oshima, T., and Nishimura, S., Biochem. Biophys. Res. Commun. 72:1137-1144 (1976).

31. Watanabe, K., Yokoyama, S., Hanske, F., Kasai, H., and Miyazawa, T., Biochem. Biophys. Res. Commun. 91:671-677 (1979).

32. Watanabe, K., Biochemistry 19:5542-5549 (1980).

33. Watanabe, K., Kuchino, Y., Yamaizumi, Z., Kato, M., Oshima, T., and Nishimura, S., J. Biochem. 86:893-905 (1979).

34. Watanabe, K., Asai, K., Oshima, T., and Kuchino, Y., J. Biochem. 90:1259-1266

EFFECT OF TEMPERATURE ON THE ACTIVITY AND STABILITY OF HIGHER PLANT CYTOCHROME c OXIDASE

Tadashi Asahi, Masayoshi Maeshima, Makoto Matsuoka,
and Ikuzo Uritani

Laboratory of Biochemistry, Faculty of Agriculture
Nagoya University
Nagoya, Japan

Physiological dysfunction of cell membranes in response to low temperature stress has been thought to be one of the most important primary events in chilling injury. Attention has been concentrated on the importance of a solidification of membrane-lipids in the membrane dysfunction. Lyons and Raison (6) found that the mitochondria from chilling-sensitive plants exhibited a downward bend at 9-12°C in the Arrhenius plot for succinate oxidase, while those from chilling-resistant plants did a continuous straight line at a temperature range from 1 to 25°C. Raison et al. (18) reported that the bend was ascribed to a phase transition of the membrane-lipids from a fluid to a solid state. These findings together with observations with other membranes have led to form the concept that cell membranes in chilling-sensitive plants undergo the phase transition at the temperature critical for chilling injury, while those in chilling-resistant plants keep their lipids fluid even at low temperatures (4,5,7.16,17). The change in the molecular architecture of membrane-lipids has been proposed to ultimately evoke injury through subsequential events including dysfunction of cell membranes and metabolic imbalance (4,7,16).

On the other hand, only a little information is available as to whether low temperature stress directly affects plant membrane-proteins to cause membrane dysfunction. Reports from our laboratory have suggested deteriorations in membrane functions by weakening in the association of proteins with membranes during chilling (25,26). In most cases, the effects of low temperatures on plant membrane-bound enzymes have been interpreted in terms of changes in the activities by the phase transition of membrane-lipids, probably through conformational changes of the enzyme proteins forced by the phase transition (4,17).

We have succeeded in purification of cytochrome c oxidase, the terminal oxidase in respiratory chain associated tightly

671

with mitochondrial inner membrane as an endo-membrane-protein, from higher plant sources (8,15), and so experiments with the enzyme as a model for membrane-proteins have been conducted to elucidate how low temperatures affect membrane-proteins as well as how phospholipids participate in the functions of membrane-bound enzymes. So far the effects of low temperatures on plant membrane-proteins have been investigated with the proteins which remain associating with, and not solubilized from, the respective membranes. Therefore, we expect to find new information from the experiments with the solubilized enzyme as to whether low temperatures directly affect membrane-proteins.

I.   COMPARISON BETWEEN SWEET POTATO AND PEA CYTOCHROME C OXIDASES WITH RESPECT TO THE MOLECULAR FORM

A.   Isolation of Cytochrome c Oxidase from Higher Plants

Cytochrome c oxidase has been isolated from many sources including various animals, fungi and prokaryotes. There are, however, only two reports from our laboratory concerning the isolation of higher plant cytochrome c oxidase (8,15). Sweet potato (Ipomoea batatas Lam. cv. Kokei No. 14) cytochrome c oxidase has been purified from the inner membrane of root mitochondria by solubilization with deoxycholate, diethylaminoethyl (DEAE)-cellulose column chromatography, and ammonium sulfate fractionation (8). Impurities in the purified preparation, if any, can be removed by sucrose density gradient centrifugation, although this procedure is accompanied by inactivation of the enzyme (8). The pea (Pisum sativum var. Alaska) enzyme has been purified from the inner membrane of shoot mitochondria by solubilization with Triton X-100 and DEAE-cellulose column chromatography (15). A pure, but inactive, preparation of this enzyme protein can be obtained by further purification with sucrose density gradient centrifugation and second DEAE-cellulose column chromatography (15). The success in purification of cytochrome c oxidase from sweet potato, a chilling-sensitive plant, and pea, a chilling-resistant plant, led us to compare between the two enzyme proteins with respect to the molecular form, in expectation that not impossibly, differences in the molecular structures between the two proteins might be related to the difference in the sensitivity to chilling.

B.   Molecular Form of Higher Plant Cytochrome c Oxidase

Animal and fungal cytochrome c oxidases consist of 6-8, probably seven, subunits, of which the three larger ones are synthesized on the mitochondrial ribosomes under the control of mitochondrial genes, while the others are on the cytoplasmic ribosomes under the control of nuclear genes (1,14,19,21). On the other hand, the enzyme from prokaryotes is composed of only 1-3 subunits (2,22,27). We have shown that both sweet potato and pea cytochrome c oxidases contain five subunits, when analyzed by electrophoresis on polyacrylamide gels containing sodium dodecylsulfate and urea (8,15). Thus higher plant cytochrome c oxidase differs from the enzyme from other sources in the subunit composition. The molecular weights of the five subunits are 39,000 (I), 33,500 (II), 26,000 (III), 20,000 (IV), and 5,700 (V) for the sweet potato enzyme, and 39,000 (I), 33,000 (II), 28,500 (III), 16,500 (IV), and 8,000-6,000 (V) for the pea enzyme (8,15). Namely, there seem to be significant differences in molecular weights of the two smaller subunits (subunits IV and V), but not in those of the other three (subunits I, II and III), between these two higher plant cytochrome c oxidases. It is of interest that the three larger subunits seem to be synthesized on the mitochondrial ribosomes, while the others seem to be synthesized on the cytoplasmic ribosomes (9,10,11).

We have separately raised antibodies against sweet potato and pea cytochrome c oxidases in rabbits by injecting the most purified preparations in complete Freund's adjuvant, and prepared immunoglobulin G fractions from the antisera (9,10,11). The antibodies each form a single precipitin line with the respective antigens in Ouchterlony double immunodiffusion tests (9,10,11). In addition, the antibody against the sweet potato enzyme also forms a single precipitin line with the pea enzyme, and vice versa the antibody against the pea enzyme does with the sweet potato enzyme (15). Either of the precipitin lines formed with the heterogeneous antigen-antibody combinations is fused with the line with the homologous combination to produce a faint spur (15), indicating that sweet potato and pea cytochrome c oxidase resemble, but differ slightly from, each other in immunological properties.

When immunoprecipitates from crude enzyme preparations from sweet potato root and pea shoot mitochondria with the respective antibodies are subjected to sodium dodecylsulfate-urea polyacrylamide gel electrophoresis, two polypeptides with molecular weights of 10,000-15,000 in addition to the five subunits are detected in both cases of sweet potato and pea (10,15). Therefore, we propose that higher plant cytochrome c oxidase is composed of five subunits and is weakly associated with two additional low-molecular-weight polypeptides in the mitochondrial inner membranes.

In conclusion, the structural outlines of sweet potato and pea cytochrome c oxidase molecules resemble each other, but there are slight differences in detailed structures between them.

II. ACTIVATION OF CYTOCHROME C OXIDASE BY PHOSPHOLIPIDS

Both the purified sweet potato and pea cytochrome c oxidase preparations contain very small amounts of phospholipids (about 2.5%, w/w) (8). Probably because of the presence of these tightly bound (annular) phospholipids, the purified preparations show some activity (we have assayed the activity by following the decrease in absorbance at 550 nm of reduced horse cytochrome c at pH 7.0: in the absence of exogenous phospholipids, the activity of the sweet potato enzyme preparation after ammonium sulfate fractionation is about 50 nmol cytochrome c oxidized per minute per mg protein).

Pre-incubation of any sweet potato or pea cytochrome c oxidase preparation (after, but not before, solubilization) with phospholipid micelles results in enhancement of the enzyme activity (8). As far as tested, all kinds of natural phospholipids (phosphatidylcholine from soybean and egg yolk, phosphatidylethanolamine from ovine brain and egg yolk, phosphatidylserine from bovine brain, and cardiolipin from bovine heart) in excess amounts stimulate the activity of the purified enzyme preparation from either sweet potato root or pea shoot mitochondria (the sweet potato enzyme preparation after ammonium sulfate fractionation or the pea enzyme preparation after first DEAE-cellulose column chromatography) to the same extent: they stimulate the activity of the sweet potato enzyme by about 3-fold (9) and that of the pea enzyme by about 2.5-fold. In other words, the same maximal activity is obtained with different kinds of phospholipids in excess amounts (the maximal activity of the sweet potato enzyme is about 150 nmol cytochrome c oxidized per minute per mg protein).

There are, however, differences in the concentration required to stimulate the activity to either the maximum or its half among different kinds of phospholipids. The reciprocal of the concentration required to stimulate the activity to the half maximum would represent the efficiency of activity-stimulation. Thus the data presented in Table I indicate that cardiolipin and phosphatidylserine are the most effective phospholipids. Phosphatidylcholine from soybean is much more effective than the same kind of phospholipid from egg yolk (Table I), indicating that the fatty acyl groups rather than the polar groups of phospholipids play an important role in the function of cytochrome c oxidase. As seen in Table I, there seems to be a correlation between the efficiency and the fatty acid composition of phospholipids: namely, there is a tendency that the more long chain-fatty acids phospholipids contain, the more effectively they stimulate the activity. Tween 80 with oleyl group, a long chain, can be substituted for phospholipids as the stimulator, but Triton X-100 with p-tert-octyl group, a very short chain, strongly inhibits the activity. This also

Table I. Stimulatory Effects and Fatty Acid Compositions of Phospholipids

Phospholipid	Amount of phospholipid[a] (μg-P)	Relative activity[b]	Fatty acid composition (%)					
			$C_{10:0}$	$C_{16:0}$	$C_{16:1}$	$C_{18:0}$	$C_{18:1 \text{ or } 2}$	C_{22}
Cardiolipin	0.075	2.33	0	1.1	1.3	3.1	94.6	0
PS	0.078	2.35	0	4.6	0	53.0	42.5	0
PC (soybean)	0.139	2.50	0	14.6	0	6.8	79.6	0
PE (ovine brain)	0.170	2.38	0	19.3	0	45.4	35.3	0
PE (egg yolk)	0.370	2.15	5.7	22.1	0	46.5	20.3	5.5
PC (egg yolk)	0.533	2.33	0	60.0	0	20.0	20.0	0

The pea enzyme preparation after DEAE-cellulose column chromatography was used. The fatty acid compositions were determined by gas chromatogrpahy of the methyl esters of the fatty acids released by hydrolysis of the phospholipids (20). PS, phosphatidyl-serine: PC, Phosphatidylcholine: PE, phosphatidylethanolamine.

a The amount of phospholipid per reaction mixture required to stimulate cytochrome c oxidase activity to half of the activity in the presence of soybean PC at a concen-tration of 1μg per reaction mixture.

b The activity in the presence of an excess amount of each phospholipid relative to the activity in its absence.

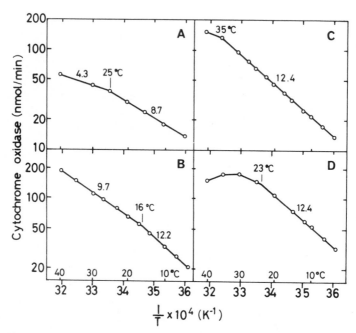

FIGURE 1. Arrhenius plots for cytochrome c oxidase in higher plant mitochondria. A: sweet potato roots. B: cassava roots. C: potato tubers. D: pea cotyledons (imbibed for 1 day). The figures near the plots indicate activation energy in kcal per mole.

suggests that long chain-fatty acyl residues of phospholipids are needed for the enzyme function.

There are also differences in the optimal pH of the stimulated activity among different kinds of phospholipids. Maximal activities are obtained around pH 7.5 with phosphatidylcholine from soybean and cardiolipin, and around pH 6.5 with phosphatidylethanolamine from ovine brain and phosphatidylserine. The optimal pH in the presence of either phosphatidylcholine or phosphatidylethanolamine from egg yolk is at pH 7.0. These results support the hypothesis that the non-polar groups rather than the polar groups of phospholipids participate in the function of cytochrome c oxidase.

Vik and Capaldi (23) have proposed that a fluid lipid environment with the non-polar parts of phospholipids is needed for the activity of cytocyrome c oxidase. Our results suggest that the length itself rather than the fluidity of the non-polar parts of phospholipids is of importance in the activity.

III.   EFFECT OF TEMPERATURE ON THE ACTIVITY
OF CYTOCHROME C OXIDASE

A.   Effect of Temperature on the Activity of Cytochrome c
Oxidase Associated with Mitochondrial Inner Membranes

Arrhenius plot for cytochrome c oxidase associated with the mitochondrial inner membrane in sweet potato roots shows discontinuity with a break point at 24-25°C (Fig. 1A) (12). A similar discontinuous plot is obtained with the enzyme in cassava (Manihot esculenta var. Crantz, cv. Ibusuki 2) root mitochondria, except that the break is at 16°C (in the case of another variety, Ibusuki 3, it is at 22°C) and the activation energy either above or below the transition temperature is higher than that in the case of sweet potato (Fig. 1B) (13). A similar downward bend in the Arrhenius plot has also been observed with cytochrome c oxidase in tomato seedling mitochondria (24). In contrast, the plot with the enzyme in potato (Solanum tuberosum Lam. cv. Rishiri) tuber mitochondria shows linearity (Fig. 1C) (12). Deviation of a plot at 40°C from the linear function seems due to inactivation of the enzyme as described later. In the case of pea cotyledon mitochondria, the plot shows a linear increase from 5 to 23°C and a convex curve above 23°C (Fig. 1D). It seems likely that the non-linearity above 23°C is caused by rapid inactivation of the enzyme. The activation energy for the enzyme in potato tuber or pea cotyledon mitochondria is similar to that below the transition temperature and is quite different from that above the temperature for the enzyme in sweet potato or cassava root mitochondria. We infer that chilling-sensitive plants exhibit discontinuity in the Arrhenius plot for cytochrome c oxidase associated with the mitochondrial inner membranes, while chilling-resistant plants do not. In the mitochondria of chilling-resistant plants, the cytochrome c oxidase seems very labile at high temperatures.

Lyons and Raison (6) reported that the mitochondria from chilling-sensitive plants exhibited a break at 9-12°C in the Arrhenius plot for succinate oxidase. Reports from our laboratory have shown that there are two breaks at 8-10 and 16-18°C in the plot for succinate oxidase of sweet potato root mitochondria (25,26). In any case, the break in Arrhenius plot for cytochrome c oxidase in sweet potato root mitochondria does not fit that for succinate oxidase. Probably, cytochrome c oxidase-step is so rapid that the step never becomes rate-limiting in succinate oxidation. In addition, the break point in Arrhenius plot for cytochrome c oxidase does not correspond to the temperature critical for chilling injury. Thus a question arises as to why chilling-sensitive plants show a bend in the Arrhenius plot for cytochrome c oxidase; in other words, as to what is the physiological meaning of discontinuity in the

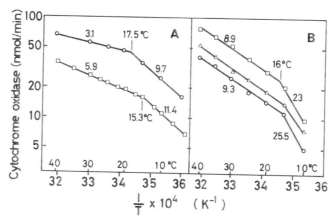

FIGURE 2. Arrhenius plots for isolated sweet potato cyto-chrome c oxidase. A: ○ , phosphatidylcholine from egg yolk;□ , Phosphatidylcholine 'from soybean. B: ○ , no phospholipids; △, dipalmitoylphosphatidylcholine;□ , dimyristoylphosphatidyl-choline.

plot. One of the speculative interpretations for this question is that in plants originating in tropical zone, the cytochrome c oxidase is able to operate at nearly a constant rate indepen-dent of temperature at a range of high temperatures. The cyto-chrome c oxidase may be adapted to a high temperature range so as to work at a similar rate to avoid metabolic imbalance.

B. Effect of Temperature on the Activity of Isolated Cytochome c Oxidase

Arrhenius plot for cytochrome c oxidase isolated from sweet potato root mitochondria (the enzyme preparation after ammonium sulfate fractionation) also shows discontinuity (Fig. 2B) (12). However, the bend is around 15°C in this case. We still have no interpretaion as to why the purified enzyme differs in the transition temperature from the enzyme associated with the mitochondrial inner membrane. As stated before, natural phos-pholipids stimulate the enzyme activity by about 3-fold, but they do not significantly alter the transition temperatures (17.5°C with phosphatidylcholine from egg yolk and 15.3°C with phosphatidylcholine from soybean) (Fig. 2A) (12). Bovine serum albumin has no effect on the Arrhenius plots.

Synthetic phospholipids, dipalmitoylphosphatidylcholine and dimyristoylphosphatidylcholine, also stimulate the activity of the purified sweet potato enzyme preparation, although they are inferior to natural phospholipids in the ability to accelerate the activity. As shown in Figure 2B, they also do not change the transition temperature (12). The phase-transition temperatures of dipalmitoylphosphatidylcholine and dimirystoylphosphatidylcholine are 40 and 23°C, respectively.

We now draw a conclusion from the above results that the bend in Arrhenius plot for cytochrome c oxidase observed with chilling-sensitive plants is not due to a change in the molecular architecture of phospholipids from a fluid to a solid structure. This conclusion coincides with the proposition stated in the preceding section that the length itself rather than the fluidity of the fatty acyl groups of phospholipids is of importance in the function of cytochrome c oxidase. Yoshida et al. (28) have suggested that a structural transition in animal cytochrome c oxidase is the primary cause of a bend in the Arrhenius plot for the enzyme in the mitochrondria. They have proposed that animal cytochrome c oxidase undergoes two forms, "hot" and "cold" types, of conformation probably through a change in the subunit arrangement. According to Kirino et al. (3), the fluctuation of adenosine 5'-triphosphatase molecule have a biphasic van't Hoff plot, in which the transition temperature is identical to the temperature at a bend in the Arrhenius plot of the enzyme activity. Therefore, we would interpret the bend in Arrhenius plot for cytochrome c oxidase observed with chilling-sensitive plants as a reflection of a conformational change in the enzyme protein. We suppose that the transition temperature for the structural change is only slightly affected by phospholipids surrounding the enzyme protein. Probably cytochrome c oxidase of chilling-resistant plants has no such ability to change its conformation. We speculate that the differences in the molecular form between sweet potato and pea cytochrome c oxidases stated before may be related to this ability.

We should make an additional comment on the interpretation of our data. There is an alternative explanation that the physical state of the annular lipid has a decisive effect on the activity of cytochrome c oxidase: it changes from a fluid to a solid phase at the observed transition temperature so that the activation energy for the enzyme activity is altered. At present, this possibility can not be excluded, but we do not think that this is very probable, because this alternative explanation raises a difficulty in understanding how exogenous phospholipids significantly stimulate the enzyme activity.

IV. THERMOSTABILITY OF CYTOCHROME C OXIDASE

As stated before, cytochrome c oxidase in the mitochrondria from chilling-resistant plants seems labile at high temperatures. The enzyme in either sweet potato root or potato tuber mitochondria is very stable at low temperatures and does not lose its activity for several minutes even at 20°C. When sweet

TABLE II.   Thermostability of Cytochrome c Oxidase

Source of mitochondria	Remaining activity of cytochrome c oxidase	
	40°C for 30 min	30°C for 30 min
Sweet potato roots	9.3%	56.9%
Sweet potato leaves	57.0%	87.4%

potato root mitochondria are kept at 30°C, the cytochrome c oxidase activity remains unchanged for the first 60-80 seconds and then begins to decline in a manner of the first-order kinetics. Under the identical conditions, the enzyme in potato tuber mitochondria holds its activity for about 40 seconds and then begins to inactivate. At 40°C, the enzyme activity in sweet potato root mitochondria remains unchanged for the first 20 seconds and then begins to decline, while that in potato tuber mitochondria decreases without such a lag period. This causes deviation of a plot at 40°C from the linearity in Arrhenius plot for the potato tuber enzyme (Fig. 1C).

Cytochrome c oxidase in sweet potato leaf mitochondria is quite stable. It loses only about half of the original activity even after incubation at 40°C for 30 minutes (Table II) (12). This suggests that the enzyme in the leaves retains active even in the daytime of the hot summer. Probably, differences in the milieu around the enzyme protein between the leaves and roots cause the difference in the thermostability of the enzyme. We have not yet compared as to the thermostability between cytochrome c oxidases in sweet potato and potato leaf mitochondria. However, we speculate that a conformation formed through a change, which is induced at the transition temperature as stated in the preceding section, may ensure the enzyme protein against inactivation, provided that the enzyme is environed with a proper milieu. We would propose that the conformational change of cytochrome c oxidase is related to the adaptation of plants to high temperatures.

V.   SUMMARY

Cytochrome c oxidase has been isolated from sweet potato root and pea shoot mitochondria. The enzyme from either source consists of five subunits and is weakly associated with two additional polypeptides in the mitochondrial inner membrane. Thus, sweet potato and pea cytochrome c oxidases resemble each other in the outlines of molecular structures. However, there are slight differences in detailed structures between them:

for instance, in the molecular weights of subunits and immuno-logical properties.

Natural phospholipids stimulate the activity of the purified enzyme preparations, and their fatty acyl groups rather than their polar groups participate in the stimulation. We suggest that the length itself rather than the fluidity of the fatty acyl groups is of importance in the enzyme function.

Arrhenius plot for cytochrome c oxidase in the mitochondrial inner membrane from either sweet potato or cassava roots, but not from pea shoots or potato tubers, shows discontinuity with a break (at 24-25°C for sweet potato roots and at 16-22°C for cassava roots). The activation energy for the enzyme from pea shoot or potato tuber mitochondria corresponds to that below the transition temperature for the enzyme from sweet potato or cassava root mitochondria. The plot for purified sweet potato cytochrome c oxidase shows a bend at about 15°C, and the transition temperature is not significantly altered by natural and synthetic phospholipids. We propose that the discontinuity in the Arrhenius plot is ascribed to a conformation change of the enzyme protein rather than a change in the fluidity of phospholipids surrounding the protein. The pea and potato enzymes may have no ability to change their conformations.

Cytochrome c oxidase in sweet potato root mitochondria seems more stable than the enzyme in potato tuber mitochondria at high temperatures. The enzyme in sweet potato leaf mitochondria is particularly stable even at high temperatures. We speculate that the thermostability of sweet potato leaf enzyme may be due to a specific environment of the enzyme protein in the membrane and related to the conformational change induced at the transition temperature.

The purpose of this work was to elucidate whether low temperatures directly affect the membrane-proteins, but the conclusion obtained is also related to the adaptation of plants to high temperatures.

REFERENCES

1. Azzi, A., Biochim. Biophys. Acta 594:231-252 (1980).
2. Hon-nami, K., and Ohshima, T., Biochem. Biophys. Res. Commun. 92:1023-1029 (1980).
3. Kirino, Y., Anzai, K., Shimizu, H., Ohta, S., Nakanishi, M., and Tsuboi, M., J. Biochem. 82:1181-1184 (1977).
4. Levitt, J., "Responses of Plants to Environmental Stresses", Vol. 1. Academic Presss, New York, 1980.
5. Lyons, J.M., Ann. Rev. Plant Physiol. 24:445-466 (1973).
6. Lyons, J.M., and Raison, J.K., Plant Physiol. 45:386-389 (1970).

7. Lyons, J.M., Raison, J.K., and Steponkus, P.L., in "Low Temperature Stress in Crop Plants", (J.M. Lyons, D. Graham, and J.K. Raison, eds.), pp. 1-24. Academic Press, New York, 1979.

8. Maeshima, M., and Asahi, T., Arch. Biochem. Biophys. 187:423-430 (1978).

9. Maeshima, M., and Asahi, T., in "Cytochrome Oxidase", (T.E. King, Y. Orii, B. Chance, and K. Okunuki, eds.), pp. 375-382.   Elsevier/North-Holland   Biomedical   Press, Amsterdam, 1979.

10. Maeshima, M., and Asahi, T., J. Biochem. 90:391-397 (1981).

11. Maeshima, M., and Asahi, T., J. Biochem. 90:398-406 (1981).

12. Maeshima, M., Asahi, T., and Uritani, I., Agric. Biol. Chem. 44:2351-2356 (1980).

13. Maeshima, M., Uritani, I., and Asahi, T., Agric. Biol. Chem. 44:2493-2494 (1980).

14. Malmstroem, B.G., Biochim. Biophys. Acta 549:281-303 (1979).

15. Matsuika, M., Maeshima, M., and Asahi, T., J. Biochem. 90:649-655 (1981).

16. Raison, J.K., in "The Biochemistry of Plants", (P.K. Stumpf and E.E. Conn, eds.), pp. 613-626. Vol. 2, Academic Press, New York, 1980.

17. Raison, J.K., in "The Biochemistry of Plants", (P.K. Stumpf and E.E. Conn, eds.), pp. 57-83. Vol. 4, Academic Press, New York, 1980.

18. Raison, J.K., Lyons, J.M., Mehlhorn, R.J., and Keith, A.D., J. Biol. Chem. 246:4036-4040 (1971).

19. Schatz, G., and Mason, T.L., Ann. Rev. Biochem. 43:51-87 (1974).

20. Schlenk, H., and Gellerman, J.L., Anal. Chem. 32:1412-1414 (1960).

21. Tzagoloff, A., Macino, G., and Sebald, W., Ann. Rev. Biochem. 48:419-441 (1979).

22. Sone, N., Ohyama, T., and Kagawa, Y., FEBS Lett. 106:39-42 (1979).

23. Vik, S.B., and Capaldi, R.A., Biochemistry 16:5575-5759 (1977).

24. Waring, A., and Glatz, P., in "Low Temperature Stress in Crop Plants" (J.M. Lyons, D. Graham, and J.K. Raison, eds.), pp. 365-374. Academic Press, New York, 1979.

25. Yamaki, S., and Uritani, I., Agric. Biol. Chem. 26:47-55 (1972).

26. Yamaki, S., and Uritani, I., Plant Cell Physiol. 15:385-393 (1974).

27. Yamanaka, T., Kamita, Y., and Fukumori, Y., J. Biochem. 89:265-273 (1981).

28. Yoshida, S., Orii, Y., Kawamoto, S., and Ikegami, A., J. Biochem. 86:1443-1450 (1979).

Index

D

M

N